Chirality at Solid Surfaces

Chirality at Solid Surfaces

Stephen J. Jenkins
University of Cambridge
Cambridge, UK

WILEY

This edition first published 2018

Registered Office(s)
John Wiley & Sons, Inc., 111 River Street, Hoboken, NJ 07030, USA
John Wiley & Sons Ltd, The Atrium, Southern Gate, Chichester, West Sussex, PO19 8SQ, UK

Editorial Office
111 River Street, Hoboken, NJ 07030, USA
9600 Garsington Road, Oxford, OX4 2DQ,

For details of our global editorial offices, customer services, and more information about Wiley products visit us at www.wiley.com.

Wiley also publishes its books in a variety of electronic formats and by print-on-demand. Some content that appears in standard print versions of this book may not be available in other formats.

Library of Congress Cataloging-in-Publication Data

Names: Jenkins, Stephen J. (Stephen John), author.
Title: Chirality at solid surfaces / by Stephen J. Jenkins.
Description: First edition. | Hoboken, NJ : John Wiley & Sons, 2018. |
Includes bibliographical references and index. |
Identifiers: LCCN 2017042876 (print) | LCCN 2017049498 (ebook) | ISBN 9781118880166 (pdf) | ISBN 9781118880142 (epub) | ISBN 9781118880128 (cloth)
Subjects: LCSH: Chirality. | Surface chemistry.
Classification: LCC QD481 (ebook) | LCC QD481 .J45 2018 (print) | DDC 541/.223–dc23
LC record available at https://lccn.loc.gov/2017042876

Cover Design: Wiley
Cover Image: Courtesy of Stephen J. Jenkins

Set in 10/12pt WarnockPro by SPi Global, Chennai, India
Printed and bound in Malaysia by Vivar Printing Sdn Bhd

10 9 8 7 6 5 4 3 2 1

To Victoria

Contents

Preface *xiii*
Acknowledgements *xxiii*

1 **Fundamentals of Chirality** *1*
1.1 Point and Space Groups *2*
1.2 Proper and Improper Symmetry *4*
1.3 Chirality in Finitude and Infinity *5*
1.3.1 Molecular Chirality *5*
1.3.2 Crystalline Chirality *8*
1.4 Routes to Surface Chirality *9*
1.4.1 Surfaces of Intrinsically Chiral Crystals *9*
1.4.2 Intrinsically Chiral Surfaces of Achiral Crystals *10*
1.4.3 Chiral Modification of Achiral Surfaces *11*
1.5 Diastereoisomerism Defined *14*
1.6 Quantifying Chirality? *15*
1.7 Enantiomeric Excess *17*
1.8 Synthesis, Separation and Sensing *19*
References *20*

2 **Surface Symmetry and Structure** *21*
2.1 Spherical Representation of Symmetry *21*
2.2 Spherical Representation of Structure *24*
2.3 Stereographic Projections: Flattening the Globe *27*
2.4 Surfaces of the Face-Centred Cubic Structure *29*
2.4.1 Reconciliation of Symmetry and Primary Structure *29*
2.4.2 Secondary and Tertiary Structure *32*
2.4.3 Commentary *34*
2.5 Surfaces of the Body-Centred Cubic Structure *36*
2.5.1 Reconciliation of Symmetry and Primary Structure *37*
2.5.2 Secondary and Tertiary Structure *39*
2.5.3 Commentary *40*
2.6 Surfaces of the Hexagonal Close-Packed Structure *42*
2.6.1 Symmetry *43*
2.6.2 Primary Structure *48*

2.6.3 Reconciliation of Symmetry and Primary Structure *52*
2.6.4 Commentary *55*
2.7 Surfaces of the Diamond Structure *56*
2.7.1 Symmetry *56*
2.7.2 Primary Structure *58*
2.7.3 Reconciliation of Symmetry and Primary Structure *59*
2.7.4 Commentary *62*
References *63*

3 Surface Energy and Surface Stress *65*
3.1 Thermodynamic Definition of Surface Energy *65*
3.2 The Tensor Nature of Surface Stress *70*
3.3 Visualisations of Surface Stress: Iconic Conics *71*
3.3.1 The Normal Stress Conic *72*
3.3.2 The Shear Stress Quartic *73*
3.3.3 The Stress Ellipse *74*
3.4 Symmetry of the Surface Stress: Eccentricity and Orientation *75*
3.4.1 Stereography and Surface Stress *77*
3.4.2 Racemic Surface Stress *79*
3.4.3 Adsorbate-Induced Asymmetry in Surface Stress *80*
3.5 Measurement of Differential Surface Stress *81*
3.5.1 Island Shape Measurement *81*
3.5.2 Contact Angle Measurement *82*
3.5.3 Cantilever Deformation *85*
3.6 Facet Formation and the Wulff Construction *86*
3.6.1 Ridge-and-Furrow Facets *86*
3.6.2 Pyramid-and-Pit Facets *88*
3.6.3 Geometrical Construction *89*
References *91*

4 Asymmetric Adsorption on Achiral Substrates *93*
4.1 Achiral Adsorbates: Gliding Through Broken Mirrors *93*
4.2 Prochiral Adsorbates: Chirality in Context *97*
4.2.1 Guanine on Au{111} *98*
4.2.2 Stilbene Derivatives on Cu{100} and Cu{110} *101*
4.2.3 Glycine on Cu{110} and Cu{311} *102*
4.2.4 Succinic and Fumaric Acids on Cu{110} *107*
4.2.5 Meso-Tartaric Acid on Cu{110} *111*
4.3 Chiral Adsorbates: Act Locally, Think Globally *112*
4.3.1 Alanine on Cu{110} and Cu{311} *112*
4.3.2 Proline on Cu{110} and Cu{311} *120*
4.3.3 Serine and Lysine on Cu{110} *125*
4.3.4 Cysteine on Cu{110} and Au{110} *128*
4.3.5 Tartaric Acid on Cu{110} *135*
4.3.6 Glutamic Acid on Ag{110} and Ag{100} *140*
4.3.7 2-Butanol on Au{111} *145*
4.3.8 Tartaric Acid on Ni{111} *146*

4.3.9 Alanine on Pd{111} *147*
4.4 Chiral Facetting: Remodelling the Surface *149*
4.4.1 Glycine, Alanine and Lysine on Cu{100} *150*
4.5 Chiral Metallorganic Frameworks: Into the Second Dimension *151*
4.5.1 Glutamic Acid on Ni/Au{111} *152*
4.5.2 Lysine on Ni/Au{111} *153*
4.5.3 Proline on Ni/Au{111} *154*
4.6 Executive Summary *156*
References *159*

5 Asymmetric Adsorption on Chiral Substrates *165*
5.1 Achiral Adsorbates on Intrinsically Chiral Substrates: Fault-Lines and Facets *165*
5.1.1 Oxygen on Cu{531} *165*
5.1.2 Cyclohexanone on Cu{643} *167*
5.1.3 NaCl on Cu{532} *168*
5.2 Prochiral Adsorbates on Intrinsically Chiral Substrates: Familiar and Strange *168*
5.2.1 Glycine on Cu{531} *169*
5.3 Chiral Adsorbates on Intrinsically Chiral Substrates: Diastereomeric Effects I *171*
5.3.1 Alanine on Cu{531} *171*
5.3.2 Serine on Cu{531} *173*
5.3.3 Cysteine on Cu{531} and Au{17 11 9} *174*
5.3.4 Tartaric Acid on Cu{531} *176*
5.3.5 Propylene Oxide and 3-Methylcyclohexanone on Cu{643} *176*
5.3.6 3-Methylcyclohexanone on Cu{531}, Cu{651} and Cu{13 9 1} *180*
5.3.7 Alanine, Serine, Lysine, Phenylalanine and Aspartic Acid on Cu{3 1 17} *182*
5.4 Chiral Adsorbates on Chirally Modified Substrates: Diastereomeric Effects II *184*
5.4.1 Propylene Oxide on 2-Butanol-Modified Pd{111} and Pt{111} *185*
5.4.2 Propylene Oxide on 2-Methylbutanoic Acid-Modified Pd{111} and Pt{111} *188*
5.4.3 Propylene Oxide on Amino Acid-Modified Pd{111} *189*
5.4.4 Glycidol on Tartaric Acid-Modified Pd{111} *190*
5.4.5 Propylene Oxide on Lysine-Modified Cu{100} *191*
5.5 Executive Summary *191*
References *193*

6 Chiral Amplification *197*
6.1 Kinetic Amplification: Surface Explosions *197*
6.1.1 Tartaric and Malic Acids on Cu{110} *200*
6.1.2 Tartaric Acid on Cu{643}, Cu{17 5 1}, Cu{531} and Cu{651} *202*
6.2 Thermodynamic Amplification: Sergeants, Soldiers and Majority Rule *206*
6.2.1 Tartaric, Succinic and Malic Acids on Cu{110} *206*

6.2.2 Heptahelicene on Cu{111}, Ag{111} and Au{111} *210*
6.2.3 Aspartic Acid on Cu{111} *215*
6.2.4 Supramolecular Assemblies on Highly Ordered Pyrolytic Graphite *217*
References *222*

7 Asymmetric Heterogeneous Catalysis *225*
7.1 Electro-Oxidation of Glucose on Pt{643} and Pt{321} *227*
7.2 Electron-Stimulated Oxidation of Methyl Lactate on Cu{643} *235*
7.3 Hydrogenation of α-Ketoesters over Platinum: The Orito Reaction *236*
7.3.1 Adsorption Geometry of Methyl and Ethyl Pyruvate *237*
7.3.2 Adsorption Geometry of Cinchonidine and its Cousins *240*
7.3.3 Binding and Reaction in the Chiral Complex *244*
7.4 Hydrogenation of β-Ketoesters over Nickel: The Izumi Reaction *247*
7.4.1 Adsorption Geometry of Methyl Acetoacetate *247*
7.4.2 Two-Dimensional Cocrystallisation: Tartaric/Glutamic Acid Modification *248*
7.4.3 Defect-Localised Oligomerisation: Modification by Aspartic Acid *250*
References *253*

8 Optical Consequences of Surface Chirality *257*
8.1 The Nature of Light *258*
8.2 Planar and Twisted Light *258*
8.2.1 Linear and Circular Polarisation *259*
8.2.2 Polarisation on a Helix *261*
8.3 Dichroic Photoemission *262*
8.4 Non-linear Optics in Chiral Systems *267*
8.4.1 Symmetry Constraints on Non-linear Optical Phenomena *267*
8.4.2 Implications for Chiral Surfaces *272*
8.4.3 Chiral SHG on Cu{111} and Au{110} *273*
8.5 Near-Field Phenomena *276*
References *277*

9 Magnetic Consequences of Surface Chirality *279*
9.1 Spin and Orbital Magnetism *279*
9.1.1 Fermions and the Dirac Equation *280*
9.1.2 Spin–Orbit Coupling *283*
9.2 Bulk Magnetocrystalline Anisotropy *285*
9.2.1 Laue Class O_h (Cubic Crystal System: O_h, T_d and O) *287*
9.2.2 Laue Class T_h (Cubic Crystal system: T_h and T) *287*
9.2.3 Laue Class D_{6h} (Hexagonal Crystal System: D_{6h}, D_{3h}, C_{6v} and D_6) *287*
9.2.4 Laue Class C_{6h} (Hexagonal Crystal System: C_{6h}, C_{3h} and C_6) *288*
9.2.5 Laue Class D_{3d} (Trigonal Crystal System: D_{3d}, C_{3v} and D_3) *288*
9.2.6 Laue Class S_6 (Trigonal Crystal System: S_6 and C_3) *289*
9.2.7 Laue Class D_{4h} (Tetragonal Crystal System: D_{4h}, D_{2d}, C_{4v} and D_4) *289*

9.2.8 Laue Class C_{4h} (Tetragonal Crystal System: C_{4h}, S_4, & C_4) *290*
9.2.9 Laue Class D_{2h} (Orthorhombic Crystal System: D_{2h}, C_{2v}, and D_2) *290*
9.2.10 Laue Class C_{2h} (Monoclinic Crystal System: C_{2h}, C_{1h}, and C_2) *292*
9.2.11 Laue Class S_2 (Triclinic Crystal System: S_2 and C_1) *293*
9.2.12 General Comments *294*
9.3 Surface Magnetocrystalline Anisotropy *295*
9.3.1 Surface MCA of Face-Centred and Body-Centred Cubic Ferromagnets *296*
9.3.2 Role of Adsorbates and Reconstruction *298*
9.4 An Aside: Vectors and Pseudovectors *299*
9.5 Spin Waves in Centrosymmetric Media *301*
9.5.1 Spin Helices *301*
9.5.2 Spin Spirals *303*
9.6 Spin Waves at a Featureless Surface *304*
9.6.1 Spin Helices *304*
9.6.2 Spin Spirals *305*
9.7 Spin Waves at Structured Surfaces *306*
9.7.1 Spin Helices at Achiral Surfaces *306*
9.7.2 Spin Helices at Chiral Surfaces *307*
9.7.3 Spin Spirals at Achiral Surfaces *307*
9.7.4 Spin Spirals at Chiral Surfaces *307*
9.8 Surface Spin Spirals Observed *308*
9.9 Skyrmions, or How to Brush a Hedgehog *309*
References *313*

10 Chiral Electronic States in Two Dimensions *315*
10.1 Dirac Cones in Graphene *316*
10.2 Dirac Cones at Half-Metal Surfaces *321*
10.3 Dirac Cones at the Surfaces of Topological Insulators *323*
10.4 Prospects for Electronic Chirality in the Chemical Context *328*
References *329*

11 Postscript *331*

A List of Abbreviations *335*

B Rules for Overlayer Periodicity Assignment *337*
B.1 Substrate Lattice *337*
B.2 Overlayer Lattice *338*
B.3 Illustrative Examples *339*
References *343*

C Further Reading *345*

Index *347*

Preface

Come on my right hand, for this ear is deaf.
The Tragedy of Julius Caesar, William Shakespeare, 1599

We shall present a formal definition of chirality by and by, but for the moment let us be guided by etymology.[1] The word 'chiral' derives from the Ancient Greek $\chi\epsilon\iota\rho$ (*cheir*) meaning 'hand'; it denotes the property of an entity (tangible or otherwise) that can (even if only hypothetically) exist in two forms related to one another in the same manner as one's right hand is related to one's left. That is, both forms are essentially identical in every regard apart from the fact that one is the mirror image of the other. Across a broad range of circumstances, one's left and right hands are functionally equivalent – in form, action and apprehension equally capable – but try to place one hand into a glove designed for the other and the distinction soon becomes clear. From this, we learn the most important lesson of chirality – that chirality only matters when chiral entities interact with other chiral entities. Chirality, therefore, is at one and the same time the most profound of characteristics and the most subtle. Chiral distinctions of crucial importance surround us, but our control over chirality (in all its forms) is limited by our access to chiral tools.

Most of us are right-handed, although a significant minority (around 1 in 10) are left-handed.[2] The origins of this disparity are not entirely understood, although genetic, epigenetic and even environmental influences have been proposed as partial explanations; the truth lies, one imagines, in some combination of all three. Whatever the whys and wherefores, right-handedness has long been considered the cultural norm, with left-handedness generally regarded as aberrant. Returning briefly to etymology, the Latin adjective *dexter* not only denotes the right-hand direction, but also connotes propriety, good fortune and (foreshadowing modern English usage) dexterity. Similarly, its antonym *sinister* denotes not only the left-hand direction, but also conveys the sense of perversity, hostility or (in its current English meaning) ill-omen. When Shakespeare has Caesar tell Antony to speak only into his right ear, he reveals not merely the dictator's infirmity, but also his hubris in cutting short talk of conspiracy – Caesar has an ear only for good news.

Practical implications of human chirality abound, from the design of tools[3] and the niceties of cricket[4] to the conventions of antique sculpture[5] and of

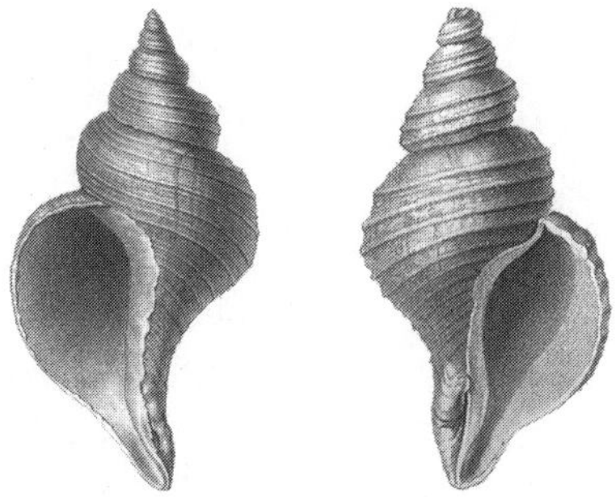

Figure 1 Fossil shells of two marine gastropod species. In panel (a) the shell of *Neptunea angulata* is a rare example that spirals anticlockwise from apex to aperture, while in panel (b) that of *Neptunea despecta* spirals in the more common clockwise sense. Reproduced from [10].

ancient writing systems.[6] Famously, the design of medieval castles is often said to have been influenced by chirality – spiral staircases typically wind clockwise when ascending, so as to favour a descending right-handed defender against a most-likely right-handed attacker.[7] Similarly, the ancient custom of travelling on the left-hand side of the road (so as to keep one's sword arm towards approaching traffic) seems to have been ubiquitous prior to the mid-eighteenth century; more recently, various jurisdictions have adopted the opposite convention in deference either to non-military practicalities or to political dogma.[8]

At any rate, chiral discrimination in the human realm is commonplace, so it is hardly surprising that examples from elsewhere in nature also abound. The shells of gastropods, for instance, typically spiral clockwise from their apex to their aperture (albeit not in all species, see Fig. 1) suggesting that some reproductive advantage is gained by individuals in a breeding population sharing the same morphological chirality.[9]

All of this suggests that chirality is not merely a particular and superficial property of the natural world, but rather a pervasive and fundamental one. We ought, therefore, to seek for the origins of chiral phenomena in living systems not merely at the macroscopic level but also at the microscopic scale of their biochemistry. The familiar double helix of DNA, for example, presents the obvious question of whether it spirals in the right- or left-handed sense. In fact, three distinct types of DNA exist, with the A and B forms exhibiting right-handed helices and the Z form a left-handed helix. There is no *a priori* reason that life could not have evolved based upon mirror-images of these molecules, but at some exceptionally early point in its development the observed asymmetry must have become locked into the genome. Other biomolecular helices are not hard to find, and these too exhibit chiral asymmetry. The protein collagen, for example, exclusively takes the form of a right-handed triple helix, the individual strands of which are themselves left-handed helical structures. The so-called α-helices that recur with great frequency in the secondary structure of proteins are, it turns out, exclusively right-handed. These latter two examples are explicable in terms of the overwhelming dominance of left-handed amino acids (the building blocks of proteins) in nature.[10]

For most practical purposes, indeed, it is essentially accurate to state that *all* amino acids found in nature are left-handed. Amongst the very rare exceptions to this rule, however, one finds that most impractical of animals, the platypus. Not

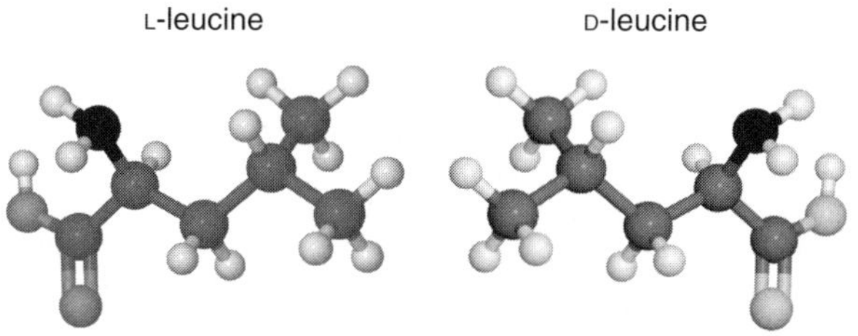

Figure 2 Left- and right-handed leucine (L-leucine and D-leucine) are mirror-image amino acids. The former is found in proteins throughout the living world, but the latter is entirely absent apart from rare cases such as the venom of the male platypus.

content with laying eggs, locating prey by electroreception and resembling a cross between an otter, a beaver and a duck, the platypus is unique amongst mammals in the possession of a venomous spur on its hind feet.[11] The venom of the platypus contains, amongst other substances, a protein incorporating the right-handed version of the amino acid leucine [11] (Fig. 2) and it may be speculated that this confers some degree of incompatibility with 'normal' biochemistry that correlates with toxicity in some way.[12] In less exotic contexts, the fact that the chirality of a chemical compound correlates with its biological activity is a fundamental pillar of pharmaceutical design.[13]

It follows, therefore, that the development of new drugs must involve careful consideration of asymmetric effects, and it is typically necessary to manufacture such compounds in chirally pure form. This may, in practice, be achieved either through asymmetric synthesis or through asymmetric separation. If one can exclusively produce only the desired chiral form, or entirely filter out the undesired form, then the biological activity can be more tightly managed; the obvious corollary is that asymmetric sensing is also mandatory, to monitor the success or otherwise of the chosen strategy. Techniques for the exertion of chiral control in homogeneous chemistry have been accorded a corresponding level of attention, and progress has been conspicuously rewarded [13]. In contrast, the realm of heterogeneous chemistry has lagged considerably, due in no small part to the constraint implied by working within two dimensions instead of three. The purpose of this book is to take stock of what has been achieved to date in the study of surface chirality, and to suggest avenues along which future efforts might usefully be directed.

With the aim of illustrating the importance of dimensionality to chirality, whilst not pre-empting the content of the book, we note that many relevant concepts can be understood most simply in terms of two-dimensional drawings of three-dimensional objects. To give this analogy some structure, we focus upon one particular corner of graphic design governed by strict rules and conventions, namely the discipline of heraldry. If one can understand the manifestations of chirality arising in this highly abstract case, the range of concrete phenomena found upon the interaction of molecules with solid surfaces ought to make just as much sense. Here, the shield will stand for the surface itself, while the objects depicted upon it, known generically as charges, will represent adsorbed molecules.

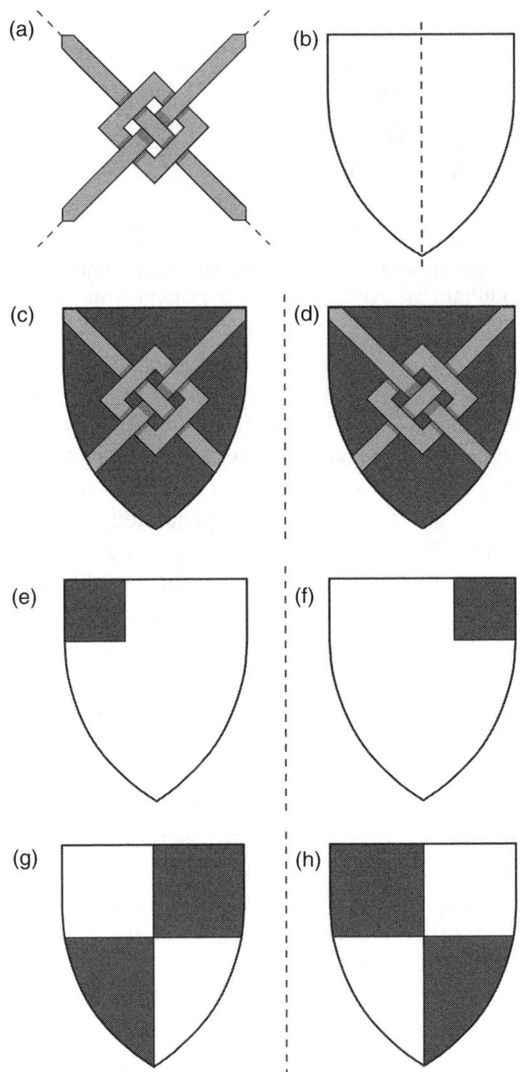

Figure 3 Panels (a) and (b) show, respectively, an isolated fret and a blank shield. In panels (c) and (d) these are combined to create two mirror-image shields. Panel (e) shows a canton in its traditional position, while panel (f) shows it in the mirror-image position. Finally, panels (g) and (h) show mirror-image quarterings of the shield. Reflection symmetry is indicated by dashed lines. (*See Plate I for the colour representation of this figure.*)

Consider the heraldic charge known as a fret – a sort of buckle formed by three interlaced laths (see Fig. 3a). When removed from its context, floating freely in space as it were, the fret is not a chiral object – it possesses multiple mirror symmetries, so reflection across an arbitrary plane consequently results in a fret that may be manhandled into exact correspondence with the original, so long as freedom of rotation in three-dimensional space is allowed. Similarly, a blank heraldic shield (Fig. 3b) is also not chiral, possessing mirror symmetry across a plane passing perpendicularly through its centre – again, reflection across an arbitrarily oriented plane can be undone by mere rotation in three-dimensional space. Place the fret upon the shield, however, and its symmetry is now constrained in just two dimensions. As oriented in Fig. 3c, the mirror symmetries of the fret do not align with that of the blank shield, and we shall assume that it is attached sufficiently strongly to the shield that this cannot be remedied by a mere rotation; the mirror-image arms shown in Fig. 3d are, therefore, genuinely (if subtly) distinct.[14]

In this way, a non-chiral object in three dimensions may acquire chirality upon being constrained to exist within a two-dimensional context. Clearly, if the fret were substituted by a device that is already chiral in three dimensions, this chirality would not in any sense be removed upon deposition onto the shield. The charges placed on shields may thus be irreducibly chiral in themselves or merely contingently so. Simpler charges, such as the canton, or square, remain stubbornly non-chiral in two dimensions as much as in three, but asymmetric placement of these objects may nevertheless lead to overall chirality, so long as we assume attachment sufficiently strong to prohibit facile translation (Fig. 3e and f).[15]

An alternative source of chirality in heraldry may be found in the quartering of shields (and indeed in other types of patterning, known collectively as variations of the field). Here, the shield itself is sub-divided into regions of differing colour, breaking the mirror symmetry and providing for two distinct mirror images (Fig. 3g and h).[16] Heraldic variations of this kind are considered fundamental to the shield itself, which is now, as it were, intrinsically chiral, independent of any chirality pertaining to wherever and whatever charges may be placed upon it.

The shields depicted in Fig. 4 thus involve no fewer than three distinct manifestations of chirality: the contingent chirality of the frets; the extrinsic chirality of the placement of these frets relative to the mirror symmetry of a blank shield; and finally the planar chirality of the quartered shield itself. All three forms of chirality are inherently two-dimensional in nature, and by their combination many subtly distinct coats of arms may therefore be constructed from the same basic elements.[17] In the first chapter of this book, we shall see how the same concepts of two-dimensional chirality – the chirality of objects *constrained* to lie upon a plane, the chirality of their *placement* on that plane, and the chirality of the *underlying* plane itself – carry over into the world of molecules and surfaces. In essence, any intrinsic chirality possessed by a given substrate may be thought of as a variation of the field; long-range chiral order in the arrangement of molecules upon the surface finds its counterpart in the asymmetric disposition of charges upon that field; and the chiralities of individual adsorbate molecules themselves may be likened to either induced or inherent asymmetry in the many different charges from which the herald may choose.

In the subsequent two chapters, we shall present a systematic treatment of the various manifestations of intrinsic chirality at the surfaces of crystalline solids, and discuss the implications of surface symmetry for the fundamental properties of surface energy and surface stress. A selective review of the literature concerning the adsorption of molecules on non-chiral and chiral surfaces will take up a further two chapters, leading to a third in which the important concept of chiral amplification is introduced. Our focus on surface chemistry will then be brought to a close with a chapter summarising current understanding of the mechanisms

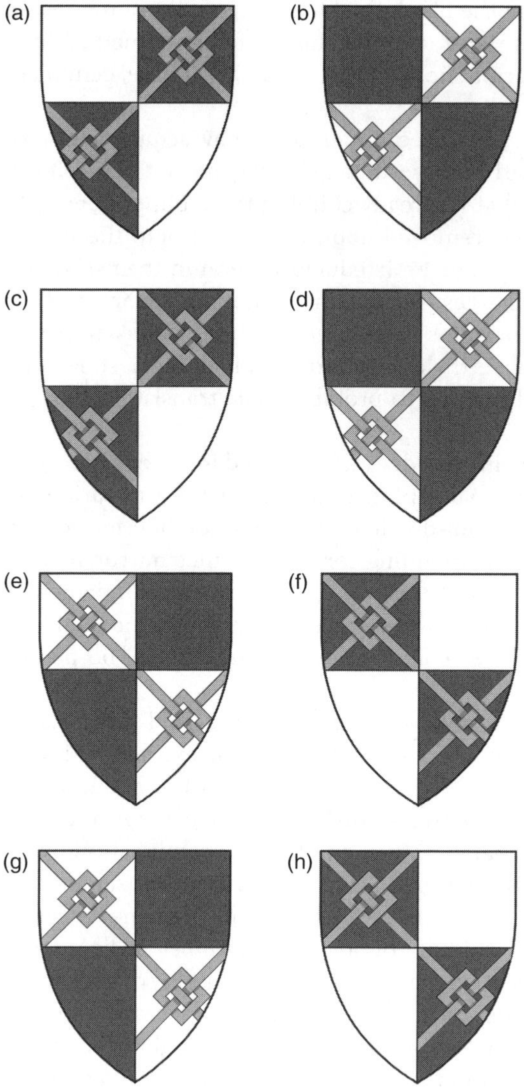

Figure 4 Variations on a coat of arms exhibiting three forms of chirality. The four shields on the left differ from the four on the right by reflection of the field; the upper four shields differ from the lower four by reflection of the fret positions; the upper two and lower two shields differ from the middle four by reflection of the frets themselves. *See Plate II for the colour representation of this figure.*

by which asymmetric heterogeneous catalysis may be achieved. In the final portion of the book, chapters on the optical, magnetic and electronic implications of surface chirality take us into the realm of surface physics. It is the firm conviction of this author that much fertile ground remains to be explored in this region, and that the interplay between physical and chemical processes at surfaces may yet prove to be the most fruitful *terra incognita* into which we might stray.

Cambridge, 23 May 2017 *Stephen J. Jenkins*

References

1 P. Cintas, *Angew. Chem. Int. Ed.* **46**, 4016 (2007).
2 D.F. Halpern and S. Coren, *Nature* **333**, 213 (1988).
3 C. Porac and S. Coren, *Lateral Preferences and Human Behaviour* (Springer-Verlag, New York, 1981).
4 M. Hirnstein and K. Hugdahl, *Br. J. Psychol.* **205**, 260 (2014).
5 S. Coren, *Am. J. Public Health* **79**, 1040 (1989).
6 J.W. Wegner and J.R. Houser Wegner, *The Sphinx that Travelled to Philadelphia: The Story of the Colossal Sphinx in the Penn Museum* (University of Pennsylvania Press, 2015).
7 A.G. Woodhead, *The Study of Greek Inscriptions*, 2nd Edition (Cambridge University Press, Cambridge, 1981).
8 A. Gardiner, *Egyptian Grammar: Being an Introduction to the Study of Hieroglyphs*, 3rd Edition, Revised (Griffith Institute, Oxford, 1957).
9 N. Guy, *Castle Stud. Group J.* **25**, 174 (2011).
10 P.H. Nyst, *Ann. Mus. R. Hist. Nat. Belg.* **3**, 1 (1878–1981).
11 A.M. Torres, I. Menz, P.F. Alewood, P. Bansal, J. Lahnstein, C.H. Gallagher and P.W. Kuchel, *FEBS Lett.* **524**, 172 (2002).
12 N. Vargesson, *Birth Defects Res. C* **105**, 140 (2015).
13 W. Knowles in *Les Prix Nobel 2001*, ed. T. Frängsmyr (Nobel Foundation, 2002); R. Noyori, *ibid.*; B. Sharpless, *ibid.*
14 G.J. Brault, *Early Blazon: Heraldic Terminology in the Twelfth and Thirteenth Centuries with Special Reference to Arthurian Heraldry*, 2nd Edition (Boydell Press, 1997).

Note

1 The origins, uses and abuses of the term are entertainingly outlined in a fascinating article by Cintas [1].
2 True ambidexterity is, apparently, extremely rare in humans (albeit the norm in most other animals) while mixed handedness (right-handed for some tasks, left-handed for others) accounts for perhaps one in a hundred people.
3 The fact that tools (interpreted broadly to include heavy machinery and motor vehicles, as well as scissors and screwdrivers) are typically designed for use by right-handed people is far from a trivial issue. One survey suggested a 2% difference in annual survival rates between left- and right-handed individuals [2], potentially accounting for an apparent drop (identified in a different survey) in the left-handed population from 13% in the 20–29 age range, and 5% in the 50–59 age range, to virtually none aged over 80 [3]. Thankfully, some of this rather alarming population drop was likely due to the outdated practice of 'training' left-handed children to be right-handed (which would have disproportionately affected older respondents) or adult handedness switching (which apparently does spontaneously occur on occasion) and some may relate to the prevalence of potentially life-limiting psychological disorders (e.g. a recent meta-analysis

of schizophrenia rates [4] found a positive correlation with non-right-handedness) but a non-negligible element seems to derive from an increased likelihood of injury when using tools maladapted to left-handed users [5].

4 There are at least three ways in which cricket is chiral. Firstly, the angle from which a bowler delivers the ball is strongly influenced by his or her handedness; secondly, the deviation of the ball's trajectory, either in flight (swing or drift) or on impact with the pitch (seam or spin) is similarly influenced by the handedness of the bowler; and finally, the batsman adopts a sideways-on stance usually determined by their handedness. Swapping the handedness of all three factors results in an approximately equivalent situation – thus, a left-hand unorthodox spinner bowling to a right-handed batsman is similar to a right-hander bowling leg-breaks to a left-handed batsman, as any schoolboy knows. In the fast-paced 20–20 variant of cricket, it has recently become common to see batsmen swap their stance to their non-dominant side, in order to confuse the opposition and/or to attempt reversed versions of their normal shots. Needless to say, 20–20 cricket is a deplorable abomination.

5 Around 10 years ago, between sessions at a conference in Paris, my colleague Stephen Driver and I found ourselves wandering the corridors of the Louvre, where we were intrigued to note that all the sphinxes we saw (there are quite a few) had tails that curled over the *right* rear paw, never the left. Despite a nagging suspicion that all the mirror-image sphinxes might somehow have found their way to the British Museum (they have not) we speculated that this was perhaps the result of a genuinely ancient artistic convention. I was pleased, therefore, to learn recently that this does indeed turn out to be the case – all sphinxes (barring 'one or two aberrant [examples from] the Ptolemaic period') do indeed curl their tails exclusively over their right rear paw [6]. More widely remarked upon is the convention in ancient Egyptian statuary for the highest status individuals to be more-or-less always represented with their *left* foot advancing. Needless to say, there is no clear consensus concerning the reason for either of these empirical observations.

6 From a Western perspective, it is tempting to view left-to-right systems of writing as somehow natural, allowing the right-handed scribe both to view his or her work in progress and to minimise the chances of smudging the drying ink. Such a narrative fails to recognise, of course, the existence of right-to-left and top-to-bottom writing systems, and indeed the fact that many of the left-to-right examples predate the widespread use of ink – one hardly supposes that smudging presented much of a problem when carving Greek letters into stone or impressing cuneiform script into clay tablets! Nevertheless, the fact remains that each writing system has typically made a decisive determination at some point and stuck to it for hundreds, if not thousands, of years. A notable exception is the archaic convention known as boustrophedon ('as the ox turns') where the direction switched between left-to-right and right-to-left for each successive line of text. Even here, however, there may be a role for chirality; in Greek boustrophedon, the familiar letters of the alphabet are themselves reversed [7], while in Egyptian

hieroglyphics the signs for people and animals face the beginning of each line [8], both examples of asymmetry making clear the direction in which the text is to be read.

7 It should be noted that this appealing theory has recently been challenged by Guy [9]. Although admitting a 20:1 ratio between clockwise and anticlockwise spirals in Norman castles, he ascribes this primarily to non-military considerations and notes that 'left-handed' staircases become increasingly common from the 1240s onward. Assertions that left-handed stairs can be associated with left-handed castle owners (e.g. the Kerr family of Ferniehirst Castle) ought to be viewed with this *caveat* in mind.

8 In Europe, for example, the adoption of right-hand traffic dates ultimately to a French revolutionary edict of 1792, which associated left-hand traffic with aristocratic manners. As with so much else, the imposition of this edict beyond the confines of France can, apparently, be attributed to Napoleon.

9 About 6 months before the composition of this preface, the UK press was briefly obsessed with the plight of 'Jeremy the Left-Handed Snail', whose search for a compatible life partner seemed to have been consummated happily after a national appeal to locate a similarly unconventional individual of the same species (see *www.bbc.co.uk/news/uk-england-nottinghamshire-37909055* for a flavour of the story). Alas, it appears that the *two* potential mates thus identified were more enamoured of one another than of Jeremy, whose search for love continues at the time of writing.

10 It is worth noting that right- and left-handed helices are defined so that the former spiral in the same sense as a standard corkscrew (i.e. one designed for use by right-handed people). For non-helical molecules, the definition of right- and left-handed versions is essentially arbitrary, although well defined by convention.

11 Strictly speaking, only the male platypus sports the venomous spur; one presumes that it is redundant to point out that only the female platypus lays eggs.

12 The occurrence of a right-handed amino acid in platypus venom seems to represent the only such example in the mammalian proteome. Other instances occur in toxins secreted by certain fungi, insect larvae, snails and tree frogs.

13 The canonical illustration of chirality's crucial influence in the action of drugs is the story of thalidomide. In the late 1950s and early 1960s, this compound was prescribed both as a sedative and in the treatment of morning sickness in pregnant women. Tragically, while the right-handed version of the drug is nominally safe the left-handed version was subsequently implicated in a wave of severe birth defects amongst children born to mothers to whom thalidomide had been prescribed; the drug was sold as a mixture of both right- and left-handed forms. In this particular case, delivery of purely right-handed molecules would not have prevented harm, however, as thalidomide readily interconverts between the two forms within the body [12].

14 It ought, perhaps, to be noted that the English system of heraldry makes no distinction between the mirror-images of frets on shields. Rendered in gold

upon red, as here, both versions of the shield are equally valid representations of the arms borne by members of the Audley family.

15 A plain white shield with red canton, whether in upper left or upper right, seems not to be associated with any particular family; the former was, however, attributed to the mythical figure of Sir Gawain in several early illustrated manuscripts of the Arthurian legends [14].

16 English heraldry does, very much, recognise the chirality of quartering. Rendered with white top left and red top right, the arms depicted are those once borne by members of the Soleigny family of Umberleigh in Devon, while the mirror image was once borne by members of the Cocke family of Broxbourne in Hertfordshire.

17 The variants shown at the upper left and immediately below are the arms borne by members of the Dutton family of Dutton in Cheshire (the mirroring of frets being immaterial in heraldry). The remainder are not, to the author's knowledge, linked with any known family. Indeed, those variants that impose yellow charges on a white background break, in so doing, one of the most fundamental rules of heraldry.

Acknowledgements

How distantly may one discern the origins of a book?

In the most immediate sense, the idea of the present volume emerged from discussions some four and a half years ago with senior commissioning editor Sarah Higginbotham at Wiley, whose advice in preparing the initial proposal was invaluable, and who, together with a variety of project editors, has been a constant and reassuring presence throughout the writing process. Taking a look down the other end of the telescope, however, it must have been Georg Held who first introduced me to the notion of chirality as a surface property, sometime around the turn of the millennium, and conversations with him over the following years have never failed to be equally enlightening. Likewise influential over a similar timespan have been Andrew Gellman, Rasmita Raval and Gary Attard, all of whom have been particularly generous with their advice and encouragement, not only in respect of my efforts in surface chirality. Overall, the surface chirality community has collectively proved to be very collegial, and it has been a pleasure to interact with a large number of people whom it would be impossible to list in full; those who go unmentioned here are appreciated nonetheless, and are requested to forgive the omission. I would, however, like to single out Karl-Heinz Ernst, not only for his kind words about the penultimate draft of this book, but also for the very helpful suggestions he made for its improvement.

Closer to home, many colleagues at Cambridge have made invaluable contributions to whatever understanding of surface chirality is displayed here. Foremost amongst these, it is beholden upon me to mention Sir David King, who not only took a keen interest in the topic at hand, but was also instrumental – almost 20 years ago – in opening the eyes of a humble surface physicist to the complexities of surface chemistry. For this, and for his unfailing support of my career as a scientist, I shall be eternally grateful. At the risk of missing someone out, particular gratitude in respect of chiral systems is also owed to Maria Blanco-Rey, Marian Bentley (née Clegg), Alex Ievins, Sakari Puisto, Nicola Scott and especially Glenn Jones, who was the first in my group to attempt calculations on chiral adsorbates. In recent years, David Madden and Israel Temprano were responsible for a wealth of experimental data from which we learnt a lot, and Marco Sacchi's careful calculations were instrumental in ensuring that we did just that; a collaboration with Jane Hinch at Rutgers stimulated much of this work, and her penetrating interrogation of unspoken assumptions was very important in honing our arguments. Special thanks, moreover, must certainly go to Stephen Driver, whose many contributions stretch back more than a decade and a half to our earliest conversations on chirality, and whose very helpful commentary on an intermediate draft of the present book raised more than a few key issues that needed attention. It is also a pleasure to record my indebtedness to Stephanie Arthey (née Pratt), with whom the stereographic approach to surface symmetry and structure was first worked out, and without whose insight at that crucial juncture this would have been a very different book, if a book at all.

Finally, my most heartfelt thanks are due to my family for their unfailing love and support, but above all others to my wife, Victoria, whose patience during the writing of this book has been phenomenal. My earnest hope is that the final product seems worthy of the sacrifices that enabled it. Meanwhile, I can only express my most sincere love and gratitude towards an unfailingly inspiring partner who is most certainly my better half.

1

Fundamentals of Chirality

And now, if e'er by chance I put
My fingers into glue
Or madly squeeze a right-hand foot
Into a left-hand shoe ....

Through the Looking-Glass, Lewis Carroll, 1871

Imagine a sophisticated three-dimensional scanner, capable of recording and encoding not only the precise shape of a physical object, but also the details of its texture, colouration, chemistry and mass distribution – in short, every aspect of its internal composition and outward appearance that one might perceive upon the most careful of inspections. Next, imagine that these encoded details are passed to a yet more sophisticated three-dimensional assembler, capable of reproducing a perfect physical replica of the original object, identical in every respect that can be observed or indeed measured. At the time of writing, such technology remains in the realm of science fiction, albeit recent advances in laser scanning and three-dimensional printing bring some superficial elements of the scheme within touching distance. Nevertheless, we can certainly conceive of the possibility of such a 'replicator', and consider its implications for our ability to distinguish between similar objects.

Now, if the replicator described above does its job perfectly, it is clear that the original object and its replica should be perfectly indistinguishable, once the former is removed from the scanner and the latter from the assembler. This much is inherent in our definition of the replicator, and should not unduly exercise our curiosity. A more interesting case may be found, however, when we make a subtle alteration to the replicator by swapping the sign of one spatial coordinate in the encoded details of the original object before passing it from the scanner to the assembler. The replica thus produced will then be identical to the original in every way *except* that it is its mirror image. Indeed, the replicator under these circumstances performs precisely the same transformation as a mirror, but unlike a mirror the 'image' it creates is substantial – capable of being picked up, moved, examined and dissected. Our question must now be whether the original and replica objects will still be indistinguishable, and our answer must, in general, be no.

Chirality at Solid Surfaces, First Edition. Stephen J. Jenkins.

The vast majority of natural objects placed into our doctored replicator will generate a replica that can be readily distinguished from the original. No amount of rotating or moving through space will succeed in making the original and the replica look precisely the same, and we shall recognise this fact by bestowing the adjective **chiral** to such objects. In fact, it is only in the case of objects possessing certain very particular symmetry properties that the original and replica *will* be indistinguishable – a situation that we shall recognise by describing such objects as **achiral**. The enumeration and categorisation of symmetry properties must therefore be our first concern.

1.1 Point and Space Groups

At the most fundamental level, symmetry refers to the property of an object whereby some **transformation** performed upon that object leaves it indistinguishable from its initial state. A plain cube, for example, may be rotated by 90° about a certain axis passing through its centre, and were it not for having observed the act in progress, it would otherwise be impossible to determine subsequently whether any action had been performed at all. This transformation may thus be said to be a **symmetry operation** of the cube, and the axis about which it occurs is said to be a **symmetry element** of the same. More specifically, this particular axis may be described as a fourfold axis, to indicate that four repetitions of the corresponding symmetry operation will return the system to its *literal* starting configuration, as opposed to a configuration that is simply *indistinguishable* from it.

To enumerate mathematically the various geometrical transformations that *may* turn out to be symmetry operations of a given object, it is convenient to employ vector notation. Let the three-dimensional vector **p** represent some specific point within the object, and the three-dimensional vector **p′** that same point after undergoing some transformation. The transformed vector is then related to the original vector by the equation

$$\mathbf{p}' = \mathbf{R}\mathbf{p} + \mathbf{T} + \mathbf{t} \tag{1.1}$$

where **R** is a 3 × 3 matrix, and both **T** and **t** are three-dimensional vectors. The matrix **R** can effect **rotation, reflection, inversion** or **scaling** (or combinations thereof) whilst the vectors **T** and **t** each effect spatial **translation**; the symbol **T** is reserved for translations corresponding to Bravais lattice vectors within a crystalline material, whereas **t** stands for any other non-Bravais translation. Applying the same geometrical transformation to all points within the object then achieves the corresponding geometrical transformation of the entire object. The act of scaling an object can never leave it apparently identical to its original state, so scaling transformations can never be symmetry operations of an object.[1] Rotation, reflection and inversion operations *can* all be symmetry operations of certain objects, with associated symmetry elements consisting of

1 Unless that object happens to be a perfect fractal, which physical objects never are.

rotation axes, **mirror planes** and **centres of inversion**, respectively, together with so-called **rotoreflection axes** corresponding (unsurprisingly) to a combination of rotation and reflection.

The special case of a null rotation (i.e. rotating through zero angular displacement) combined with null translation (i.e. translating through zero linear displacement) is described as the **identity** operation, and leaves the object entirely unmoved. More general transformations for which the net translational vector $\mathbf{T} + \mathbf{t}$ is null are known as **point operations**, because they leave at least one point in space entirely unmoved, while those for which the net translational vector is finite may be described as **space operations**. Clearly, the distinction between the two categories can be highly dependent upon one's choice of origin. Space operations that can be reduced to point operations by a suitable choice of origin are known as **symmorphic** operations. Space operations that cannot be reduced to point operations by a suitable choice of origin are known as **non-symmorphic** operations, of which the **glide** symmetry operation (a combination of reflection across a plane with translation through half of a Bravais lattice vector lying within that plane) and the **screw** symmetry operation (a combination of rotation about an axis with translation by some fraction of a Bravais lattice vector lying along that axis) are the commonly recognised examples.

The set of all possible symmetry operations for a given object is known as its **space group**, and may in general contain both point and space operations. For a spatially finite object (such as a molecule) all of these operations are necessarily symmorphic, but for a spatially periodic object (such as a perfect crystal) both symmorphic and non-symmorphic space operations may arise. A space group containing *only* symmorphic operations is said to be a symmorphic space group, while one that contains even a single non-symmorphic operation is said to be a non-symmorphic space group.

The set of transformations obtained by simply *omitting* the translational component from each symmetry operation of the space group of an object is known as the **point group of the space group**. In the case of a spatially finite object, the point group of the space group is *identical* to the space group for one particular choice of origin. For a spatially periodic object, however, this will only be true if the space group is symmorphic; otherwise, the point group of the space group will contain transformations distinct from the symmetry operations of the space group, whatever the choice of origin. In either case, the point group of the space group consists of a set of exclusively point operations. Furthermore, in the case of a spatially periodic object, it is this point group and not the space group that will constrain the macroscopic shape (and all other macroscopic properties) of that object.

It is worth noting very carefully that the point group of the space group of a crystal is not necessarily synonymous with either 'the set of all point symmetry operations pertaining to the lattice' or 'the set of all point symmetry operations pertaining to the crystal'. If the space group contains glide or screw operations, then the point group of the space group will contain reflection or rotation symmetry operations that are not symmetries of the crystalline structure; conversely, the point group of the space group may lack certain symmetry operations that are indeed symmetry operations of the lattice. Despite the scope for ambiguity, however, the point group of the space group is generally referred to as the point group *of the crystal*, and we shall follow this practice in all subsequent chapters.

1.2 Proper and Improper Symmetry

Amongst all the various symmetry operations that may be found within the point or space group of an object, those for which the matrix **R** has a determinant equal to unity possess an important special property – they can actually be performed upon a rigid physical body in the real world, not just as a thought experiment or as a reflection in a mirror. Examples include, but are not limited to, the identity, rotation, translation and screw operations. Such operations are collectively described as **proper** symmetry operations.

In contrast, operations including, but again not limited to, the reflection, inversion and glide operations *cannot* be performed upon a rigid physical body in the real world. These are operations that can only be performed as a reflection in a mirror, or as a thought experiment, and they are collectively described as **improper** symmetry operations. The determinant of the matrix **R** in these cases will be minus unity.

Sequential application of successive proper symmetry operations to an object achieves a resultant operation that is itself also a proper operation; the transformation matrix obtained by multiplying several matrices of unit determinant together will itself have a determinant of unity. Similarly, a sequence of operations in which an even number is improper will also result in a proper composite operation, since multiplying together their individual transformation matrices will again yield a matrix with unit determinant. Concatenating a sequence of operations for which the number of improper operations is odd, however, results in an improper combined operation, described by a transformation matrix whose determinant is minus unity.

The crucial significance of these definitions in our present considerations is that the mirror image of an object can be manipulated by a series of proper symmetry operations to look identical to the original object if, and only if, its space group contains at least one improper symmetry operation. Referring back to our earlier definition of chirality, this may be re-stated as:

> An object whose space group contains no improper symmetry operations is necessarily chiral; the presence of even a single improper symmetry operation in the space group renders an object achiral.

This statement is precise and complete. It embodies the relatively obvious fact that an object possessing a mirror symmetry operation in its space group is achiral (since this implies straightforwardly that it is unchanged when reflected across a suitably oriented mirror plane) but also the somewhat less obvious fact that inversion or glide symmetries would be equally capable of ensuring that the object is achiral (even in the absence of mirror symmetry). Enumeration of the space group operations of an object, followed by their categorisation into proper and improper types, provides a foolproof test for chirality and is to be favoured over attempts at visualising the effects of reflection and subsequent manipulation.

1.3 Chirality in Finitude and Infinity

As with all aspects of symmetry, the chirality of an object is intimately tied up with its dimensionality. Objects that may be achiral in zero-dimensional isolation may become chiral upon repetition within a two- or three-dimensional lattice, and vice versa. Since we shall be interested, ultimately, in chirality at surfaces, where zero-dimensional molecules meet three-dimensional solids, it makes sense to describe the chirality of those cases first, before tackling the two-dimensional realm of the surface itself.

1.3.1 Molecular Chirality

As we have mentioned already, the space group of a spatially finite object, such as a molecule, is practically synonymous with its point group. That is, for a particular choice of origin, only symmetry operations having no translational component need be considered. Permissible symmetry operations therefore comprise rotations, reflections and inversions alone (or their combinations). A point group containing only rotations (including the identity operation) lacks improper symmetry operations, and therefore indicates that the object is chiral; one that includes either reflections or inversions (or both) indicates an achiral object. Real molecules collectively exhibit a wide range of different point groups of both the chiral and the achiral varieties.

Consider, for example, the pyruvic acid molecule depicted in Fig. 1.1a. The point group of this molecule includes a reflection across a mirror plane (the plane containing the carbon and oxygen atoms) and therefore indicates that the molecule is achiral. In contrast, the lactic acid molecules depicted in Fig. 1.1b, which may be obtained from pyruvic acid by hydrogenation, each entirely lack reflection and inversion symmetry operations in their point groups, demonstrating that they are chiral. As a consequence of this chirality, lactic acid molecules

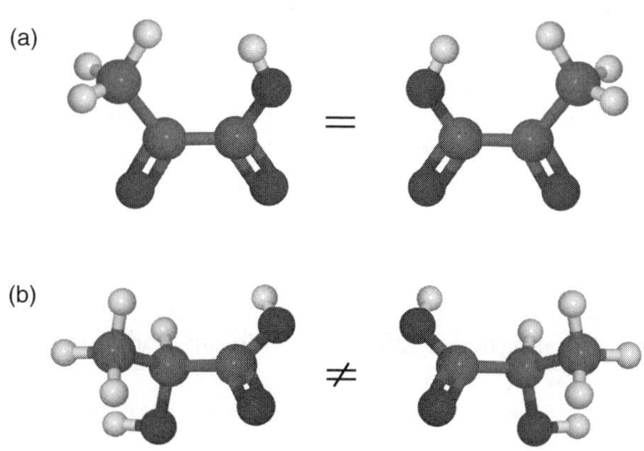

Figure 1.1 Mirror images of (a) pyruvic acid and (b) lactic acid molecules. In the case of pyruvic acid, the reflected molecule is identical to the original, to within a sequence of rotations; for lactic acid, however, the original molecule and its mirror image intrinsically differ and cannot be reconciled by any sequence of proper operations.

can be found in two forms, known as **enantiomers** (i.e. stereoisomers related to one another as mirror images). Enantiomers are necessarily energetically degenerate with one another and should arise in equal proportion in any process that is not in some sense asymmetric in nature.

Now, the symmetry properties of a molecule are clearly properties possessed by the molecule *as a whole*, not by some fragment of the molecule in isolation. It follows, therefore, that the chirality or otherwise of a molecule is also a property of the whole molecule, and we would do well to keep this important point always in mind. Nevertheless, it is clear in the case just described that chirality appeared when a particular kind of change was effected at one particular carbon atom, where hydrogenation occurs. In pyruvic acid, this carbon atom forms three bonds to different atoms or groups of atoms, and these bonds lie within the mirror plane of the molecule. In lactic acid, on the other hand, this carbon atom now forms four bonds to different atoms or groups of atoms, and these can no longer all lie within the same plane. This carbon atom is said to be a **chiral centre**, because it is the vital change in geometry at this site in lactic acid that breaks the improper symmetry found in pyruvic acid. In general, any atom (usually carbon) that forms bonds to four *chemically or geometrically distinct* atoms or groups of atoms will be a chiral centre; if fewer than four bonds are present,[2] or if the bonded atoms or groups of atoms are not all distinct, then the central atom is not a chiral centre. Molecules, such as pyruvic acid, featuring an atom that can be converted into a chiral centre in a single reaction step are designated **prochiral**, and are often of importance in synthetic pathways leading to chiral products.

One further advantage to identifying the chiral centre of a molecule arises in the matter of nomenclature. Clearly, the two enantiomers of lactic acid shown in Fig. 1.1b are distinct from one another, but normal chemical naming conventions are incapable of specifying which of the two we might be discussing at any given moment. The Cahn–Ingold–Prelog convention, however, provides an entirely robust procedure for assigning identifying labels to the enantiomers, as follows:

1) Identify the chiral centre of each enantiomer, assuming for the moment that there is only one;
2) Assign priorities to the groups surrounding the chiral centre, according to a more or less arcane set of rules based on the atomic numbers and connectivities of the constituent atoms;
3) Viewing the chiral centre from the direction precisely opposite to its lowest priority group, the label *R* is assigned if one sees a clockwise arrangement of the remaining groups from high to low priority, and the label *S* if the order is anticlockwise.

There is certainly some devilry in the detail of Step 2, which we shall not pursue further here; the interested reader is referred to the original description [1]. For our present purposes, it is sufficient merely to recognise that the labels *R* and *S* may be assigned unambiguously for any individual chiral centre. The letters *R* and

2 For this purpose, a well-defined lone pair may be considered as a dangling bond that counts to the same degree as a non-dangling bond.

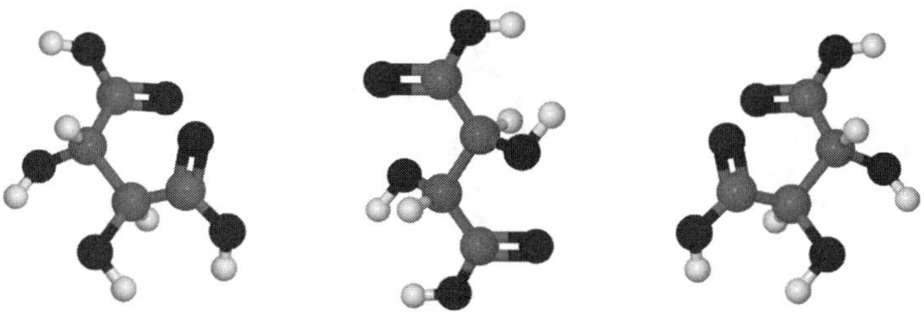

Figure 1.2 Chiral tartaric acid (*R,R* and *S,S* enantiomers) and the meso isomer (*R,S*-tartaric acid, equivalent to *S,R*-tartaric acid).

S stand, respectively, for the Latin words *rectus* and *sinister* – meaning 'Right' and 'Left' – but clearly the assignment is essentially arbitrary (albeit well defined) and does not correspond in any way to any essential quality of 'rightness' or 'leftness'. Indeed, closely related molecules, differing only by a relatively minor substitution (e.g. replacement of an alcohol group by a thiol group) may acquire opposing *R/S* labels.

An alternative labelling convention, commonly applied when dealing with relatively simple chiral molecules, assigns labels D and L by making comparison with the enantiomers of glyceraldehyde. The *R* enantiomer of glyceraldahyde happens to effect a clockwise rotation in the plane of linearly polarised light (and is therefore said to be 'dextrorotary') whilst the *S* enantiomer causes an anticlockwise rotation (and is correspondingly said to be 'laevorotary'). The prefixes *dextro-* and *laevo-* are, once again, of Latin origin and again refer to 'right' and 'left'. Enantiomers (of other species) whose chiral centre may be synthesised through achiral chemistry from the dextrorotary form of glyceraldehyde are assigned the label D, while those that may be similarly synthesised from the laevorotary form are assigned the label L. Note that, once again, the procedure for assignment of labels is well defined (although more limited in applicability than the *R/S* convention) but apart from the case of glyceraldehyde itself, there is no link between the D/L labels and the property of dextro-/laevo-rotation.[3]

Although useful, the concept of a chiral centre must be invoked with care, as can be seen most clearly when a molecule contains two or more chiral centres. Consider the tartaric acid molecules depicted in Fig. 1.2, where each chiral centre has been labelled according to the Cahn–Ingold–Prelog convention. In the *R,R* and *S,S* cases, the point group of each molecule is devoid of reflection and inversion symmetries, and is thus chiral as we might have expected. In the *R,S* and *S,R* cases, however, each molecule possesses an inversion symmetry and is therefore achiral; in fact, these 'two' molecules are actually precisely the *same* molecule viewed from different perspectives. Tartaric acid can thus occur in a

3 Indeed, it is important to distinguish the D and L labels described in the text, which do *not* imply the sense of optical rotation, from the labels *d* and *l* which explicitly *do* indicate the dextrorotary and laevorotary forms. In view of the potential for confusion, the usage of *d*/*l* to indicate rotary sense is now deprecated, and ± labels used instead to represent the same quality. It should go without saying that the *R*/*S* labels of the Cahn–Ingold–Prelog notation are no more indicative of rotary sense than are the D/L labels.

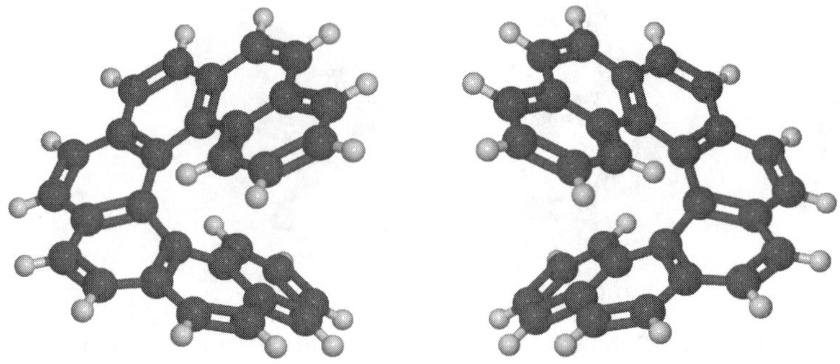

Figure 1.3 Heptahelicene enantiomers, conventionally labelled M and P (i.e. minus and plus).

chiral form, with two enantiomers labelled *R,R* and *S,S*, or in an achiral form in which the two chiral centres effectively cancel each other out. An achiral molecule in which two chiral centres negate one another is known as a **meso isomer**, and can only occur when the two chiral centres exist in otherwise identical parts of the molecule. More generally, of course, a molecule may have two chiral centres that exist in distinct parts of the molecule, in which case the *R,R* and *S,S* forms comprise one pair of enantiomers, while the *R,S* and *S,R* forms comprise a second pair of enantiomers not degenerate with the first.

Finally, it is very important to understand that a molecule may be chiral *without* possessing any chiral centres at all. A classic example may be found in the helicene molecules, where contiguous aromatic rings are connected so as to form a helical structure. No single atom can be identified as a chiral centre, yet these molecules are plainly chiral (see Fig. 1.3).

1.3.2 Crystalline Chirality

Turning to extended solid materials, it is customary to begin by ignoring the existence of any surfaces; in effect, one considers such a material to be infinite in extent. In the case of an amorphous solid, it is clear from the outset that all improper symmetry elements must be lacking, and hence that the object is chiral. This is, however, an essentially trivial form of chirality, since the exact mirror-image form of an amorphous solid is just one amongst an infinity of virtually degenerate alternative arrangements of the constituent atoms, rather than a practically realisable configuration. Non-trivial chirality in solids can only be spoken of when there is long-range order, so that a lack of improper symmetry really does imply the existence of enantiomeric configurations that can realistically be obtained and studied.

Let us, therefore, consider the range of possible space groups available to three-dimensional crystals. It turns out that there are precisely 230 of these, but the majority include improper symmetry operations and hence lead to crystals that are non-chiral. Only 65 of the space groups lack all improper symmetry operations, and crystals conforming to one of these so-called **Sohncke** space groups will necessarily be chiral.[4]

Examples of chiral solids include cases, such as crystalline tartaric acid, where a chiral molecule forms the basis of the unit cell. Here, the chirality of the situation is demanded by the molecule, and the long-range ordering may or may not reflect the chirality (i.e. chiral molecules may be distributed through space within a non-chiral lattice). On the other hand, one also finds crystals, such as quartz, where there are no identifiably chiral sub-units, and in these cases it is precisely the long-range ordering that is responsible for chirality.

1.4 Routes to Surface Chirality

In considering surface chirality, it is worth dismissing (as above) those trivial cases where chirality is *formally* present due to the absence of improper symmetry, but *practically* absent because one is dealing with an amorphous or otherwise disordered system. We shall restrict ourselves, therefore, to the surfaces of crystalline materials, and to adsorption that leads to at least partial (if not complete) ordering of adsorbates. Even within these restrictions, there are several ways in which chirality may arise.

1.4.1 Surfaces of Intrinsically Chiral Crystals

Perhaps the most obvious way in which surface chirality may arise occurs at the surfaces of chiral crystals. Clearly, if the underlying bulk material is itself already chiral, the creation of a surface (whether by cleavage, cutting or growth) cannot possibly *add* any of those improper symmetry elements whose presence would be needed to render the surface achiral. In this way, the surfaces of crystals such as quartz or tartaric acid, mentioned in the previous section, would necessarily be chiral, whatever orientation the surface takes. In terms of possible applications for chiral surfaces, one feature here is that the particular enantiomer to be used may be chosen at the level of the bulk material. Less conveniently, however, one is limited to only those materials that happen to crystallise in these forms; to date, the surface of quartz has dominated work in this area [2]. If one's interest lies in asymmetric heterogeneous catalysis, for example, then the absence of chiral crystalline polymorphs for reactive transition metals would be problematic. It may, of course, be possible to use chiral crystals as support materials, in the hope that they may template the formation of reactive metals into asymmetric thin-films or nanoparticles, but this avenue has yet to be explored in depth.

4 Confusingly, the term **chiral space group** is reserved *not* for the full set of 65 Sohncke space groups, but only for a subset of 22 space groups whose symmetry elements are not preserved upon applying an improper symmetry operation *to the space group*. It is self-evident that a space group whose formal structure is itself chiral can only give rise to a chiral crystal; the same is true, however, for the other 43 Sohncke space groups, whose symmetry elements are related to one another in an achiral fashion, but which simply do not include any that correspond to improper operations.

1.4.2 Intrinsically Chiral Surfaces of Achiral Crystals

Whenever the three-dimensional structure of a bulk crystal is terminated by a surface plane, there exists the possibility that certain of the symmetry operations of the bulk may be removed. The most trivial example would be where the surface plane lies parallel to a bulk mirror plane, in which case the mirror plane most certainly would not remain as a symmetry operation of the truncated material. More generally, all types of symmetry operation are potentially susceptible to being broken by the existence of the surface – which ones actually *are* broken in any given case will be determined by the space group of the bulk crystal and the particular orientation of the surface. Assuming the bulk crystal to be achiral, the possibility of surface chirality will be realised whenever the surface orientation is such as to remove all of the improper symmetry elements present in the bulk. In contrast to the examples in the preceding subsection, this form of surface chirality may be achieved, at least in principle, for *any* crystalline material, including those taking the highly symmetric face-centred cubic, body-centred cubic and hexagonal close-packed structures favoured by (chemically interesting) transition and coinage metals. Another advantage is that the same parent bulk material may be used to create surfaces of opposing chirality, simply by changing the orientation at which the surface is cut. The link between orientation and chirality will be thoroughly discussed in the next chapter, but for the moment we merely present a couple of examples to illustrate our point.

The fact that surfaces of achiral materials may be chiral by virtue of their intrinsic structure seems to have been explicitly discussed first by Gellman and co-workers [3] in a 1996 paper concerning the {643} surfaces of silver. They noted that surfaces possessing regularly kinked step-edges must be chiral if the length of step either side of each kink differs (see Fig. 1.4a). Furthermore, they introduced the idea of labelling the chirality of such surfaces within a convention inspired by the Cahn–Ingold–Prelog rules; the priority of groups surrounding a chiral centre was replaced with that of steps and terraces surrounding each chiral kink, arbitrarily defined in the (descending) order long-step/short-step/terrace. A little later, Ahmadi *et al.* [4] pointed out that the kinked surfaces of face-centred cubic materials would be chiral even if the step lengths were equal (e.g. in the {531} surface type, see Fig. 1.4b) and Attard subsequently proposed a modified version of the Cahn–Ingold–Prelog convention that has become the most widely used in the field [5]. This revised scheme is rooted in the observation that each kink site is necessarily surrounded by microfacets of {111}, {100} and {110} type, arbitrarily assigned (descending) priority in that order; if the microfacets cycle clockwise around the kink atom, in descending order of priority, the surface is labelled R, and conversely the designation S correlates with the anticlockwise sense. Although unambiguous, and relatively easy to use, this convention is limited when applied to materials of other crystalline structures. For example, some surfaces of body-centred cubic materials (e.g. the {321} surface type, see Fig. 1.4c) may be chiral despite the complete absence of kink sites. For this reason, an alternative notation has been proposed [6], in which the labels D and L are used, based upon the symmetry of crystallographic directions, rather than

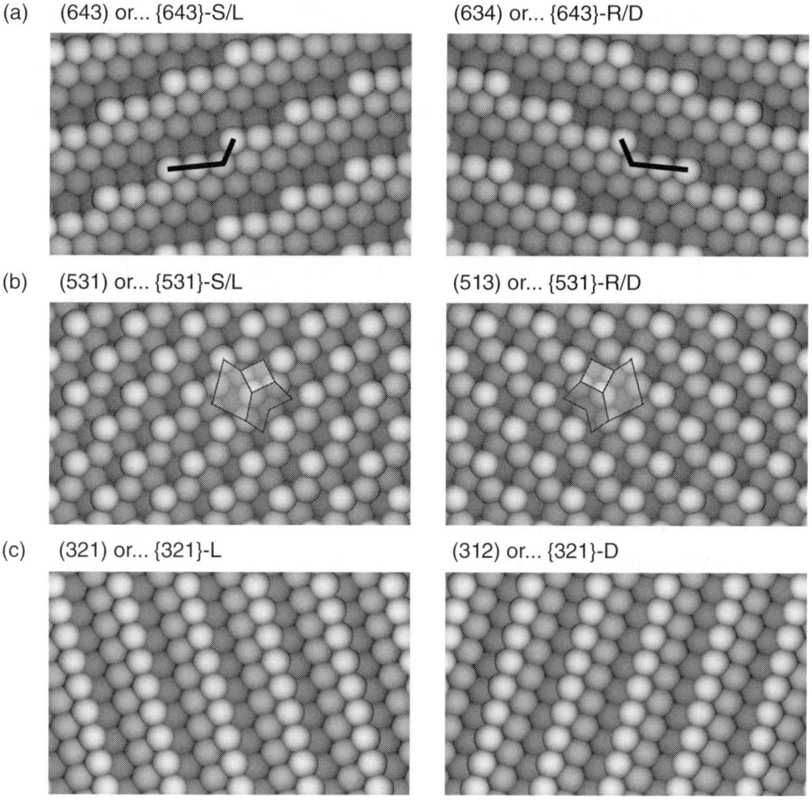

Figure 1.4 Enantiomorphically related surfaces of (a) *fcc*-{643} type, (b) *fcc*-{531} type, and (c) *bcc*-{321} type. In (a) an L-shaped mark highlights unequal step lengths either side of a kink atom, while in (b) the four-, five- and six-sided shapes indicate {100}, {110} and {111} microfacets, respectively. For the *fcc* cases, both the R/S and D/L notations are given, but only the latter is suitable for the *bcc* case.

the structure of microfacets; for face-centred cubic materials, the D and R labels correspond with one another, as do the L and S labels, but the approach is more widely applicable to other crystal structures, as we shall see in the next chapter.

1.4.3 Chiral Modification of Achiral Surfaces

Finally, we note the possibility that a surface whose structure is achiral when clean may be rendered chiral by the adsorption of chiral molecules. If these molecules arrange themselves in disordered fashion on the surface, then the long-range chirality thus obtained will be of a trivial nature, but one may still speak of non-trivial chirality at short length-scales (of the order of the molecular size); we might term this **irreducible molecular chirality** (as distinct from the **contingent molecular chirality** that may be induced in adsorbed achiral molecules). In such cases, the surface may now react in an asymmetric manner towards further chiral stimuli – surface chemical reactions, for instance, may become **enantioselective**[5] – but the mechanism must be explicable at a purely local level. On the other hand, it may be that adsorbed chiral molecules arrange themselves on the surface in a highly ordered fashion, and if they do so then it is further possible (though by no means certain) that they may adopt chiral long-range order; we might term this **intrinsic lattice chirality** if it is dictated by the overlayer lattice *per se* and **extrinsic lattice chirality** if it arises from the orientation of an otherwise achiral overlayer lattice relative to the underlying substrate lattice. If either of these situations arises globally, the surface ought once again to display an asymmetric response to chiral stimuli, but the origin of the asymmetry may lie in either the irreducible molecular chirality or the intrinsic/extrinsic lattice chirality (or some combination of both). Upon adsorption of an even-handed mixture of chiral molecules, domains of opposing local lattice chirality may occur, but the global behaviour will fail to display asymmetry, owing to degeneracy – mirror-image domains will arise in essentially equal proportion. On the other hand, even the adsorption of achiral molecules may result in locally lattice-chiral domains, once again displaying degeneracy and hence a lack of global asymmetry.

Furthermore, we ought also to recognise the possibility that adsorption of chiral molecules may induce a chiral reconstruction in the underlying substrate. Such a reconstruction might be ordered at long range (i.e. associated with a chiral lattice) or only at short range (e.g. the surface might break up into a series of facets, each locally ordered at the nanoscale but of varying size and shape at the mesoscale). In the case of a facetted surface, it may occur that some facet orientations are found to correspond to intrinsically chiral surfaces, in which case they may be said to display **planar substrate chirality** within themselves; nevertheless, even achiral facets may meet along low-symmetry edges or at low-symmetry vertices, in which case these may be viewed as loci of **linear or pointlike substrate chirality**. The range of possible behaviours for adsorbed chiral molecules will be discussed in depth across several later chapters, but for the moment we simply emphasise the key distinctions between the length-scales and dimensionality pertaining to different forms of surface chirality. We summarise these as follows:

Irreducible Molecular Chirality
Chirality associated with a single molecule, and which cannot be removed without breaking at least one covalent bond. Such chirality in an adsorbate would typically be present even if the molecules were abstracted from the surface (e.g. alaninate is chiral not only when adsorbed on a surface, but also in solution).

Contingent Molecular Chirality
Chirality associated with a single molecule, but which can be removed without breaking even a single covalent bond. Such chirality in an adsorbate is typically related to internal deformation that would vanish if the molecules were abstracted from the surface (e.g. glycinate may adopt a chiral conformation when adsorbed on a surface, but is achiral in solution).

5 We define an enantioselective reaction as one which proceeds with different rates for the two different enantiomers of the same chiral species.

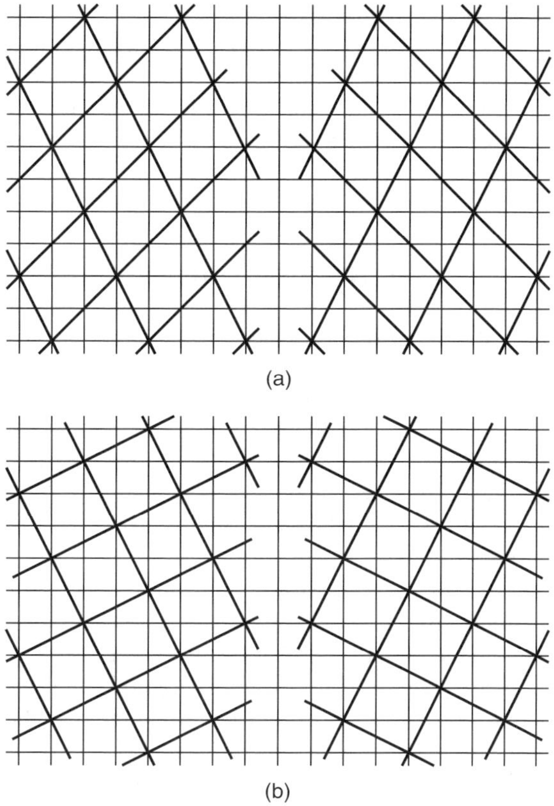

Figure 1.5 Intrinsic and extrinsic lattice chirality. In panel (a) the overlayer lattice, indicated with bold lines, is oblique in nature and would exhibit two mirror-image forms regardless of its relationship with the underlying substrate lattice, shown in feint lines. In panel (b) the overlayer lattice is square in nature, and would be achiral in isolation; chirality here derives only from the overlayer lattice's misalignment with the substrate lattice.

Intrinsic Lattice Chirality
: Chirality associated with the intrinsic nature of the two-dimensional lattice adopted by adsorbates. Such chirality (see Fig. 1.5a) is only present when the overlayer lattice is of oblique type (i.e. not of square, triangular, rectangular or rhombic symmetry).

Extrinsic Lattice Chirality
: Chirality associated with the registry between the two-dimensional lattice adopted by adsorbates and that of the underlying substrate. Such chirality (see Fig. 1.5b) may occur for any type of overlayer lattice apart from the oblique type (for which lattice chirality is intrinsic).

Planar Substrate Chirality
: Chirality associated with the arrangement of substrate atoms across a two-dimensional facet (which may or may not lie parallel to the macroscopic surface plane). Such chirality is dictated only by the orientation of the facet and the three-dimensional structure of the parent bulk material (see Chapter 2).

Linear Substrate Chirality
: Chirality associated with the arrangement of substrate atoms along a one-dimensional line where two facets meet (either along a ridge or a furrow). Such chirality is dictated only by the orientations of the two facets and the three-dimensional structure of the parent bulk material (see Section 3.6.1).

Pointlike Substrate Chirality
Chirality associated with the arrangement of substrate atoms at a zero-dimensional point where three facets meet (either at the apex of a pyramid or the bottom of a pit). Such chirality is dictated only by the orientations of the three facets and the three-dimensional structure of the parent bulk material (see Section 3.6.2).

1.5 Diastereoisomerism Defined

The categorisation of surface chirality introduced in the preceding section makes it clear that a given system may, taken as a whole, display chirality in multiple aspects simultaneously. Throughout this book, we shall be interested in situations where the combination of two chiral aspects results in an **enantiospecific** outcome, defined as one that distinguishes in some way between the two different enantiomers of one of these chiral aspects.[6] This already rather general concept may, in turn, be further generalised by considering the labelling of chiral entities in the most abstract sense.

Let us imagine, therefore, that some particular system displays chirality in two distinct aspects. These might be, for instance, planar substrate chirality and irreducible molecular chirality, but for our present purpose we need only recognise them as separate from one another and keep track of their chiral labels independently. If each chiral aspect is individually labelled either D or L, then the four possible combinations may be listed as DD, LL, DL and LD, where the first-listed symbol should be taken to label the chirality of the first chiral aspect of the system, and the second-listed symbol to perform the same function for the second. Clearly, the combinations labelled DD and LL are precise enantiomers of one another and must therefore represent degenerate situations; similarly, the DL and LD combinations form another degenerate enantiomeric pair of situations. It is evident, however, that the DD and LL combinations are related to neither the DL nor the LD combination by mirror symmetry. This type of non-degenerate and non-enantiomeric relationship is said to be **diastereomeric** in nature, and the entire pattern of enantiomeric and diastereomeric pairs (Fig. 1.6) constitutes

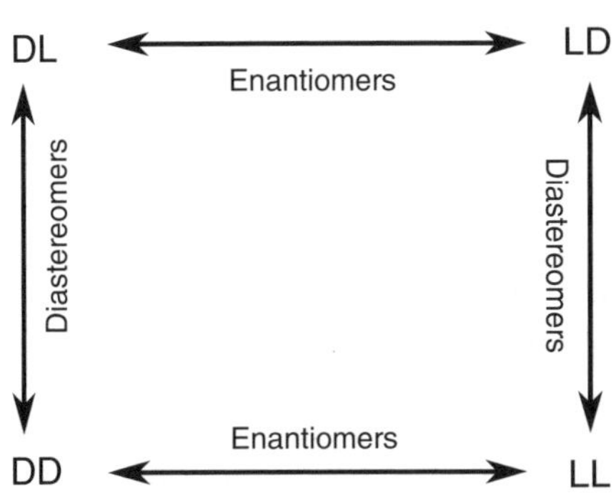

Figure 1.6 Pattern of diastereoisomerism, whereby DL and LD combinations form a degenerate enantiomeric pair, as do DD and LL combinations. In contrast, DL and DD combinations form a non-degenerate diastereomeric pair, as do the LD and LL combinations.

6 Strictly speaking, the more restrictive term 'enantioselective' (previously defined) only really applies to the concept of enantiospecificity in the context of chemical reaction rates.

the phenomenon of **diastereoisomerism**.[7] The origin of all enantiospecificity lies in the diastereomeric difference pertaining to some property of the system in question – purely enantiomeric differences are discernable only by other chiral systems, whereas diastereomeric differences are relevant to all manner of properties that are not themselves inherently chiral, such as thermodynamic and/or kinetic parameters.

1.6 Quantifying Chirality?

Defined as the absence of improper symmetry elements, the state of being chiral is absolutely binary. It is no more possible to be 'a little bit chiral' than it is to be 'a little bit pregnant'. An entity is either chiral or it is not, and it ought therefore to be readily apparent that any effort to *quantify* the degree of chirality such an entity possesses must surely be doomed to failure – or at least ambiguity. This consideration has not, however, prevented a number of authors from proposing 'figures of merit' that purport to reveal how chiral an object might be. At best, these quantities may correlate with *some* aspect of chirality, but considerable caution should be exercised in their interpretation.

The essential point to grasp is this: while it is true that a particular chiral entity may behave as if it is 'more chiral' than another, according to the outcome of some specific type of experiment, it is always possible that the apparent 'order of chirality' may be reversed when studied via some other type of experiment. It is only ever possible to gauge the 'degree of chirality' in terms of the system's diastereomeric response to some well-defined 'probe of chirality'.[8]

To give a concrete example, consider two entirely different species whose degree of chirality is first assigned by means of their optical rotary power (i.e. their ability to rotate the linear polarisation vector of light in transmission); let us suppose that we test enantiopure samples of each species that happen to rotate the polarisation vector in the same sense, and naively insist that the species with the higher optical rotary power is somehow 'more chiral' than the other. If the same two species are subsequently tested as to their ability to act as asymmetric catalysts in some particular reaction, there is absolutely no reason to suppose that the species with the greater optical rotary power will be more effective in producing an excess in one of the product enantiomers over the other; indeed, there is no reason a priori to suppose that the two species will produce asymmetry amongst the products in the same sense as each other, despite their optical rotations being in the same sense. It is clear, therefore, that

7 We have already come across one example of this type of chiral combinatronics in the case of molecules possessing two chiral centres (see Section 1.3) where we noted that the *R*,*R* and *S*,*S* forms were enantiomers of one another but distinct from *R*,*S* and *S*,*R* forms that constitute a second enantiomeric pair.

8 Harris *et al.* [7] have rigorously confirmed this principle within the framework of group theory, showing that not only may there be infinitely many chiral parameters that characterise a single chiral system equally well, but also that these parameters can vary independently of one another under a continuous deformation of that system. Indeed, it is even possible for a particular parameter to switch sign independently of all the others, so that the apparent handedness of a system may swap from the viewpoint of one probe, but not from that of another!

the 'quantity' and 'sign' of chirality differs between physical and chemical probes of the chirality, according to the diastereoisomerism exhibited by the chirality of the sample coupled with that of the probe. And that is even without considering that one might devise more than one type of physical probe (based, for example, on stress or magnetism) and more than one type of chemical probe (based on a different chemical reaction). The quantification of chirality by means of a particular chiral probe is of use only in the context of *that specific probe,* and hence is of no *practical* use whatsoever in predicting behaviour in other contexts.

For the reasons outlined above, there have been a few efforts to devise schemes for quantifying chirality based upon geometric considerations alone. These efforts miss the point, however, since they may equally be considered (in an abstract sense) as tests of chirality based not upon a physical or chemical probe but upon a geometric probe. It is, of course, entirely possible to define geometric tests to quantify the extent to which one enantiomer differs from its mirror image, but there is no reason to suppose that any scale of chirality based upon such a test should have any bearing upon the behaviour of those enantiomers when subjected to asymmetric physical or chemical stimuli. Even restricting ourselves entirely to the realm of geometry, inconsistency is bound to arise, as may be seen by extending the classic example of hands and gloves to include feet and shoes.

The origin of asymmetric response in chiral systems is often illustrated by means of analogy with the idea of fitting hands into gloves.[9] Imagine that a left hand and a right hand correspond to the two enantiomers of a chiral species. So long as those two hands experience only non-chiral environments, their behaviours will be indistinguishable. Tasked with catching a spherical ball, for instance, either hand can do so equally well. Faced with a chiral environment, however, the two hands may be readily distinguished. If provided with a right-handed glove, for example, the right hand will be able to slip inside easily, whereas the left hand will not. This much is obvious, and implies that the 'glove-shaped' geometric probe registers the hands as being 'very chiral' (i.e. the diastereomeric difference in geometry for the hand–glove system is large). Extending the analogy to include feet, one imagines a left foot and a right foot, equally capable of kicking a spherical ball, but only the right foot will slip easily into a right-footed shoe. Here, the 'shoe-shaped' geometric probe registers the feet as being 'very chiral' (i.e. the diastereomeric difference in geometry for the foot–shoe system is large).

The ambiguity of this approach is revealed, however, when the 'shoe-shaped' geometric probe is used to test the chirality of hands; here, both hands will fit into a right-footed shoe with approximately equal ease – the probe will register the hands as being only 'slightly chiral' (i.e. the diastereomeric difference in geometry for the hand–shoe system is small). Similarly, if the 'glove-shaped' geometric probe is used to test the chirality of feet, both feet will fit into a right-handed glove almost equally badly – the probe will register the feet as being only 'slightly chiral' (i.e. the diastereomeric difference in geometry for the foot–glove system is small). The moral of the story is that there is no such thing as an absolute quantitative measure of chirality, even when restricting oneself to

9 See the preface to this book.

the realm of pure geometry, let alone when concerned with physical or chemical phenomena. Chirality simply *appears* to be strong when the probe that is in use happens to match, in some sense, the nature of the chiral entities under test. It is quite possible for an object to yield a very strong asymmetric response under one set of circumstances, despite showing only a very weak asymmetric response under another. In this book, we shall endeavour to refrain from making absolute statements about the degree of chirality evident at a surface, as opposed to specific statements about the asymmetry of response to a specific situation, reflected in the diastereomeric difference of some quantity of interest.

1.7 Enantiomeric Excess

Notwithstanding the admonition of the preceding section, there is at least one aspect in which chirality can be meaningfully quantified: when a *mixture* of chiral entities is present, one may sensibly express the dominance of one or the other enantiomer by means of that quantity known as the **enantiomeric excess**. Given the symbol *ee*, the enantiomeric excess is defined as

$$ee = \frac{N_{(+)} - N_{(-)}}{N_{(+)} + N_{(-)}} \tag{1.2}$$

where $N_{(+)}$ and $N_{(-)}$ are, respectively, the number of dextrorotary and laevorotary entities (or number of moles of such entities) present in the mixture. Despite arguments [8, 9] that this definition introduces needless complexities that would be avoided by use of a simple enantiomeric ratio (i.e. $er = N_{(+)}/N_{(-)}$) the enantiomeric excess remains the *de facto* standard for quantifying the asymmetry of chiral mixtures.[10] We note here two special cases, namely: (i) the so-called **racemic** case, where the enantiomeric excess is precisely zero, and (ii) the **enantiopure** case, where the enantiomeric excess is either plus or minus unity. A sample with arbitrary enantiomeric excess can always be obtained by combining a **racemate** (i.e. a racemic mixture) with an appropriate quantity of enantiopure material.

Extending the concept from the description of chiral *entities* to the description of chiral *processes*, we can define an **enantiomeric excess in the rates** of a reaction. That is, we specify

$$ee_r = \frac{r_{\parallel} - r_{\perp}}{r_{\parallel} + r_{\perp}} \tag{1.3}$$

where we shall take $r_{\parallel}$ to be the rate when the reaction takes place between entities whose chiral labels (i.e. D or L) happen to match, and $r_{\perp}$ to be the rate when the chiral labels do not match. The $\parallel$ and $\perp$ symbols thus identify the two non-degenerate states of the diastereomeric situation; positive values of ee_r imply that the reaction occurs fastest when the chiral labels of the involved entities match, while negative values would imply a faster reaction when they differ.

---

10 It is, of course, essentially trivial to convert between enantiomeric excess and enantiomeric ratio. The necessary equations are simply $ee = (er - 1)/(er + 1)$ and $er = (1 + ee)/(1 - ee)$.

Furthermore, we may write these rates as

$$r_{\parallel} = k_{\parallel} f_{\parallel} \tag{1.4}$$

and

$$r_{\perp} = k_{\perp} f_{\perp} \tag{1.5}$$

where $k_{\parallel}$ and $k_{\perp}$ are rate 'constants' varying only with temperature, and where $f_{\parallel}$ and $f_{\perp}$ are functions representing all other aspects of the current state of the system; in the context of a gas-phase reaction, these latter functions will, in general, depend upon the concentrations of any reactants; for a surface reaction, the relevant variables may be the coverage of reactants and/or the areal density of active sites. With these definitions in mind, we may now introduce the quantity

$$ee_k = \frac{k_{\parallel} - k_{\perp}}{k_{\parallel} + k_{\perp}} \tag{1.6}$$

which we shall term the **enantiomeric excess in the rate constants**. Note that this last quantity must be carefully distinguished from the enantiomeric excess in the rates – identical rate constants may give rise to rather different actual rates depending upon the current coverages and/or availability of active sites.

Now, rate constants for activated processes can generally be written in the form

$$k = \nu e^{-E/k_B T} \tag{1.7}$$

and doing so in the preceding expression yields

$$ee_k = \frac{\nu_{\parallel} e^{\alpha} - \nu_{\perp} e^{-\alpha}}{\nu_{\parallel} e^{\alpha} + \nu_{\perp} e^{-\alpha}} \tag{1.8}$$

with

$$\alpha = (E_{\perp} - E_{\parallel})/2k_B T \tag{1.9}$$

where $E_{\parallel}$ and $E_{\perp}$ are the activation energies, $\nu_{\parallel}$ and $\nu_{\perp}$ the corresponding pre-exponential factors, k_B the Boltzmann constant and T the temperature.

If we make the assumption that the pre-exponential factors are not sensitive to chirality (i.e. $\nu_{\parallel} = \nu_{\perp}$) this reduces to the simple form

$$ee_k = \tanh \alpha \tag{1.10}$$

which varies monotonically from plus or minus unity at absolute zero temperature to zero at infinite temperature. In fact, the same is equally true for the general expression also.[11] We should, however, stress that no such simple behaviour may be inferred for the enantiomeric excess in the rates, for the reason alluded to above – the rates are a product of the rate constants with functions expressing the current state of the system, and these latter will, in turn, depend upon

11 Differentiating Eq. (1.6) with respect to temperature and setting the result to zero, one finds $k_{\parallel} k'_{\perp} = k_{\perp} k'_{\parallel}$ (where primed quantities represent temperature derivatives) as the criterion for a stationary value of ee_k. Making use of Eq. (1.7), it becomes clear that this criterion is satisfied only in the trivial case where the activation energies $E_{\parallel}$ and $E_{\perp}$ are equal. In all other cases, the enantiomeric excess in the rate constants can have no stationary values and must, therefore, vary monotonically. Since the values at absolute zero and infinite temperatures may readily be shown to be unity and zero, the statement in the text is proved.

the history of the system. In comparing the enantiomeric excess in the rates, it matters greatly whether a given experiment is performed, for example, at constant temperature or at constant rate of temperature change.

Finally, we note that all three of these enantiomeric excess quantities may be recast in percentage form, namely, $\%ee = ee \times 100\%$ and similar expressions for $\%ee_r$ and $\%ee_k$. In this form, the range of each is transformed to $\pm 100\%$, and the racemic situation corresponds to a value of 0%.

1.8 Synthesis, Separation and Sensing

To conclude this introduction, let us reflect that whilst studying chiral systems may be intellectually stimulating in its own right, the true test of our mastery of any subject lies in our ability to make use of it for predefined ends. If we truly understand it, we ought to be able to apply it. From the standpoint of chemistry, at least, three target applications for surface chirality may readily be identified and summarised as follows:

Asymmetric Synthesis

Ideally, one could imagine a heterogeneous analogue of the catalytic processes currently carried out in the homogeneous realm, where chiral catalysts skew the product distribution of a particular reaction to favour a marked excess of one enantiomer over its mirror image. The advantage over a homogeneous catalytic reaction would be the ability to run continuously, rather than in batch mode, but the major challenge lies in achieving a sufficiently high enantiomeric excess. Applications for chiral molecules (e.g. in the pharmaceutical industry) typically require very high levels of enantiomeric purity, so if asymmetric synthesis alone is to be relied upon its selectivity must be extreme.

Asymmetric Separation

Failing to synthesise a particular chiral product with sufficient enantioselectivity, one might resort to *post hoc* separation of a racemic (or even just slightly enantio-impure) mixture. Again, this can already be achieved by homogeneous means, but the efficiencies of heterogeneous processing are very attractive. The essence of asymmetric separation lies in binding one enantiomer in excess over the other, but in addition to the inherent requirement for extreme selectivity, a further challenge lies in the need to avoid poisoning the adsorbent; an ideal solution might hypothetically be found in a switchable system capable of binding first one enantiomer and then the other, so that temporal cycling can be employed to maintain an active surface.

Asymmetric Sensing

Under this heading, one imagines a device capable of responding to the chemical asymmetry of two enantiomers by changing some readily measurable physical property. For example, such a device might respond to an enantiomeric excess by exhibiting a proportional (or at least monotonic) change of reflectivity, magnetism or conductivity. In this way, it would be possible to monitor the chiral composition of a flowing product stream in real time, confirming that asymmetric synthesis or separation processes are achieving the desired enantiomeric excess.

References

1 R.S. Cahn, C. Ingold and V. Prelog, *Angew. Chem. Int. Ed. Engl.* **5**, 385 (1966).
2 J.W. Han and D.S. Sholl, *Phys. Chem. Chem. Phys.* **12**, 8024 (2010).
3 C.F. McFadden, P.S. Cremer and A.J. Gellman, *Langmuir* **12**, 2483 (1996).
4 A. Ahmadi, G.A. Attard, J. Feliu and A. Rodes, *Langmuir* **15**, 2420 (1999).
5 G.A. Attard, *J. Phys. Chem. B* **105**, 3158 (2001).
6 S.J. Jenkins and S.J. Pratt, *Surf. Sci. Rep.* **62**, 373 (2007).
7 A.B. Harris, R.D. Kamien and T.C. Lubensky, *Rev. Mod. Phys.* **71**, 1745 (1999).
8 R.E. Gawley, *J. Org. Chem.* **71**, 2411 (2006).
9 B.C. Gibb, *Nat. Chem.* **4**, 237 (2012).

2

Surface Symmetry and Structure

[Nature] is an infinite sphere, the centre of which is everywhere, the circumference nowhere.

Pensées, Blaise Pascal, 1670 (tr. William F. Trotter, 1910)

In the previous chapter, we considered the various ways in which surfaces might be chiral, (i) through the intrinsic chirality of the bulk material; (ii) arising due to the choice of surface crystallographic plane; (iii) bestowed by adsorbed molecules. Now we will look in more detail at the middle option, clarifying the circumstances in which chirality does and does not occur. In order to do so, we will first introduce a useful method of visualising the totality of all possible surface planes for a given bulk crystal structure. This will allow us to identify families of surfaces sharing similar structural or symmetry properties.

2.1 Spherical Representation of Symmetry

To begin, let us imagine an infinitely large sphere crafted from a single perfect crystal of some given material. The crystalline nature of the sphere implies that it possesses well-defined crystallographic axes, which we shall suppose ourselves able to discover by some means, and hence it is possible to label points on the surface of the sphere through the use of Miller indices.[1] Indeed, microscopic analysis of the sphere's surface at a point labelled by some particular Miller indices would reveal an infinitesimally small patch of *approximately* planar surface whose structure locally matches that of the corresponding Miller-index surface that would be formed by slicing the crystal on the same plane to form an *extended* planar surface. The surface of the sphere as a whole, therefore, represents all possible surface orientations available to a crystal of the particular structure chosen. In Fig. 2.1a, the locations of a set of high-symmetry surfaces, belonging to the {111}, {100} and {110} types, are displayed for the case of a sphere constructed from a material

1 The crystallographic axes imply the existence of a real-space lattice, which in turn implies the existence of a reciprocal-space lattice; the Miller indices are simply integer multipliers for the primitive reciprocal-lattice vectors, defining the surface-normal direction and hence the orientation of the corresponding surface plane [1].

Chirality at Solid Surfaces, First Edition. Stephen J. Jenkins.

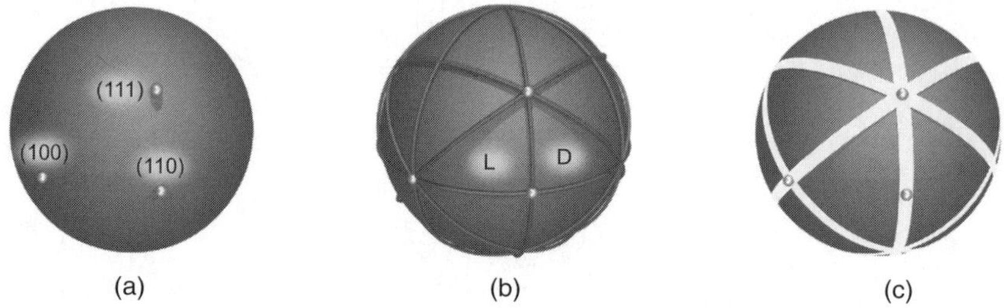

Figure 2.1 Spherical representations of surface symmetry and structure. In panel (a) the points on an infinitely large single-crystal sphere that correspond to patches of (111), (100) and (110) surfaces are highlighted. In panel (b) the 'mirror zones' corresponding to surfaces possessing at least one mirror symmetry are marked; regions lying off the mirror zones contain chiral surfaces, labelled either D or L within each triangle. In panel (c) the 'primary structural zones' are marked, corresponding to surfaces possessing at least one direction in which close-packed atomic chains lie within the surface plane.

conforming to a cubic crystal structure (either simple cubic, face-centred cubic or body-centred cubic). In the language usually employed in describing such a representation, we shall call these points on the surface of our sphere the **poles** of the corresponding Miller-index surfaces.

We can now make use of the sphere to catalogue Miller-index surfaces sharing particular qualities. Sticking with the example of a cubic crystalline material, Fig. 2.1b now highlights all those Miller-index surfaces whose structure features at least one plane of mirror symmetry. Again adopting the usual language, we shall call the great circles thus formed **zones** and label each one by the crystallographic direction (the **zone axis**) lying perpendicular to the plane defined.[2] Moreover, since the zones in question denote the existence of mirror symmetry, we shall be even more specific and describe them as **mirror zones**.

In fact, if we restrict ourselves, for the moment, to the consideration of *ideal* surfaces (i.e. we ignore the possibility of surface relaxation or reconstruction) we do not even need to scrutinise the structure of each Miller-index surface individually in order to construct a picture of the mirror zones. Consideration of the bulk symmetry found in simple cubic, face-centred cubic and body-centred cubic crystals (which share identical point groups) leads one to note that all feature mirror planes oriented perpendicular to the ⟨100⟩ and ⟨110⟩ crystallographic directions. Simply marking the intersection of the sphere's surface with a set of all such planes (each passing through the sphere origin) is sufficient to identify the Miller-index surfaces through which these bulk mirror planes cut vertically; these, and *only* these, Miller-index surfaces will display the corresponding mirror symmetry in their geometry. It is, therefore, precisely the ⟨100⟩ and ⟨110⟩ zones that we must identify as the mirror zones in this case.

---

2 Many authors take the zone axis to be the fundamental defining element of the zone, describing the corresponding great circle as merely the 'trace' of the zone. Semantics aside, both the zone axis and the great circle do nothing more than represent the same set of planes – the zone itself – and we shall simplify matters by referring to both as if they *were* the zone for which they stand. That is, references to a zone should always be regarded as referring fundamentally to a specific set of planes, but whether these are represented by a zone axis or a great circle should be inferred from context.

The surface symmetries possible for simple cubic, face-centred cubic and body-centred cubic crystals are thus *almost* fully represented by a set of ⟨100⟩ and ⟨110⟩ mirror zones marked on the surface of our sphere. No other mirror symmetry operations exist, the fact that these crystal structures feature only a single atom per primitive unit cell precludes the possibility of glide symmetry, and true three-dimensional inversion symmetry is impossible at a surface. Only the possibility of rotational symmetry is missing from the representation, but for the purposes of investigating chirality such proper surface symmetry is not relevant and the absence of it from our representation is no great loss.

Examination of the mirror zones in Fig. 2.1b leads us now to a number of conclusions – each individually relatively unremarkable to anyone with experience of the commonly studied surfaces of cubic crystals, but nevertheless noteworthy here because they achieve the level of proof rather than anecdote. Firstly, we note that there are three, and *only* three, types of Miller-index surface having more than one mirror plane, namely the {100} surfaces (which have four mirror planes), the {111} surfaces (which have three) and the {110} surfaces (which have two). Secondly, we note that there are infinitely many Miller-index surfaces that possess but a single mirror plane, namely all those (excepting the three types listed above) whose Miller indices feature zeroes or repeated magnitudes, such as the {210} surface (whose indices include a zero) and the {311} surface (whose indices include two of the same magnitude). And thirdly, we note that that there are infinitely many Miller-index surfaces whose structures feature no mirror planes at all and hence (in the certain absence of inversion symmetry) are necessarily chiral, namely those whose Miller indices are all non-zero and of differing magnitudes, such as the {531} surface.

Regarding those chiral surfaces, whose poles lie *within* the triangles defined by the mirror zones, it is worth remarking here that precisely *one* instance of each chiral {*hkl*}-type surface will be found in each such triangle.[3] All will be energetically degenerate with one another, but the instance found in any particular triangle will be of opposite chirality to those instances found in all adjacent triangles. So, for example, the (531) instance of the cubic {531}-type surfaces lies in a triangle adjacent to others encompassing the (513), (351) and ($53\bar{1}$) instances, all three of which share identical structures – the mirror image of that belonging to (531). It follows generally, therefore, that one may arbitrarily assign a chiral label to all the surfaces whose poles lie in one particular triangle, knowing then that the opposite label must apply to all surfaces whose poles lie in adjacent triangles. It remains only to decide upon a suitable arbitrary rule by which to make the first assignment.

One particularly attractive scheme for arbitrarily assigning a chirality label to any given triangle is based purely upon consideration of the symmetry found at that triangle's vertices. In general, it will be found that the surfaces whose poles lie at the vertices of a triangle will be of three differing types, with three differing

3 As per standard usage, the notation {*hkl*} implies the (hkl) surface and all other surfaces related to it by symmetry (either proper or improper). In general, the {*hkl*} set is obtainable from (hkl) by permutation of the order and sign of its Miller indices, *consistent with the point group of the space group*. We shall refer to each individual surface thus obtained as an **instance** of the {*hkl*} type.

symmetries. Moreover, it will often be found that these three symmetries can readily be placed in rank order from least to most symmetric.[4] Tracing a path around the perimeter of any given triangle from the least symmetric vertex, through the vertex of median symmetry, to the most symmetric vertex (and, by implication, on to the least symmetric vertex again) then involves a well-defined sense of rotation, either clockwise or anticlockwise, which may be invoked to assign the chirality label for that triangle. If the sense of rotation is clockwise, we assign the label 'D' to the triangle (and indeed to all surfaces whose poles lie within that triangle), whereas if it is anticlockwise we assign the label 'L'. In the *fcc* case described above, the (531) surface would lie within a triangle labelled L, and can hence be considered to be an instance of a cubic {531}-L surface; the (513), (351) and $(53\bar{1})$ surfaces, in contrast, would all lie within triangles labelled D, and are thus instances of cubic {531}-D.

2.2 Spherical Representation of Structure

We have seen how certain symmetry properties found at the Miller-index surfaces of a crystal may conveniently be summarised by plotting zones associated with corresponding symmetry properties of the bulk material (its mirror planes, in the case described above). Although illustrated for cubic crystals, the approach is entirely general and would work equally well if one wished to represent the symmetries found at the surface of, say, an orthorhombic crystal – the particular zones would differ, but the basic ideas would remain the same. Moreover, there is really no need to restrict ourselves to treating only the *symmetry* properties of Miller-index surfaces in this way. How then might we extend the idea to purely *structural* aspects of the Miller-index surfaces?

One reasonably obvious starting point is to consider the existence of close-packed chains of atoms at the surface. Such a feature will not only be readily apparent when one examines the detailed structure of the surface, but also carries with it important implications regarding the stability and reactivity of surfaces, as we shall shortly discuss. Let us, therefore, take notice of the directions within the bulk material in which atoms are found to be close-packed, and plot zones onto the surface of our sphere lying perpendicular to these directions; these **primary structural zones** (we shall meet secondary and tertiary examples later) will then indicate all surfaces featuring at least one surface-parallel direction in which the atoms are close-packed. To give a concrete example, we take the case of an *fcc* crystal, where the close-packed directions in the bulk material are exclusively the ⟨110⟩ directions. The corresponding zones are shown on the surface of our sphere in Fig. 2.1c, whereupon a number of observations may be made.

---

4 That is, a surface displaying three mirror planes is of higher symmetry than one displaying only two, for example. In cases where mirror and glide planes are both present, we can still apply this method, but rank according to mirror symmetries first, considering the number of glide planes only as a 'tie-breaker' if necessary. For the *fcc* crystal structure, the rank order of vertex poles runs from the {100} type, through the {111} type, to the {110} type.

Firstly, amongst the surfaces of an *fcc* crystal, only two types feature close-packed chains of atoms lying in two or more directions within the surface plane, namely the {111} surfaces (in which there are three such directions) and the {100} surfaces (in which there are two); surfaces of this kind necessarily display rather weakly corrugated structures, and so may usefully be described as **flat**. Secondly, there are an infinite number of surfaces featuring close-packed chains of atoms lying in a single direction within the surface plane, namely those *fcc* surfaces (other than those mentioned above) having two Miller indices of identical magnitude, such as the {110} or {311} surfaces; these surfaces are typically more corrugated than the flat surfaces, and may sensibly be regarded as **stepped**. And thirdly, there will also be an infinite number of possible surfaces in which *no* close-packed chain of atoms runs across the surface in any direction, such as the {210} and {531} surfaces; in these cases any chain of close-packed atoms lying *close* to the surface plane must necessarily 'kink' at regular intervals, and indeed one may reasonably describe such surfaces as being **kinked**. Examples of flat, stepped and kinked *fcc* surfaces may be examined in Fig. 2.2, whereupon the suitability of these descriptions ought to become evident.[5]

Now, the most important advantage of adopting this taxonomy of surface structure is revealed when one considers the coordination numbers of atoms in the surface layer. So long as the underlying bulk crystal structure involves just one atom per primitive unit cell, the least coordinated atoms of a kinked surface will possess precisely half the number of nearest neighbours as an atom in the bulk would have. The least coordinated atoms of a stepped surface (of the same material) would possess precisely one more nearest neighbour than this, due to the presence of chains of close-packed atoms along one direction within the surface plane. Finally, the least-coordinated atoms of a flat surface (again of the same material) would possess at least two more nearest neighbours than the least coordinated atoms in the kinked surface. For example, the kinked surfaces of an *fcc* crystal would all have a coordination number of 6 for their least coordinated atoms (precisely half the bulk coordination number for this structure), whereas the least coordinated atoms in the stepped surfaces all have a coordination number of 7; the flat surfaces of *fcc* crystals display coordination numbers of 8 (for the {100} surface) and 9 (for the {111} surface) for their least coordinated atoms. Similar trends in coordination number will be seen for all materials, although for crystals whose bulk structure involves more than one atom per primitive unit cell the governing rules are not, in general, so simply stated.

The consequence of this predictability in coordination number is quite profound when it comes to the stability of surfaces against thermal roughening. The removal of one of the least coordinated atoms from a kinked surface (of a material having one atom per primitive unit cell in the bulk) involves the loss of a certain

5 We should, however, recognise that the term 'flat' is often used in surface science to indicate morphology over a rather longer range, in contradistinction to the term 'rough'. Similarly, the term 'stepped' may sometimes refer to regular steps of the kind described here, but at other times to irregular steps between mesoscopic terraces that may themselves be either flat, stepped or kinked at the atomistic level. To avoid confusion, this book will adopt the term 'planar' to indicate a non-rough surface, and 'terraced' to indicate the presence of planar regions at differing heights; the boundaries between adjacent planar regions will then be referred to as 'terrace edges' rather than steps.

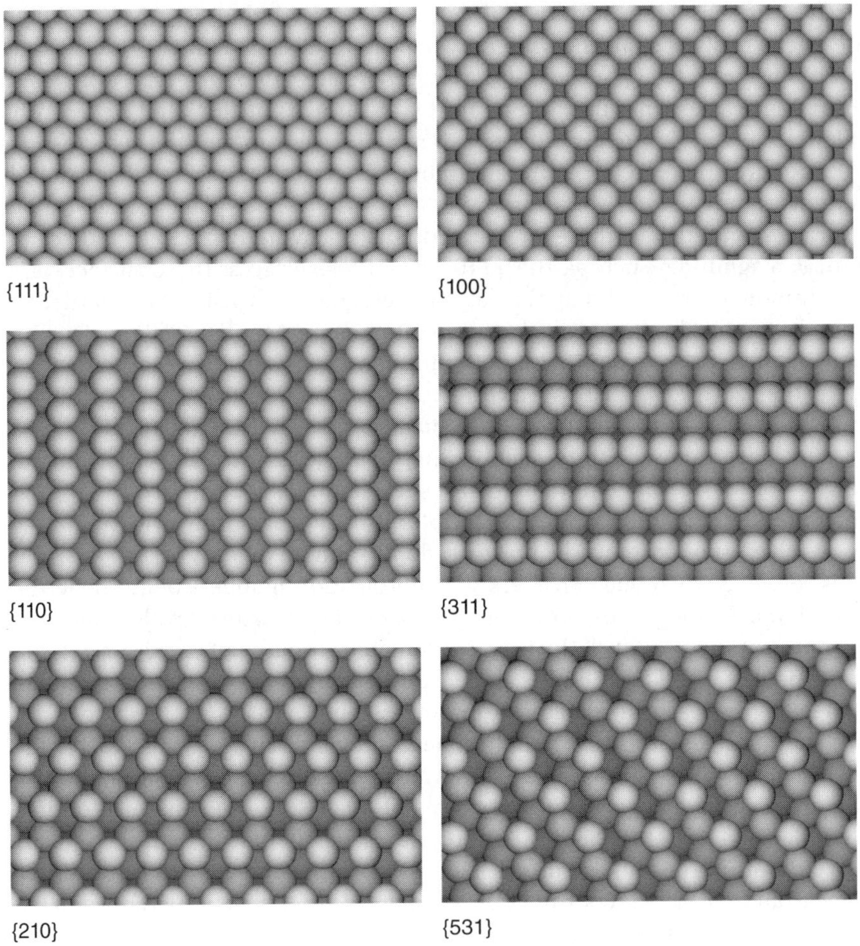

Figure 2.2 Flat ({111} and {100}), stepped ({110} and {311}) and kinked ({210} and {531}) surfaces of the face-centred cubic structure.

number of nearest-neighbour interactions, but the absence of intra-layer nearest neighbour interactions means that depositing that same atom onto a pristine section of surface some distance from its original site will result in the formation of *exactly the same number* of new nearest-neighbour interactions. To the extent that the cohesive energy of a solid mainly comprises its nearest-neighbour interactions, the formation of an adatom–vacancy pair in this way will be approximately thermoneutral. At any finite temperature, therefore, one may reasonably anticipate that entropy should drive the surface towards *repeated* and *random* adatom–vacancy pair formation, resulting in a surface that is rough at the atomic scale (see, e.g. the case of Cu{531} [2]). On a stepped surface of the same material, in contrast, the number of nearest-neighbour interactions lost upon removal of an atom to form a vacancy will always exceed (by two) the number gained when that atom is deposited on a distant patch of pristine surface; the entropic driving force towards disorder will consequently not succeed in producing a rough surface unless some critical temperature is exceeded. On the other hand, *concerted* adatom–vacancy pair creation leading to the formation of mesoscopic facets *can* be near-thermoneutral on both kinked *and* stepped surfaces, and thus may occur either spontaneously (relatively rare) or in response to adsorption (more common on kinked surfaces, but not unheard of on stepped ones). By similar arguments, the flat surfaces of this same material will be yet more stable against thermal roughening. Although defined in purely geometric terms, and in relation to the unrelaxed surface at that, the appellations flat, stepped and kinked thus turn out to have a strong connection to physical behaviour of real surfaces.

Indeed, looking beyond the propensity to adopt mesoscopically rough surface structures, the issue of coordination number ought also to be relevant to the chemical reactivity of surfaces. In general, the more undercoordinated an atom at the surface, the more strongly we should expect it to interact with adsorbed molecules.

2.3 Stereographic Projections: Flattening the Globe

Before proceeding to look in more detail at the surface symmetry and structure of specific crystalline materials, it must be admitted that the spherical representations introduced above, whilst attractive and useful, are not readily constructed without the aid of a computer. Put simply, three-dimensional objects are unsuited to being sketched on the back of either an envelope or a napkin, and so we might reasonably ask whether the same information might be adequately summarised in a two-dimensional representation instead. Fortunately, just such a representation does exist, in the form of a so-called **stereographic projection**.

The concept of a stereographic projection is not by any means unique to the description of surface symmetry and structure. In fact, such projections are widely used in the fields of materials science and crystallography to represent a variety of features of bulk materials (e.g. slip planes, etc). Even more broadly, the stereographic projection is just one of many different projections devised over the years to map the surface of a three-dimensional sphere onto a flat two-dimensional plane.[6]

To understand the stereographic projection, first imagine the spherical representation described above to be pushed up against a flat semi-transparent plane, as in Fig. 2.3. The point of contact between plane and sphere we shall call the **projection pole**, while the point on the sphere furthest from the plane will be termed the **projection antipole**. Next, imagining the sphere itself to be translucent but the poles and zones on its surface to be opaque, we mentally install a point light source at the projection antipole and consider the shadow that it casts upon the plane; clearly, this maps all the poles and zones onto a convenient

6 The Mercator projection has long been the *de facto* standard for mapping the entirety of the terrestrial globe onto the pages of an atlas, albeit many alternatives have been proposed to ameliorate is many deficiencies. The stereographic projection, however, has become probably the most common approach in mapping the Earth's polar regions.

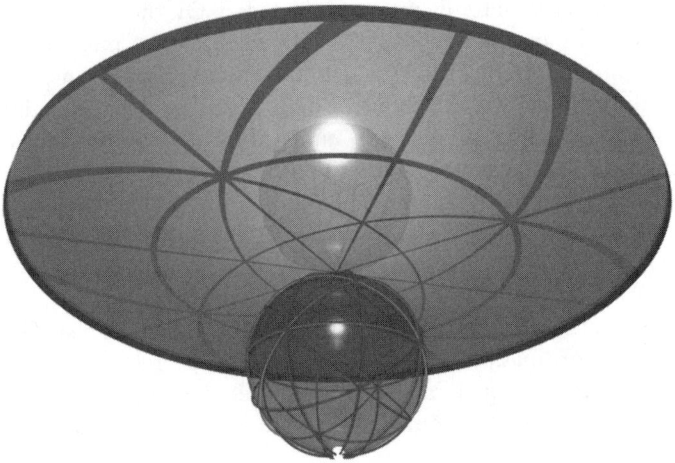

Figure 2.3 Construction of a stereographic projection, showing the mirror zones of face-centred and body-centred cubic materials.

two-dimensional representation. Conventionally, however, we imagine ourselves 'viewing' this representation from the other side of the semi-transparent plane, thus avoiding a reversal of chiral sense that would otherwise occur.

Notice that any zone passing through the projection pole will project onto the plane as a straight line extending to infinity, whereas zones that do not pass through the projection pole will project as circles. A notable special case, of course, is a zone whose identifying pole happens to be the projection pole itself, leading to a projected circle of radius twice that of the sphere, centred on the centre of the projection. Indeed, if we imagine such a zone (regardless of whether it turns out to be a mirror or structural zone of the material in question) the circle it projects onto contains within it the projections of all the poles and zones lying in the upper half of the sphere; the projections lying outside of this circle derive exclusively from poles and zones lying in the lower half of the sphere. It is usual, therefore, to display only the portion of the projection lying within the projected equatorial zone, since that portion lying outside may simply be inferred if necessary.

Before proceeding, it is perhaps wise to emphasise one particularly important feature of the stereographic projection – namely, that it is an 'equal angle' projection, but not an 'equal distance' one. That is, one can be confident that two zones meeting at some specific angle on the surface of the sphere will map onto lines in the projection that also meet at the very same angle. For example, the three mirror zones that converge upon each {111} surface in the spherical representation for the *face-centred cubic* crystal structure are separated by angles of 60°, and the same is true of their projections onto the two-dimensional plane. Conversely, distances measured along the zones of the spherical representation are *not* preserved upon projection; a zone segment of some given length will project to a greater length on the two-dimensional plane if it lies far from the projection pole than if it lies close to the projection pole.

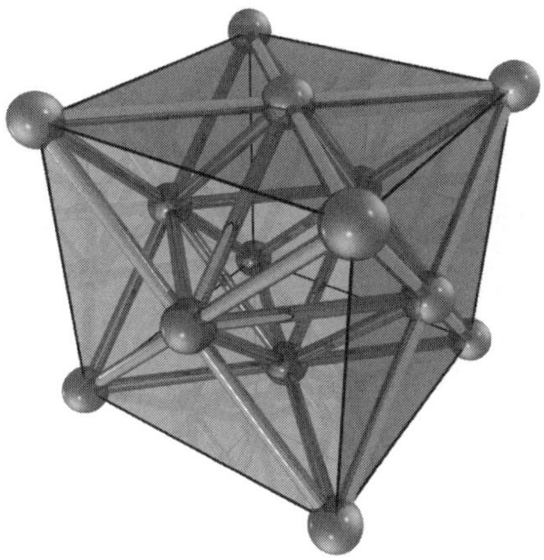

Figure 2.4 Conventional unit cell of the face-centred cubic crystal structure.

2.4 Surfaces of the Face-Centred Cubic Structure

We here turn in more detail to the surfaces of several particularly important crystal structures, starting with the face-centred cubic structure adopted by many metals. Indeed, this is the most common crystal structure amongst the transition elements (i.e. those having partially filled d orbitals) and the coinage metals (i.e. copper, silver and gold). The primitive unit cell is a parallelopiped containing a single atom, although a more symmetric conventional unit cell containing four atoms is more frequently depicted (Fig. 2.4). As mentioned in Sections 2.1 and 2.2, the bulk structure features mirror planes perpendicular to the $\langle 100 \rangle$ and $\langle 110 \rangle$ crystallographic axes, the latter also corresponding to the close-packed directions, and these observations permit the identification of a variety of symmetry and structure classifications. Let us therefore begin by revisiting these, slightly more formally, focussing particularly upon the intersection between structure and symmetry revealed by the spherical representation (or, equivalently, the stereographic projection).

2.4.1 Reconciliation of Symmetry and Primary Structure

In Fig. 2.5, we show the same mirror and primary structural zones plotted previously in Fig. 2.1b and c, but now combined onto the same sphere (or projection) rather than presented separately. Immediately, a number of facts become clear: (i) that only the flat surfaces of $\{111\}$ and $\{100\}$ type, have more than two mirror planes; (ii) that the $\{110\}$ surfaces are the most highly symmetric of the stepped surfaces; (iii) that all other stepped surfaces have precisely one mirror plane and (iv) that all chiral surfaces are kinked, although not all kinked surfaces are chiral. Furthermore, we can construct a table (Fig. 2.6) highlighting all possible combinations of structure and symmetry that can arise for the surfaces of *fcc* crystals. Here, we distinguish between surfaces having differing numbers of mirror planes by introducing the term **reflexive** to indicate a surface having a single mirror plane, while **doubly reflexive, triply reflexive** and **quadruply reflexive** (collectively **multiply reflexive**) indicate two, three and four mirror planes respectively.

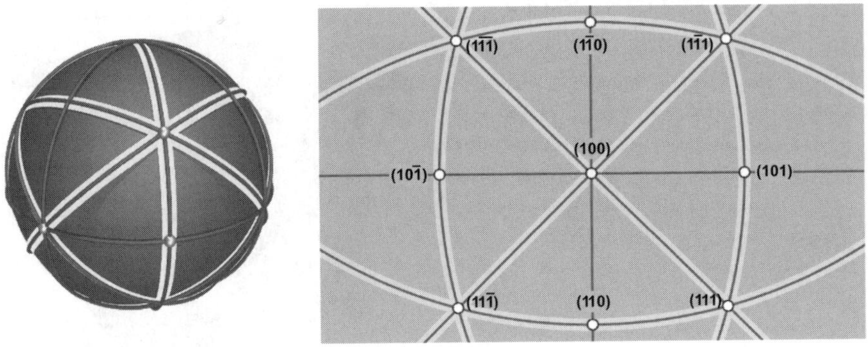

(a) (b)

Figure 2.5 Combined representations of symmetry and structure for face-centred cubic materials; in panel (a) the spherical representation is given, while in panel (b) the same information is projected onto a flat plane, using the (100) surface as the projection pole. (*See Plate III for the colour representation of this figure.*)

Symmetry \ Structure	Flat	Stepped	Kinked
Chiral			✓
Reflexive		✓	✓
Doubly reflexive		✓	
Triply reflexive	✓		
Quadruply reflexive	✓		

Figure 2.6 Permitted combinations of structure and symmetry for the surfaces of face-centred cubic materials.

In this way, we identify just *six* possible combinations of structure and symmetry that may arise at these surfaces:

Quadruply Reflexive Flat Surfaces

This category contains only a single type of surface, namely the six {100} surfaces. Two of the four mirror planes lie perpendicular to ⟨100⟩ crystallographic axes within the surface plane, while the other two lie perpendicular to a pair of surface-parallel ⟨110⟩ directions. These latter directions also define close-packed chains of atoms within the surface plane. The coordination number of the least coordinated atoms is 8.

Triply Reflexive Flat Surfaces

This category again contains only a single type of surface, namely the eight {111} surfaces. Here, the three mirror planes all lie perpendicular to ⟨110⟩ crystallographic axes within the surface plane, which once more also define close-packed chains of atoms. The coordination number of the least coordinated atoms is 9.

Doubly Reflexive Stepped Surfaces

Once again, this category contains only a single type of surface, namely the 12 {110} surfaces. In each of these, one mirror plane lies perpendicular to a surface-parallel ⟨100⟩ crystallographic axis, while the other lies perpendicular to a ⟨110⟩ crystallographic axis within the surface plane. It is this latter axis that also defines the close-packed direction corresponding to the surface steps. The coordination number of the least coordinated atoms is 7.

Reflexive Stepped Surfaces

This category contains an infinite number of different surface types, conforming to the constraint that precisely *two* of their Miller indices share the same magnitude while *none* are zero (each such type arising in 24 different instances). In each case, the single mirror plane lies perpendicular to the ⟨110⟩ crystallographic axis that defines the close-packed direction of the surface steps. The coordination number of the least coordinated atoms is 7.

Reflexive Kinked Surfaces

This category again contains an infinite number of different surface types, conforming this time to the constraint that precisely *one* of the Miller indices is zero and *none* share the same magnitude (each such type arising in 24 different instances). Here, the single mirror plane lies perpendicular to a ⟨100⟩ crystallographic axis, but there is *no* close-packed direction within the surface plane. The coordination number of the least coordinated atoms is 6.

Chiral Kinked Surfaces

Once again, this category contains an infinite number of different surface types, defined by the constraint that all *three* Miller indices are of differing magnitude and *none* are zero (each such type will arise in 48 different instances – 24 of each enantiomer). Here, we find no mirror plane (hence the chirality) and no close-packed chains lying within the surface plane (hence the surface's kinked nature). The coordination number of the least coordinated atoms is 6.

At this level of analysis, perhaps the most striking observation is that it is impossible to conceive of an intrinsically chiral *fcc* surface that is not also kinked. As we argued in Section 2.2, there are strong grounds upon which to expect kinked surfaces to be particularly prone to atomic-scale roughening and/or facet formation. Thus we may be equally sure that all chiral *fcc* surfaces are likely to be prone to the same phenomena, with all the associated implications for our ability to study them in detail via standard surface science techniques.[7]

Incidentally, it was the strong link between kinks and chirality at the surfaces of *fcc* crystals that led Attard to the first (and still most widely used) general scheme for assignment of chirality labels to intrinsically chiral surfaces [3]. The foundation of this approach, it will be recalled, is the observation that any kink on an *fcc* surface may be thought of as the intersection of three micro-facets

7 Rough surfaces are not readily amenable to techniques such as low-energy electron diffraction, for example, since it relies upon the existence of long-range order. Spectroscopic techniques that simply average over surface sites (rather than relying on constructive interference between them) fare better, but even in these cases inhomogeneity of the surface can broaden spectral features to the point of uselessness. Surface imaging techniques, such as scanning probe microscopy, are best-placed to cope with rough and/or facetted surfaces.

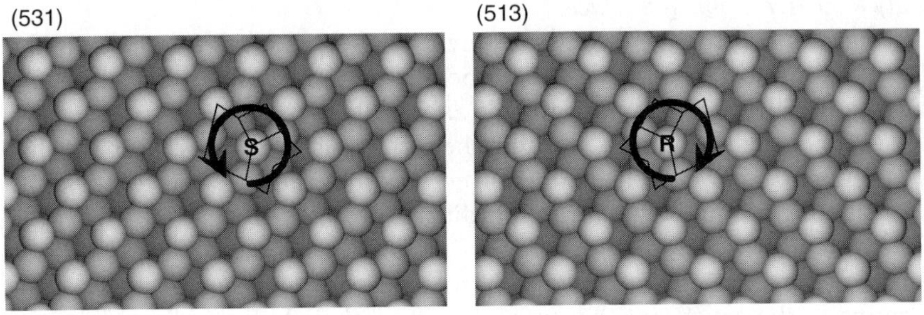

Figure 2.7 The Attard convention for labelling the chirality of kinked surfaces, exemplified by the (531) and (513) surfaces. Visiting the {111}, {100} and {110} microfacets in that order, one rotates anticlockwise on an S surface, but clockwise on an R surface.

of {110}, {100} and {111} type (Fig. 2.7). As mentioned briefly in Chapter 1, if visiting these cyclically in the order of decreasing density of surface-layer atoms (i.e. {111} → {100} → {110} and back to the beginning again) involves traversing a clockwise path around the kink, then the surface is labelled ‘R’, whereas an anticlockwise path results in the surface being labelled ‘S’. For the surfaces of *fcc* crystals, a surface labelled ‘R’ according to this scheme would always turn out to be labelled ‘D’ according to the scheme outlined in Section 2.1, which was based upon the symmetry of vertex poles in the spherical representation (or, indeed, the stereographic projection). Conversely, a surface labelled ‘S’ according to the kink/micro-facet scheme would always turn out to be labelled ‘L’ according to the vertex/symmetry scheme. The two schemes are thus practically interchangeable (barring a trivial R ↔ D and S ↔ L relabelling) with little to choose between them.[8] The vertex/symmetry scheme truly comes into its own, however, for crystal structures where it is possible to find chiral surfaces that have no kinks, as we shall see for the *bcc* case below.

2.4.2 Secondary and Tertiary Structure

Before turning to the body-centred crystal structure, we first complete our discussion of the face-centred crystal structure by pushing the idea of structural zones just a little further. Thus far, we have considered primary structural zones, corresponding to the ⟨110⟩ close-packed directions. It is possible, however, to define additional zones corresponding to more complex structural features. One way of doing this, for example, would be to look for directions in which one can find zig–zags consisting of close-packed *pairs* of atoms (rather than close-packed *chains*) alternating in direction *ad infinitum*. In the face-centred cubic structure, such features can be found lying in the ⟨100⟩ and ⟨211⟩ crystallographic directions; we shall call the zones associated with these directions the **secondary structural zones**. Adding these to the stereographic representation is immediately enlightening, as they intersect with the primary

8 One might, however, note that it is arguably more satisfying to label the chirality (a symmetry property of the system) by means of a scheme based upon the symmetry of vertices rather than the structure of micro-facets.

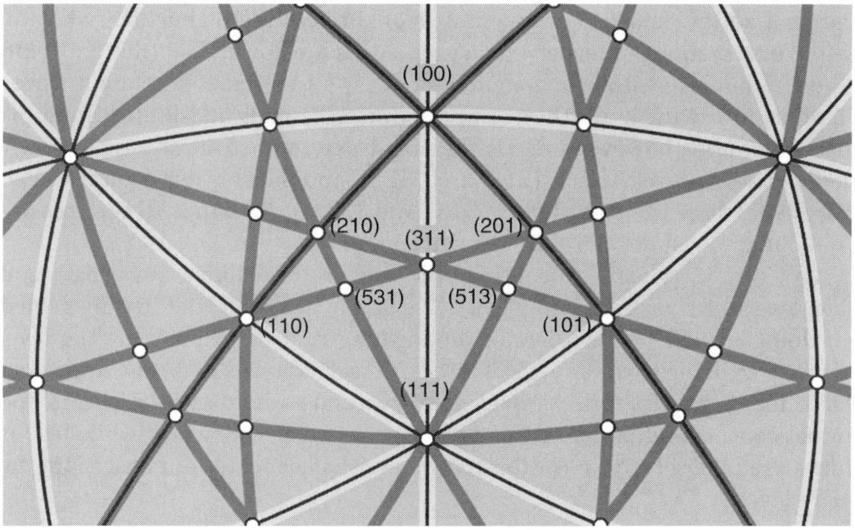

Figure 2.8 Stereographic projection onto the (311) pole, showing mirror zones together with primary and secondary structural zones for surfaces of face-centred cubic materials. (*See Plate IV for the colour representation of this figure.*)

structural zones, and with each other, at poles corresponding to a small number of highly significant surface types (Fig. 2.8). Firstly, we may note that the {111} and {100} surfaces, highlighted already at the intersection of primary structural zones, are picked out once again by the secondary structural zones. Secondly, however, the {110} surfaces, which have *not* already been highlighted at the purely structural level (recall that they are 'merely' the most symmetric of the stepped surfaces) are now picked out at the intersection between a primary structural zone with no less than *three* secondary structural zones; they may thus now also be considered the most structurally simple of the stepped surfaces, in addition to their special symmetry status as the only doubly reflexive surfaces. Thirdly, the {311} surfaces, which have also not been highlighted at the level of primary structure, are now picked out as the most structurally simple of the (singly) reflexive stepped surfaces. Fourthly, and finally, the {210} and {531} surfaces are picked out at the intersections of *three* and *two* secondary structural zones respectively; in a sense, therefore, they respectively represent the most structurally simple of the achiral and chiral kinked surfaces. The full set of surfaces highlighted at primary and secondary level just happens to be the set of surfaces chosen to illustrate the definitions of flat, stepped and kinked surfaces in Fig. 2.2.

Further insight may, of course, be gained by extending the structural zones to a tertiary level, and the natural way to do this is to consider directions in which the structure repeats not with every adjacent atom (the close-packed chains of primary structure) nor with every alternate atom (the zig–zag chains of secondary structure) but with every third atom. In the face-centred cubic structure, such features are found lying in the ⟨310⟩, ⟨111⟩ and ⟨321⟩ crystallographic directions, and we shall describe the corresponding zones as **tertiary**

structural zones. Rather a large number of surfaces are highlighted when all possible intersections of tertiary structural zones are considered (not only with each other, but also with primary and secondary structural zones) and it would be tedious to list them all (they can be found in Jenkins and Pratt [4]). Some of particular note, however, include the only three stepped surface types newly picked out at this level ({331}, {211} and {511}) and the only two achiral kinked surface types similarly picked out ({310} and {320}). A further 16 surfaces are picked out amongst the chiral kinked category.

It is, of course, possible to continue the exercise of defining quaternary and higher levels of structural complexity. Inevitably, however, the insight gained from doing so will follow a law of diminishing returns. Already by the level of tertiary structure, we have recognised 27 surface types as being somewhat notable for their structural simplicity; to continue would clearly lead to the eventual conclusion that *all* surfaces are somewhat special, which while true is hardly a useful observation. For this reason, we shall conclude our analysis at the tertiary level.

2.4.3 Commentary

One further point might usefully be made here, regarding the chiral nature of certain surfaces. Describing the surface structure by means of a two-dimensional Bravais lattice, supplemented with a semi-infinite motif of atoms extending from the top-layer down into the bulk, we can effect a slightly different categorisation than that outlined above. In this alternative approach, we may first classify according to the symmetry of the two-dimensional lattice, and then note whether any change in symmetry is introduced by the atomic basis.

Now, in two dimensions, there are only five symmetrically distinct lattice types, namely: square, triangular, rectangular, rhombic and oblique (see Fig. 2.9). Of these, the first four are unambiguously achiral, possessing mirror symmetry within the two-dimensional plane. The final lattice type – oblique – may be considered chiral, so long as 'above' and 'below' the plane are distinct (as is certainly true when the lattice describes a surface). Accordingly, it is trivially certain that any surface having an oblique two-dimensional lattice must necessarily be chiral, since no conceivable atomic basis can introduce improper symmetries that are absent from the underlying lattice. A more interesting question, however, is whether all chiral surfaces must necessarily have oblique lattices; if any may be found that do not, then *these* must be chiral only due to their atomic basis.

To address this issue, it is worth reviewing each of the two-dimensional lattice types in turn. Starting with the square lattice, it is trivial to assert that this underlies the structure of *fcc*-{100} surfaces, but less so to demonstrate that it underlies no other type of surface; nevertheless, the latter statement may be proven to be true [5]. Similarly, the triangular lattice evidently underlies the structure of *fcc*-{111} surfaces, but once again it is not obvious that it underlies no other type of surface, even though this statement too may be proven to be true [5].

Since all surfaces whose normals lie on a single mirror zone (i.e. those surfaces whose Miller indices include either a single pair of equal magnitude or a single index of zero magnitude) possess reflection symmetry, it is certain that their

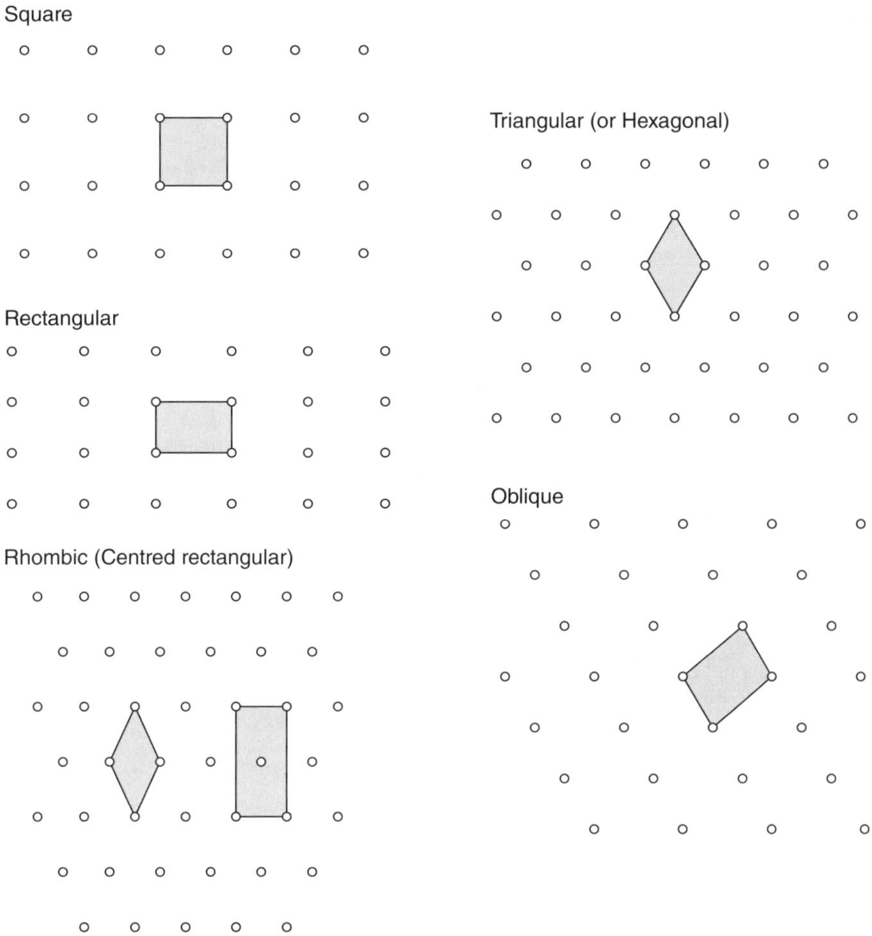

Figure 2.9 The five symmetrically distinct lattice types in two dimensions, with primitive unit cells (of the maximum possible symmetry) marked.

two-dimensional lattices must themselves possess mirror symmetry within the plane. This limits the possibilities to the four achiral lattices, but given that two of these (the triangular and square) are ruled out for all but the quadruply and triply reflexive surfaces ({100} and {111}) we are, in effect, left with only the rectangular and rhombic possibilities. One can, in fact, show that for the face-centred cubic surfaces of $\{ppq\}$ type, rhombic lattices are associated with those having both p and q odd, whereas for those of $\{pq0\}$ type, rhombic lattices are associated with either p or q being odd but not both; when these conditions are not met, the lattice is necessarily rectangular [5].

As for those surfaces whose normals do not lie on mirror zones (i.e. those that are chiral) it turns out that indeed there *are* several with rectangular and rhombic two-dimensional lattices, in addition to the more prevalent examples with oblique lattices. Although one can derive rules to distinguish between these

Table 2.1 Proportions of chiral *fcc* surfaces having rhombic, rectangular and oblique two-dimensional lattices.

Maximum Index	Rhombic (%)	Rectangular (%)	Oblique (%)
8	8	17	75
16	6	11	83
32	4	7	89
64	3	4	93

All such surfaces having Miller indices up to and including the stated maximum magnitude are classified.

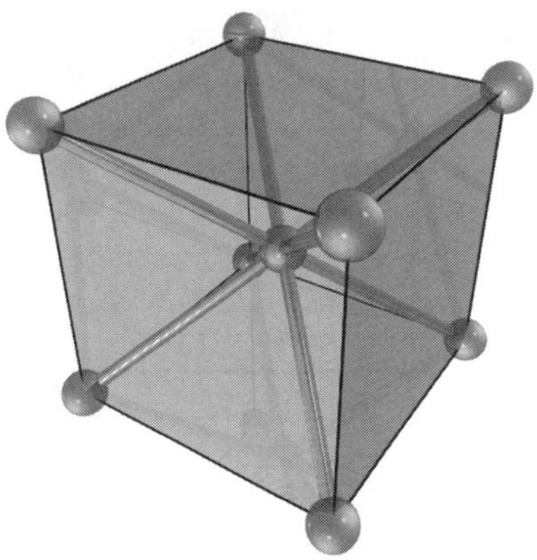

Figure 2.10 Conventional unit cell of the body-centred cubic crystal structure.

types based on their Miller indices, these rules are far from straightforward; the interested reader is referred to a forthcoming publication for details [5]. For our present purposes, we simply note that as one allows Miller indices of progressively greater magnitude, the proportion of rhombic lattices drops relative to rectangular lattices, and both drop relative to oblique lattices (see Table 2.1). Amongst surfaces having rhombic lattices, those of {531} type have the smallest primitive unit cell, while for surfaces having rectangular lattices, this distinction belongs to the {421} type; the smallest primitive unit cell for any surface having an oblique two-dimensional lattice is found for those of {321} type.

2.5 Surfaces of the Body-Centred Cubic Structure

The body-centred cubic crystal structure is most notably adopted by a number of highly reactive transition metals, including iron, molybdenum and tungsten. Unlike the face-centred cubic structure discussed above, it is not a close-packed structure and the coordination number of atoms in the bulk is only 8, rather than 12 (Fig. 2.10). Nevertheless, the bulk structure still features directions in which atoms form close-packed chains; these are directions of ⟨111⟩ type, however, not

⟨110⟩ as in the *fcc* case. The consequences of this difference for the analysis of surface structure and symmetry are quite profound, as we shall shortly see.

2.5.1 Reconciliation of Symmetry and Primary Structure

Since the body-centred cubic crystal structure shares the same point group as the face-centred cubic structure, its mirror zones are precisely the same as we have seen before; we now combine them, however, with a set of primary structural zones corresponding to the ⟨111⟩ close-packed directions of the *bcc* structure (Fig. 2.11). Once again, the structural zones allow us to rigorously define flat, stepped and kinked surfaces according to whether the corresponding pole lies upon more than one, precisely one, or no structural zones. For the *bcc* case, only the {110} surfaces may be regarded as flat, having two ⟨111⟩ close-packed directions lying within the surface plane. Stepped surfaces, for which only one such direction lies within the surface plane, are those for which the magnitudes of two Miller indices add up to that of the third (e.g. the (321) surface). And all other surfaces, for which there are no surface-parallel ⟨111⟩ axes, are necessarily kinked. As before, the designations of kinked, stepped and flat relate rather directly to the coordination numbers of atoms in the uppermost layer. In the kinked case, the least coordinated atoms have a coordination number of 4, which is precisely half that of the bulk. In close analogy with the *fcc* case, this latter fact implies that removal of an atom from the surface layer, forming a vacancy, necessitates the loss of exactly the same number of nearest-neighbour interactions as would be gained upon deposition of that atom onto a patch of pristine surface distant from its original location. Formation of an adatom – vacancy pair is thus approximately thermoneutral, so kinked *bcc* surfaces are likely to be just as rough, on an atomic scale, as are kinked *fcc* surfaces. The *bcc* stepped surfaces, on the other hand, all have coordination numbers of 5 for their least coordinated atoms, which exceeds by 2 the number of nearest-neighbours that an adatom would possess,

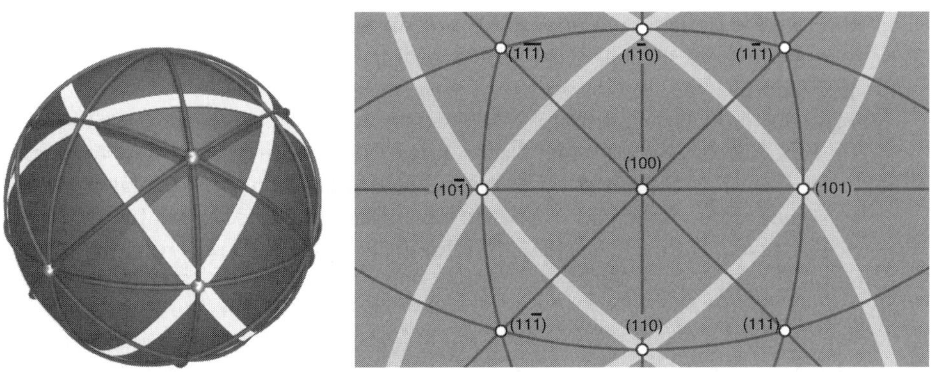

(a) (b)

Figure 2.11 Combined representations of symmetry and structure for body-centred cubic materials; in panel (a) the spherical representation is given, while in panel (b) the same information is projected onto a flat plane, using the (100) surface as the projection pole. (*See Plate III for the colour representation of this figure.*)

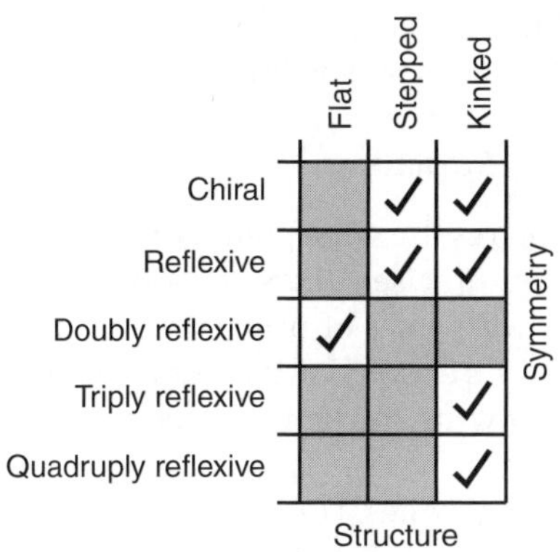

Figure 2.12 Permitted combinations of structure and symmetry for the surfaces of body-centred cubic materials.

indicating a significant energy penalty for atomic-scale roughness. Similarly, the flat *bcc* surfaces feature least coordinated atoms with a coordination number of 6, which is higher by 3 than that of an adatom, making them even less prone to roughness.

The combination of mirror and structural zones now allows us to make a number of observations for the *bcc* case, analogous to those noted above for the case of *fcc*, namely: (i) that the flat surfaces of {110} type, possess only two mirror planes; (ii) that the {211} surfaces are the most highly symmetric of the stepped surfaces; (iii) that all other stepped surfaces are chiral and (iv) that all other chiral surfaces are kinked, although not all kinked surfaces are chiral.

The table in Fig. 2.12 summarises the *seven* different combinations of structure and symmetry that may arise for surfaces of *bcc* crystals:

Quadruply Reflexive Kinked Surfaces

This category contains only a single type of surface, namely the six {100} surfaces. Two of the four mirror planes lie perpendicular to ⟨100⟩ crystallographic axes within the surface plane, while the other two lie perpendicular to a pair of surface-parallel ⟨110⟩ directions, but there are no close-packed directions lying within the surface plane. The coordination number of the least coordinated atoms is 4.

Triply Reflexive Kinked Surfaces

This category again contains only a single type of surface, namely the eight {111} surfaces. Here, the three mirror planes all lie perpendicular to ⟨110⟩ crystallographic axes within the surface plane, but once more there are no close-packed chains of atoms within the surface plane. The coordination number of the least coordinated atoms is 4.

Doubly Reflexive Flat Surfaces

Once again, this category contains only a single type of surface, namely the 12 {110} surfaces. In each of these, one mirror plane lies perpendicular to a surface-parallel ⟨100⟩ crystallographic axis, while the other lies perpendicular to a ⟨110⟩ crystallographic axis within the surface plane. In addition, two ⟨111⟩ axes lie within the surface plane, implying two directions in which there are close-packed chains of atoms. The coordination number of the least coordinated atoms is 6.

Reflexive Stepped Surfaces
This category once more contains only a single type of surface, namely the 24 {211} surfaces. Here, the single mirror plane lies perpendicular to a ⟨110⟩ crystallographic axis, which lies perpendicular to a ⟨111⟩ axis defining the close-packed direction of the surface steps. The coordination number of the least coordinated atoms is 5.

Reflexive Kinked Surfaces
This category contains an infinite number of different surface types, conforming either to the constraint that precisely *one* of the Miller indices is zero with *none* sharing the same magnitude, *or* to the constraint that precisely *two* of their Miller indices share the same magnitude while *none* are zero (each such type arising in 24 different instances). Here, the single mirror plane lies perpendicular to a ⟨100⟩ crystallographic axis, but there is *no* close-packed direction within the surface plane. The coordination number of the least coordinated atoms is 4.

Chiral Stepped Surfaces
Once again, this category contains an infinite number of different surface types, defined by the constraint that all *three* Miller indices are of differing magnitude, *none* are zero, and the magnitudes of two sum to the magnitude of the other (each such type will arise in 48 different instances – 24 of each enantiomer). Here, we find no mirror plane (hence the chirality), but nevertheless there is a single close-packed ⟨111⟩ axis within the surface plane, corresponding to the direction of the surface steps. The coordination number of the least coordinated atoms is 5.

Chiral Kinked Surfaces
This final category again contains an infinite number of different surface types, defined by the constraint that all *three* Miller indices are of differing magnitude, *none* are zero, and the summed magnitudes of no pair of indices equals that of the other (each such type will arise in 48 different instances). Here, we find no mirror plane (hence the chirality) and no close-packed chains lying within the surface plane (hence the surface's kinked nature). The coordination number of the least coordinated atoms is 4.

2.5.2 Secondary and Tertiary Structure

Just as we supplemented our discussion of the surfaces possible for face-centred cubic materials by examining secondary and tertiary structural zones, so we can augment our understanding of the surfaces possible for body-centred materials in a similar manner. Once again, we define secondary structural features to be zig–zags formed by pairs of close-packed atoms, which in this case are found to extend in the ⟨100⟩ and ⟨110⟩ crystallographic directions. Plotting the corresponding secondary structural zones onto a stereographic representation of the structure and symmetry (Fig. 2.13) highlights (in addition to the {110} surfaces picked out at the level of primary structure) the {211} surfaces (hitherto only notable as the most symmetric of the stepped surfaces) and the {111} surfaces (not previously picked out at all). The {211} surfaces now emerge, therefore, as the structurally simplest of the stepped surfaces, not merely the

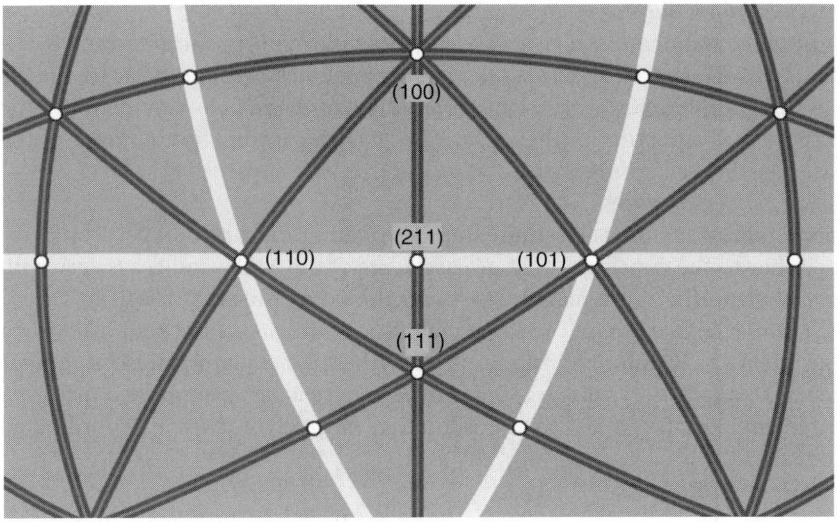

Figure 2.13 Stereographic projection onto the (211) pole, showing mirror zones together with primary and secondary structural zones for surfaces of body-centred cubic materials. (*See Plate IV for the colour representation of this figure.*)

most symmetric, and the {100} and {111} surfaces are seen to be the structurally simplest of the achiral kinked surfaces (lying at the intersections of three and two secondary structural zones, respectively).

As before, we can go further still by introducing the concept of tertiary structural features, in which the alternation of direction between adjacent pairs of atoms repeats only every third atom; such features occur in the body-centred cubic structure along the ⟨331⟩ and ⟨311⟩ crystallographic directions. Plotting the corresponding tertiary zones onto the relevant stereographic representation of structure and symmetry now yields newly highlighted intersections with primary structural zones at {321} poles, which are thus recognised as the structurally simplest of the chiral stepped surfaces.

New intersections with secondary structural zones are found at the {611}, {411}, {433}, {332} and {310} poles, now seen to be the most structurally simple achiral kinked surface types. The chiral kinked {521}, {631}, {653} and {543} surface types round out the list of surfaces highlighted at the level of tertiary structure, and following the precedent set when discussing the surfaces of face-centred cubic materials, we draw our analysis of the present case to a close at this point.

2.5.3 Commentary

Arguably the most significant of the structure–symmetry categories for the body-centred cubic materials is that of the chiral stepped surfaces, exemplified in Fig. 2.14 by the {532}, {321} and {431} varieties. The lack of kink sites implies that they will be stable against thermal roughening, in contrast to *every* chiral *fcc* surface. Such surfaces might, therefore, be ideal test-beds for asymmetric

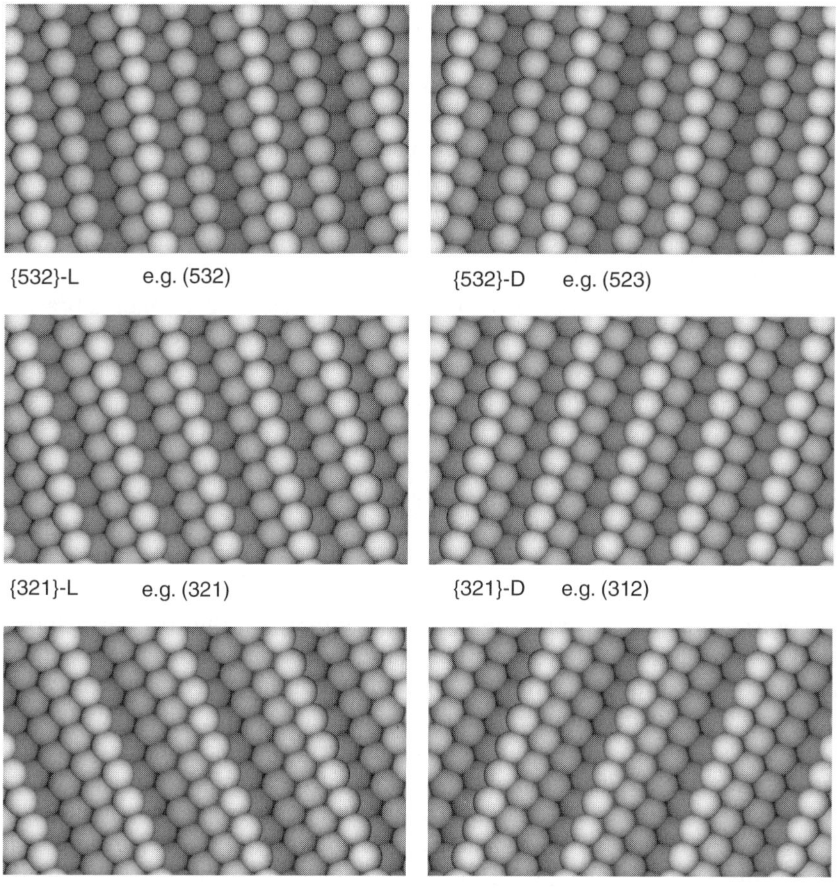

Figure 2.14 Exemplars of chirality without kinks, the {532}, {321} and {431} surfaces of body-centred cubic materials exist in mirror image forms.

surface science. Admittedly, the high reactivity of *bcc* metals argues against their usefulness as substrates for the adsorption of complex organic molecules (which will likely dissociate on the surface, in whole or in part) but the physical manifestations of chirality (e.g. mechanical, optical or magnetic properties) ought to be rather more readily accessible.

In addition, these *bcc* surfaces illustrate the advantage of applying a convention for chiral labelling based upon the symmetry of vertices in the spherical (or stereographic) representation, rather than one based upon recognising microfacets in the vicinity of a kink atom. Quite simply, the chiral stepped surfaces have no kink atoms, so the structure-based definition fails to identify suitable labels (R or S). The symmetry-based definition, however, allows us to assign labels D and L to each triangle in the *bcc* representation, and indeed to the individual surfaces whose poles lie within them, just as easily as we did for the *fcc* case.

2.6 Surfaces of the Hexagonal Close-Packed Structure

The hexagonal close-packed structure is adopted by a number of transition metals, including cobalt, ruthenium and rhenium. As the name suggests, it entails close-packing of atoms, with an ideal packing density equal to that of the *fcc* structure.[9] The structure (see Fig. 2.15) is based upon a hexagonal Bravais lattice, but with a basis of *two* atoms per primitive unit cell – a fact that ensures its surface symmetries and structures are far more complex than for either of the two cases discussed so far.

It is common (though not universal) practice to describe the hexagonal Bravais lattice by means of *four* primitive lattice vectors, rather than the three used for cubic systems. Here, we shall define $\mathbf{a}_1$, $\mathbf{a}_2$ and $\mathbf{a}_0$ to be a set of coplanar primitive lattice vectors lying within the basal plane and separated by angles of 120°; clearly, *one* of these three can always be considered as redundant, since a translation in the direction of one can always be equally well represented as a linear combination of translations along the other two. A fourth primitive lattice vector, which we shall denote $\mathbf{a}_3$, lies along the sixfold rotational axis, perpendicular to the other three.

Directions within the hexagonal lattice may then be defined in four-index notation as $[uvtw]$, just as we would define them as $[uvw]$ within the three-index notation for non-hexagonal systems; the redundancy amongst the basal lattice vectors must be ensured, however, by demanding that the sum $u + v + t$ be zero. By analogy with the usual three-index notation, angle brackets $\langle uvtw \rangle$ will imply the complete set of directions related by symmetry to the $[uvtw]$ direction by operations within the point group of the crystal. In the case of *hcp*, the set $\langle uvtw \rangle$ may be obtained from $[uvtw]$ by simply essaying all possible combinations of the following operations: (i) interchange of the u, v and t indices, (ii) change of sign of the w index and (iii) change of sign of all four indices. Close-packed chains of atoms in the *hcp* structure lie along the $\langle 11\bar{2}0 \rangle$ directions, while in the $\langle 20\bar{2}3 \rangle$ directions we find intermittent close-packing (i.e. pairs of adjacent atoms separated by gaps equivalent to five times the close-packed spacing).

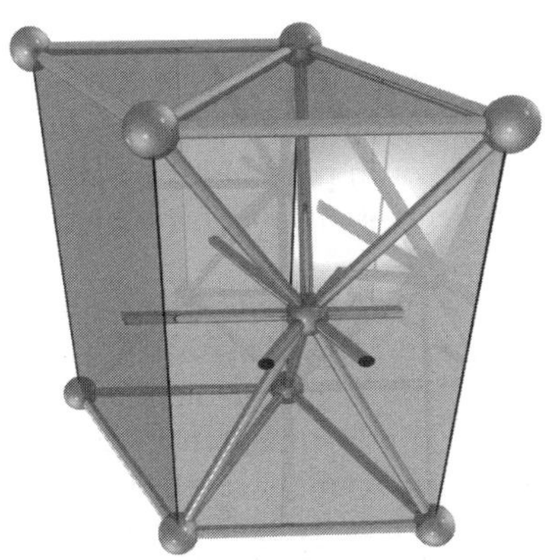

Figure 2.15 Primitive unit cell of the hexagonal close-packed crystal structure.

9 That is, packing hard spheres in the *hcp* structure leads to the same packing density as *fcc*; in practice, relaxing the hard-sphere approximation permits a slight variation.

Planes within the hexagonal lattice are defined by first obtaining the reciprocal lattice vectors $\mathbf{b}_1$, $\mathbf{b}_2$ and $\mathbf{b}_3$ (satisfying the relation $\mathbf{a}_i \cdot \mathbf{b}_j = 2\pi\delta_{ij}$) and then expressing the normal to the plane in the form $h\mathbf{b}_1 + k\mathbf{b}_2 + l\mathbf{b}_3$. The plane may then either be represented within the three-index notation as (hkl) or within a four-index notation as $(hkil)$, where i is a redundant index chosen to ensure that the sum $h + k + i$ is zero.[10] Curly brackets $\{hkil\}$ are, not unexpectedly, used in the four-index notation to indicate the complete set of planes related to the $(hkil)$ plane by operations within the point group of the crystal. In the same way as outlined above for directions, the $\{hkil\}$ set in the *hcp* case may be obtained from $(hkil)$ by combinations of (i) interchange amongst the first three indices, (ii) change of sign of the fourth index and (iii) change of sign of all indices. With these preliminaries under our belt, let us now consider the structure and symmetry of *hcp* surfaces.

2.6.1 Symmetry

As ever, we begin by indicating the mirror zones of our underlying Bravais lattice within a spherical representation and/or stereographic projection (Fig. 2.16a); these lie perpendicular to the $\langle 0001 \rangle$, $\langle 11\bar{2}0 \rangle$ and $\langle 10\bar{1}0 \rangle$ directions. Any surface whose pole lies on a single mirror zone will possess precisely one mirror plane, while any lying on multiple mirror zones will possess the corresponding number of mirror zones. We can, furthermore, label triangles either D or L according to the same rule applied above for cubic systems.[11] Whilst Fig. 2.16a accurately represents the symmetry of the hexagonal Bravais lattice, however, we immediately detect problems in using it to understand the symmetry of *hcp* surfaces. For instance, the *hcp*-$\{0001\}$ surfaces possess not six but merely three mirror planes, while the *hcp*-$\{11\bar{2}0\}$ surfaces possess not two but merely one. What has gone wrong?

The answer, of course, is that we are ignoring the fact that the *hcp* structure involves *two* atoms within its primitive unit cell. These are placed in highly symmetric locations, but nevertheless they *do* reduce the symmetry of the crystal structure relative to its underlying Bravais lattice. Specifically, the mirror planes lying perpendicular to $\langle 10\bar{1}0 \rangle$ directions are reduced to glide planes. Within each glide plane, one may identify infinitely many glide vectors – translation of the crystal through any one of these results in a structure related to the original by reflection across the corresponding glide plane. In consequence of this reduction in symmetry, surfaces whose poles lie on $\langle 10\bar{1}0 \rangle$ zones do *not* display a corresponding mirror symmetry. Instead, they *may* display glide symmetry, but only if one of the glide vectors for the specific $\langle 10\bar{1}0 \rangle$ zone in question happens to lie

10 Note that although the three-index notation for a plane may be obtained from the four-index notation simply by omitting the third index, it is *not* generally possible to obtain the three-index notation for a direction from its four-index notation in the same similar way. Note also that the $[hkil]$ direction is *not*, in general, perpendicular to the $(hkil)$ plane. Conversely, the $[uvtw]$ direction *will* be parallel to the $(hkil)$ plane if the sum $uh + vk + ti + wl$ is zero.

11 One may adopt the convention applied within the International (Hermann–Mauguin) system of symmetry classification, which (arbitrarily) considers the $\langle 10\bar{1}0 \rangle$ poles of a hexagonal system to be of nominally higher symmetry than the $\langle 11\bar{2}0 \rangle$ poles.

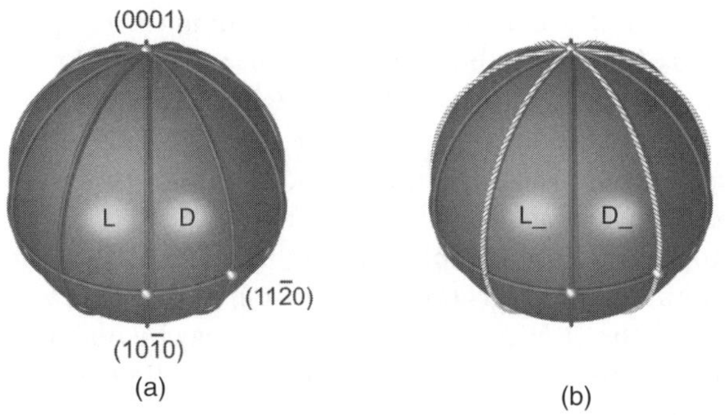

Figure 2.16 Spherical representations of surface symmetry for materials of hexagonal close-packed structure. In panel (a) the mirror zones of the underlying simple hexagonal lattice are marked; in panel (b) those that reduce to glide zones within the full crystal structure are shown striped. Chiral labels are given for two triangles, based in each case on the lattice symmetry alone; for the crystal symmetry, we anticipate an additional chiral label depending upon the relative orientations of the lattice, motif and surface plane.

within the surface plane. We will therefore indicate $\langle 10\bar{1}0\rangle$ zones (which we shall now call **glide zones**) on our spherical representation (or stereographic projection) by means of a striped line (see Fig. 2.16b) reminding ourselves that these no longer imply mirror symmetry in the surface, but rather the *possibility* of glide symmetry. One can, in fact, show that only those surfaces whose indices take the form $\{p\,p\,\overline{2p}\,2q\}$ (with p and q positive integers) satisfy the conditions for glide symmetry, having poles lying on a glide zone *and* a glide vector of the relevant glide plane lying within the surface plane. We shall call such surfaces **glissadic** (Fig. 2.17).

As a result of these modifications to our spherical representation (or stereographic projection) we see that the $\{0001\}$ surfaces are indeed now indicated to possess just three mirror planes rather than six. Similarly, the $\{11\bar{2}0\}$ surfaces clearly possess just one, not two, although they also display glide symmetry (i.e. they are **reflexive-glissadic**). The $\{10\bar{1}0\}$ surfaces, on the other hand, genuinely do possess two mirror planes (i.e. they are doubly reflexive) and are represented accordingly. All other surfaces possess either a single mirror plane (reflexive), a single glide symmetry (glissadic), or no mirror/glide symmetry at all. Since glide symmetry, like mirror symmetry, is improper, it is only this last group of surfaces (i.e. those displaying neither mirror nor glide symmetry) that are in some sense chiral.

Now, the potential for an unreconstructed surface to display glide symmetry is one consequence of the bulk crystal structure having more than one atom per primitive unit cell. In fact, if one were to label half the atoms in the bulk *hcp* structure α and the other half β, in such a way that all α atoms occupy one hexagonal Bravais lattice and all β atoms another, then each layer in the glissadic surfaces would comprise equal numbers of α and β atoms. All the other non-glissadic surfaces lying on glide zones (i.e. those without the necessary in-plane glide vectors to display glide symmetry) would have alternating layers of purely α and purely β character. Upon inspection, it turns out that *these* surfaces are chiral (as the

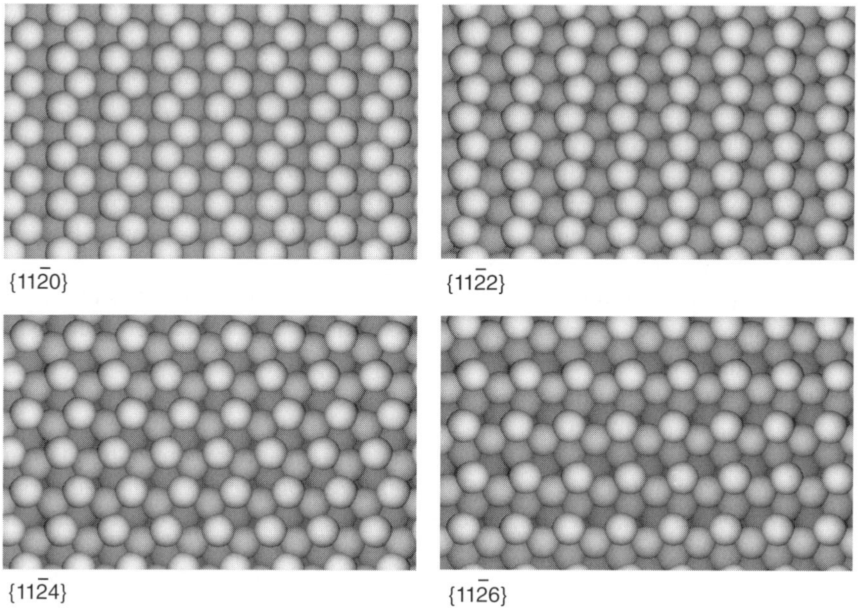

Figure 2.17 Examples of glissadic surfaces derived from the hexagonal close-packed crystal structure; in each case the defining glide plane is aligned up the page.

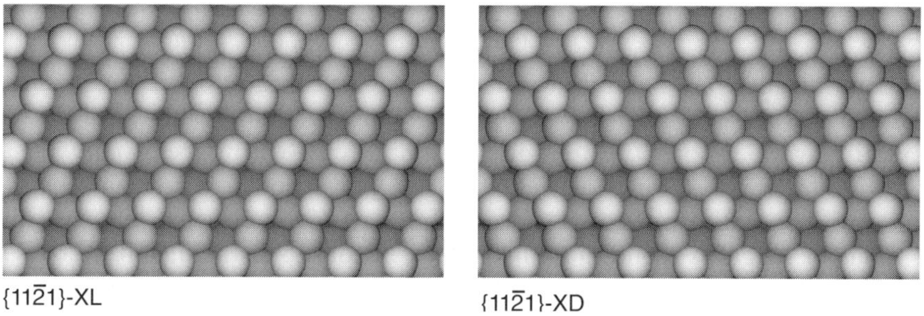

Figure 2.18 As an example of a racemic surface, the *hcp*-{11$\bar{2}$1} orientation supports two distinct terminations, each the mirror image of the other; chiral labels are assigned as described in the text, and addition or removal of one complete layer of atoms would interconvert between one enantiomorph and the other.

absence of mirror and glide symmetry would indicate) but terminating the surface on a layer of α atoms yields the mirror image structure to that obtained upon terminating the same surface on a layer of β atoms; we shall call such surfaces **racemic** (Fig. 2.18).

Unlike the *fcc* and *bcc* crystals, where swapping the surface chirality necessitated changing to a different set of Miller indices (i.e. shifting to a different triangle), here it can be achieved for racemic surfaces merely by terminating the surface at a different layer. Since the two possible terminations are energetically degenerate, a real surface would likely display terraces of both terminations, separated by monoatomic terrace edges,[12] in equal proportion. For the glissadic surfaces, on the other hand, terminating at different levels yields identical results, with no alternation in chirality (because such surfaces are not chiral).

Indeed, this observation suggests a further type of symmetry that we have not yet considered explicitly, namely the property of a surface being identical or non-identical upon addition/removal of a single atomic layer. We may introduce, therefore, the so-called **truncation** symmetry operation, which consists of changing the height at which the surface is terminated by one atomic layer.[13] In and of itself, truncation is a proper symmetry operation, whose presence or absence in the space group of the surface carries no implications for chirality. The glissadic surfaces possess truncation symmetry (i.e. they are unchanged upon addition or removal of a single atomic layer) but are achiral only because of their glide symmetry. The racemic surfaces, on the other hand, have neither reflection nor glide symmetries, but *do* possess what one might describe as **reflexotrunction** symmetry (i.e. they are symmetric under the combination of truncation *and* reflection operations applied together). Incidentally, the {0001} surfaces are the sole examples of *hcp* surfaces possessing **rototruncation** symmetry, which is to say that they are symmetric under the combination of truncation *and* (threefold) rotation operations applied together; in view of this fact, we describe them as **bayonet** surfaces, alluding to the combined push–twist associated with a so-called bayonet mechanism.[14]

More generally, however, we can identify a class of *hcp* surfaces possessing truncation symmetry, which we shall describe as **uniterminated**, recognising that they each have a single unique termination. All have indices of the form $\{3p+q\ q\ \overline{3p+2q}\ 2s\}$ with p, q and s positive integers,[15] and all surfaces with indices of this form are uniterminated. Surfaces not conforming to this pattern necessarily display two differing terminations upon addition/removal of a single atomic layer. Amongst the non-glissadic uniterminated surfaces, we find just two possibilities: those for which either q or s is zero possess a single mirror plane (**uniterminated-reflexive**) while all others possess none (**singly chiral**). The modifier 'singly' is required in this latter category to distinguish from those non-uniterminated surfaces that lie off the mirror and glide zones; as we shall see, *these* surfaces may justifiably be described as **doubly chiral**.

Before tackling the doubly chiral surfaces, however, let us reflect upon the sources of chirality in the cases described above. We began by noting that even the underlying hexagonal lattice is capable of supporting a chiral surface, if cut to expose a plane whose pole lies off all of its mirror zones. The addition of a

12 We utilise the term 'terrace edge' to indicate the (probably irregular) boundary between two terraces described by the same set of Miller indices but terminated at different heights. The term 'step' would be more usual, but is problematic here because we are using this already to refer to the regular close-packed chains of atoms running across certain surfaces with particular Miller indices.

13 The surfaces of any crystal structure having only a single atom per primitive unit cell must all necessarily display truncation symmetry; only for those crystals having multiple atoms per primitive unit cell does truncation symmetry become interesting.

14 The bayonet mechanism is nowadays most familiar from the bayonet light-bulb fitting, but does, of course, derive its name from the method by which bayonets are attached to firearms.

15 The general form of indices for the glissadic surfaces, given above, may be obtained from the unitermination form simply by setting $p = 0$ before relabelling $q \rightarrow p$ and $s \rightarrow q$.

basis to the crystal structure can reduce the overall symmetry, but it can never increase it. It follows, therefore, that any surface of the *hcp* structure whose pole lies off all of its mirror and glide zones must necessarily be chiral due to the way in which the lattice is cut by the surface. On the other hand, we have seen that chirality can exist for surfaces whose poles lie on the glide zones of the *hcp* structure, albeit alternating from one layer to the next, but this chirality cannot possibly be attributed to the cutting of the lattice alone; neither can the atomic basis alone be responsible, since it is in itself achiral. Clearly, the chirality emerges from the interplay between the lattice symmetry, the symmetry of the atomic basis, and the particular choice of surface plane; this constitutes a second source of chirality, in addition to that related purely to the cutting of the lattice. In the racemic surfaces, therefore, we see the case where the first source of chirality (which we shall call the **one-chirality**) is inactive; the chiralities of the two possible surface terminations are thus solely determined by the second source of chirality (which we shall call the **two-chirality**). The singly chiral surfaces, in contrast, correspond to the case where it is the two-chirality that is inactive (because unitermination implies that both atoms in the atomic basis are equivalently oriented with respect to the plane of the surface) and the chirality arises solely due to the one-chirality. Accordingly, we may label each instance of a singly chiral surface either DX or LX, where in each case the first symbol (D or L) is assigned to the one-chirality with reference to the chirality labels corresponding to the underlying hexagonal lattice (Fig. 2.16) and the second symbol (X) indicates that the two-chirality is inactive. Similarly, we can label the two terminations of each racemic surface XD or XL, where the X now indicates that it is the one-chirality that is inactive, with the D and L symbols defining the two-chirality according to some suitable convention that we shall settle upon shortly. Surfaces with either mirror or glide symmetry must, of course, be labelled XX, since both sources of chirality must be inactive in these cases.

Returning now to the case of surfaces failing to conform to bayonet, reflexive-glissadic, doubly reflexive, uniterminated-reflexive, reflexive, glissadic, racemic or singly chiral type, we conclude that these must be *doubly* chiral due to a combination of both one-chirality and two-chirality acting together. By analogy with doubly chiral molecules, therefore, we ought to anticipate finding four versions of each such surface: an enantiomorphically related pair labelled DD and LL, and a second enantiomorphically related pair labelled DL and LD (and diamorphically related to the first pair). This is, indeed, what we find. Taking the $\{21\bar{3}1\}$ surfaces as archetypal examples of this category, we see that there are two non-degenerate terminations possible for the $(21\bar{3}1)$ surface, which we can label DD and DL since they both lie in a D triangle of the hexagonal Bravais lattice's spherical (or stereographic) representation but differ in their two-chirality. Within a neighbouring L triangle, however, the $(3\bar{1}\bar{2}1)$ surface also has two non-degenerate terminations, one of which is the (degenerate) mirror image of the DD case and can thus sensibly be labelled LL, and the other of which is the (degenerate) mirror image of the DL case and should be labelled LD.

To underline the degeneracy of the DD/LL pair, we will adopt the practice of appending a subscript τ to both, yielding a **composite symmetry symbol** of the

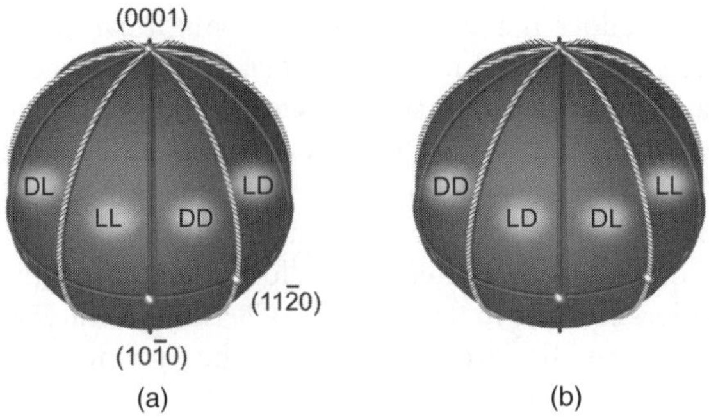

Figure 2.19 Chirality labelling scheme for surfaces of hexagonal close-packed materials. In panel (a) the labels used for the α termination are shown, while in (b) the labels are for the β termination.

form DD_τ or LL_τ. The degeneracy of the DL/LD pair is highlighted by appending a subscript $\bar{\tau}$, giving us $\mathrm{DL}_{\bar{\tau}}$ and $\mathrm{LD}_{\bar{\tau}}$. In essence, the subscript τ is just a marker to indicate a surface for which the one-chirality and two-chirality match, and conversely $\bar{\tau}$ simply indicates that they are opposed. Although this may seem somewhat redundant at this stage, it does give us a useful shorthand for referring to *either* the DD or the LL surface without specifying *which* (or similarly referring to either one of the DL or LD surfaces). We shall retain the $\tau/\bar{\tau}$ notation for non-uniterminated XX surfaces also, but replace *both* with u for general uniterminated surfaces (hence DX_u and LX_u for the singly chiral surfaces; XX_u for uniterminated-reflexive surfaces) or with o for non-glissadic surfaces whose poles lie on a glide zone, where both terminations are degenerate (hence XD_o and XL_o for racemic surfaces and XX_o for *both* terminations of the bayonet surface). For the glissadic surfaces, we append a subscript ø, pronounced 'ur', to indicate that both o and u could equally be applied to such a surface (hence $\mathrm{XX}_{\o}$ in these cases).

Finally, to fulfil the promise of a convention for assigning labels to the two-chirality of surfaces, let us (arbitrarily) define that the α and β atoms of the bulk structure are located at vector positions [0000] and $\frac{1}{6}[20\bar{2}3]$ relative to each lattice point; we can then (again arbitrarily) assign two-chirality labels as shown in Fig. 2.19 for the α- and β-terminated surfaces. According to this convention, for example, it is the α-terminated $(21\bar{3}1)$ surface that is labelled DD, whereas the identical $(2\bar{3}11)$ surface is β-terminated.[16] Similarly, considering the racemic $(11\bar{2}1)$ surface, it is the α termination that is labelled XD and the β termination that is labelled XL; the reverse would be true for the $(2\bar{1}\bar{1}1)$ surface.

2.6.2 Primary Structure

Following the same logic applied for the *fcc* and *bcc* crystal structures, let us now construct a spherical representation (in lieu of a stereographic projection) of the primary structural features of the *hcp* surfaces. As before, we identify the directions corresponding to close-packed chains of atoms in the bulk structure (in this case, the $\langle 11\bar{2}0\rangle$ directions) and plot the zones for which these are the

16 Note that α and β atoms are, of course, identical. We make a distinction between them only in order to provide ourselves with a convenient convention to label surfaces with active two-chirality in an unambiguous (albeit essentially arbitrary) manner.

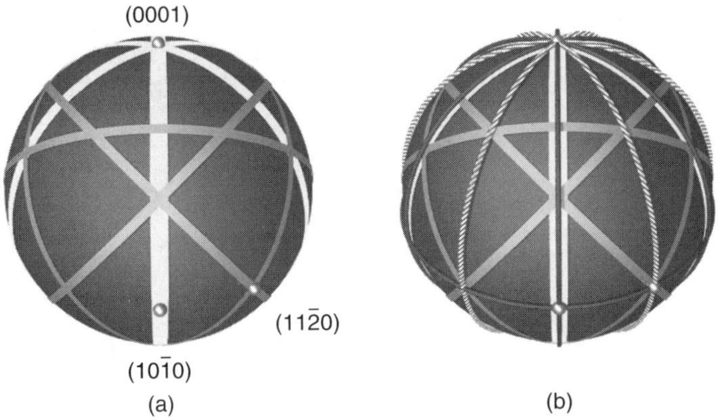

Figure 2.20 Spherical representation of surface structure for hexagonal close-packed materials. In panel (a) the primary (broad) and interrupted primary (narrow) structural zones are shown; in panel (b) the mirror and glide zones of the symmetry representation are overlaid. (*See Plate III for the colour representation of this figure.*)

poles (Fig. 2.20). Here, however, we can also usefully take note of those directions for which an intermittent form of close-packing is observed (the $\langle 20\bar{2}3\rangle$ directions) and plot the zones associated with these too, on the same diagram. Since we describe the zones associated with true close-packed directions as primary structural zones, we shall describe the zones related to the intermittent close-packed directions as **interrupted primary zones**. Consideration of the various intersections between these two types of zone allows us to identify six different structural categories amongst the *hcp* surfaces.

First, we note the existence of flat surfaces, exclusively of $\{0001\}$ type, where three primary structural zones intersect. Next, there are the stepped surfaces, with poles lying on a single primary structural zone (but *not* on an interrupted primary zone). And completing the set of surfaces similar to the *fcc* and *bcc* surfaces, we find the kinked surfaces, whose poles lie on neither primary nor interrupted primary zones. In addition, however, there are a further three surface structural categories that we have not met above.

A surface whose pole lies on a single interrupted primary zone but not on a primary structural zone displays the interesting feature that the surface contains pairs of close-packed atoms (because an interrupted close-packed chain lies within the surface plane). We shall describe such surfaces as **geminal** to reflect this pairing[17] (Fig. 2.21).

There are only two types of surface with poles lying at the intersection of *two* interrupted primary zones but not on a primary structural zone, namely the $\{11\bar{2}0\}$ and $\{11\bar{2}2\}$ surfaces. In both cases, the surface is notable for a zig–zag chain of atoms (due to the presence of two intersecting interrupted close-packed chains) and we describe such surfaces as being of **meandering row**

17 The term 'geminal' is intended to reference the mythological twins, Castor and Pollux, transformed by Zeus into the constellation of Gemini.

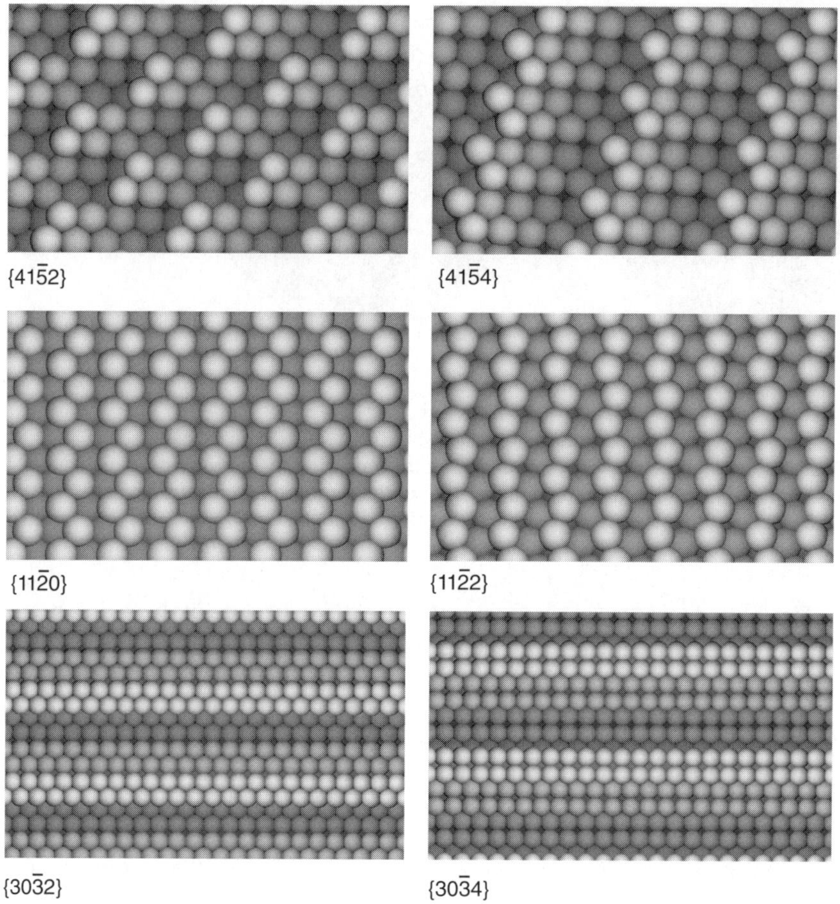

Figure 2.21 Geminal ($\{41\bar{5}2\}$ and $\{41\bar{5}4\}$), meandering row ($\{11\bar{2}0\}$ and $\{11\bar{2}2\}$) and interrupted flat ($\{30\bar{3}2\}$ and $\{30\bar{3}4\}$) surfaces of the hexagonal close-packed structure.

type (Fig. 2.21). In the $\{11\bar{2}0\}$ surfaces, the angle internal to the zig–zag satisfies the equation

$$\cos\zeta = \left(\frac{4-3r^2}{4+3r^2}\right) \tag{2.1}$$

where $r = c/a$ with $a = |\mathbf{a}_1| = |\mathbf{a}_2| = |\mathbf{a}_0|$ and $c = |\mathbf{a}_3|$. For the $\{11\bar{2}2\}$ surfaces, the corresponding internal angle is instead given by

$$\cos\eta = -\left(\frac{2+3r^2}{4+3r^2}\right) \tag{2.2}$$

and for the ideal c/a ratio of $\sqrt{8/3}$ we find $\zeta = 109.47°$ and $\eta = 146.44°$.

Finally, we note that the $\{30\bar{3}4\}$ and $\{30\bar{3}2\}$ surfaces have poles located at the intersection of a primary structural zone with one or two interrupted primary zones respectively. In each case, therefore, we find a single surface direction in

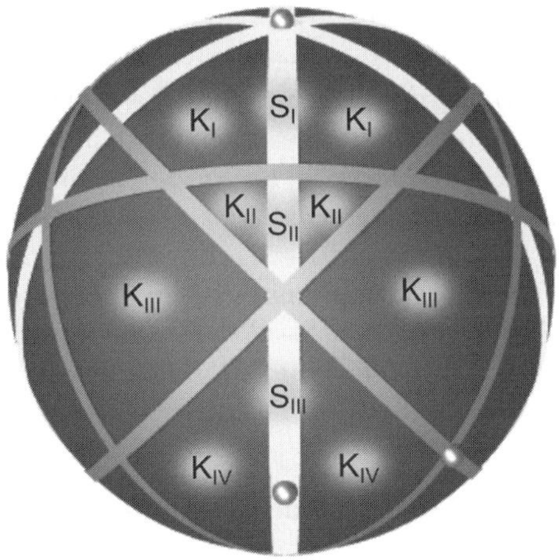

Figure 2.22 Regions, discussed in the text, relevant to the coordination numbers of the least coordinated atoms in surfaces of hexagonal close-packed materials.

which close-packed chains of atoms occur, akin to a stepped surface, but instead of each step-edge consisting of a single such chain, here we find two adjacent chains at the same height. The effect is to create what looks like a very narrow terrace of effectively 'flat' surface, so we shall call these **interrupted flat** surfaces (Fig. 2.21) in recognition of this appearance.

Identifying the coordination numbers of the least coordinated atoms in surfaces of face-centred cubic and body-centred cubic materials was, it will be recalled, straightforward; for the face-centred cubic case, the relevant value was 9 for the flat {111} surfaces, 8 for the flat {100} surfaces, 7 for all stepped surfaces, and 6 for all kinked surfaces; for the body-centre cubic case, the relevant value was 6 for the only flat surfaces (i.e. {110}), 5 for all stepped surfaces, and 4 for all kinked surfaces. In the case of hexagonal close-packed materials, the situation is much more complex, with the coordination numbers varying not only with the type of surface (i.e. flat, stepped, kinked, etc.) but also with the region of the stereogram in which its pole is found. It will be convenient to define distinct types of stepped surfaces, denoted S_n, where the numerical value of the subscript indicates unity plus the number of interrupted flat poles that must be crossed when travelling from the {0001} pole to the pole in question along the primary structural zone; the stepped surfaces are thus broken down into S_I, S_{II} and S_{III} categories (see Fig. 2.22). The kinked surfaces may be similarly sub-divided, now taking the numerical value of the subscript as unity plus the minimum number of interrupted primary zones that must be crossed when drawing a path from the {0001} pole to the pole in question, leading to K_I, K_{II}, K_{III} and K_{IV} categories (see Fig. 2.22). These sub-divisions have a bearing upon the coordination numbers in each case. A full description is given by Jenkins and Pratt [4] but here we summarise the basic points by stating that the coordination number of the least coordinated atoms is....

Flat – Nine for both terminations;
Meandering-Row – Seven (both atoms in the unique termination);
Interrupted-Flat {30$\bar{3}$2} – Eight (both atoms in the unique termination);

Interrupted-Flat $\{30\bar{3}4\}$ – Seven & eight (two atoms in the unique termination);
Geminal – Six and seven (two atoms in the unique termination);
S_I *(General)* – Seven for both terminations;
S_I *(Uniterminated)* – Seven (both atoms in the unique termination);
S_{II} & S_{III} *(General)* – six for one termination, eight for the other;
S_{II} & S_{III} *(Uniterminated)* – six and eight (two atoms in the unique termination);
K_I & K_{III} *(General)* – Six for both terminations;
K_I & K_{III} *(Uniterminated)* – Six (both atoms in the unique termination);
K_{II} & K_{IV} *(General)* – Five for one termination, seven for the other;
K_{II} & K_{IV} *(Uniterminated)* – Five & seven (two atoms in the unique termination).

2.6.3 Reconciliation of Symmetry and Primary Structure

Putting together the mirror and structural zones for *hcp*, we arrive once more at the point where it is possible make some general observations. In this case, we may note (i) that the only flat surfaces are of $\{0001\}$ type; (ii) that the most highly symmetric stepped surfaces are of $\{10\bar{1}0\}$ type; (iii) that all other stepped surfaces possess a single mirror plane, as do the interrupted flat surfaces; (iv) that geminal surfaces are always singly chiral, but not all singly chiral surfaces are geminal; (v) that surfaces of meandering row type are always glissadic, but aside from these all glissadic surfaces are kinked and (vi) that all racemic and doubly chiral surfaces are also kinked.

Regarding the overall categorisation of surfaces, we find a total of fourteen different combinations of symmetry and structure (Fig. 2.23) that may arise for surfaces of *hcp* crystals:

Bayonet Flat Surfaces

This category contains only a single type of surface, namely the two $\{0001\}$ surfaces. Each possesses three mirror planes, perpendicular to the $\langle 11\bar{2}0\rangle$ crystallographic axes, but the defining symmetry is a that of rototruncation (i.e. addition or removal of a layer of atoms, combined with rotation about a vertical axis – in this case, by 60°). Close-packed chains of atoms run in the $\langle 10\bar{1}0\rangle$ directions.

Reflexive-Glissadic Meandering Row Surfaces

This category contains only the six surfaces of $\{11\bar{2}0\}$ type. Each is uniterminated (i.e. displays pure truncation symmetry) and possesses a mirror plane perpendicular to the [0001] crystallographic axis, together with a glide symmetry comprising reflection across a plane perpendicular to one of the $\langle 10\bar{1}0\rangle$ axes combined with translation through half a lattice spacing in a $\langle 11\bar{2}0\rangle$ direction parallel to that plane. A meandering row with internal angle ζ (see above) runs in the same $\langle 11\bar{2}0\rangle$ direction.

Figure 2.23 Permitted combinations of structure and symmetry for the surfaces of hexagonal close-packed materials.

Symmetry \ Structure	Flat	Interrupted flat	Stepped	Geminal	Meandering row	Kinked
Doubly chiral						✓
Singly chiral				✓		✓
Racemic						✓
Glissadic					✓	✓
Reflexive			✓			✓
Uniterminated-reflexive		✓	✓			✓
Doubly reflexive			✓			
Reflexive-glissadic					✓	
Bayonet	✓					

Doubly Reflexive Stepped Surfaces

This category contains only the six surfaces of $\{10\bar{1}0\}$ type. Each possesses a mirror plane perpendicular to the [0001] crystallographic axis, and another perpendicular to one of the $\langle 11\bar{2}0\rangle$ axes. Close-packed step edges run parallel to the latter. There is no truncation symmetry, so the two terminations of each surface are non-degenerate.

Uniterminated-Reflexive Interrupted Flat Surfaces

This category contains two types of surface, namely the six $\{30\bar{3}2\}$ and the six $\{30\bar{3}4\}$ surfaces. In all cases, the surface exhibits a single mirror plane perpendicular to one of the $\langle 11\bar{2}0\rangle$ crystallographic axes, and a close-packed chain of atoms runs parallel to the same axis. Owing to pure truncation symmetry, each layer is identical and features narrow strips of triangular or square local structure, for surfaces of $\{30\bar{3}2\}$ and $\{30\bar{3}4\}$ type, respectively.

Uniterminated-Reflexive Stepped Surfaces

This category contains an infinite number of surfaces whose indices conform to the pattern $\{3p\ 0\ \overline{3p}\ 2q\}$ with p and q being positive integers and $q > 2$. For each type, there will be six distinct instances. Once again, each exhibits a single mirror plane perpendicular to one of the $\langle 11\bar{2}0\rangle$ crystallographic axes, and a close-packed chain of atoms runs parallel to the same axis. Truncation symmetry implies that all layers are identical.

Uniterminated-Reflexive Kinked Surfaces

This category contains an infinite number of surfaces whose indices conform to the pattern $\{3p+q\, q\, \overline{3p+2q}\, 0\}$ with p and q being positive integers. For each type, there will be 12 distinct instances. Each exhibits a single mirror plane, perpendicular to the [0001] crystallographic axis, but there are no close-packed pairs or chains of atoms lying within the surface plane. Once again, truncation symmetry implies that all layers are identical.

Reflexive Stepped Surfaces

This category consists of an infinite number of surfaces whose indices contain one of zero magnitude amongst the first three without falling into one of the categories outlined above. Each such surface type will occur in six distinct instances, exhibiting a single mirror plane perpendicular to one of the $\langle 11\overline{2}0\rangle$ crystallographic axes. A close-packed chain of atoms runs parallel with the same axis. There is no truncation symmetry, so the two terminations available to each surface will, in general, be non-degenerate.

Reflexive Kinked Surfaces

This category consists of an infinite number of surfaces whose fourth index is zero without falling into one of the categories outline above. Each such surface type will occur in 12 distinct instances, exhibiting a single mirror plane perpendicular to the [0001] crystallographic axis. There are no close-packed pairs or chains of atoms lying within the surface plane, and lack of truncation symmetry implies that each such surface possesses two possible non-degenerate terminations.

Glissadic Meandering Row

This category contains only the six surfaces of $\{11\overline{2}2\}$ type. Each displays a glide symmetry comprising reflection across a plane perpendicular to one of the $\langle 10\overline{1}0\rangle$ directions combined with translation through half a lattice spacing in a $\langle 11\overline{2}0\rangle$ direction parallel to that plane. A meandering row with internal angle η (see above) runs in the same $\langle 11\overline{2}0\rangle$ direction. In common with other glissadic surfaces, truncation symmetry implies that only one distinct termination exists.

Glissadic Kinked

This category contains an infinite number of surfaces whose indices conform to the pattern $\{p\, p\, \overline{2p}\, 2q\}$ with p and q being positive integers and $q > 1$. Each type occurs in six different instances. All display both truncation symmetry (i.e. they are uniterminated) and a glide symmetry combining reflection across a plane perpendicular to one of the $\langle 10\overline{1}0\rangle$ axes combined with translation through half a lattice spacing in a $\langle 11\overline{2}0\rangle$ direction parallel to that plane. Each atom is devoid of nearest neighbours within its own layer.

Racemic Kinked

This category consists of an infinite number of surfaces whose indices conform to the pattern $\{p\, p\, \overline{2p}\, 2q \pm 1\}$ with p and q being positive integers. Each type occurs in six different instances. The defining symmetry of these surfaces is that of reflexotruncation (i.e. addition or removal of a layer of atoms, combined with reflection across a vertical plane – in this case, perpendicular to one of the $\langle 10\bar{1}0 \rangle$ axes). The resulting two enantiomerically related terminations are labelled D or L, according to their two-chirality. Once again, each atom is devoid of nearest neighbours within its own layer.

Singly Chiral Geminal

This category contains an infinite number of surfaces whose indices conform to the pattern $\{3p + q\ q\ \overline{3p + 2q}\ s\}$ with p and q being positive integers and s equal to $p, p + q$ or $2p + q$. Each type occurs in 12 different instances - 6 of each enantiomer, labelled D or L according to their one-chirality. Such surfaces possess only pure truncation symmetry in addition to the translational periodicity common to all crystalline surfaces, implying that only a single distinct termination exists for each. Every atom has a single nearest neighbour within its own layer.

Singly Chiral Kinked

This category of surface contains an infinite number of surfaces whose indices conform to the pattern $\{3p+q\ q\ \overline{3p+2q}\ s\}$ with p and p being positive integers and s not equal to p, $p+q$ or $2p+q$. Each type occurs in 12 different instances - 6 of each enantiomer, labelled D or L according to their one-chirality. Once again, such surfaces possess only pure truncation symmetry in addition to the translational periodicity common to all crystalline surfaces, implying that only a single distinct termination exists for each. No atom possesses any nearest neighbours within its own layer.

Doubly Chiral Kinked

Finally, this category contains an infinite number of surfaces whose indices do not conform to any of the special cases outlined above. Each type occurs in 12 different instances - 6 with one-chirality D and 6 with one-chirality L. Such surfaces display no symmetry apart from the translational periodicity common to all crystalline surfaces, implying that each can exist with two non-degenerate terminations, labelled D or L according to their two-chirality. No atom possesses any nearest neighbours within its own layer.

2.6.4 Commentary

Clearly, one of the key differences between the surfaces of hexagonal close-packed materials and those of the face-centred and body-centred cubic materials is to be found in the number of distinct terminations that may exist for the same set of indices. The maximum number of such terminations is simply equal to the number of atoms within the primitive unit cell of the parent bulk material: one in the two cubic cases, versus two in the hexagonal case. Granted, the existence of uniterminated surfaces means that the maximum number of terminations is not always attained, but the non-uniterminated situation is certainly not limited to exotic surfaces. The $\{0001\}$ surfaces, as we have seen above, show two terminations related to one another by rotational symmetry; they are degenerate, though not identical, and consequently both co-exist in roughly equal proportions on real surfaces. The $\{10\bar{1}0\}$ surfaces, to give another example, can have two distinct terminations that are definitely *not* degenerate; the coordination number of the least-coordinated atoms is eight for one of the terminations and six for the other, meaning that the latter is overwhelmingly favoured.[18]

To generalise for *all* surfaces of hexagonal close-packed materials, it can be shown that within a model that counts only nearest-neighbour interactions (and assumes all of these to be equal) the termination for which the coordination number in the outermost layer is maximised corresponds to the lowest surface energy. Whenever two terminations have distinctly different coordination numbers in their outermost layers, therefore, we should expect a clear energetic preference for one over the other, as we have indeed seen for the $\{10\bar{1}0\}$ surfaces. Where two terminations share the same coordination number in the outermost layer, however, their surface energies are likely to be very similar, even though they may have very different structures. A good example of this may be seen in the case of the $\{2\bar{1}\bar{3}1\}$ surface type; this comes from the K_{III} region of the stereogram, and hence both of its two distinct terminations have atoms in the outermost layer displaying a coordination number of 6. Pleasingly, density functional theory (DFT) calculations for Co, Ru and Re have revealed, respectively, differences of only 1.2%, 0.9% and 0.3% between the surface energies of the two terminations [6]. These surface energies are similar enough to anticipate co-existence of both terminations at room temperature in each case, suggesting the likelihood of a rough surface unless the domains are large – clearly, the energy attributable to domain boundaries will play a key role. The crucial point is that the state of affairs where this must be considered is readily predictable according to the criteria presented in this commentary.

2.7 Surfaces of the Diamond Structure

The diamond structure (see Fig. 2.24) is adopted not only by crystalline carbon, but also by silicon and germanium, making it one of the most important crystal structures in the study of semiconductors. From the point of view of the spherical/stereographic approach adopted above, it shares a number of features with the *hcp* structure, notably the presence of glide symmetry and of directions with only intermittent close-packing. Indeed, there are *no* directions in which infinitely extended close-packed chains of atoms occur, which sets it aside from *all* of the cases discussed thus far. On the other hand, the cubic symmetry of the diamond structure is quite familiar from the cases of *fcc* and *bcc*. For all of these reasons, it makes an interesting additional topic for study.

2.7.1 Symmetry

As ever, we begin by plotting the mirror zones of the crystal structure's underlying Bravais lattice in a spherical representation (or, equivalently, onto a stereogram),

18 Indeed, the favoured termination of the $\{10\bar{1}0\}$ surfaces of the hexagonal close-packed structure bears a striking resemblance to the face-centred cubic $\{110\}$ and body-centred cubic $\{211\}$ surfaces, reflecting the fact that each represents the maximally reflexive stepped surfaces of their respective structures (i.e. the only stepped surfaces with two mirror planes for the hexagonal close-packed and face-centred cubic structure, and the only stepped surface with even a single mirror plane for the body-centred cubic structure).

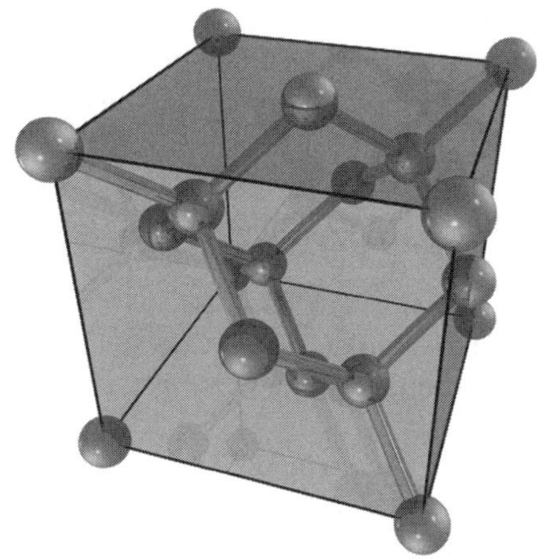

Figure 2.24 Conventional unit cell of the diamond crystal structure.

which in the case of the diamond structure means revisiting the symmetry of the *fcc* lattice. The required mirror zones are just those perpendicular to the $\langle 100\rangle$ and $\langle 110\rangle$ axes of the bulk lattice. As before, (see Fig. 2.1b), these delineate a set of triangles, within each of which the chirality of the surfaces of *fcc* crystals can be assigned (according to an arbitrary but well-defined rule). The fact that the diamond structure contains two atoms per unit cell, however, means that the $\langle 100\rangle$ mirror planes of the lattice do not apply to the crystal; instead, they are glide planes, within which it is possible to find glide vectors corresponding to the translational components of glide operations involving reflection across the plane. By analogy with the *hcp* case, therefore, we can identify two sources of chirality: what we shall call the one-chirality derives from the truncation of the underlying lattice by the surface plane, whereas the two-chirality arises due to the interplay between the lattice symmetry, the symmetry of the atomic basis and the particular choice of surface plane.

Let us, as before in the case of *hcp* crystals, label half of the atoms in the diamond structure α and the other half β. More specifically, we will identify the α atoms with a vector displacement of [000] relative to each lattice point, and the β atoms with vector displacement $\frac{1}{4}[111]$. We can then assign labels for the two-chirality of surfaces with α and β termination as shown in Fig. 2.25.

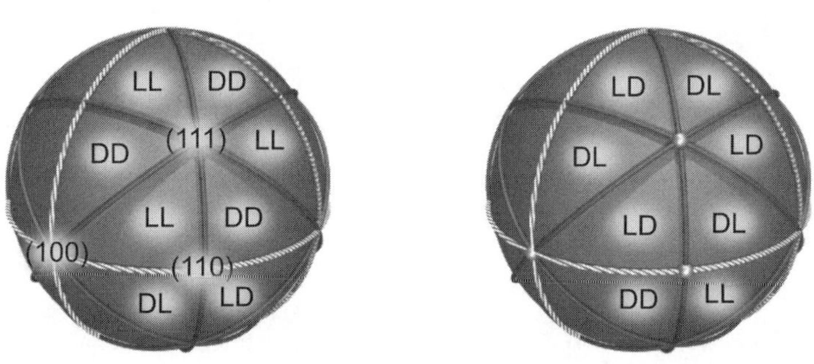

Figure 2.25 Chirality labelling scheme for surfaces of diamond structure materials. In panel (a) the labels used for the α termination are shown, while in (b) the labels are for the β termination.

Following the same logic as for the hexagonal system, we also specify that any surface termination labelled either DD or LL will be designated τ, while any labelled either DL or LD will be labelled $\overline{\tau}$. In this way, we unambiguously assign labels to all of the doubly chiral surfaces of the diamond structure.

Once again, however, we can predict a variety of special symmetry cases [7]. For instance, it is clear that for any surface whose pole lies on a mirror zone, both the one- *and* two-chirality must be inactive, so the appropriate composite symmetry symbol will, in general, be XX_τ and $XX_{\overline{\tau}}$ for non-degenerate terminations, or XX_u for uniterminated-reflexive surfaces. Where unitermination does not coincide with a mirror or glide zone, the result is a singly chiral surface, with composite symmetry symbol DX_u or LX_u, while a non-uniterminated surface whose pole lies on a glide zone will be racemic (i.e. have alternating chirality upon addition or removal of a single atomic layer) and its two degenerate terminations will have composite symmetry symbols XD_o and XL_o. Where a surface whose pole lies on a glide zone is uniterminated, in contrast, it must have $XX_\emptyset$ as its composite symmetry symbol, indicating that it is achiral due to the presence of glide symmetry (i.e. glissadic).

2.7.2 Primary Structure

The main difference between the metallic crystal structures (i.e. face-centred cubic, body-centred cubic and hexagonal close-packed) and the diamond crystal structure adopted by several elemental semiconductors and insulators (i.e. carbon, silicon, germanium, α-tin) is the lack of extended close-packed chains of atoms in the latter. In the stereographic analysis presented here, there are thus no primary structural zones to define flat and stepped surfaces in quite the same manner. Instead, we find only interrupted close-packed chains – lying in the $\langle 111 \rangle$ crystallographic directions – which we shall associate with corresponding interrupted primary zones (see the similar treatment of interrupted features in the structure of hexagonal close-packed materials, described above).

At the intersection of the interrupted primary zones (see Fig. 2.26) we find the poles corresponding to surfaces of $\{110\}$ type, which, for materials taking

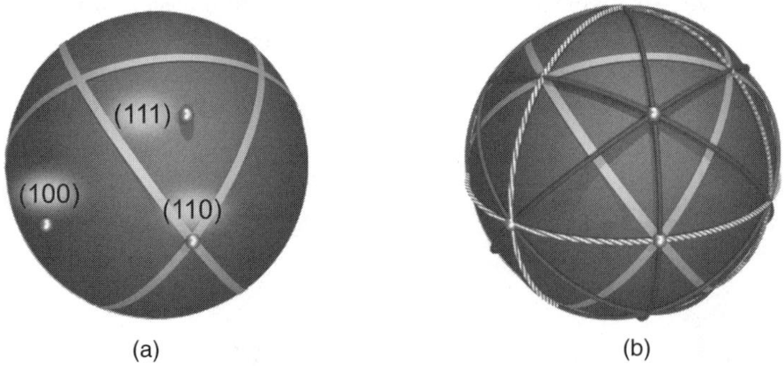

Figure 2.26 Spherical representation of surface structure of diamond structure materials. In panel (a) the interrupted primary structural zones are shown; in panel (b) the mirror and glide zones of the symmetry representation are overlaid. (*See Plate III for the colour representation of this figure.*)

the diamond structure, are the only exemplars of the meandering row class. In this case, the meandering rows exhibit an internal angle of $\cos^{-1}(-1/3)$ or 109.47°. In contrast, geminal surfaces (featuring only close-packed pairs of atoms) can be found whenever the pole lies on a zone perpendicular to one of the ⟨111⟩ crystallographic directions, while the epithet 'kinked' applies to all other surfaces [7].

Just as we were able to identify the coordination numbers of atoms in the outermost layer of hexagonal close-packed surfaces, so we are again able to do so for the surfaces of materials taking the diamond crystal structure. As before, the results break down according to the various structural classifications introduced above. The kinked surfaces are subdivided into K_I and K_{II} categories, where the numerical value of the subscript now indicates unity plus the minimum number of interrupted primary zones that must be crossed in order to reach the pole in question from a pole of {100} type. The coordination number of atoms in the outermost layer is then:

Meandering-Row – Three (both atoms in the unique termination);
Geminal – Two and three (two atoms in the unique termination);
K_I *(General)* – Two for both terminations;
K_I *(Uniterminated)* – Two (both atoms in the unique termination);
K_{II} *(General)* – One for one termination, three for the other;
K_{II} *(Uniterminated)* – One and three (two atoms in the unique termination).

2.7.3 Reconciliation of Symmetry and Primary Structure

Putting together the mirror and structural zones for the diamond structure, we arrive once more at the point where it is possible to make some general observations. In this case, we may note (i) that there are no truly flat or stepped surfaces, due to an absence of close-packed directions in the bulk; (ii) that the closest-to-flat surfaces are those of meandering row type, namely the {110} surfaces and (iii) that the most highly symmetric geminal surfaces are those of {211} type, which have a single mirror plane.

Regarding the overall categorisation of surfaces, we find a total of 11 different combinations of symmetry and structure (Fig. 2.27) that may arise for surfaces of diamond structure crystals:

Bayonet Kinked Surfaces
This category contains only a single type of surface, namely the six {001} surfaces. Each possesses two mirror planes, perpendicular to ⟨110⟩ crystallographic axes, but the defining symmetry is that of a rototruncation (i.e. addition or removal of a layer of atoms, combined with rotation about a vertical axis – in this case, by 90°). There are no close-packed pairs or chains of atoms lying within the surface plane.

Reflexive-Glissadic Meandering Row Surfaces
This category contains only the 12 surfaces of {110} type. Each is uniterminated (i.e. displays pure truncation symmetry) and possesses a mirror plane perpendicular to one of the ⟨110⟩ crystallographic axes, together with a glide

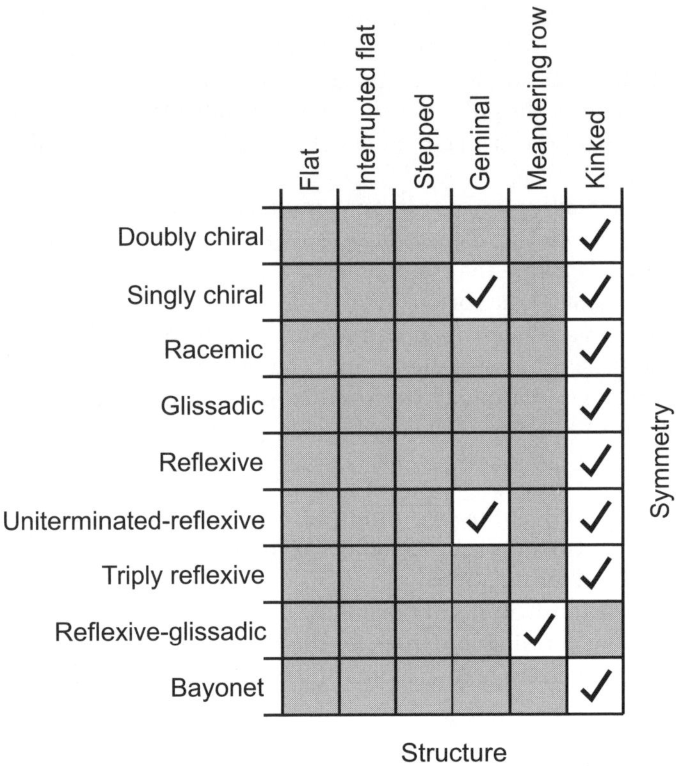

Figure 2.27 Permitted combinations of structure and symmetry for the surfaces of diamond structure materials.

symmetry comprising reflection across a plane perpendicular to one of the ⟨100⟩ axes combined with translation through half a lattice spacing in a ⟨110⟩ direction parallel to that plane. A meandering row with tetrahedral internal angle runs in the same ⟨110⟩ direction.

Triply Reflexive Kinked Surfaces

This category contains only the eight surfaces of {111} type. Each possesses three mirror planes, perpendicular to ⟨110⟩ crystallographic axes, but there is no truncation symmetry so the two possible terminations are non-degenerate. There are no close-packed chains of atoms lying within the surface plane in either termination.

Uniterminated-Reflexive Geminal Surfaces

This category contains only the 24 surfaces of {112} type. Each possesses a single mirror plane, perpendicular to one of the ⟨110⟩ crystallographic axes, and truncation symmetry implies that all possible terminations of the surface are equivalent. There are no close-packed chains of atoms lying within the surface plane, but each atom does possess a single nearest neighbour within the same layer.

Uniterminated-Reflexive Kinked Surfaces

This category consists of an infinite number of surfaces whose Miller indices conform to the pattern $\{2p + 1\ 2p + 1\ 2q\}$ with p and q both integers (excepting the special cases of {110} and {211} surfaces dealt with above). Each type occurs in 24 distinct instances, obtained by permutation of the index signs and ordering. All exhibit a single mirror plane, perpendicular to one of the ⟨110⟩ crystallographic axes, and truncation symmetry implies that each possesses only a single distinct termination. There are no close-packed pairs or chains of atoms lying within the surface plane.

Reflexive Kinked Surfaces

This category contains an infinite number of surfaces whose Miller indices conform to the pattern $\{p\,p\,2q + 1\}$ with p and q integers (and excepting the special cases of {001} and {111} surfaces dealt with above). Once again, each type will occur in 24 distinct instances, in accordance with the crystalline symmetry. All exhibit a single mirror plane, perpendicular to one of the ⟨110⟩ axes, and the absence of truncation symmetry implies that each instance possesses two possible non-degenerate terminations. No close-packed pairs or chains of atoms lie within the surface plane.

Glissadic Kinked Surfaces

This category consists of an infinite number of surfaces whose Miller indices conform to the pattern $\{2p + 1\ 2q + 1\ 0\}$ with p and q unequal integers. Each type occurs in 24 different instances, obtained by permuting the order and signs of the indices. All such surfaces display truncation symmetry (i.e. they are uniterminated) in addition to glide symmetry that combines reflection across a plane perpendicular to one of the ⟨110⟩ axes with translation through half a lattice spacing in the ⟨001⟩ direction parallel to that plane. Each atom is devoid of nearest neighbours within its own layer.

Racemic Kinked Surfaces

This category contains an infinite number of surfaces whose Miller indices conform to the pattern $\{2p + 1\ 2q\ 0\}$ with p and q both integers. Each type occurs in 24 different instances. The defining symmetry is that of reflexotruncation (i.e. addition or removal of a layer of atoms, combined with reflection across a vertical plane – in this case, perpendicular to one of the ⟨110⟩ axes). The resulting two enantiomerically related terminations are labelled D or L, according to their two-chirality. There are no close-packed pairs or chains of atoms lying within the surface plane.

Singly Chiral Geminal Surfaces

This category contains an infinite number of surface types whose Miller indices conform to the pattern $\{p\,q\,p + q\}$ with p and q different positive integers. Each such type lacks mirror symmetry and will arise in 48 different instances – 24 of each enantiomer, labelled D or L according to their one-chirality. There is no close-packed chain direction lying within the surface plane, but each atom does have a single nearest neighbour within the same layer. Only a single distinct termination is possible for each instance of such a surface.

Singly Chiral Kinked Surfaces
This category comprises an infinite number of surfaces whose Miller indices (i) are all of differing magnitude, (ii) are all non-zero and (iii) conform to the pattern $\{2p + 1\, 2q + 1\, 2r\}$ with p, q and r integers. Each such surface type arises in 48 different instances – 24 of each enantiomer, labelled D or L according to their one-chirality – and displays no mirror symmetry. Each atom is devoid of nearest neighbours within its own layer, and there is only a single distinct termination possible.

Doubly Chiral Kinked
Finally, this category contains an infinite number of surfaces whose Miller indices do not conform to any of the special cases outlined above. Each type occurs in 48 different instances – 24 with one-chirality D and 24 with one-chirality L. Such surfaces display no symmetry apart from the translational periodicity common to all crystalline surfaces, implying that each can exist with two non-degenerate terminations, labelled D or L according to their two-chirality. No atom bonds to any nearest neighbours within its own layer.

2.7.4 Commentary

Finally, we close this chapter on intrinsic surface chirality by noting that we have limited ourselves thus far to materials comprising atoms of only a single element. Clearly, materials composed of more than one element are likely to exhibit lower symmetry in their bulk structures, and correspondingly are more likely to exhibit chiral surface structures. It is, of course, possible to extend the stereographic approach adopted here to such surfaces, but inevitably the analysis will become ever more complex. Alternatively, we might seek to describe multi-element materials in terms of simpler single-element structures, including chirality due to the chemical identity of certain atoms in an ad hoc manner. To take a simple example, one might treat the zincblende crystal structure as a variant upon the diamond crystal structure, noting only that symmetry operations that interchange α and β positions are no longer valid. This would, in turn, mean that diamond-structure surfaces whose achirality is due only to glide symmetry (e.g. surfaces of $\{310\}$ type) will be chiral in the case of the zincblende structure. Another important difference is that the distinction between α-terminated and β-terminated surfaces is physically meaningful in the zincblende case, relating to the difference between cation-terminated and anion-terminated surfaces. It follows, therefore, that the criterion for unitermination is effectively synonymous with the criterion for the surface of a zincblende material to be non-polar; since polar surfaces are strictly unstable against reconstruction, it further follows that only the reflexive-glissadic, uniterminated-reflexive, glissadic and singly chiral categories should be stable in unreconstructed form.

References

1 N.W. Ashcroft and N.D. Mermin, *Solid State Physics* (Saunders College Publishing, 1976).

2 M.L. Clegg, L. Morales de la Garza, S. Karakatsani, D.A. King and S.M. Driver, *Top. Catal.* **54**, 1429 (2011).

3 G.A. Attard, *J. Phys. Chem. B* **105**, 3158 (2001).

4 S.J. Jenkins and S.J. Pratt, *Surf. Sci. Rep.* **62**, 373 (2007).

5 S.J. Jenkins (submitted for publication).

6 S.J. Jenkins, *Surf. Sci.* **604**, 1688 (2010).

7 S.J. Jenkins, *Langmuir* **33**, 12820 (2017).

3

Surface Energy and Surface Stress

Work consists of whatever a body is obliged to do...
The Adventures of Tom Sawyer, Mark Twain, 1876

Arguably, the most fundamental property of any system is its energy, since this dictates both its stability and its capacity to do useful work. By considering virtual work at equilibrium, one obtains the stress that acts upon the system, and from this it is possible to predict how the system will evolve in response to external perturbation. When studying a surface, therefore, it is essential to begin with a rigorous definition of its energy and to investigate how virtual work associated with its deformation defines the stress implicit in its very existence. To do so, we had best go all the way back to first principles.

3.1 Thermodynamic Definition of Surface Energy

Consider an isolated system of volume V, comprising a solid body surrounded by fluid,[1] held at constant temperature, T, and pressure, p. At equilibrium, the internal energy of the complete system may be written as

$$U = U_1 + U_2 \tag{3.1}$$

where U_1 is the internal energy of the solid (including its surfaces) and U_2 that of the fluid (including the regions in close proximity to the solid). These two energies may be written as

$$U_m = TS_m - pV_m + \sum_n \mu_n N_n^m \tag{3.2}$$

where S_m and V_m represent, respectively, the entropy and volume associated with each component of the system. The symbol N_n^m represents the number of particles of species n in component m, and the chemical potential for each species is denoted by μ_n.

1 The fluid may be liquid or gas, and in the latter case may even be taken to the tenuous limit of a vacuum. We shall refer to all two-dimensional properties derived here as 'surface' properties, rather than 'interface' properties, but the difference is really only semantic.

Chirality at Solid Surfaces, First Edition. Stephen J. Jenkins.

Now, imagine cleaving the solid in two, creating a total area A of new surface that did not previously exist. The total internal energy of the system now becomes

$$U = U_1 + U_2 + U_s \tag{3.3}$$

where U_s is the internal energy associated with the newly created surface.

Let us assume, first of all, that the work, W, done in creating the new surface is proportional to the surface area, so that

$$W = \gamma A \tag{3.4}$$

and hence

$$U_s = TS_s + \gamma A + A \sum_n \mu_n \Gamma_n \tag{3.5}$$

where S_s is the entropy associated with the new surface.[2] The quantity Γ_n here represents the number of particles per unit area that may have accumulated at the new surface (either due to segregation of atoms from the bulk of the solid, or due to adsorption from the bulk of the fluid). We shall call the property γ the **specific surface energy**, and investigate below what consequences its value may imply.

Clearly, we must have

$$\sum_m N_n^m + \Gamma_n A = N_n \tag{3.6}$$

where N_n is the total number of particles of species n contained overall within the system . Any non-zero Γ_n consequently implies a change in either N_n^1 or N_n^2 (or both) and a corresponding change in U_1 and/or U_2.

Now, differentiating our expression for internal energy, we obtain the following fundamental thermodynamic relations for the bulk

$$dU_m = TdS_m - pdV_m + \sum_n \mu_n dN_n^m \tag{3.7}$$

and surface

$$dU_s = TdS_s + \gamma dA + Ad\gamma + A \sum_n \mu_n d\Gamma_n \tag{3.8}$$

parts of our system. Note carefully the crucial distinction that the derivative Vdp does not occur in the bulk expression, because the pressure is held constant, whereas both γdA and $Ad\gamma$ appear in the surface expression. It is quite possible to imagine cutting our original bulk solid at two different angles, creating the same area of new surface in each case but with differing internal energies due to anisotropy in the bulk structure; such a difference must be captured by different values of the specific surface energy, γ, so the second of our two derivative terms cannot be left out in this case.[3]

---

2 The volume associated with the surface is considered to be zero.

3 There has, it must be admitted, been a vigorous (occasionally ill-tempered) debate in the surface science literature over recent years concerning the validity of this argument. A small group of authors [1–7] insist that the derivative of a thermodynamic potential can never contain terms of the form $XdY + YdX$, and that the term $Ad\gamma$ ought, therefore, to be omitted in deference to γdA. Others argue [8–13] that this view is based upon a mistakenly inflexible interpretation of the fundamentals

From the internal energies defined above, it is straightforward to deduce the corresponding Gibbs free energy, since

$$G_m = U_m + pV_m - TS_m = \sum_n \mu_n N_n^m \tag{3.9}$$

and

$$G_s = U_s - \gamma A - TS_s = A \sum_n \mu_n \Gamma_n \tag{3.10}$$

yielding

$$G = G_1 + G_2 + G_s = \sum_n \mu_n N_n \tag{3.11}$$

for the total Gibbs free energy of the system. Its derivative takes the form

$$dG = \sum_n (\mu_n dN_n + N_n d\mu_n) \tag{3.12}$$

and since we are dealing with an isolated system at equilibrium with constant temperature and pressure, we know that $dN_n = dG = 0$, implying that

$$\sum_n N_n d\mu_n = 0 \tag{3.13}$$

for this system.

In similar fashion, we obtain the Helmholtz free energy for the bulk components

$$F_m = U_m - TS_m = -pV_m + \sum_n \mu_n N_n^m \tag{3.14}$$

and surface component

$$F_s = U_s - TS_s = \gamma A + A \sum_n \mu_n \Gamma_n \tag{3.15}$$

yielding

$$F = F_1 + F_2 + F_s = -pV + \gamma A + \sum_n \mu_n N_n \tag{3.16}$$

for the total Helmholtz free energy of the system. Its derivative takes the form

$$dF = -pdV + \gamma dA + Ad\gamma \tag{3.17}$$

where, once again, both terms in the derivative of γA must be retained. Note that terms in $\mu_n dN_n$ vanish, since we are dealing explicitly with an isolated system; those in $N_n d\mu_n$ do likewise, owing to the equilibrium condition on the Gibbs free energy.

---

of thermodynamics. It seems to this author that the argument of the former group is rather obviously flawed, since their fundamental thermodynamic relation, omitting the $Ad\gamma$ term, is incapable of predicting a change in the internal energy as a result of area-preserving deformations – hence implying (incorrectly, of course) that shear strains may be accomplished without work!

The significance of these free energies is, of course, their relationship to various kinds of work done on (or by) the system under certain conditions. Specifically, the Gibbs free energy quantifies the maximum *useful* work that may be extracted from a system (zero, in this case, because it is isolated) while the Helmholtz free energy dictates the *total* work involved in perturbing the system subject to the condition that the temperature starts and ends at the temperature of its surroundings. Focusing upon this latter property, we therefore have

$$dW = \left(\frac{\partial F}{\partial A}\right)_T dA \tag{3.18}$$

for the infinitesimal work done in stretching our new surface to increase its area by an amount dA. Note that this is a rather different way to increase the amount of surface present than the process of cleavage employed at the start of this discussion. Inserting Eq. (3.16) and assuming that the overall volume of our system remains essentially constant while stretching the surface, we obtain

$$dW = \left[\gamma + A\left(\frac{\partial \gamma}{\partial A}\right)\right] dA \tag{3.19}$$

which expresses the fact that increasing the surface area generally requires work both in creating a new surface at the original specific surface energy (the first term on the right-hand side above) *and* altering the specific surface energy across the whole surface (the second term on the right-hand side).

We can, however, be a little more precise than this, by specifying the perturbation of the surface not as a simple change of area but instead as a combination of strains. Following the usual conventions, we set $d\epsilon_{11}$ and $d\epsilon_{22}$ as $\delta x_1/x_1$ and $\delta x_2/x_2$ for normal strain components in the x_1 and x_2 directions, with $d\epsilon_{12}$ and $d\epsilon_{21}$ as the shear strains $\delta x_1/x_2$ and $\delta x_2/x_1$ (see Fig. 3.1). We hence deduce that

$$\frac{\partial A}{\partial \epsilon_{ij}} = A\delta_{ij} \tag{3.20}$$

from which we obtain the result

$$\sum_{ij}\left(\frac{\partial \gamma}{\partial \epsilon_{ij}}\right) d\epsilon_{ij} = \sum_{ij}\left(\frac{\partial \gamma}{\partial A}\right)\left(\frac{\partial A}{\partial \epsilon_{ij}}\right) d\epsilon_{ij} = A\left(\frac{\partial \gamma}{\partial A}\right) dA \tag{3.21}$$

by means of the chain rule.

Noting also that

$$\sum_{ij} \gamma \delta_{ij} d\epsilon_{ij} = \gamma dA \tag{3.22}$$

we obtain

$$dW = \sum_{ij} \sigma_{ij} d\epsilon_{ij} \tag{3.23}$$

where the tensor quantity

$$\sigma_{ij} = \gamma\delta_{ij} + \frac{\partial \gamma}{\partial \epsilon_{ij}} \tag{3.24}$$

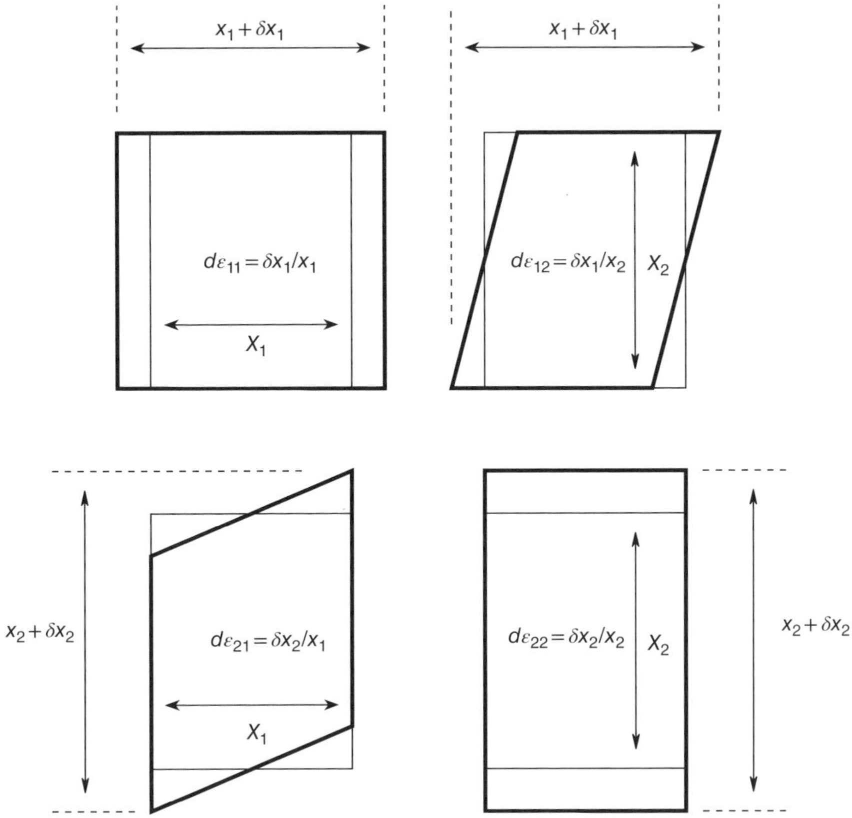

Figure 3.1 Definition of surface strain components in terms of the infinitesimal deformation (δx_1, δx_2) of a small element of area $A = x_1 x_2$.

clearly plays the role of **surface stress** (defined as per Fig. 3.2). That is, the work done in deforming a surface is simply the tensor product of the stress tensor with the strain tensor.

This last equation is the so-called Shuttleworth equation, the most important consequence of which is that the surface stress is, in general, anisotropic, due to the second term on its right-hand side. The specific surface energy, in general, changes by different amounts when the surface is strained in different senses, so the corresponding surface stress components vary likewise. In this respect, the stress at a solid surface differs qualitatively from that of a liquid surface, where the second term on the right-hand side of the Shuttleworth equation necessarily vanishes.[4] For liquids, the surface stress is, therefore, simply equivalent to the product of the specific surface energy with the unit matrix; both surface stress and specific surface energy are then typically referred to as the **surface tension** – a term that has no well-defined meaning for a solid surface.

4 Any deformation of a liquid that varies the surface area will result in flow of particles between the bulk and the surface, such that the density of particles at the surface (and hence the specific surface energy) remains constant.

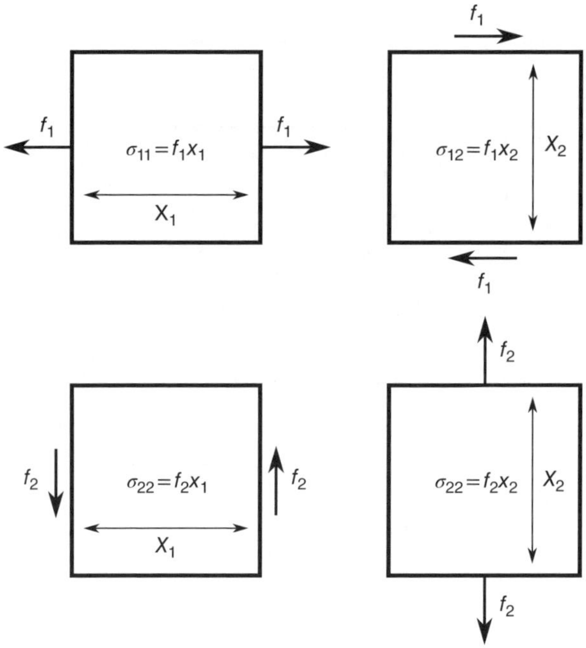

Figure 3.2 Definition of surface stress components in terms of forces (f_1, f_2) acting upon a small element of area $A = x_1 x_2$.

3.2 The Tensor Nature of Surface Stress

It would appear, from the definition of surface stress introduced above, that a full consideration of its effects requires us to consider a total of four components – two so-called **normal stress components**, σ_{11} and σ_{22}, and two **shear stress components**, σ_{12} and σ_{21}. The first two correspond to stresses associated with stretching/compressing the surface, while the second pair represents stresses associated with deformations of the surface that do not change the surface area.[5] If the shear stress components were to differ from one another, however, one would find unbalanced forces tending to rotate the surface, implying that at equilibrium these last two components must be equal. It is usual, therefore, to omit σ_{21} from explicit discussion, dealing only with σ_{12} as one of just three *independent* stress components (there being no similar restriction on the values of σ_{11} and σ_{22}).

It should also be apparent, however, that a simple change of axes would change the values of all our stress components (whether we consider all four, or merely the three independent ones). That is, a set of stress components σ_{ij} defined relative to a set of axes $(\hat{\mathbf{x}}_1, \hat{\mathbf{x}}_2)$ will, in general, differ from stress components σ'_{ij} defined relative to a set of axes $(\hat{\mathbf{x}}'_1, \hat{\mathbf{x}}'_2)$. Nevertheless, it must surely be the case that these components refer to the *same* physical property of the surface, since

5 It is a little unfortunate that the word 'normal' is used here in an entirely different sense to its usage in referring to the 'normal' direction of the surface, but the problem is essentially unavaoidable. In both cases, the intent is to imply perpendicularity, but whereas for the 'surface normal' the referent is the surface plane, for the 'normal stress' the referent is the boundary of the surface area element acted upon in Fig. 3.2.

our choice of axes is essentially arbitrary. It follows, therefore, that the variation of stress components upon changing our coordinate system must be well defined according to some fairly simple mathematical rule. In fact, if the $(\hat{\mathbf{x}}_1, \hat{\mathbf{x}}_2)$ axes are rotated anticlockwise (when viewed from above) by an angle θ relative to the $(\hat{\mathbf{x}}'_1, \hat{\mathbf{x}}'_2)$ axes, that rule reduces to

$$\sigma_{11} = \sigma'_{11}\cos^2\theta + \sigma'_{22}\sin^2\theta + 2\sigma'_{12}\sin\theta\cos\theta \tag{3.25}$$

$$\sigma_{22} = \sigma'_{11}\sin^2\theta + \sigma'_{22}\cos^2\theta - 2\sigma'_{12}\sin\theta\cos\theta \tag{3.26}$$

$$\sigma_{12} = (\sigma'_{22} - \sigma'_{11})\sin\theta\cos\theta + \sigma'_{12}(\cos^2\theta - \sin^2\theta), \tag{3.27}$$

which happens to be the usual transformation rule for a two-dimensional rank-two tensor, proving that the surface stress is itself just such a tensor.

Now, one important consequence of the tensor nature of the surface stress is that it is possible to diagonalise its matrix to obtain eigenvalues (known as the **principal stress components**) and eigenvectors (the **principal stress axes**). That is, for some particular coordinate system with $\hat{\mathbf{x}}'_1 = \hat{\mathbf{p}}$ and $\hat{\mathbf{x}}'_2 = \hat{\mathbf{q}}$, one finds $\sigma'_{11} = \sigma_p$, $\sigma'_{22} = \sigma_q$ and $\sigma'_{12} = 0$, with $(\hat{\mathbf{p}}, \hat{\mathbf{q}})$ being the principal stress axes and σ_p, σ_q the corresponding principal stress components. We can, therefore, conclude that

$$\sigma_{11} = \sigma_p\cos^2\theta + \sigma_q\sin^2\theta \tag{3.28}$$

$$\sigma_{22} = \sigma_p\sin^2\theta + \sigma_q\cos^2\theta \tag{3.29}$$

$$\sigma_{12} = (\sigma_q - \sigma_p)\sin\theta\cos\theta, \tag{3.30}$$

where θ should now be interpreted as the angle (measured anticlockwise when viewed from above) that the general axes $(\hat{\mathbf{x}}_1, \hat{\mathbf{x}}_2)$ are rotated relative to the principal stress axes $(\hat{\mathbf{p}}, \hat{\mathbf{q}})$. The principal stress components and their corresponding axes thus contain sufficient information to reconstruct the stress components relative to any arbitrary coordinate system one may desire. In matrix notation, we may represent the same realisation more succinctly as

$$\boldsymbol{\sigma} = \mathbf{R}^T\boldsymbol{\Sigma}\mathbf{R} \tag{3.31}$$

where

$$\boldsymbol{\Sigma} = \begin{pmatrix} \sigma_p & 0 \\ 0 & \sigma_q \end{pmatrix} \tag{3.32}$$

is the diagonal matrix of the principal stress components, and

$$\mathbf{R} = \begin{pmatrix} \cos\theta & -\sin\theta \\ \sin\theta & \cos\theta \end{pmatrix} \tag{3.33}$$

is the standard two-dimensional rotation matrix ($\mathbf{R}^T$ being its transpose).

3.3 Visualisations of Surface Stress: Iconic Conics

The tensor nature of surface stress lends itself to a range of opportunities for the construction of useful visual representations. Here, we discuss three reasonable options, each with its own strengths and weaknesses. The **normal stress conic** and **shear stress quartic** approaches have the advantage of simplicity, but each displays only one aspect of the surface stress; the **stress ellipse** approach can include both normal and shear stress components within a single figure, but is much less straightforward to interpret.

3.3.1 The Normal Stress Conic

Consider a conic section whose semi-major and semi-minor axes are aligned with the principal stress axes of a particular surface, with radii in these directions equal to the reciprocal square roots of the corresponding principal stress component magnitudes. Such a curve satisfies the equation

$$\sigma_p p^2 + \sigma_q q^2 = \pm 1 \tag{3.34}$$

where p and q are coordinates measured along the $\hat{\mathbf{p}}$ and $\hat{\mathbf{q}}$ axes. In the case where both principal stresses share the same sign, the resulting curve is an ellipse, whereas with differing signs two conjugate hyperbolae are found[6] (see Fig. 3.3).

Setting

$$p = r(\theta)\cos\theta \tag{3.35}$$

$$q = r(\theta)\sin\theta, \tag{3.36}$$

where $r(\theta)$ is the radius measured from the origin of the conic, we find (by substitution into Eq. (3.34)) that

$$r(\theta) = |\sigma_p \cos^2\theta + \sigma_q \sin^2\theta|^{-1/2} \tag{3.37}$$

for a radius measured at an angle θ anticlockwise from the $\hat{\mathbf{p}}$ axis, and

$$r(\theta + \pi/2) = |\sigma_p \sin^2\theta + \sigma_q \cos^2\theta|^{-1/2} \tag{3.38}$$

for one measured at an angle θ anticlockwise from the $\hat{\mathbf{q}}$ axis.

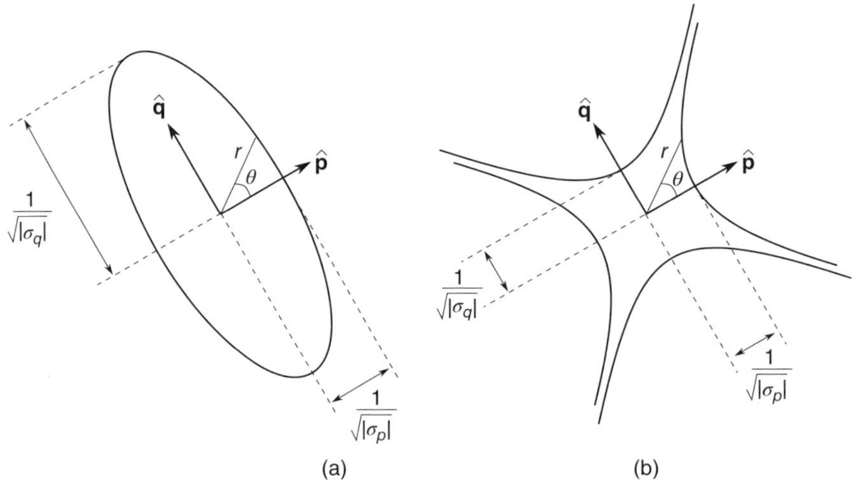

Figure 3.3 Normal stress conics, for cases with (a) both principal stress components positive and (b) one positive and one negative principal stress component.

---

6 We restrict our solutions to the real (p, q) plane.

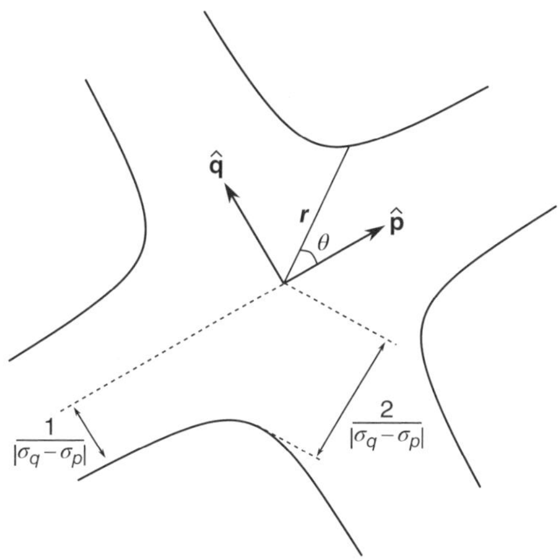

Figure 3.4 The shear stress quartic.

But these last two equations have right-hand sides reciprocal to the square roots of the magnitudes of Eqs. (3.28) and (3.29), introduced above to describe the normal stress components, σ_{11} and σ_{22}, along axes $(\hat{\mathbf{x}}_1, \hat{\mathbf{x}}_2)$ rotated anticlockwise by an angle θ relative to the principal stress axes $(\hat{\mathbf{p}}, \hat{\mathbf{q}})$. In other words, the conic radius at any given angle θ is equal to the reciprocal square root of the normal stress component's magnitude at that angle. For this reason, a conic thus defined may sensibly be described as the **normal stress conic**. It provides a neat visual representation of how the normal components of surface stress vary azimuthally, between the extremal values found along the principal stress axes.

3.3.2 The Shear Stress Quartic

Consider the cruciform curve defined by the quartic equation

$$(\sigma_q - \sigma_p)^2 p^2 q^2 = p^2 + q^2 \tag{3.39}$$

where once again p and q are coordinates measured along the $\hat{\mathbf{p}}$ and $\hat{\mathbf{q}}$ axes. Such a curve (Fig. 3.4) approaches the origin most closely when p and q are equal, at which points the radius, $r(\theta)$, takes the value $2/|\sigma_q - \sigma_p|$. Conversely, the radius tends to infinity along the $p = \pm 1/(\sigma_q - \sigma_p)$ and $q = \pm 1/(\sigma_q - \sigma_p)$ asymptotes.

Setting $p = r(\theta)\cos\theta$ and $q = r(\theta)\sin\theta$ as before, we find (by substitution into Eq. (3.39)) that

$$r(\theta) = \frac{1}{|\sigma_q - \sigma_p| \sin\theta\cos\theta} \tag{3.40}$$

for a radius measured at an angle θ anticlockwise from either the $\hat{\mathbf{p}}$ or the $\hat{\mathbf{q}}$ axis. Since this is simply the reciprocal of the magnitude of the right-hand side of Eq. (3.30), the radius of this particular cruciform curve is just the reciprocal of the magnitude of the shear stress component, σ_{12}. Just as the normal stress ellipse gave us a neat way to visualise the normal stress components, so also the **shear stress quartic**, as we shall henceforth call it, performs much the same function for the shear stress components.

3.3.3 The Stress Ellipse

The two curves introduced above provide visualisations for the normal and shear stress components in a relatively straightforward manner, but neither shows *both* types of stress on the same diagram. For cases where both principal stress components are positive, however, a third curve can do just this, albeit at the expense of a slightly more involved interpretation.

Let us begin with an ellipse whose semi-major and semi-minor axes are aligned with the principal stress axes of a particular surface, with radii in these directions equal to the corresponding principal stress components (cf. the normal stress conic, above, where the *reciprocal square roots* of principal stress magnitudes were used). Such an ellipse satisfies the equation

$$\frac{p^2}{\sigma_p^2} + \frac{q^2}{\sigma_q^2} = 1 \tag{3.41}$$

where p and q are, as usual, coordinates measured along the principal stress axes $(\hat{\mathbf{p}}, \hat{\mathbf{q}})$. In anticipation of the following section, we note that such an ellipse may be characterised by its **eccentricity**, defined as whichever of $\sqrt{1 - \sigma_q/\sigma_p}$ or $\sqrt{1 - \sigma_p/\sigma_q}$ is real.

As before, we may set $p = r(\theta)\cos\theta$ and $q = r(\theta)\sin\theta$, but substituting these into Eq. (3.41) does not lead to any obviously useful insights in this case. We merely find

$$r(\theta) = \sigma_p\sigma_q(\sigma_q^2\cos^2\theta + \sigma_p^2\sin^2\theta)^{-1/2} \tag{3.42}$$

for a radius measured at an angle θ anticlockwise from the $\hat{\mathbf{p}}$ axis. We can make progress, however, by introducing a second angle, ϕ, such that

$$p = \sigma_p \cos\phi \tag{3.43}$$

$$q = \sigma_q \sin\phi \tag{3.44}$$

defines a direction that intercepts a circle of radius σ_p at a distance p along the $\hat{\mathbf{p}}$ axis and a circle of radius σ_q at a distance q along the $\hat{\mathbf{q}}$ axis, as depicted in Fig. 3.5. The angles θ and ϕ are clearly related by

$$|\sigma_q| \tan\phi = \sigma_p \tan\theta \tag{3.45}$$

and become equal when the radius lies along either the $\hat{\mathbf{p}}$ or the $\hat{\mathbf{q}}$ axis.

Now, consider the projection of $r(\theta)$ along the directions parallel and perpendicular to that defined by the angle ϕ. After some trigonometry (i.e. elimination of θ using Eq. (3.45)) we obtain

$$r(\theta)\cos(\phi - \theta) = \sigma_p\cos^2\phi + \sigma_q\sin^2\phi \tag{3.46}$$

and

$$r(\theta + \pi/2)\cos(\phi - \theta) = \sigma_p\sin^2\phi + \sigma_q\cos^2\phi \tag{3.47}$$

with

$$r(\theta)\sin(\phi - \theta) = (\sigma_q - \sigma_p)\sin\phi\cos\phi \tag{3.48}$$

which have right-hand sides that should, by now, be starting to look familiar (albeit we have seen them, in Eqs (3.28)–(3.30), with θ in place of ϕ).

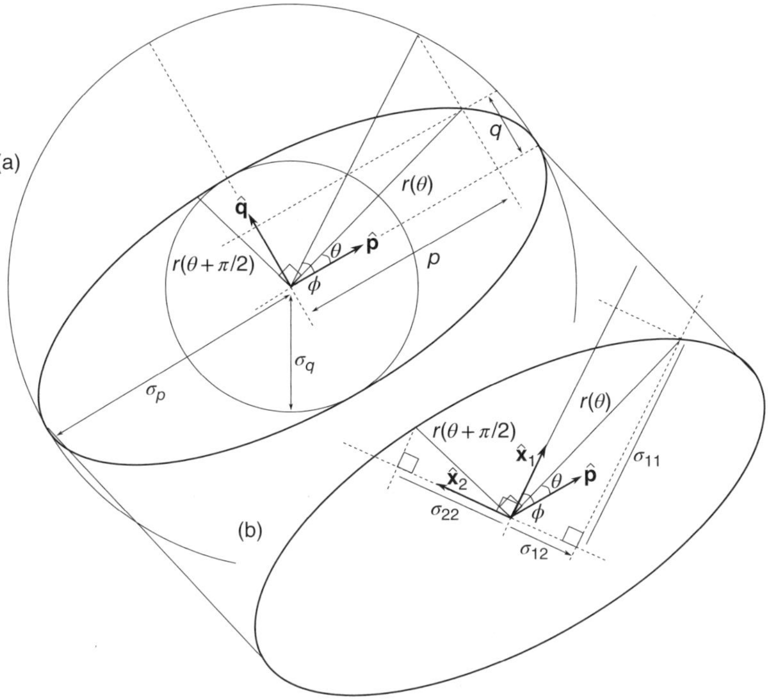

Figure 3.5 The stress ellipse, showing (a) construction of the auxiliary angle ϕ and (b) resolution of the radius at angle θ to give normal and shear stress components in the $(\hat{\mathbf{x}}_1, \hat{\mathbf{x}}_2)$ coordinate system.

The upshot of this analysis is that, should we wish to read off from our diagram the normal and shear stress components relative to a set of axes $(\hat{\mathbf{x}}_1, \hat{\mathbf{x}}_2)$ rotated by an angle ϕ relative to the principal stress axes $(\hat{\mathbf{p}}, \hat{\mathbf{q}})$, we need first to ascertain the angle θ, related to ϕ via Eq. (3.45) (or, equivalently, obtained by construction, as in Fig. 3.5) and take the projections of the ellipse radii $r(\theta)$ and $r(\theta + \pi/2)$ *along* and *perpendicular* to $\hat{\mathbf{x}}_1$ and $\hat{\mathbf{x}}_2$. In this way, the ellipse defined here, with semi-major and semi-minor axes equal to the principal stress components, can be taken to represent *all* of the information held within the stress tensor. Accordingly, we shall refer to it as the **stress ellipse**.

3.4 Symmetry of the Surface Stress: Eccentricity and Orientation

As the foregoing discussion should have made clear, the tensor nature of the surface stress means that, unlike the scalar surface energy, it is intimately related to the *directional* properties of the surface. This makes it particularly diagnostic of surface symmetry, in ways which we shall now explore.

First, let us consider the case of an unstrained surface[7] having three or more mirror or glide planes. The surface stress curves described in the preceding section (i.e. the normal stress conic, the shear stress quartic and the stress ellipse) must all respect the symmetry of the surface.[8] In this instance, that means the curves must themselves share the same three (or more) mirror or glide symmetries of the surface. Now, the only ellipse that displays at least three mirror planes is, in fact, a circle; no ellipse with unequal semi-major and semi-minor axes will do so. Consequently, we can deduce that the stress ellipse of such a surface must perforce be a stress circle. We can furthermore deduce from this fact that the normal stress of the surface will be the same when measured with respect to *any* set of axes, so that the normal stress conic is necessarily a normal stress circle. Moreover, the shear stress will always be zero and, indeed, the shear stress quartic must consequently have an infinite radius in any direction. The upshot of all these considerations is that the stress of this surface is actually entirely isotropic and devoid of any shear component. Even though a crystalline surface structure only displays a finite number of mirror planes, the surface stress displays an infinite number and is only ever purely tensile or purely compressive; the only property of the surface stress ellipse *not* constrained by symmetry is its area.

Turning now to the case of a surface (again unstrained) having either one or two mirror or glide planes,[9] we again insist that the various stress curves must respect the surface symmetry. The difference, of course, is that it is now possible to conceive of a non-circular ellipse possessing two mirror planes, and any such ellipse would satisfy our symmetry requirement, so long as its mirror symmetries coincide with those of the surface. That is, for a surface with two mirror or glide planes, the principal axes of the stress ellipse must align with those mirror or glide planes; for a surface with but a single mirror or glide plane, one principal axis of the stress ellipse must align with it and the other must necessarily lie perpendicular to it. The non-circular nature of the ellipse means, however, that not only can the area of the stress ellipse vary unconstrained by symmetry, as in the isotropic case above, but so too can its aspect ratio (the ratio between the principal stress components) and hence eccentricity. The surface stress, in other words, may be anisotropic; in extreme cases, it is even possible for the normal stress along one principal axis to be tensile while that along the other is compressive. Accordingly, the normal stress conic can now be either elliptical or hyperbolic in nature, while the shear stress quartic is no longer pushed out to infinite radius at all angles. In fact, the latter curve reveals a shear stress that disappears only when measured

---

7 The symmetry constraints upon surface stress described in this section are all susceptible to modification in the light of stress induced by means of an externally imposed strain; clearly, deformation of the sample can break any or all symmetries of the unstrained surface.

8 The relationship between the symmetry of a system and that of its measurable properties is dictated by Neumann's principle. Essentially, the measurable properties must display *at least* the symmetry of the system as a whole. In certain cases, a particular measurable property may show *more* symmetry, which in fact is the case for the surface stress, but it can never show *less*.

9 If a surface possesses precisely two mirror planes, it must be the case that they are perpendicular to one another.

relative to a set of axes aligned with the principal stress axes, reaching a maximum magnitude for axes aligned precisely midway between these.

Finally, for an unstrained surface having no mirror or glide planes (i.e. a chiral surface) the surface stress is entirely unconstrained by surface symmetry. The surface stress ellipse must remain elliptical and can certainly be non-circular, but now it can adopt an arbitrary orientation with respect to the crystallographic directions of the surface; the orientation now joins the area and aspect ratio of the stress ellipse as an unconstrained parameter. The normal stress conic and shear stress quartic, it should be noted, must share the same orientation as the stress ellipse, and the maximum shear stress (smallest radius of the shear stress quartic) still occurs for axes precisely midway between the principal axes of stress.

3.4.1 Stereography and Surface Stress

To illustrate the points raised above, let us now consider a few case studies in which the surface stress of chiral systems has been calculated from first principles. We begin with the work of Blanco-Rey *et al.* [14], in which a series of body-centred cubic surfaces were investigated. The series was chosen to comprise a zone perpendicular to one of the $\langle 111 \rangle$ directions, and hence all the surfaces were of stepped type (see extensive discussion in Chapter 2). All such surfaces are necessarily chiral, for materials taking the body-centred cubic crystal structure, excepting those of $\{211\}$ type. Accordingly, one should expect the surface stress ellipse to have no particular orientation defined by symmetry. Indeed, the question tackled by this work was whether one should anticipate *any* degree of predictability in the orientation of the surface stress ellipse at all.

The sequence of surfaces (211) – (532) – (321) – (431) – (110) was studied and the surface stress calculated by first-principles density functional theory (DFT) in each case, for body-centred cubic iron, molybdenum and tungsten (see Fig. 3.6) [14]. Results for the latter two elements were qualitatively similar to one another: the stress ellipse displayed only moderate anisotropy for all surfaces (eccentricities in the range 0.2–0.8) and the principal stress axes rotated anticlockwise smoothly through a total angle of 60° in going from the (211) to the (110) surface; note that all these surfaces are designated L and that the rotation would proceed in the opposite sense for a sequence of D surfaces. In contrast, the results for iron showed a rather greater degree of anisotropy for all surfaces (eccentricity in excess of 0.8) and the principal stress axes were found to rotate clockwise smoothly through a total angle of 30° in going from the (211) to the (110) surface; again the sense of rotation would be reversed if D surfaces were studied in place of L surfaces. The contrarotation of the surface stress ellipse in the case of iron, relative to the cases of molybdenum and tungsten, means that whilst all three metals display a semi-major stress ellipse axis that bisects the close-packed $\langle 111 \rangle$ directions of the (110) surface, the semi-major axis on the (211) surface lies *along* the $[1\bar{1}\bar{1}]$-oriented step-edge in the cases of molybdenum and tungsten, but *perpendicular* to it in the case of iron. It is tempting to view this as a classic example of anomalous behaviour in the first row of each block in the periodic table, but as yet there is insufficient data to be certain; subsequent calculations [15] breaking down the surface stress into

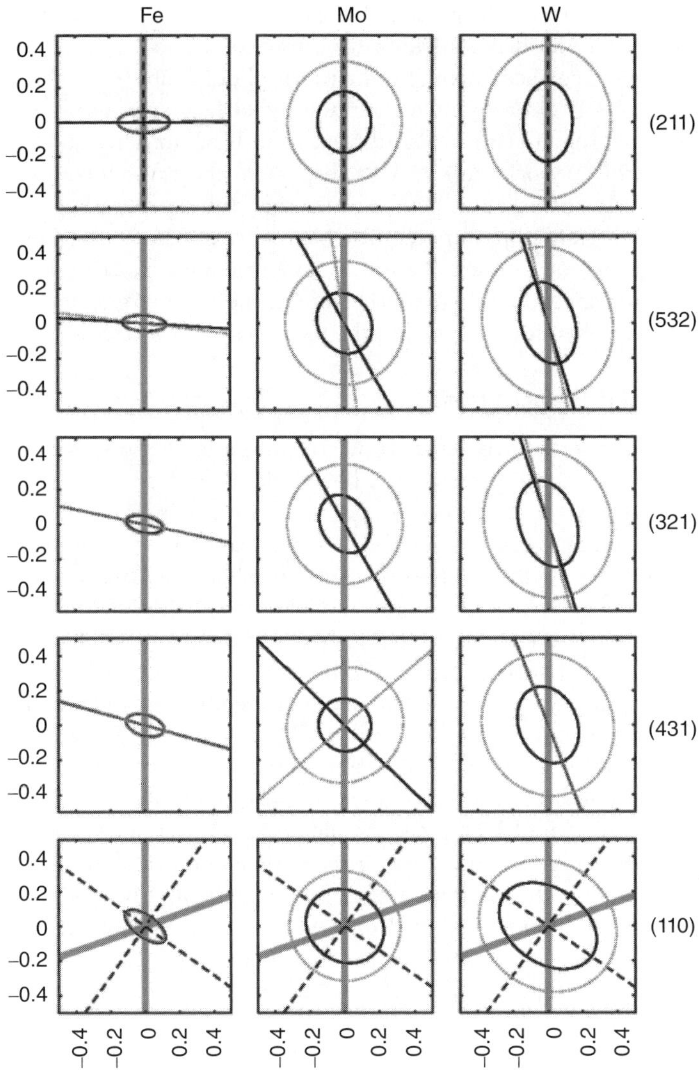

Figure 3.6 DFT-calculated stress ellipses for sequences of iron, molybdenum and tungsten surfaces (units are eV Å^{-2}). The larger ellipse in each case represents the stress of the ideal unrelaxed surface, while the smaller corresponds to the relaxed surface. Full straight lines indicate the orientation of the regular steps found on all these surfaces and dashed lines indicate mirror planes where these exist. Reproduced with permission from [14]. Copyright (2009) American Physical Society.

contributions from electronic kinetic energy, from exchange–correlation energy and from electrostatics, suggest that the picture may well be more complicated than this.[10]

10 In general, the elements found in the first row of each block of the periodic table exhibit anomalous behaviour with respect to elements further down their group. That is, those elements having partially occupied 1s, 2p, 3d or 4f orbitals are unusual, which is usually attributed to the lack

3.4.2 Racemic Surface Stress

A case where the symmetry properties of the surface stress ellipse turns out to have potentially profound implications for surface physics and chemistry is to be found in the work of Etman *et al.* [16]. Here, the surface stress was calculated, by first-principles DFT, for both terminations (XL_o and XD_o) of the racemic $\{11\bar{2}1\}$ surface of hexagonal close-packed rhenium. The stress ellipse was found to be highly anisotropic, with the maximum normal stress amounting to 351 meVÅ^{-2} (5.62 Jm^{-1}) and the minimum normal stress being 94 meVÅ^{-2} (1.51 Jm^{-1}) – an eccentricity of 0.96, exceeding even the eccentricities reported for the iron surfaces mentioned above. Crucially, the chirality of the terminations ensured that the principal axes of the surface stress ellipse were not aligned with any high-symmetry crystallographic axis. The two-dimensional lattice of the $\{11\bar{2}1\}$ surface is rectangular, with a primitive unit of that shape oriented with its long side along the $[11\overline{26}]$ direction; the surface stress ellipse of the XL_o termination was found to have its semi-major axes rotated by 15° clockwise from this direction, while the surface stress ellipse of the XD_o termination had *its* axis rotated by the same amount in the anticlockwise direction (see Fig. 3.7).

The practical upshot of this is that an unstrained $\{11\bar{2}1\}$ surface, which should exhibit terraces of both terminations in essentially equal proportion, will be subject to macroscopic surface stress aligned to respect the symmetry of the rectangular two-dimensional lattice, but its microscopic surface stress will vary from one asymmetric orientation to its mirror image on going from one terrace to another. Furthermore, it was estimated that the application of just 1% tensile surface strain in the direction that maxmimises the surface stress difference between terraces, could potentially skew the energetic balance (in favour of either the XL_o or the XD_o termination) by an amount comparable with the energy differences typically involved in driving surface reconstruction. This raises the intriguing possibility that the surface might be amenable to reversibly switching between XL_o dominance and XD_o dominance – all under the control of an externally

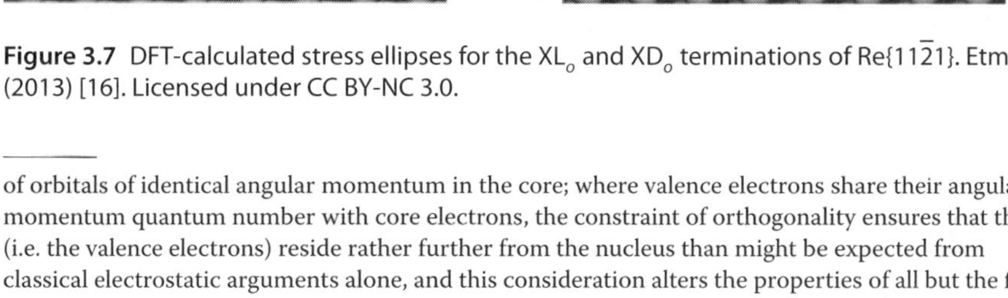

Figure 3.7 DFT-calculated stress ellipses for the XL_o and XD_o terminations of Re$\{11\bar{2}1\}$. Etman (2013) [16]. Licensed under CC BY-NC 3.0.

---

of orbitals of identical angular momentum in the core; where valence electrons share their angular momentum quantum number with core electrons, the constraint of orthogonality ensures that they (i.e. the valence electrons) reside rather further from the nucleus than might be expected from classical electrostatic arguments alone, and this consideration alters the properties of all but the first element in each group.

applied strain. Even more remarkable, if the two terminations could be shown to display enantioselective chemistry, then the external strain could actually be capable of directing asymmetric chemistry. At present, these exciting possibilities have yet to be fully explored.

3.4.3 Adsorbate-Induced Asymmetry in Surface Stress

The examples discussed thus far have all involved the intrinsic stress of clean surfaces, but it ought also to be noted that adsorption can modify this stress[11] – potentially in asymmetric ways. To illustrate this, we focus upon adsorption of the amino acid alanine on Cu{110}, as studied (via first-principles DFT calculations) by Blanco-Rey and Jones [18]. We shall describe the adsorption of amino acids, including alanine, on Cu{110} at some length in the following chapter (Section 4.3.1) but for the moment let us simply summarise that upon annealing at moderate temperatures the molecules self-assemble into well-ordered two-dimensional networks held together by hydrogen bonds. These networks form a (3×2) overlayer, in which each primitive unit cell contains two molecules; each molecule attaches to the surface in a chiral **footprint** that takes the form of a right-angled triangle.[12] Since half the footprints in this dense overlayer are mirror images of the other half, it follows that any asymmetry in the surface stress must arise from subtle differences in the adsorption geometry and intermolecular interactions caused by the chirality of the alanine molecules. In fact, the authors find that although the surface stress is strongly modified by the presence of alanine (becoming overall compressive instead of tensile) the principal axes are rotated by only around 5° relative to the mirror planes of the clean surface – clockwise for L-alanine, and anticlockwise for D-alanine. Incidentally, a much more asymmetric result (misaligned by 15° from the mirror planes) is obtained when only half the molecules are included in the model (and hence only one footprint enantiomer is occupied); in this case, the sense of rotation matches that of the sequence adjacent–opposite–hypotenuse with respect to the acute angle of the footprint triangle. Clearly, the surface stress modification brought on by adsorption is sensitive to the adsorbate–substrate interaction itself, as well as to any interadsorbate interactions. Disentangling these contributions, in order

---

11 A particularly vivid illustration of adsorption-induced stress modification – albeit an achiral one – may be found in the case of Au{111}. Here, the so-called 'herringbone' reconstruction of the clean surface arises from a densification in the uppermost atomic layer; 23 gold atoms are squeezed into an area nominally large enough for only 22, and the consequent mismatch with the second layer results in a series of linear rumples (known as solitons) that zig–zag across the surface in pairs. Upon exposure to NO_2, molecules are found to adsorb preferentially in the spaces between pairs of solitons, modifying the local surface stress and deforming the bounding solitons in the process [17]. As islands of NO_2 grow, the local stress field (as inferred from the soliton deformation) induces a distinct mesoscopic ordering that breaks down only when adjacent solitons are forced so close together that they coalesce and eject nanoclusters of gold atoms.

12 The importance of a molecule's footprint, in relation to its intrinsic handedness, was memorably emphasised by the image of the so-called 'Frankenfoot' introduced by Forster *et al.* [19] – picture a hand grafted onto a foot, and one arrives at the (disturbing) conclusion that the misguided scientist responsible will have faced a choice between four distinct creations: left hand on right foot, right hand on right foot, left hand on left foot and right hand on left foot. The relationship between these forms will, of course, follow the pattern of diastereoisomerism illustrated more formally in Fig. 1.6.

to interpret experimental measurement, is likely to require careful consideration of calculations such as these. In principle, however, they provide a means by which the asymmetric adsorption may be monitored through measurement of stress changes and/or any associated strain that may occur.

3.5 Measurement of Differential Surface Stress

The direct measurement of absolute surface stress is not at all a simple business. Essentially, one needs to find a way in which to strain the surface to a quantifiable degree whilst simultaneously recording the work expended in doing so. Even in isotropic cases, this is not straightforward, but for anisotropic cases (and particularly chiral cases, where the principal axes of stress are not known a priori) the difficulties are considerable. The author is not aware of any measurement of absolute surface stress carried out in a chiral case.

In contrast, the measurement of surface stress differences (or surface stress changes) is rather more amenable to experimental study. This is because such differences (or changes) can manifest as deformations in the macroscopic shape of a sample, or the microscopic shape of some observable surface feature. Here, we outline a few of the ways in which differential[13] surface stress either has been measured, or could hypothetically be measured.

3.5.1 Island Shape Measurement

One of the most beautiful measurements of differential surface stress has been achieved in the case of Ge{001}. This surface typically reconstructs by the formation of tilted dimers, which minimise the number of dangling bonds and hence the specific surface energy. Distinct rows of adjacent dimers run across the surface in the direction perpendicular to the dimers themselves. Unsurprisingly, given the discussion above, the principal axes of stress are aligned with the mirror planes of the surface. More remarkably, however, the principal stress parallel to the dimer rows is tensile, while that perpendicular to the dimer rows is compressive.

Now, within the nomenclature introduced in Chapter 2, the Ge{001} surface falls into the bayonet category, meaning that addition or removal of a single layer of atoms results in a structure identical to the original surface but rotated by some angle (in this instance 90°). A monoatomic step on this surface therefore separates terraces on which the dimer rows run in perpendicular directions and within which the principal surface stress components are interchanged. A monoatomic island surrounded by much larger terraces (or, in the case studied by Middel *et al.* [20], a vacancy island) would therefore be subject to unbalanced stresses were it to remain unstrained.

Consider, for example, an island (or vacancy island) of square shape, whose edges align with the dimer rows of either the terrace within the island or that lying outside of the island (see Fig. 3.8). Let us denote the compressive principal

13 Here we mean differential in the sense of a difference, not a derivative.

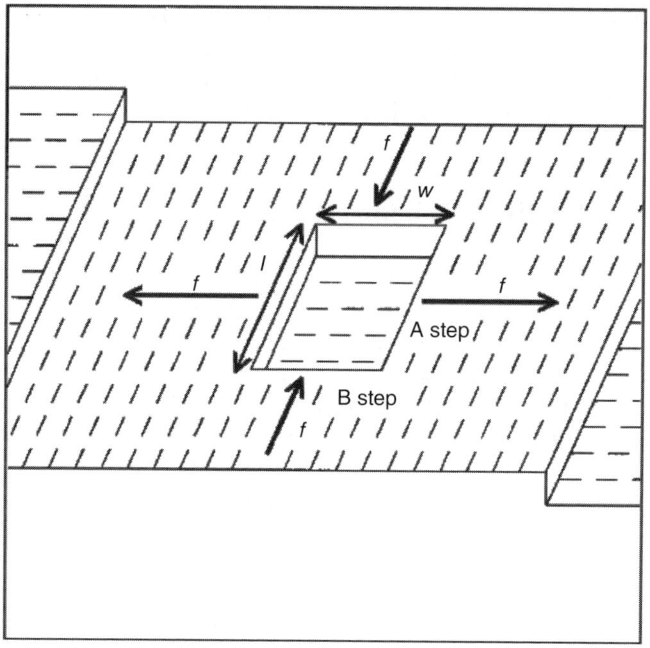

Figure 3.8 Schematic view of forces acting upon a vacancy island at the Ge{001} surface, leading to a rectangular island morphology ($w \neq l$). Reproduced with permission from [20]. Copyright (2002) American Physical Society.

stress component of an idealised Ge{001} surface as σ_c and the tensile principal stress component as σ_t. At one pair of island edges (those lying parallel to the island's dimer rows) the unbalanced surface stress amounts to $\sigma_t - \sigma_c$, while at the other pair (those lying perpendicular to the island's dimer rows) the unbalanced surface stress is $\sigma_c - \sigma_t$. Treating the island as an elastic object in its own right, one would consequently expect that it ought to deform in response to these unbalanced stresses, attaining a distinctly rectangular shape at equilibrium. Analysing island aspect ratios observed in scanning tunnelling microscopy (STM), along the lines first set out by Li *et al.* [21], the authors were able to deduce a value of $|\sigma_t - \sigma_c|$ equal to 80 meV for this surface.

3.5.2 Contact Angle Measurement

An alternative method that is often purported to measure surface stress is founded upon consideration of the **contact angle** between a liquid droplet and the solid surface in question. Strictly speaking, however, this method technically only yields differential stress, and any inference of the absolute stress based upon this technique necessarily involves the invocation of various assumptions of questionable veracity. The essence of the method is readily explained with reference to the diagram presented in Fig. 3.9, where the angle ξ is the eponymous contact angle. At the point of contact itself, three different surface stresses act, namely: (i) the surface stress of the clean surface beyond the edge of the droplet, (ii) the surface stress of the surface lying beneath the droplet and (iii) the stress

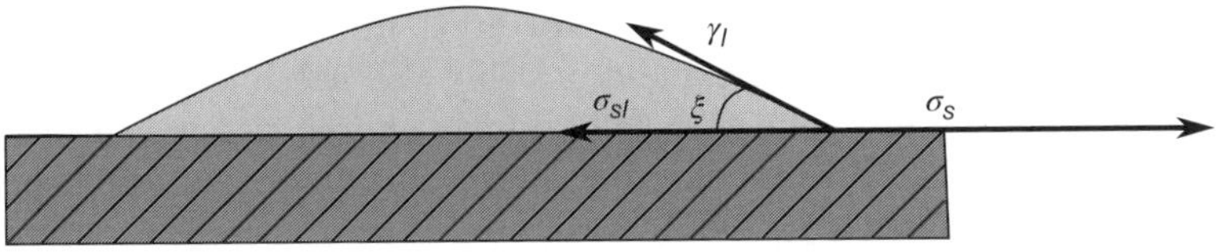

Figure 3.9 Schematic view of forces acting upon a liquid droplet.

of the liquid surface of the droplet itself. The latter stress is synonymous with the surface tension of the liquid and hence is isotropic. The other two stresses, each deriving from a solid surface, may in general be anisotropic, but for simplicity we will initially consider only the isotropic case.

Resolving the various surface stresses in any direction parallel to the plane of the surface now yields the so-called **Young equation** for the equilibrium condition

$$\gamma_l \cos \xi = \sigma_s - \sigma_{sl} \qquad (3.49)$$

where σ_s is the surface stress of the clean surface, σ_{sl} is the surface stress associated with the solid/liquid interface and γ_l is the surface tension of the liquid. For a liquid of known surface tension, therefore, it is possible to deduce the difference between σ_s and σ_{sl} from the measurement of the contact angle ξ. Most commonly, a series of measurements are made for a homologous sequence of different liquids, and the cosine of the contact angle plotted against the surface tension of the liquid; this so-called **Zisman plot** generally yields a well-defined (if not necessarily straight) line, and the intercept with $\cos \xi = 1$ allows one accurately to determine $\sigma_s - \sigma_{sl}$ directly from the value of γ_l at that point.[14]

Strictly speaking, the value of $\sigma_s - \sigma_{sl}$ deduced from the Zisman plot is no more and no less than the **critical surface tension** at which a liquid will *just* wet the solid surface. One occasionally finds it asserted that this value also represents 'the surface energy' or 'the surface stress' of the solid, but either statement would evidently be entirely incorrect; only if σ_{sl} were close to zero would such an assignment be even approximately valid, but there is no reason to suppose this always to be the case,[15] and the Zisman plot cannot, in itself, provide support for this notion. With rather more justification, however, a number of simple models have been put forward to partition σ_{sl} into various components, depending upon the nature of the interaction between the solid and the liquid. Within the context of such models, therefore, it may be possible to extract an approximate absolute value for σ_s, based on these assumptions about the functional form of σ_{sl}. Nevertheless, it would be wise to treat such measurements of absolute surface stress with due caution.

14 Note that the less homologous the liquids used, the less well defined will be the line in the Zisman plot; the observed correlation is essentially empirical and not strongly constrained by fundamental considerations.

15 If one *assumes* that the specific work of adhesion between the liquid and solid (i.e. the work required per unit area in order to separate them) is simply the geometric mean of the solid and liquid specific surface energies, then the interfacial energy can indeed be shown to vanish at the critical point of the Zisman plot, but this assumption is likely true only in the absence of strong liquid–solid interactions.

Where the contact angle method *could* be of great utility, however, is in resolving the anisotropy of surface stress and possibly also its chirality. Let us, therefore, consider the case of a solid surface having an entirely general surface stress tensor $\boldsymbol{\sigma}$ when clean, and some different interfacial surface stress tensor $\boldsymbol{\sigma}'$ when covered with a particular liquid. Next, we shall re-write the Young equation by resolving stresses along an in-plane direction lying at an angle of θ, measured anticlockwise from the major principal axis of $\boldsymbol{\sigma}$ when viewed from above. The normal stress of the clean surface at this angle is simply

$$\sigma_{11} = \sigma_p \cos^2\theta + \sigma_q \sin^2\theta \tag{3.50}$$

as per Eq. (3.28), with σ_p and σ_q being the major and minor principal stress components of the clean surface. Now, if the major principal axis of $\boldsymbol{\sigma}'$ is rotated by an angle ϕ anticlockwise from that of $\boldsymbol{\sigma}$ (again, when viewed from above) it follows that the normal stress of the liquid-covered surface at the angle θ is

$$\sigma'_{11} = \sigma'_p \cos^2(\theta - \phi) + \sigma'_q \sin^2(\theta - \phi) \tag{3.51}$$

where σ'_p and σ'_q are the major and minor principal stress components of $\boldsymbol{\sigma}'$, respectively. We must therefore have

$$\gamma_l \cos\xi = \sigma_p \cos^2\theta + \sigma_q \sin^2\theta - \sigma'_p \cos^2(\theta - \phi) - \sigma'_q \sin^2(\theta - \phi) \tag{3.52}$$

which may be simplified to

$$\gamma_l \cos\xi = (\sigma_p - \sigma'_{11})\cos^2\theta + (\sigma_q - \sigma'_{22})\sin^2\theta + 2\sigma'_{12} \sin\theta \cos\theta \tag{3.53}$$

with

$$\sigma'_{11} = \sigma'_p \cos^2\phi + \sigma'_q \sin^2\phi \tag{3.54}$$

$$\sigma'_{22} = \sigma'_p \sin^2\phi + \sigma'_q \cos^2\phi \tag{3.55}$$

$$\sigma'_{12} = (\sigma'_q - \sigma'_p) \sin\phi \cos\phi \tag{3.56}$$

analogous to Eqs (3.28)–(3.30).

In the event that $\phi = 0$ or $\phi = \pi$ (i.e. that the principal stress axes for liquid-covered and clean surfaces share the same orientation), Eq. (3.53) describes either an ellipse or two conjugate hyperbolae (depending upon the relative values of their principal stress components). In the general case, however, plotting $\cos\xi$ versus θ should yield a more complicated curve. Differentiating Eq. (3.53) and setting the result equal to zero, we find that the radius of this curve should display extremal values at angles satisfying

$$\tan 2\theta = \frac{2\sigma'_{12}}{(\sigma_p - \sigma_q) - (\sigma'_{11} - \sigma'_{22})} \tag{3.57}$$

and alternating between minima and maxima.

Now, in general, the relationship between the $\cos\xi$ curve and the surface stress components is sufficiently complicated that it may not be possible to extract the latter from the measurement of the former in a single case. By using a range of different liquids, however, it should be possible, through trial and error, to hit upon one for which the curve of $\cos\xi$ versus θ turns out to be either an ellipse or a pair of conjugate hyperbolae; if, and only if, this condition is met, the axes of this conic section must be aligned perfectly with those of the stress tensor of the clean surface. Contact angle measurements, therefore, potentially offer a direct method by which the orientation of the stress tensor may be obtained, providing a unique test of calculations of this property at chiral surfaces.

3.5.3 Cantilever Deformation

Perhaps the most commonly applied method to measure differential surface stress is that based upon the deflection of a cantilever, pioneered by Ibach and co-workers [22, 23]. Various cantilever geometries may be imagined, although the simplest takes the form of a narrow rectangle, clamped at one end. The surface stresses on either side of the cantilever are initially equal, but adsorption of material preferentially on one side causes the corresponding surface stress to change. The induced difference in surface stress between one side and the other then creates a deformation resulting in a displacement of the cantilever tip, which can be measured by any convenient method (most frequently, by measuring the deflection of a laser beam reflected from the cantilever tip).

In the case of a rectangular cantilever, Godin *et al.* [24] have shown that a uniform and isotropic differential surface stress, $\Delta\sigma$, may be calculated from the cantilever's radius of curvature, R, via

$$\Delta\sigma = \frac{2l^3 k}{3R(1-\nu)wt} \tag{3.58}$$

where ν and k are the Poisson ratio and spring constant of the cantilever, respectively, and where l, w and t are the cantilever length, width and thickness. Alternatively, an expression due to Stoney [25] approximates the (uniform and isotropic) differential surface stress as

$$\Delta\sigma = \frac{Et^2}{6R(1-\nu)} \tag{3.59}$$

where E is the Young's modulus of the cantilever.

Evidently, use of non-rectangular cantilevers would necessitate revision of the equations quoted above, and it has been argued that even for the rectangular case there ought to be corrections to account for practical details, such as the method of clamping one end of the cantilever when in use. Tamayo *et al.*, for example, have derived modified equations to deal with just this circumstance [26]. Arguably, the ideal method of analysis would involve finite-element modelling of the deformation, and this would likely be essential in cases where the surface stress is anisotropic and not necessarily aligned with the symmetry of the cantilever.

The cantilevers used for surface stress measurements to date have typically been polycrystalline, and the emphasis has very much been on stress induced by molecules adsorbing in rather disordered fashion. The surface stress tensor involved in such cases is essentially isotropic, so the analysis may proceed as outlined above. If one were to use a single-crystal cantilever, however, it is entirely possible for the surface stress tensor to become appreciably anisotropic. In such cases, the anisotropy will be constrained by the symmetry of the substrate, and for non-chiral surfaces will, therefore, possess principal axes of stress that align with particular crystallographic axes. Assuming the cantilever to have been manufactured with high-symmetry crystallographic directions symmetrically disposed with respect to its macroscopic shape, the resulting deformation of the cantilever is likely to be a pure bend, as before. Indeed, in experiments using single-crystal cantilevers with exposed {111} faces, deformations have been analysed perfectly adequately on precisely this basis [22]. If, however, the surface were chiral, there would be no reason to expect the stress tensor to align with the macroscopic shape, and the cantilever ought not only to bend but also to twist. Measuring the twist entails practical difficulties for the experimentalist, but ought not to be fundamentally impossible. Elucidating the relationship between the twist and the stress tensor, however, would be very difficult without the use of finite-element simulations. Nevertheless, the fact remains that cantilever-based techniques represent a very promising route towards the quantitative sensing of chirality in adsorbate molecules and the structures that they form upon surfaces.

3.6 Facet Formation and the Wulff Construction

One of the most spectacular ways in which a surface may respond to having high surface energy is to be found in the formation of facets. Here, the original planar surface breaks up into mesoscale features with local normals aligned at some angle to the macroscopic surface normal. The result will necessarily be a greater surface area, but if the specific surface energy of the facets is sufficiently low compared with that of the original surface plane an overall reduction in the total surface free energy may result. We consider two cases in detail: (i) where the facets form a ridge-and-furrow structure and (ii) where they create a landscape of pyramids and pits.

3.6.1 Ridge-and-Furrow Facets

Consider a planar surface of infinite two-dimensional extent, having specific surface energy γ_0 and no surface excess of any atomic species (i.e. $\Gamma_j = 0$ for all j). Over an area A_0, the surface Helmholtz free energy, F_0, will therefore simply be $\gamma_0 A_0$.

Next, let us imagine the same surface having broken up into several facets, each with their normal vector at an angle of either θ_1 or θ_2 to that of the original surface, such that the facets form a series of aligned ridges and furrows across the surface (Fig. 3.10). In order that the macroscopic orientation of the surface remains the same overall, we can deduce that

$$A_1 \sin\theta_1 = A_2 \sin\theta_2 \tag{3.60}$$

where A_1 and A_2 are the total areas of the two facet types. Moreover, since the projection of the facetted surface onto the original surface plane must match the original area, we can also state that

$$A_1 \cos\theta_1 + A_2 \cos\theta_2 = A_0 \tag{3.61}$$

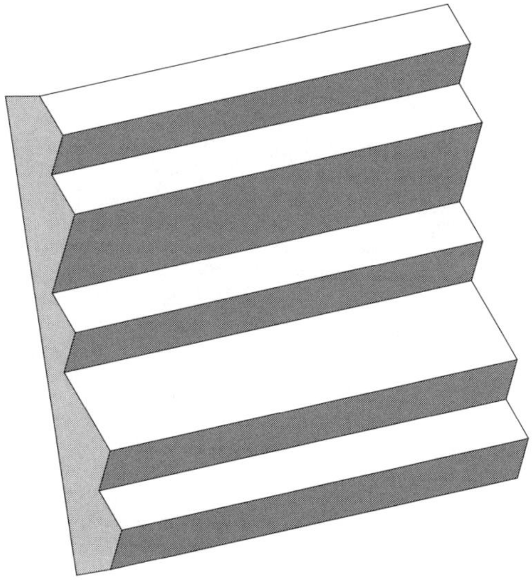

Figure 3.10 Top-down view of ridge-and-furrow facets.

and note that the surface energy of the facetted surface, F_{12}, must be given by

$$F_{12} = \gamma_1 A_1 + \gamma_2 A_2 \tag{3.62}$$

where γ_1 and γ_2 are the specific surface energies of the two facets, and we have neglected any contribution from the very tops of the ridges and bottoms of the furrows.

From the first two of these equations, we may obtain

$$A_1 \sin\theta_1(\cot\theta_1 + \cot\theta_2) = A_0 \tag{3.63}$$

and

$$A_2 \sin\theta_2(\cot\theta_1 + \cot\theta_2) = A_0 \tag{3.64}$$

giving

$$F_{12} = A_0 \left(\frac{\gamma_1 \csc\theta_1 + \gamma_2 \csc\theta_2}{\cot\theta_1 + \cot\theta_2} \right) \tag{3.65}$$

for the surface energy of the facetted surface. Facetting will then occur, of course, whenever F_0 exceeds F_{12}, leading to the inequality

$$\gamma_0 \leq \frac{\gamma_1 \csc\theta_1 + \gamma_2 \csc\theta_2}{\cot\theta_1 + \cot\theta_2} \tag{3.66}$$

as the condition for stability of the original planar surface.

With regard to the question of chirality, it should be noted that an initially chiral surface orientation cannot become achiral due to the occurrence of facetting, but the dimensionality of the chirality may be altered. While the chirality of the non-facetted surface would be described (in the terms introduced in Chapter 1) as displaying planar substrate chirality, it may be that the new facets happen to be achiral (although this need not necessarily be so) in which case the only remaining chirality in the system must be sought along the linear ridges and furrows where those facets meet.

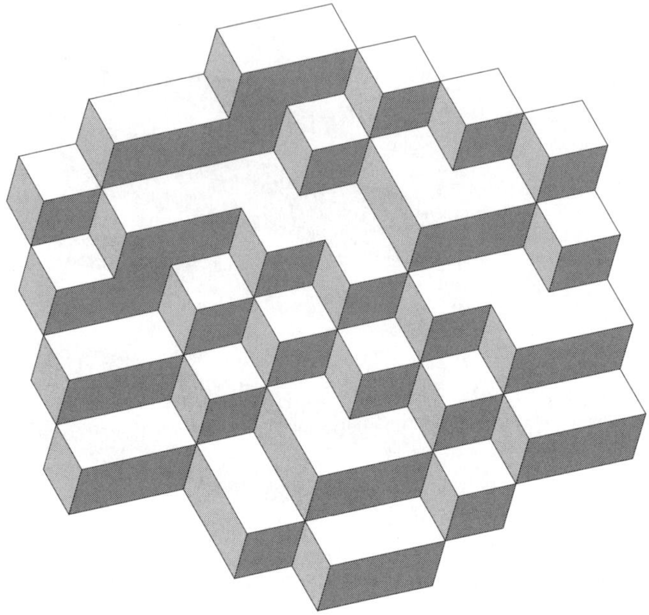

Figure 3.11 Top-down view of pyramid-and-pit facets.

3.6.2 Pyramid-and-Pit Facets

The discussion of facetting presented above for the case of two facet orientations (i.e. ridge-and-furrow facetting) may readily be extended to the case of three facet orientations. Here, the facets form triangular pyramids and correspondingly inverted pits (see Fig. 3.11). Once again, the conservation of macroscopic surface orientation provides one constraint, allowing us to assert that

$$A_1 \sin\theta_1 = A_2 \sin\theta_2 = A_3 \sin\theta_3 \tag{3.67}$$

where now A_1, A_2 and A_3 are the total areas of each facet types, while θ_1, θ_2 and θ_3 are the angles between the respective facet normal vectors and the macroscopic surface normal. Equating the total projected area of these facets with the area of the original planar surface, A_0, provides another constraint,

$$A_1 \cos\theta_1 + A_2 \cos\theta_2 + A_3 \cos\theta_3 = A_0 \tag{3.68}$$

while the surface energy of the facetted surface, F_{123}, is given by

$$F_{123} = \gamma_1 A_1 + \gamma_2 A_2 + \gamma A_3 \tag{3.69}$$

where γ_1, γ_2 and γ_3 are the specific surface energies of the three facet types.

From the first two equations, we derive the expressions

$$A_1 \sin\theta_1(\cot\theta_1 + \cot\theta_2 + \cot\theta_3) = A_0 \tag{3.70}$$

$$A_2 \sin\theta_2(\cot\theta_1 + \cot\theta_2 + \cot\theta_3) = A_0 \tag{3.71}$$

and

$$A_3 \sin\theta_3(\cot\theta_1 + \cot\theta_2 + \cot\theta_3) = A_0 \tag{3.72}$$

for the individual facet areas, and so the surface energy of the facetted surface may simply be written as

$$F_{123} = A_0 \left(\frac{\gamma_1 \csc\theta_1 + \gamma_2 \csc\theta_2 + \gamma_3 \csc\theta_3}{\cot\theta_1 + \cot\theta_2 + \cot\theta_3} \right) \tag{3.73}$$

and the inequality that implies stability for the original surface against facetting then becomes

$$\gamma_0 \leq \frac{\gamma_1 \csc\theta_1 + \gamma_2 \csc\theta_2 + \gamma_3 \csc\theta_3}{\cot\theta_1 + \cot\theta_2 + \cot\theta_3} \tag{3.74}$$

in rather close analogy with the corresponding expression for the ridge-and-furrow case.

In a similar analogy, we also note that pyramid-and-pit facetting can again alter the dimensionality of surface chirality. If the chirality of the ideal surface may be said to be planar in nature, then the chirality expressed at the apex of each pit or pyramid must be thought of as pointlike. Even when the individual facets themselves become achiral, pointlike and/or linear, chirality must necessarily remain if the original non-facetted surface was chiral.

3.6.3 Geometrical Construction

In fact, Eqs (3.66) and (3.74) suggest a convenient geometrical construction to determine the stability of surfaces against facetting. Let us begin by imagining a plane oriented parallel to the original planar surface, positioned so that its distance of closest approach to the origin is proportional to its specific surface energy, γ (see Fig. 3.12). Next, we imagine either two or three planes parallel to the potential facet orientations we wish to consider, positioned so that *their* distances of closest approach to the origin are also proportional to their respective specific surface energies (with the same constant of proportionality as before). Now, one can show that, in the ridge-and-furrow case the line of intersection between the two facet planes falls *on or beyond* the plane of the original surface if, and only if, Eq. (3.66) is satisfied. Similarly, in the pyramid-and-pit case the point of intersection between the three facet planes is likewise located beyond the original plane if, and only if, Eq. (3.74) is satisfied. In other words, the original surface plane is stable against facetting only if the intersection between any pair or trio of potential facet planes falls on or beyond the original surface plane – each plane being drawn at a perpendicular distance from the origin proportional to its specific surface energy.

This observation permits us to deduce the equilibrium shape of a crystallite, by means of the celebrated Wulff construction.[16] We begin by considering the set of all possible surface orientations, represented by planes placed at distances from the origin determined (as above) by their specific surface energies. The equilibrium shape is then simply identified as the region that can be accessed from the origin without crossing any of the planes. Each of the planes that bound this region must necessarily lie closer to the origin than any intersection of multiple

16 Although proposed by Wulff in 1901, the most satisfying treatment is arguably that provided by Herring 50 years later [27].

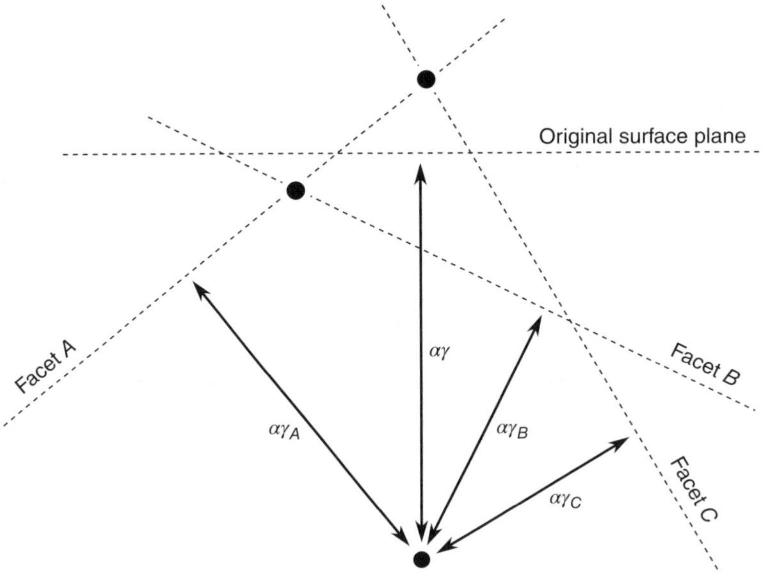

Figure 3.12 Construction for the stability of facet combinations, with planes plotted such that their closest approach to the origin is proportional to their specific surface energy. The original surface plane indicated will be unstable against decomposition into a combination of facet orientations *A* and *B*, because the intersection point of these planes lies within the original plane. The original surface plane will be stable against decomposition into a combination of facet orientations *A* and *C*, however, since *their* planes intersect beyond the original plane.

planes, and must therefore be stable against facetting. Any variation of this shape must involve an increase in free energy and will hence be disfavoured.

Calculations of surface free energy for *fcc* metals generally predict (and experiments confirm) low-temperature equilibrium crystallite shapes comprising primarily {111} facets, a smaller surface area of {100} facets, and a very small proportion of higher-index facets. At higher temperatures, these additional facets start to become important, leading to rounding of the crystallites, but still the close-packed flat facets are generally dominant (see Fig. 3.13); the same

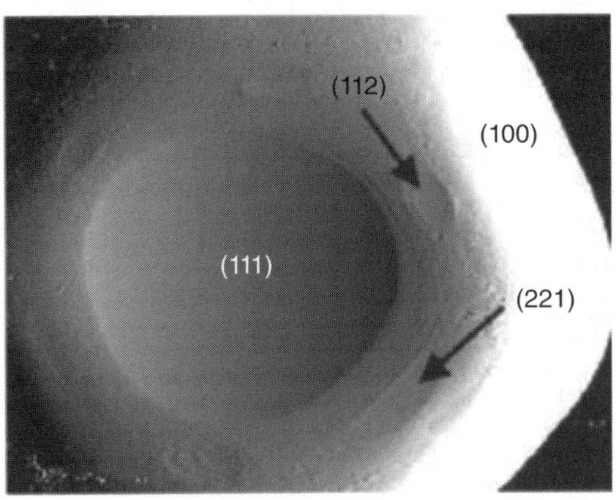

Figure 3.13 STM image of a lead crystallite equilibrated at 353 K; image size approximately $970 \times 750\ nm^2$. Reproduced with permission from [28]. Copyright (2007) Springer.

dominance of close-packed flat surfaces also holds for *bcc* and *hcp* metals. Deviations from this rule are typically only observed when adsorbates are introduced. It follows, therefore, that one might reasonably expect high-index planar surfaces of metals to be prone to break up into facets of {111}, {100} or {110} orientation. That this does *not* always occur must, presumably, be attributable to kinetic considerations. We have hitherto ignored the free energy cost associated with boundaries between facets (i.e. the very tops of the ridges and pyramids, or the bottoms of the furrows and pits). This is reasonable when the facets become very large, as the boundaries make up a vanishingly small proportion of the overall surface area, but in order to reach the condition of large facets, the surface must pass through intermediate stages where the facets are much smaller. Accordingly, it is possible for the system to become kinetically trapped in a higher free energy state than might be expected. Certainly, it is possible to obtain samples of high-index metal surfaces that seem not to be prone to facetting, at least when clean, that one might not expect to be stable based upon the Wulff construction. Many of these have been studied in connection with their chirality, or the chirality of adsorbates attached to them, as we shall see over the next few chapters.

References

1 E.M. Gutman, *J. Phys. Condens. Matter* **7**, L663 (1995).
2 D.J. Bottomley and T. Ogino, *Phys. Rev. B* **63**, 165412 (2001).
3 V.A. Marichev, *Surf. Sci.* **600**, 4527 (2006).
4 V.A. Marichev, *Surf. Sci.* **602**, 1131 (2008).
5 D.J. Bottomley, L. Makkonen and K. Kolari, *Surf. Sci.* **603**, 97 (2009).
6 D.J. Bottomley, L. Makkonen and K. Kolari, *Surf. Sci.* **603**, 2356 (2009).
7 D.J. Bottomley, L. Makkonen and K. Kolari, *Surf. Sci.* **604**, 2066 (2009).
8 D. Kramer and J. Weissmüller, *Surf. Sci.* **601**, 3042 (2007).
9 D. Kramer and J. Weissmüller, *Surf. Sci.* **602**, 1133 (2008).
10 H. Ibach, *Surf. Sci.* **603**, 2352 (2009).
11 J.C. Eriksson and A.I. Rusanov *Surf. Sci.* **603**, 2348 (2009).
12 J.C. Eriksson and A.I. Rusanov *Surf. Sci.* **604**, 1062 (2010).
13 J.C. Eriksson and A.I. Rusanov *Surf. Sci.* **604**, 2069 (2010).
14 M. Blanco-Rey, S.J. Pratt and S.J. Jenkins, *Phys. Rev. Lett.* **102**, 026102 (2009).
15 M. Blanco-Rey and S.J. Jenkins, *J. Phys. Condens. Matter* **22**, 135007 (2010).
16 H. Etman, G. Held, S.J. Jenkins and R.A. Bennett, *Phys. Chem. Chem. Phys.* **15**, 20823 (2013).
17 S.M. Driver, T. Zhang and D.A. King, *Angew. Chem. Int. Ed.* **46**, 700 (2007).
18 M. Blanco-Rey and G. Jones, *Phys. Rev. B* **81**, 205428 (2010).
19 M. Forster, M.S. Dyer, M. Persson and R. Raval, *J. Am. Chem. Soc.* **133**, 15922 (2011).
20 M.T. Middel, H.J.W. Zandvliet and B. Poelsema, *Phys. Rev. Lett.* **88**, 196105 (2002).
21 A. Li, F. Liu and M.G. Lagally, *Phys. Rev. Lett.* **85**, 1922 (2000).

22 A. Grossmann, W. Erley, J.B. Hannon and H. Ibach, *Phys. Rev. Lett.* **77**, 127 (1996).
23 H. Ibach, *Surf. Sci. Rep.* **29**, 195 (1997).
24 M. Godin, V. Tabard-Cossa, P. Grütter and P. Williams, *Appl. Phys. Lett.* **79**, 551 (2001).
25 G.G. Stoney, *Proc. R. Soc. London, Ser. A* **82**, 172 (1909).
26 J. Tamayo, J.J. Ruz, V. Pini, P. Kosaka and M. Calleja, *Nanotechnology* **23**, 475702 (2012).
27 C. Herring, *Phys. Rev.* **82**, 87 (1951).
28 H.P. Bonzel, D.K. Yu and M. Scheffler, *Appl. Phys. A* **87**, 391 (2007).

4

Asymmetric Adsorption on Achiral Substrates

Linnaeus and Cuvier have been my two gods, though in very different ways, but they were mere schoolboys to old Aristotle.

Letter to W. Ogle, Charles Darwin (22 February 1882)

The adsorption of molecules on crystalline surfaces can, of course, occur in either ordered or disordered fashion. Whilst disordered adsorption necessarily breaks all point symmetries of an otherwise achiral surface, including its improper symmetries, the resulting long-range configuration can hardly be said to be chiral in anything other than the most trivial sense; the mirror-image enantiomer of a specific disordered adsorption geometry is just one out of countless essentially degenerate alternative geometries, and its specific significance to our understanding of the system is hence vanishingly small. It is, of course, possible that the environment immediately surrounding an individual adsorbate may be rendered locally chiral, even in the case of disordered adsorption.

In the case of ordered adsorption, by way of contrast, the point symmetry of an otherwise achiral surface *may* be lowered, but this is by no means a universal occurrence; furthermore, any reduction in symmetry *may* remove all improper symmetry that the clean surface might have had, but again this is not necessarily so. In the event that all improper symmetry of an otherwise achiral substrate *is* broken by the adsorption of an ordered overlayer, however, the concept of enantiomeric overlayer geometries is well defined and it is natural to describe the situation as sensibly chiral. In the discussion that follows, we shall focus predominantly upon the necessary conditions leading to long-range globally chiral networks. It is important to stress, from the outset, that this discussion is emphatically not intended as a comprehensive catalogue of all the work done in this highly active area – examples have been carefully selected to illustrate specific types of phenomena, rather than included solely for completeness, and the omission of any particular study should not be taken to imply anything other than a practical decision in the context of a finite page count.

4.1 Achiral Adsorbates: Gliding Through Broken Mirrors

It is worth pointing out from the start that any reduction in symmetry caused by adsorption onto an achiral surface need not necessarily be driven by any

Chirality at Solid Surfaces, First Edition. Stephen J. Jenkins.

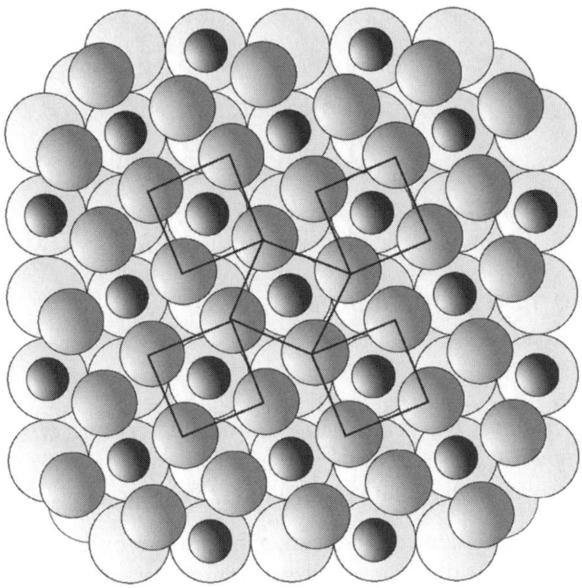

Figure 4.1 Clock reconstruction of Ni{100} induced by nitrogen adatoms (lateral displacements exaggerated for clarity). Reproduced with permission from Ref. [3]. Copyright (2016) Elsevier.

complexity in the structure or symmetry of the individual adsorbates. A classic example is to be found in the so-called 'clock' reconstruction of the Ni{100} surface under the influence of carbon or nitrogen adatoms [1–3]. In this system, the adsorption of N_2 is dissociative, so the actual adsorbates are simply individual N adatoms – their structure and symmetry could hardly be simpler; nevertheless, the four metal atoms bound to each adatom are observed to rotate within the surface plane, alternately clockwise and anticlockwise from one adsorption site to the next (Fig. 4.1). Viewed from the vantage point of a single N adatom, the resulting structure *appears* to be chiral, since the local metal atoms have rotated in one specific sense, but taking the surface as a whole this is clearly not the case. In fact, the ⟨001⟩ mirror planes of the clean-surface space group have become glide planes in the space group of the modified surface, but none have vanished without trace, let alone all. The surface symmetry has been lowered (if we consider glide to be inferior to reflection) but improper symmetry operations in the space group have simply been replaced with other improper symmetry operations, so clearly the result remains achiral. Indeed, the point group of the surface is unaltered.[1]

Another example in which mirror symmetry is reduced to glide may be found upon the adsorption of CO at the Pt{110}(1×2) surface. Here, the clean surface exhibits a 'missing row' reconstruction, corresponding to an ordered ridge-and-furrow structure that exposes {111}-type microfacets. The ridges lie along the in-plane $\langle\bar{1}10\rangle$ direction and are highly conducive to the upright adsorption of CO, which decorates alternate atop sites at coverages up to 0.5 ML (Monolayers). As the coverage approaches 1.0 ML, however, the density of CO molecules along the ridges increases towards a limit of one molecule per

1 Recall that both mirror and glide planes in the space group yield mirror planes in the point group (strictly, the point group of the space group).

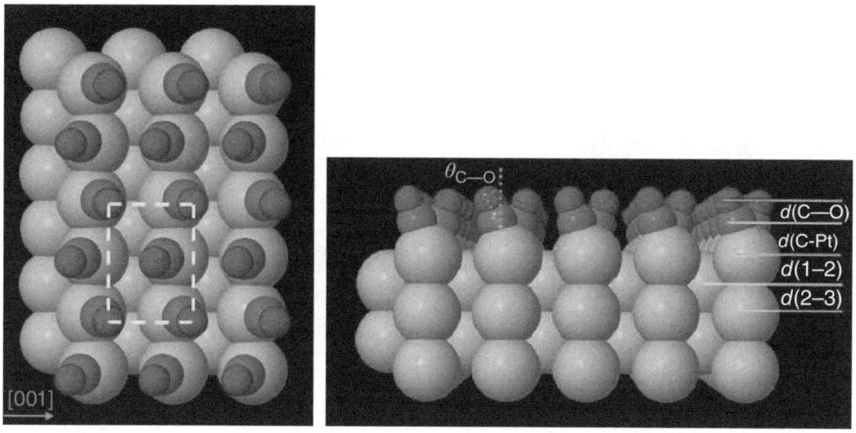

Figure 4.2 Alternating tilts amongst CO molecules adsorbed on Pt{110}(1×2). Reproduced with permission from Ref. [4]. Copyright (2012) Elsevier.

atop site. As a consequence of this, the molecules take on a marked tilt within the ⟨001⟩ azimuth [4], alternating from one site to the next to minimise steric repulsion (Fig. 4.2). The result is that the $\langle\bar{1}10\rangle$ mirror plane is removed, whilst that lying perpendicular to ⟨001⟩ is converted to a glide plane. Once again, we see that a highly symmetric molecule can readily lower the symmetry of an achiral surface, albeit in the two examples thus far discussed some improper symmetry is preserved and the system remains achiral.

From the observation that highly symmetric adsorbates can reduce mirror symmetry to glide, it is a small step to the realisation that in some cases the improper symmetry may be removed altogether. One of the clearest examples of this may be found in the adsorption of benzene on Ni{111}, which at saturation achieves a coverage of 1/7 ML in a $(\sqrt{7}\times\sqrt{7})$R19.1° overlayer [5]. The high packing density of molecules in this phase requires that the hydrogen atoms of neighbours interlock in the manner of meshed gears, necessitating a rotation of each adsorbate by c.30° about the surface normal. When the molecular rotation occurs in the clockwise sense, the $(\sqrt{7}\times\sqrt{7})$ unit cell is rotated by 19.1° clockwise relative to the (1×1) unit cell of the clean surface (Fig. 4.3) but crucially there exists also a degenerate structure in which the molecules and unit cell are rotated anticlockwise. These two structures are each extrinsically lattice-chiral (see Section 1.4.3) as the overlayer breaks completely the mirror symmetries of the underlying substrate. The degeneracy of the two enantiomeric structures, however, necessitates that both must arise on the surface with equal probability. Accordingly, the surface will display regions of each structure in essentially equal proportion. Such regions may be termed **mirror domains**.

It should be stressed that it is by no means necessary to involve molecules of the size of benzene to obtain overlayers exhibiting mirror domains. Even such simple molecules as CO and NO_2 can yield similar behaviour, for example, as has been found upon coadsorption of these species on the Au{111} surface. In this case, the CO molecules are only very weakly bound to the substrate, but

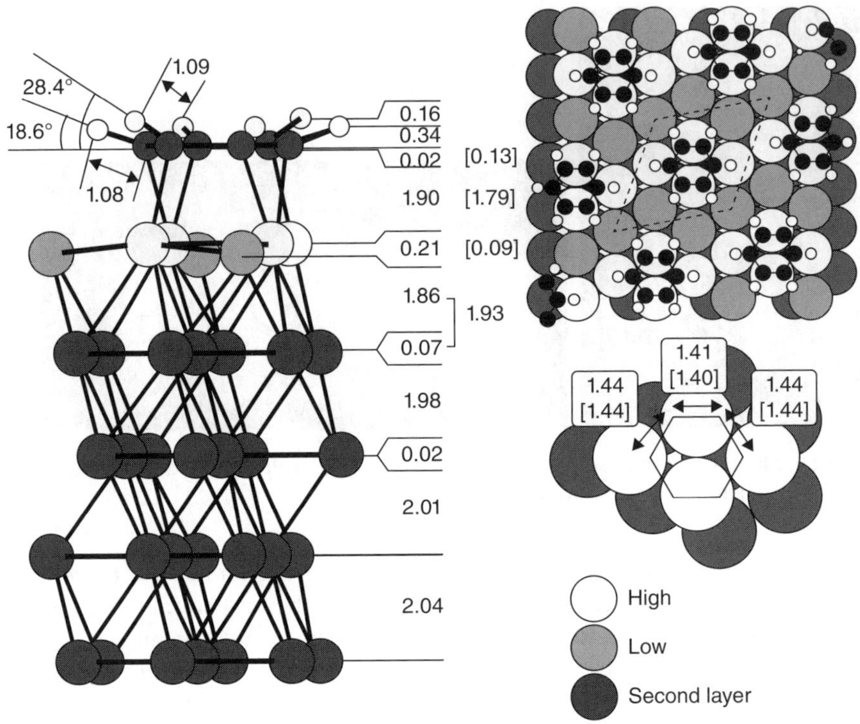

Figure 4.3 Benzene in its clockwise-rotated ($\sqrt{7} \times \sqrt{7}$) phase on Ni{111}. Reproduced with permission from Ref. [5]. Copyright (2001) American Institute of Physics.

are stabilised by a complex combination of electrostatic interactions with the more strongly bound NO_2 molecules [6]. Each CO molecule binds upright in an atop site, surrounded by six bidentate NO_2 molecules that bridge between two atop sites; viewing the central CO molecule as a hub, the surrounding NO_2 molecules exhibit a 'pinwheel' pattern of either clockwise or anticlockwise sense (Fig. 4.4). The long-range order again displays a ($\sqrt{7} \times \sqrt{7}$)R19.1° unit cell, with the clockwise pinwheel corresponding to a clockwise rotation of the unit cell and anticlockwise pinwheel to the anticlockwise unit cell rotation.[2]

In all these cases, it is notable that only minimal distortion of the individual molecular adsorbates is involved, and indeed that the interactions between adsorbates are neither highly specific nor directional. That they nevertheless exhibit asymmetry within mirror domains suggests strongly that mere geometry is sufficient to drive this tendency. An interesting take on this has been provided by Kasperski and Szabelski [9–11], who have conducted Monte Carlo simulations of highly idealised adsorbates having either tripodal or cruciform shape (on triangular and square lattices, respectively). These calculations demonstrate a clear

2 The chirality of this arrangement was highlighted by Zhang *et al.* [7], but one may also note earlier comments by Gellman [8] on *fcc*-{111} ($\sqrt{7} \times \sqrt{7}$)R19.1° overlayers in general, not to mention *fcc*-{100} ($\sqrt{5} \times \sqrt{5}$)R26.6° overlayers.

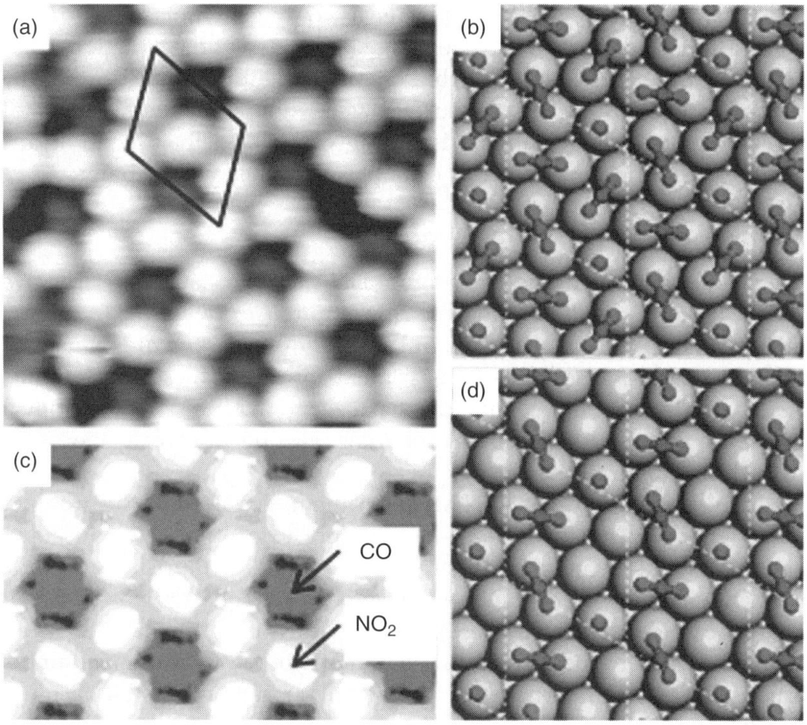

Figure 4.4 Coadsorption of CO and NO_2 on Au{111}, showing in (a) and (c) experimental and simulated STM images, with the anticlockwise pinwheel model in (b) and a disfavoured alternative in (d). Reproduced with permission from Ref. [6]. Copyright (2005) American Physical Society.

tendency towards the formation of domains possessing either intrinsic or extrinsic lattice chirality, even when realistic 'chemistry' is entirely absent from the model. Furthermore, the porosity of the chiral networks is found to vary in a predictable manner according to the length of the tripod/cross arms and the imposed packing density [9]. By 'activating' different combinations of certain sites within the model adsorbates (i.e. simply by altering their inter-molecular interaction energies relative to those of the other sites) it turns out to be possible to tune the self-assembled structures between close-packed, porous, rod-like and thread-like variants [10, 11].

4.2 Prochiral Adsorbates: Chirality in Context

Prochiral molecules are those that, whilst themselves strictly achiral, can react to form chiral products in a single step; global symmetry implies, however, that both enantiomers will be produced in equal quantity, unless some chiral modifier is involved as a catalyst. When a prochiral molecule is adsorbed at a surface, the geometric constraint thus applied may be sufficient to result in a chiral adsorption configuration, even when both adsorbate and substrate are achiral to begin with – a situation that we have described as contingent molecular chirality (see Section 1.4.3).[3] Needless to say, the global symmetry of the situation again demands that both enantiomers of such a chiral configuration must ordinarily occur in equal proportion across the surface as a whole,[4] but the existence of extended domains of a single chirality is certainly possible, albeit by no means certain. Here, we enumerate some clear examples that illustrate the point.

4.2.1 Guanine on Au{111}

The molecule guanine, one of the four natural DNA bases, is a purine derivative based upon a fused pyrimidine-imidazole ring system (*see Plate XIV*).[5] Its planar geometry renders it strictly achiral, but several of its carbon atoms may be described as prochiral centres – hydrogenation at any one of these, for example, would result in a chiral product. Similarly, the adsorption of the ring system in a flat-lying surface geometry would break the molecule's only mirror plane, again leading to a chiral conformation.

Scanning tunnelling microscopy (STM) experiments reported by the group of Besenbacher [15] show that upon room-temperature adsorption onto a Au{111} surface, guanine spontaneously self-assembles into islands comprising sub-units of four individual molecules, dubbed 'quartets'. Each quartet possesses an approximate fourfold rotational symmetry, but no mirror symmetry, and indeed quartets of two distinct types are observed and attributed to two enantiomers of a chiral structure. The internal structure of the quartets can be glimpsed in high-resolution images, matching well to two proposed mirror-image structures calculated by density functional theory (DFT) to be particularly stable; these DFT calculations do not include the substrate–adsorbate interaction, but this is expected to be very small, as evidenced by the fact that adsorption does not lift the well-known herringbone reconstruction of the Au{111} surface. Each molecule within a quartet binds to two neighbours via two pairs of hydrogen

3 The chirality of an adsorbed prochiral molecule is, of course, bistable. That is, it may flip from one handedness to the other, either spontaneously or under the influence of an external perturbation. In some cases, the activation barrier may be sufficiently small that the situation is essentially fluxional at the temperature of interest, whilst in others there may be considerable thermal stability. A good example of perturbation-driven handedness change may be found in the case of propene adsorption on Cu{211}, where Parschau *et al.* [12] have induced flipping by means of inelastic electron tunnelling (IET) within an STM experiment.

4 The exception to this general rule is, of course, the case when careful steps have been deliberately taken to prepare the prochiral adsorbate from a chiral precursor in such a way that some degree of asymmetry is preserved, as in the very recent work of Stetsovych *et al.* [13] (see also the comments of Ernst [14]).

5 For ease of reference, molecular adsorbates discussed in this book are collated in the colour plates. Simple inorganic molecules are omitted, as are species used only as solvents. Most of the remaining cases have been classified according to their dominant functional groups, according to the following hierarchy: (i) acids/esters, (ii) ketones/aldehydes, (iii) alcohols/epoxides. Where a particular molecule falls into two or more categories, it has been placed within only the higher-order grouping, to avoid repetition. A few of the more complicated molecules have been separately gathered under the general headings of either sugars or alkaloids, leaving only a small number of orphans to form a miscellaneous category of their own. Sugars have been represented in their linear form, rather than the cyclic pyranose forms found more typically in solution, simply to avoid the excessive display of the several distinct isomers in which the latter forms occur.

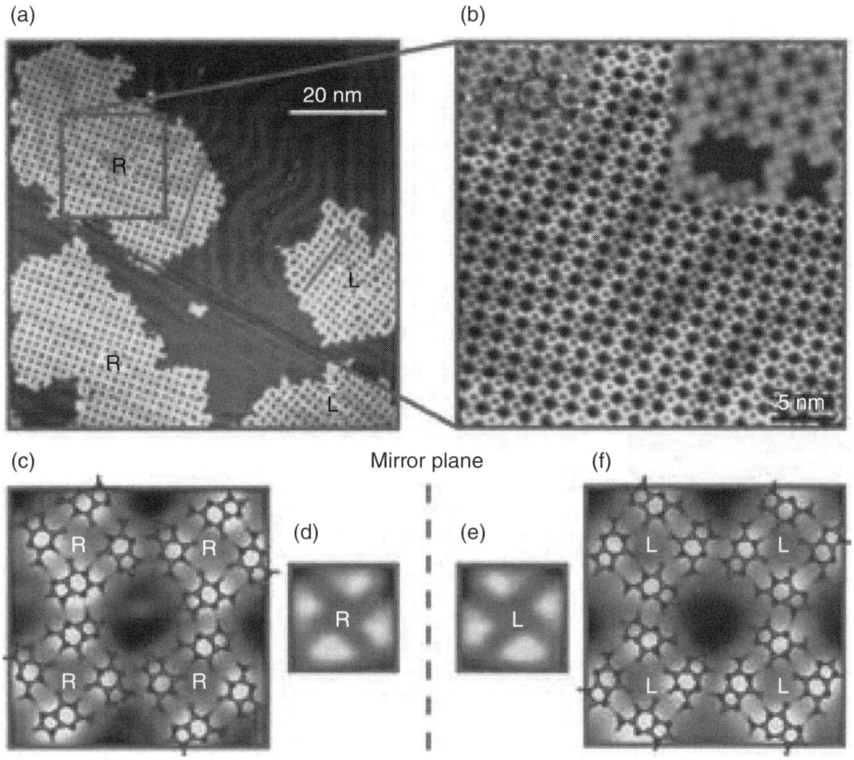

Figure 4.5 STM images of homochiral domains of guanine on Au{111}. Note the herringbone reconstruction visible in bare patches of panel (a) and just perceptible as a modulation beneath the overlayer in panel (b). Panels (c) and (d) show quartets exhibiting shared orientation, while panels (e) and (f) depict the mirror-image situation. Reproduced with permission from Ref. [15]. Copyright (2009) Wiley.

bonds implying eight intra-quartet hydrogen bonds in all. Adjacent quartets can then bind to one another via further pairs of inter-quartet hydrogen bonds allowing the formation of a network of quartets extending across tens of nanometres. In the room-temperature experiments, islands comprising exclusively one or the other quartet enantiomer were observed (Fig. 4.5). That is, the system exhibits spontaneous segregation into domains that one might describe as **homochiral** (i.e. containing structural motifs of a single handedness[6]). Unlike the cases described above for benzene on Ni{111} or CO/NO_2 coadsorption on Au{111}, the chirality here is not simply attributable to the lateral orientation of the adsorbates, but also to the face that each adsorbate orients down onto the surface; self-assembly here requires that each molecule contributing to a quartet (and indeed to an entire homochiral domain) be the 'same way up'. Were it possible, therefore, to enact a hydrogenation reaction within a single such domain, it might be expected that a single enantiomer of the chiral product would result.

---

6 In this case, the motif in question is the quartet of guanine molecules; in other cases, the motifs may simply be individual molecules.

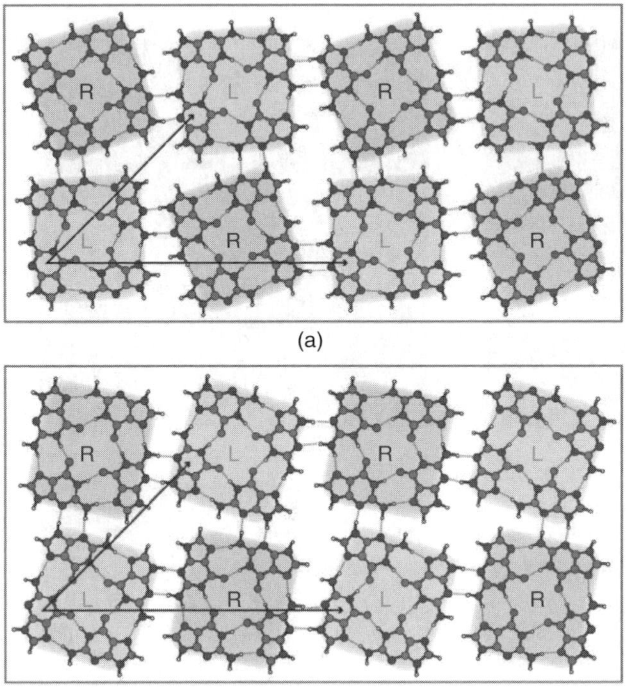

(a)

(b)

Figure 4.6 Schematic depictions of possible heterochiral domains of guanine on Au{111}. Reproduced with permission from Ref. [15]. Copyright (2009) Wiley.

Upon annealing to 400 K, Xu *et al.* [15] report that the homochiral domains are replaced by a **heterochiral** structure in which each quartet retains a distinct chirality but now quartets of *both* enantiomeric forms combine within each domain (Fig. 4.6). Indeed, the resulting domains show strict alternation between quartet enantiomers, and they can hence be considered to be intimately racemic at the level of the quartets. In fact, however, it is possible to propose *two* degenerate mirror-image models that fit the STM images equally well, suggesting that at the supra-quartet level of organisation there remains some degree of chirality. The authors' DFT calculations indicate a very small energy difference between the heterochiral and homochiral structures, but favouring the latter; the fact that the former is observed experimentally upon annealing is, therefore, attributed to configurational entropy associated with disorder between the two mirror-image forms of the structure containing both quartet enantiomers. That is, the heterochiral domain is suggested to contain a strict alternation of the quartet enantiomers, but the exact orientation of these quartets varies locally between two degenerate supra-quartet arrangements in a more or less random manner. The guanine/gold system thus shows a complex relationship[7] between local and long-range symmetry, encompassing, under various conditions,

7 Replacing guanine with adenine results in an even more complex system, described in detail by Lukas *et al.* [16], but it would be disproportionate to summarise it here.

local chirality, homochiral segregation, intimately racemic heterochirality and hierarchical chirality.

4.2.2 Stilbene Derivatives on Cu{100} and Cu{110}

Two further examples will suffice to illustrate the possibility of switching between homochiral and heterochiral domains, both involving molecules derived from *trans*-stilbene adsorbed on copper surfaces.[8] In the first case, the transition is driven by temperature, as per the example of guanine discussed above, but in the second it is the surface coverage that appears to be the crucial factor.

We begin, then, with the adsorption of *trans*-4,4′-dicarboxystilbene (*trans*-DCSB, *see Plate XVIII*) on Cu{110}, as reported by Cortés *et al.* [20]. Upon deposition at room temperature, the adsorbed moieties self-assemble in islands revealed by STM to be ordered in what the authors describe as a 'parquet pattern'. Although the DCSB molecule displays improper symmetry in isolation, this is lost upon adsorption, implying that two distinct chiral forms are possible on the surface. Close examination of the parquet-pattern islands indeed reveals moieties of two mirror-image forms, but only one type is present in any one island. That is, the parquet-pattern islands formed at room temperature constitute homochiral domains. Upon deposition at around 395 K, however, the adsorbed moieties no longer form parquet-pattern islands, but instead assemble into a so-called 'herringbone pattern'. Detailed scrutiny of this high-temperature phase reveals that although two mirror-image types of herringbone-pattern island can be discerned, each comprises individual moieties of *both* chiralities. In other words, the islands are internally heterochiral, even though they do possess an overall chirality associated with their packing – a clear manifestation of the distinction between contingent molecular chirality and (in this case, intrinsic) lattice chirality.

Turning now to the adsorption of *trans*-2-(4-pyridyl)-4-vinylbenzoic acid (*trans*-PVBA, *see Plate XVIII*) on Cu{100}, STM experiments conducted by Vidal *et al.* [21] reveal parquet-pattern islands formed at low coverage upon deposition at 300 K followed by brief annealing at 400–450 K. Once again, the surface removes the improper symmetry found in the isolated molecules, leading to adsorbed moieties possessing contingent molecular chirality; each parquet-pattern island is found to comprise only one or other molecular handedness, implying the existence of mirror-image homochiral domains. As the coverage increases beyond about 1/20 ML, however, the molecular ordering changes abruptly, switching to a single heterochiral phase in which each unit cell contains adsorbed moieties of both mirror-image types in a complex 'butterfly pattern'. In both this case and that of DCSB, the subtle balance between intermolecular interactions and the substrate–overlayer registry has been invoked to explain the apparent contextual sensitivity that drives the observed transition from homochiral to heterochiral domains [20, 21].

8 A wider variety of systems could, of course, be adduced here (see, for example, Bohringer *et al.* [17], Gopakumar *et al.* [18], or Yang *et al.* [19]) but there is merit in offering just two, rather similar, systems to compare and contrast.

4.2.3 Glycine on Cu{110} and Cu{311}

The immediately preceding examples concerned cases where the adsorbed molecules retained the symmetry of their gas-phase counterparts, aside from a single mirror plane and/or centre of inversion removed by the surface. Now we consider an alternative situation, where the adsorbed molecules may be strongly distorted in order to conform to the available adsorption sites of the surface. In doing so, we shall turn to one of the classic systems of the field, namely the adsorption of glycine on Cu{110}. As the simplest of the amino acids (and indeed the only achiral member of that class, *see Plate XI*) glycine serves not only as an interesting case study in its own right, but also as an important baseline for studies of more complex (and chiral) amino acids. The Cu{110} substrate proves to be an amenable support for the formation of two-dimensional hydrogen-bonded networks, neither binding too weakly to allow annealing nor too strongly to permit ready diffusion and self-assembly. Recent work on the Cu{311} surface, which differs subtly in its symmetry and structure from Cu{110}, provides an insightful counterpoint to the earlier work on the higher symmetry substrate, shedding light on the murky relationship between the adsorbate symmetry, the symmetry of its binding site and the symmetry of the resulting hydrogen-bonded network.

As an isolated molecule, glycine adopts a strictly achiral ground state configuration, in which the plane of its acid group lies within the mirror plane of the carbon backbone. Under different environmental conditions the molecule can variously become deprotonated at the acid group, forming the anion known as glycinate, or protonated at the amine group, forming the cation; when both modifications are present, the resulting neutral species is described as a zwitterion. Upon room-temperature adsorption onto the Cu{110} surface, it is generally agreed that deprotonation occurs, so that the molecule binds in its anionic form. Low-exposure reflection absorption infrared spectroscopy (RAIRS) experiments reported by Barlow *et al.* [22] reveal (amongst other details) a clear absorption peak at 1417 cm^{-1} associated with the symmetric —CO_2 stretch mode of the glycinate species (Fig. 4.7a and b). A corresponding peak associated with the antisymmetric —CO_2 stretch mode is conspicuously absent, implying that the vector between the two oxygen atoms lies parallel to the surface.[9] Similarly missing is any sign of the —NH_2 scissor mode, although a band attributed to the —NH_2 wag mode is identified at 1084 cm^{-1}; taken together, these facts confirm that the amine group has not gained a proton, and that it lies with its plane nearly parallel to the surface. At higher exposures, both —NH_2 wag and scissor modes are observed, at 1105 cm^{-1} and 1578 cm^{-1}, respectively, albeit the latter only relatively weakly, indicating that some small twist in the amine group has occurred (Fig. 4.7c and d). At the same time, the —CO_2 symmetric stretch mode shifts slightly to 1422 cm^{-1} and a new mode attributed to the asymmetric —CO_2 stretch appears quite strongly at 1630 cm^{-1}. It seems, therefore, that the acid group of at least some adsorbed glycinate moieties rotates signficantly as the surface coverage increases.

---

9 The surface selection rule for RAIRS tells us that only modes having an out-of-plane component of dynamic dipole moment will give rise to absorption peaks.

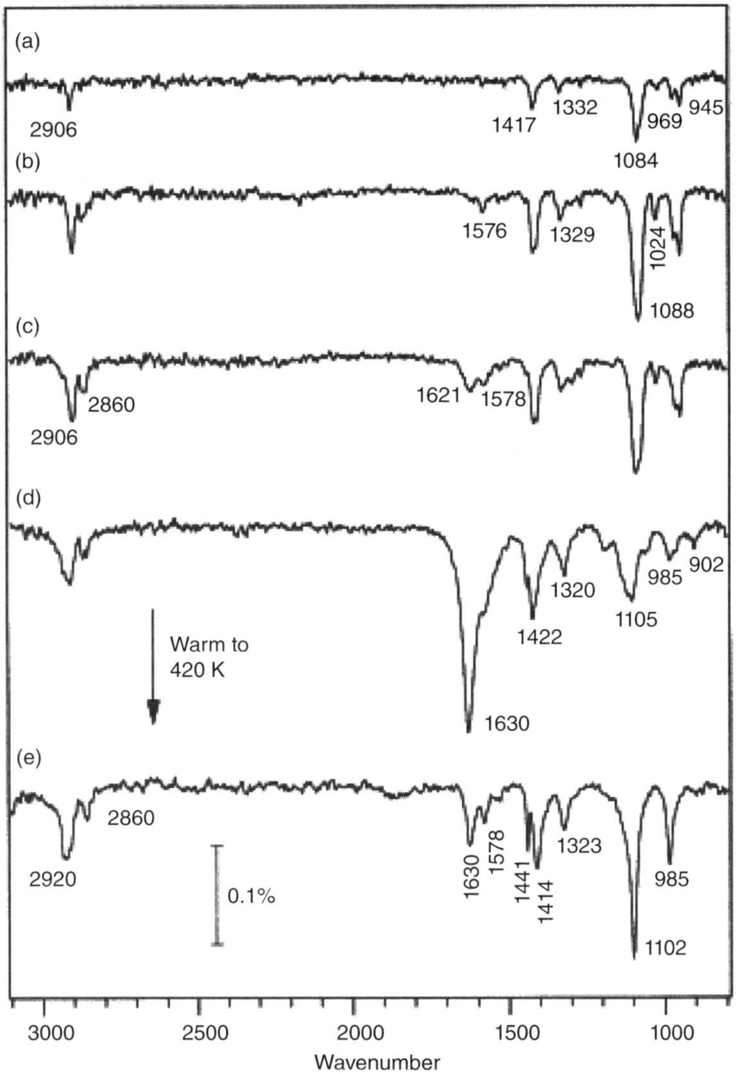

Figure 4.7 RAIR spectra obtained upon 300 K exposure of Cu{110} to increasing doses of glycine (spectra (a)–(d)) and after heating to 420 K (spectrum (e)). Reproduced with permission from Ref. [22]. Copyright (1998) Elsevier.

Upon heating the adsorbed overlayer at 420 K, the high-coverage features in the RAIR spectra weaken but do not altogether vanish (Fig. 4.7e) suggesting that some dissociation or desorption occurs. A (3×2) low-energy electron diffraction (LEED) pattern, poorly resolved at room temperature, sharpens considerably at the same time. Examination of this pattern reveals systematically missing spots, consistent with the existence of glide symmetry in the overlayer structure. Since glide symmetry implies that the unit cell contains an even number of entities, one may readily conclude that there must be two glycinate moieties per unit cell

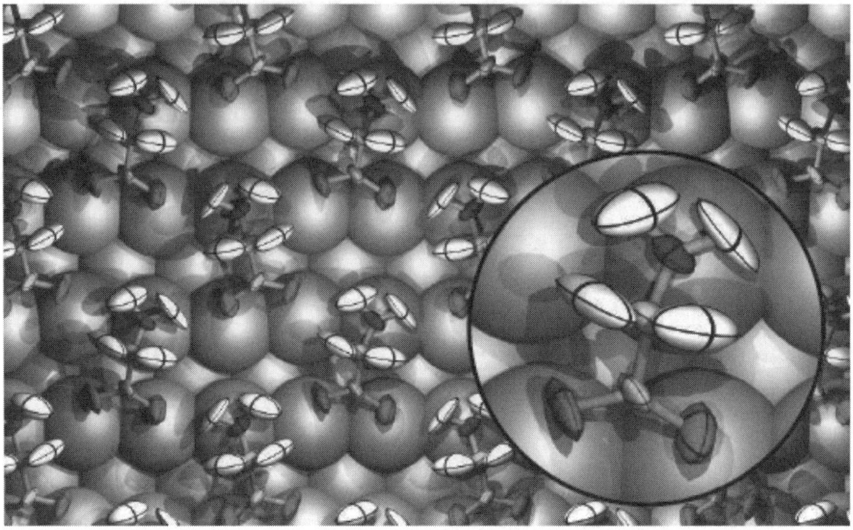

Figure 4.8 Calculated structure of glycinate adsorbed on Cu{110} in a heterochiral (3×2) structure.

(there being insufficient space for four or more). The accepted model for this phase consists of a heterochiral pair of distorted glycinate molecules bound to the surface in a μ_3 configuration (i.e. three-point binding via two oxygen atoms and one nitrogen atom) and to each other via hydrogen bonds (Fig. 4.8). The twist in the molecular backbone favours one strong and one weak hydrogen bond, rather than two of intermediate strength, but more importantly it also presents the electronegative atoms of the molecule to a right-angled triangle of copper atoms that constitutes its footprint on the surface (Fig. 4.9a). Assuming

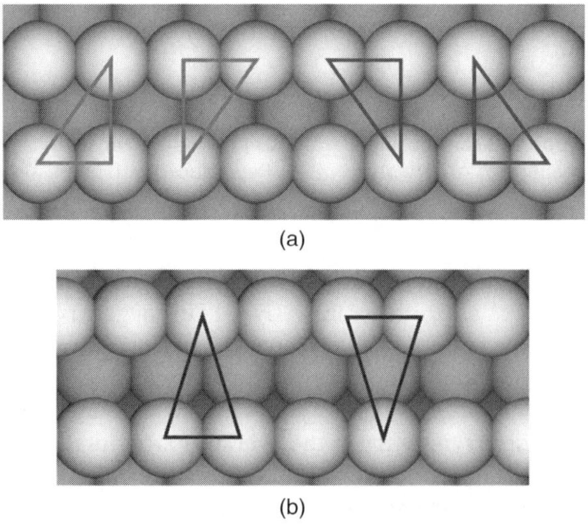

Figure 4.9 Triangular footprints on (a) Cu{110} and (b) Cu{311}. Reproduced under Creative Commons Attribution 4.0 from Ref. [23].

this footprint to be particularly favourable in terms of binding energy, one realises immediately the fact that the Cu{110} surface possesses only right- and left-leaning footprints, practically ensuring that adsorbed glycinate molecules must adopt a locally chiral geometry. Subsequent DFT calculations also confirmed that this heterochiral structure is energetically favoured [24] and X-ray photoemission spectroscopy (XPS) experiments show that all oxygen atoms experience identical chemical environments [25] consistent with μ_3 binding. The RAIR spectra obtained at higher exposures suggest that some conversion to a μ_2 configuration takes place as the coverage increases, presumably in order to free-up additional adsorption sites for extra glycinate molecules to occupy. Some STM images [26] have been interpreted as revealing homochiral domains[10] but the evidence is by no means clear cut and tip artefacts may play a role; certainly, quantitative photoelectron diffraction (PhD) [27] and LEED [28] analyses have pretty much unequivocally supported the standard μ_3 heterochiral model for coverages up to 1/3 ML.

In many respects, the Cu{311} surface is very similar to Cu{110} (see Section 2.4) but it differs in a couple of crucial features, the most important of which is that triangles of copper atoms suitable for μ_3 adsorption of glycinate are isosceles in form, rather than right-angled (see Fig. 4.9b). That is, the footprint of a glycinate molecule adsorbed on Cu{311} will necessarily be achiral, rather than chiral. A second difference is, of course, the presence of only a single mirror plane in Cu{311}, as opposed to two in the case of Cu{110}. This means that footprints pointing in different directions along the $\langle\bar{1}33\rangle$ axis are non-degenerate and must be taken separately into consideration when developing structural models. Experiments reported by the present author's own group [23, 29] show the formation at low exposures of a sharp (2,1;1,2) LEED pattern[11] upon adsorption of glycine at room temperature; the corresponding surface unit cell is large enough to contain only a single glycinate moiety in the μ_3 binding configuration, suggesting immediately that removing the necessity to occupy pairs of chiral footprints has simplified the structure relative to that found on Cu{110}. RAIR spectra [23] taken in the low-exposure regime reveal absorption bands at 1097 and 1414 cm^{-1}, highly reminiscent of features observed previously on Cu{110} and similarly attributed predominantly to —NH_2 wag and symmetric —CO_2 stretch modes, respectively; both the asymmetric —CO_2 stretch and any pure —NH_2 scissor mode are absent from the spectra, further supporting a μ_3 binding geometry (Fig. 4.10). DFT calculations for a model based upon this assumption yield vibrational frequencies in excellent agreement with the experimental spectra. Higher exposures lead to the emergence of distinct streaks in the LEED pattern, maintaining the overall

10 In this context, homochiral should be taken to imply that all glycinate moieties within a single domain are chirally distorted in the same sense. Likewise, heterochiral would imply glycinate moieties that are chirally distorted in two opposing senses within a single domain.
11 Once the overlayer periodicity is sufficiently complex as to require matrix notation, it is inevitably the case that multiple equivalent ways can be found to indicate the same physical structure. Throughout this book, all periodicities described in the literature have been translated (where necessary) to conform to a single logical convention that emphasises the symmetry of the overlayer and its relationship with that of the underlying substrate. This convention is described in Appendix B.

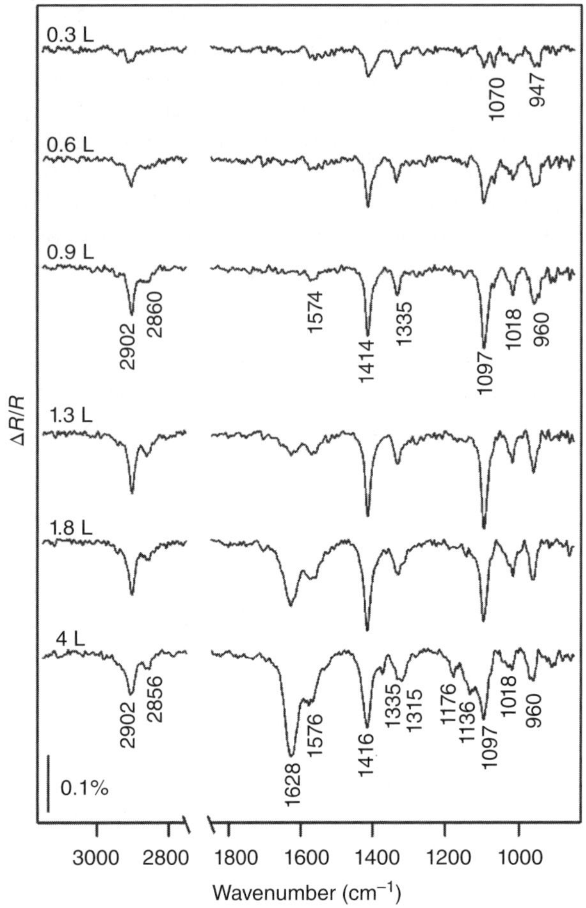

Figure 4.10 RAIR spectra obtained upon 300 K exposure of Cu{311} to increasing doses of glycine. Reproduced under Creative Commons Attribution 4.0 from Ref. [23].

symmetry but indicative of disorder in two azimuthal directions, coincident with the growth of an absorption feature at 1628 cm^{-1} indicative of a new μ_2 binding configuration.

STM images obtained after dosing at 300 K in the low-exposure regime reveal a highly ordered array of molecules arranged in accordance with the (2,1;1,2) LEED pattern, but annealing to 410 K in the high-coverage regime reveals something radically different (Fig. 4.11). Under conditions consistent with the existence of μ_2 species (as per the RAIR spectra) one finds a disordered array of dark zig–zag lines crossing the surface and failing to respect its mirror symmetry, leaving only rather small strips of (2,1;1,2) structure between them. These dark lines do, in fact, contain distinct features, suggesting that they represent moieties bound in a different conformation, rather than the absence of adsorbates altogether. The natural conclusion is that they correspond to the μ_2 species, while the bright spots within the remaining ordered regions correspond to the μ_3 species; certainly, the relative preponderance of the two types of feature seems to correlate nicely with the RAIR spectra. The zig–zag meandering of these μ_2 boundaries is also consistent with the bi-directionality of the streaks observed in LEED

Figure 4.11 STM image (200 × 200 Å^2) obtained upon annealing a saturated glycinate overlayer on Cu{311} to 410 K. Dark zig–zag trenches are believed to harbour μ_2 species, while the intervening spots of intermediate brightness are consistent with (2, 1; 1, 2) periodicity and contain μ_3 species; the brightest spots are thought to correspond to alaninate contamination. Reproduced with permission from Ref. [29]. Copyright (2014) Elsevier.

patterns obtained at high coverage, so a consistent picture seems to be emerging. The high-symmetry phase that pertains at coverages up to 1/3 ML is entirely achiral, with each individual glycinate moiety maintaining a mirror-symmetric local bonding configuration, unlike the intimately racemic situation found on Cu{110}; this difference is attributable to the different footprints available in the two cases. At coverages above 1/3 ML, the conversion of some glycinate moieties from μ_3 to μ_2 binding necessarily breaks this symmetry on a local level. The zig–zag features associated with μ_2 binding create locally chiral regions that extend beyond a single molecule, however, as the zigs and the zags may each run for around 10 molecules before a change in direction. They may, in effect, be considered to be one-dimensional domains of extrinsic lattice chirality, threading between intervening regions where achirality appears to be maintained.

4.2.4 Succinic and Fumaric Acids on Cu{110}

Moving away from amino acid adsorption, another heavily studied class of adsorbates is that of the dicarboxylic acids. Simple achiral examples of this class of compound include succinic acid and fumaric acid (*see Plate XII*) which have both been examined in some detail on the Cu{110} surface. Added complexity, compared with the case of glycine, is found in the possibility that either one or both of the acid groups may be deprotonated, leading to either mono- or bi-carboxylate species. As above, RAIRS experiments have proved vital in determining which types of species are present under different conditions, and their combination with LEED and STM experiments is once again extremely powerful.

Upon room-temperature adsorption of succinic acid on Cu{110}, Humblot *et al.* [30] report the presence at the lowest coverages of an absorption band at 1425 cm^{-1}, attributed in part to a symmetric —CO_2 stretch mode. A corresponding band at 1712 cm^{-1}, attributable to the asymmetric —CO_2 stretch mode, only appears at slightly higher coverages. This behaviour is interpreted as revealing initial adsorption in the bisuccinate form, with the monosuccinate appearing only as the surface becomes more densely occupied. The purely bisuccinate regime is characterised by overlaid $(1, \bar{1}; 6, 3)$ and $(1, 1; 6, \bar{3})$ LEED patterns.[12] Although neither periodicity implies intrinsic lattice chirality (they each correspond to a rectangular lattice) they are not aligned with the high-symmetry directions of the underlying lattice and thus exhibit extrinsic lattice chirality. When the monosuccinate is present, a further c(4×2) LEED pattern is overlaid, although so long as the coverage is not *too* high the continued presence of the chiral LEED patterns indicates that this coexists with the previous bisuccinate phase. Increasing the coverage beyond an estimated 1/6 ML, however, leads to the loss of the chiral LEED patterns, and the formation of a (presumed) purely monosuccinate phase. Conversely, annealing at 400 K leads to the loss of both the c(4×2) LEED pattern *and* the 1712 cm^{-1} absorption band, implying the generation of a purely bisuccinate phase existing in the aforementioned chiral domains. The group of Ernst subsequently reported that annealing at 580 K results in a further restructuring, resulting in overlaid $(1, 1; 5, \bar{2})$ and $(1, \bar{1}; 5, 2)$ LEED patterns[13] [31].

STM images obtained by Humblot *et al.* [30] bear a very interesting, and somewhat non-trivial, relationship to the order revealed by LEED. In the pure monosuccinate phase, characterised by a c(4×2) LEED pattern, for example, the images recorded by STM actually display c(8×2) periodicity. The authors propose that individual monosuccinate moieties are dispersed in a c(4×2) arrangement, but that the free acid groups of neighbouring pairs interact via hydrogen bonds to create dimers in an effective c(8×2) arrangement (Fig. 4.12). The STM is then held to image the dimer superstructure, while LEED is supposed to register only the underlying disposition of the individual molecules. The argument is reasonably plausible, although it is a little surprising that LEED seems to be insensitive to the molecular distortions implied by dimer formation. If the free acid groups are sufficiently mobile to be disordered, and hence invisible to LEED, one wonders why the STM images are quite so apparently well ordered. In the absence of any alternative hypothesis, however, this one will have to suffice for the moment.

In similar fashion, the pure bisuccinate phase, characterised by overlaid $(6, \bar{3}; \bar{1}, \bar{1})$ and $(6, 3; \bar{1}, 1)$ LEED patterns, yields STM images of chiral phases having apparent $(7, \bar{2}; \bar{2}, \bar{2})$ and $(7, 2; \bar{2}, 2)$ periodicity.[14] Here, detailed examination of the STM images reveals the existence of trimers, despite the absence of any possibility of hydrogen-bonded interaction between neighbouring bisuccinate moieties. Furthermore, it is far from clear why the LEED patterns should indicate a periodic repeat along the (1,1) surface direction at half the apparent spacing revealed by STM; it has been speculated [32] that LEED is either sensitive primarily to the carboxylate groups, or that some reconstruction of the surface with the smaller period so dominates the LEED pattern as to swamp the higher order periodicity of the adsorbates. Again, neither explanation entirely convinces, but no obvious alternatives present themselves.

---

12 In the original paper, these are described as $(9,0; \bar{1}, 1)$ and $(1, 1; \bar{9}, 0)$, which fails to emphasise the rectangular symmetry of the overlayer lattice.

13 Originally described as $(1, 1; \bar{7}, 0)$ and $(7,0; \bar{1},1)$.

14 Originally described as $(2, 2; \bar{9}, 0)$ and $(9,0; \bar{2}, 2)$.

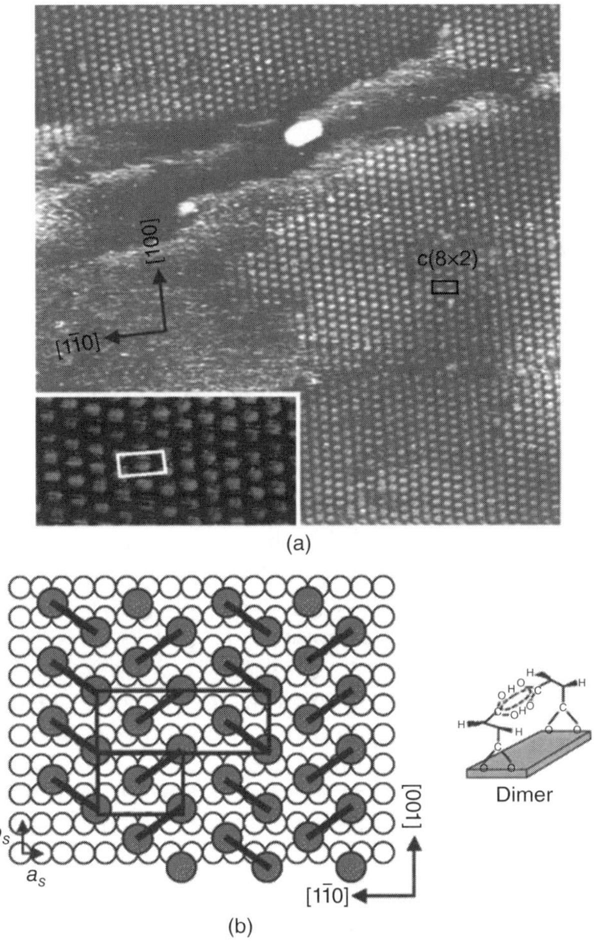

Figure 4.12 The monosuccinate phase on Cu{110}: (a) STM image (500 × 500 Å^2) showing c(8×2) periodicity; (b) schematic indicating the relationship between this lattice and the c(4×2) periodicity revealed by LEED. Reproduced with permission from Ref. [30].Copyright (2004) American Chemical Society.

An interesting counterpoint to these adsorption studies on succinic acid is to be found in the work of Karageorgaki *et al.* [33] on the adsorption of fumaric acid. The two compounds differ in that the latter is unsaturated, with the consequence that it possesses another potential site for interaction with the surface (i.e. its —C=C— bond). Despite this, several aspects of its structure on Cu{110} are highly reminiscent of succinic acid, but the interpretation of these must necessarily be somewhat different.

Upon room-temperature adsorption of fumaric acid on Cu{110}, only a poorly ordered c(2×4) phase is formed at saturation coverage, but its LEED pattern sharpens upon annealing to 373 K; further annealing to 423 K yields a (2×4) LEED pattern with systematic absences indicative of two orthogonal glide planes. An absorption band at around 1420 cm^{-1} is interpreted as corresponding

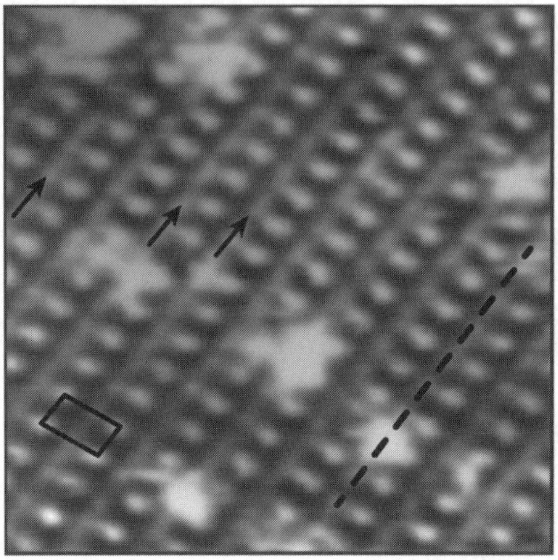

Figure 4.13 STM image (169×169 Å^2) of the $(1, \bar{1}; 5, 2)$ phase observed on Cu{110} after exposure to fumaric acid. The marked unit cell indicates the periodicity of the bright features, believed to be molecular in nature; the arrows highlight rows of fainter features, interpreted as copper adatoms, which are believed to repeat at twice the spatial frequency of the molecules (although this cannot be discerned from the image shown here). The dashed line marks an antiphase domain boundary. Reproduced with permission from Ref. [33]. Copyright (2014) Elsevier.

to the —CO_2 symmetric stretch mode of a carboxylate group, while a broad OH stretch band centred on 3159 cm^{-1}, together with a strong —C=O band at 1700 cm^{-1}, suggests that at least one acid group remains intact; hence these phases have been supposed to comprise monofumarate species [33]. This conclusion is seemingly at odds with the existence of two glide planes in the (2×4) phase, which would require each molecule to terminate identically at each end. It is conceivable that the ordered phase in fact contains only bifumarate species, while the spectroscopic signature of the monofumarate derives from a disordered phase invisible to LEED (or vice versa). Alternatively, one may suggest that LEED in this case is primarily sensitive to the metal atoms, and relatively less so to the details of the molecular structure.

At sub-saturation coverages, three different phases comprising chiral domains have been reported upon annealing at 423 K [33]. The highest coverage of these, occurring at 81% of saturation coverage, entails overlaid $(6, \bar{3}; \bar{1}, \bar{1})$ and $(6, 3; \bar{1}, 1)$ LEED patterns, similar to those seen upon adsorption of succinic acid.[15] At 78% of saturation coverage, however, the overlaid LEED patterns are $(1, 1; 5, \bar{2})$ and $(1, \bar{1}; 5, 2)$,[16] while at 70% of saturation coverage they are $(1, \bar{1}; 3, 2)$ and $(1, 1; 3, \bar{2})$.[17] STM images of the $(1, \bar{1}; 5, 2)$ and $(6, \bar{3}; \bar{1}, \bar{1})$ phases (as identified by LEED) reveal doubled periodicity along the (1,1) surface direction (see Fig. 4.13) as observed previously for succinic acid. In this case, however, patches of surface devoid of molecules have been imaged between the adsorbate islands; notably, these contain features, interpreted as copper adatoms, that display the periodicity of the LEED patterns, and it is supposed that the same reconstruction continues beneath the adsorbate islands too. The implication, yet again, is that LEED in these systems is primarily sensitive to the disposition of the metal atoms, and that higher order periodicity in the disposition of the adsorbates, which should strictly

15 Originally described as $(1, 1; \bar{6}, 3)$ and its enantiomer.

16 Originally described as $(1, 1; \bar{5}, 2)$ and $(1, \bar{1}; 4, 3)$.

17 Originally described as $(1, \bar{1}; 3, 2)$ and its enantiomer.

give rise to additional fractional order spots in the LEED pattern, is somehow suppressed. One may speculate that significant flexural motion of the molecular backbone may be responsible, but at present this point remains obscure. It is clearly, however, of great potential importance when interpreting LEED data, and in particular the presence or absence of glide symmetry.[18]

4.2.5 Meso-Tartaric Acid on Cu{110}

Another dicarboxylic acid of considerable interest is tartaric acid, which differs from succinic acid by the substitution of a hydroxyl group in place of one hydrogen from each of the two central carbon atoms (*see Plate XII*). This substitution converts each of these carbon atoms into a chiral centre, leading to three possible isomers in all. When both chiral centres share the same chirality within the Cahn-Ingold-Prelog notation, the resulting molecule is itself chiral and may be denoted as *R,R* or *S,S*; these forms are also referred to as L-tartaric acid and D-tartaric acid, respectively. Experimental studies of the adsorption of these chiral forms will be outlined in Section 4.3.5. When the two chiral centres have opposite chirality, however, the resulting *R,S* molecule is actually achiral (or prochiral[19]) and known as *meso*-tartaric (or *m*-tartaric) acid.

Upon adsorption on Cu{110}, *m*-tartaric acid has been reported, on the basis of LEED experiments conducted by the group of Ernst [34, 35], to display three different ordered structures as a function of coverage. At a coverage estimated as around 0.1 ML, a (13×5) LEED pattern is observed, while at saturation coverage (estimated as around 0.25 ML) a c(8×2) pattern is reported. Neither of these provides any indication of chiral structure. At intermediate coverages, however, estimated as around 0.16 ML, the LEED pattern can only be interpreted as a superposition of the patterns emanating from domains of the intrinsically chiral $(1, \bar{2}; 6, 1)$ and $(1, 2; 6, \bar{1})$ lattices.[20]

Analysis of XPS data sheds light on the chemical nature of the adsorbed molecules, with peaks in the C 1s region of the spectra at 288.6 and 287.8–288.0 eV associated with carbon involved in intact acid and deprotonated carboxylate groups, respectively. On this basis, the saturated c(8×2) phase is identified as a monotartrate [35], in close analogy to the saturated monosuccinate c(4×2) phase described in the preceding section. The phase comprising chiral domains, on the other hand, is interpreted as deriving from bitartrate adsorbates, in just the same way that the analogous phase for succinic acid was identified with bisuccinate moieties. A model for the chiral domain structure was proposed [35] in which intramolecular hydrogen bonding between hydroxyl and carboxylate groups precludes the possibility of intermolecular hydrogen bonding; as we shall see, the same will not be true when chiral tartaric acid is considered, leading to rather different chiral phases, but that discussion we defer for the moment.

18 See, especially, the case of alanine adsorbed on Cu{110}, described in Section 4.3.1.

19 In this case, prochirality relates to the possibility of switching *off* the asymmetry of one chiral centre, leaving the other intact to confer chirality on the molecule as a whole.

20 Originally described as $(6, 1; \bar{1}, 2)$ and $(6, \bar{1}; \bar{1}, \bar{2})$.

4.3 Chiral Adsorbates: Act Locally, Think Globally

The common feature of all those systems described above is, of course, that the adsorbate molecules involved are achiral (or prochiral) when considered in isolation. They may give rise to chiral domains upon adsorption, but domains of opposite chirality must necessarily arise in equal proportion. The same is *not* true, however, when chiral adsorbates are involved, as we shall now discuss.

4.3.1 Alanine on Cu{110} and Cu{311}

One of the most intensively studied of all chiral molecular adsorbates is the amino acid alanine, and of all the substrates upon which its adsorption has been examined, Cu{110} has received by far the most attention. As in the case of the similar, but achiral, amino acid glycine, however, adsorption on Cu{311} provides an interesting point of comparison, and this much less-studied system will be described here also.

Given the strong structural similarities between alanine and glycine (*see Plate XI*) it is perhaps unsurprising that enantiopure alanine adsorbed on Cu{110} displays many of the traits exhibited by its achiral cousin; a classic study from the group of Raval makes this very clear [36, 37]. Once again, the presence of a RAIRS absorption peak attributed to the symmetric $—CO_2$ stretch mode (1411 cm^{-1}) clearly indicates deprotonation of the acid group upon room-temperature adsorption, while the absence of the $—NH_2$ scissor mode (c. 1576 cm^{-1}) shows that the amine group lies more or less parallel to the surface (see Fig. 4.14). Similarly absent, at least at low coverages, is any hint of the asymmetric $—CO_2$ stretch mode, confirming that the molecule must adopt a μ_3 binding configuration, analogous to that found previously in the case of glycinate; XPS data is also consistent with this conclusion. Predictably, a strong asymmetric $—CO_2$ stretch band emerges at 1626 cm^{-1} as the coverage is increased towards saturation, revealing the growing presence of μ_2 moieties as previously observed for glycinate. Whereas a weak (3×2) LEED pattern was noted in the case of glycinate, however, no discernible room-temperature LEED pattern was reported for alaninate at either low or high coverage, implying a lack of long-range order. STM images obtained from these phases also lack long-range order, but do reveal some local organisation in the saturated overlayer, suggestive of short chains (and double chains) of adsorbates lying roughly along either the (1, 1) surface direction, in the case of L-alaninate, or the $(1, \bar{1})$ surface direction, in the case of D-alaninate. Clearly, these chains do not respect the mirror symmetry of the substrate, but they are too short and randomly arranged to give rise to a well-defined chiral LEED pattern.

Upon annealing the overlayer at c.400–430 K, the authors of this study report the development of intrinsically chiral $(2, \bar{2}; 5, 3)$ and $(2, 2; 5, \bar{3})$ LEED patterns when using L-alanine or D-alanine, respectively.[21] Interestingly, however, the corresponding STM images reveal considerable variability within the overlayer network, albeit based upon just two basic motifs: hexamers and octomers, highly

21 Originally described as $(2, \bar{2}; 5, 3)$ and $(5, \bar{3}; 2, 2)$.

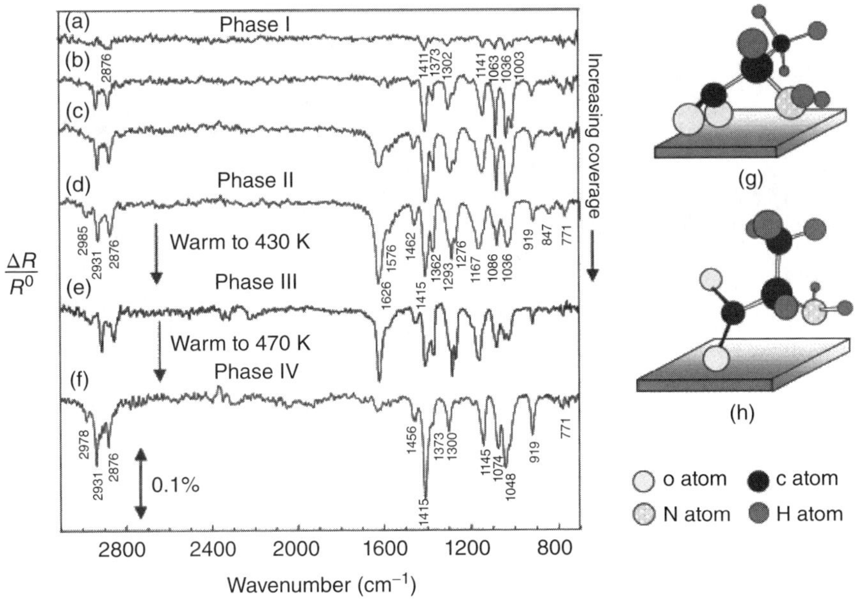

Figure 4.14 RAIR spectra obtained upon exposure of Cu{110} to increasing doses of L-alanine at 300 K (spectra (a)–(d)) and after annealing the saturated overlayer to 430 K (spectrum (e)) and 470 K (spectrum (f)). Schematics (g) and (h) depict μ_3 and μ_2 models for adsorption of L-alaninate. Reproduced with permission from Ref. [36]. Copyright (2005) Elsevier.

reminiscent of the double chains seen prior to annealing [36, 37]. Domains are observed consisting entirely of either hexamers or octomers (never both) and whilst they extend without evident limit in one dimension, their width is limited to the length of a single hexamer/octomer in the other (see Fig. 4.15). Hexamer and octomer domains are interspersed aperiodically on the surface, giving rise to a structure that cannot, strictly, be described by a single two-dimensional lattice. In effect, two different lattices (neither individually consistent with the observed LEED pattern) are separated by domain boundaries running along a common crystallographic direction. It is worth emphasising that when domains are sufficiently small, diffraction patterns cannot simply be expected to comprise superposed versions of the patterns that would pertain were the domains infinite in extent.

Further annealing their sample at 470 K, the authors report suppression of the $—CO_2$ asymmetric stretch mode, interpreted as being due to loss of the μ_2 species, simultaneous with transformation of the LEED pattern to one of (3×2) symmetry [36]. A subsequent study clarifies, however, that this phase is actually subject to a distortion taking the form of a subtle displacement of some LEED spots from their expected positions within a 'true' (3×2) pattern [38]; one plausible interpretation offered in that work is that the molecules predominantly adopt a (3×2) structure on the surface (analogous with that adopted by glycinate moieties at temperatures in excess of 420 K) but subject to some longer-range superstructure whose periodicity must be convoluted with the short-range

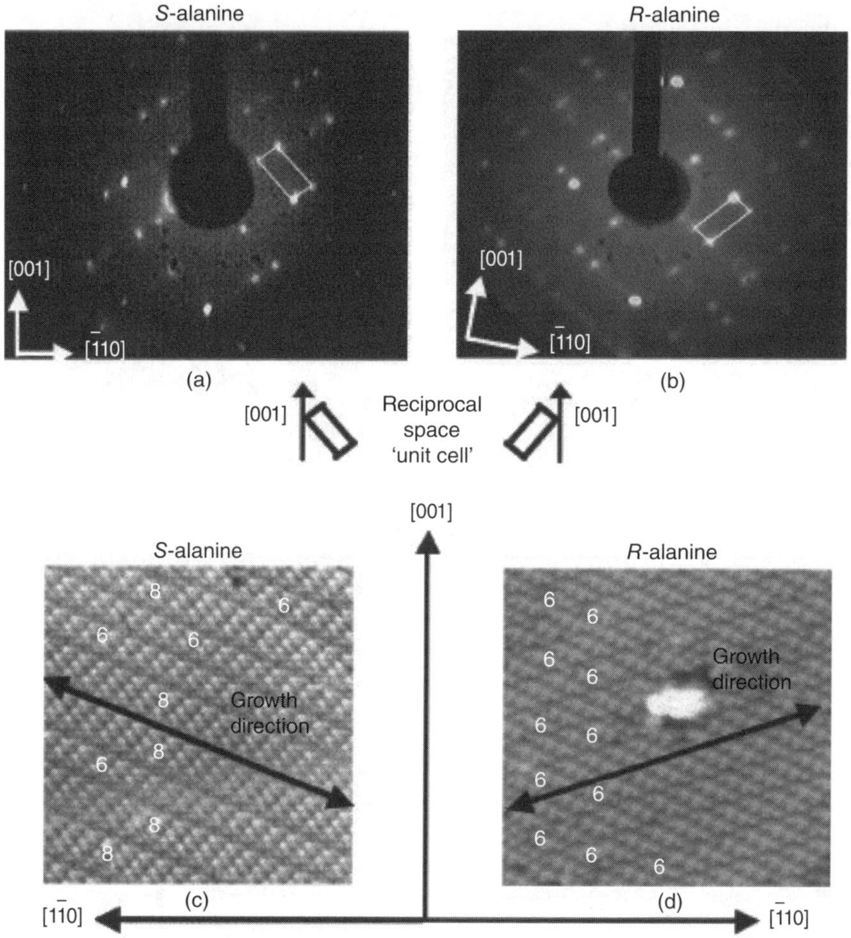

Figure 4.15 LEED patterns (52 eV) and STM images (165 × 160 Å^2) acquired upon annealing alaninate overlayers on Cu{110} at c. 400–430 K. Panels (a) and (c) show results obtained by dosing L-alanine, while panels (b) and (d) were obtained using D-alanine. Numerals 6 and 8 on the STM panels indicate rows comprising hexamers and octomers respectively. Reproduced with permission from Ref. [36]. Copyright (2005) Elsevier.

order to produce a 'complex' LEED pattern that approximates to a 'distorted' (3×2) pattern. Interestingly, although the deposition of racemic alanine yields very similar results in most regards, it has been found possible to obtain a 'true' (3×2) LEED pattern when both enantiomers are present on the surface at saturation coverage, although a 'distorted' pattern is still observed at slightly lower coverage [39]; this coverage dependence suggests that the presence of low-density domain boundaries may be responsible for the distortion, and such a model would appear to be consistent with recent hydrogen- and helium-scattering experiments reported by Seifert *et al.* [40, 41]. It should also be noted that the various (3×2)-esque LEED patterns all exhibit systematic absences (i.e. missing spots) ordinarily diagnostic of the existence of glide symmetry in the overlayer [36, 38, 39]. For the racemic case, this is uncontroversial, since an intimate mixture of L- and D-alaninate can readily adopt a glide-symmetric configuration. In the enantiopure cases, however, it is strictly impossible for a truly glide-symmetric structure to form, unless racemisation occurs upon annealing; an alternative scenario is that what appear to be systematic absences when referenced to a 'distorted' (3×2) LEED pattern may simply not form part of the 'complex' LEED pattern that the 'distorted' pattern merely approximates [38].

The first computational study of alaninate on Cu{110} was reported by Rankin and Sholl [42] whose DFT calculations (constrained to perfect (3×2) periodicity) strongly supported μ_3 models for both enantiopure and racemic cases, similar to that calculated previously by the same authors for glycinate [24]; data from single-enantiomer PhD experiments performed and analysed by Woodruff and co-workers [43] are consistent with the lowest energy enantiopure structure, subject to only minor modifications. Later, DFT calculations reported by Jones *et al.* [38] emphasised the myriad local energy minima accessible to the overlayer at room temperature (see Fig. 4.16) suggesting in particular that the orientation of methyl groups is likely to be both varied and dynamic; orbital energies from these calculations were used to aid in the interpretation of near-edge X-ray absorption fine structure (NEXAFS) data, which once again proved consistent with the μ_3 consensus in both outline and detail. Topological analysis of DFT-calculated electron density was subsequently used by Jones *et al.* [44] to argue that whereas compact glycinate overlayers are slightly stabilised by attractive substrate-mediated interactions over-and-above the influence of hydrogen bonding, the opposite is true for alaninate and hydrogen bonds must actively overcome substrate-mediated repulsion between neighbouring moieties.

Turning to adsorption of alanine on the Cu{311} surface, one again finds a system showing many similarities with adsorption on Cu{110} but with some telling differences, just as we saw above in the case of glycine adsorption. As before, the important structural distinction between Cu{110} and Cu{311} is that, while the former features left- and right-leaning triangular footprints for the μ_3 adsorption of amino acids, the latter features isosceles triangles (Fig. 4.9). In preliminary work, Clegg *et al.* [45] reported the formation of a (2, 1; 1, 2) LEED pattern upon deposition of enantiopure alanine at room temperature up to around 1/3 ML coverage; at higher coverages, the LEED pattern proved to be quite sensitive to the precise conditions both of preparation and of the experiment itself (i.e. electron flux, electron energy and surface temperature during exposure) and were not readily reconcilable with the STM images reported. It was speculated that regions of intrinsically chiral (2, 1; 4, 6) and extrinsically chiral (2, 1; 4, 7) order[22] might coexist in the case of D-alanine (or the mirror-image (1, 2; 6, 4) and (1, 2; 7, 4) phases in the case of L-alanine[23]) in variable proportion and with disordered domain boundaries. Subsequent detailed study has entirely borne out this interpretation, whilst providing additional spectroscopic insight [46] and considering also the adsorption of racemic alanine [29].

22 The (2, 1; 4, 7) phase actually has a rectangular unit cell.
23 Originally described as (6, 4; 1, 2) and (7, 4; 1, 2).

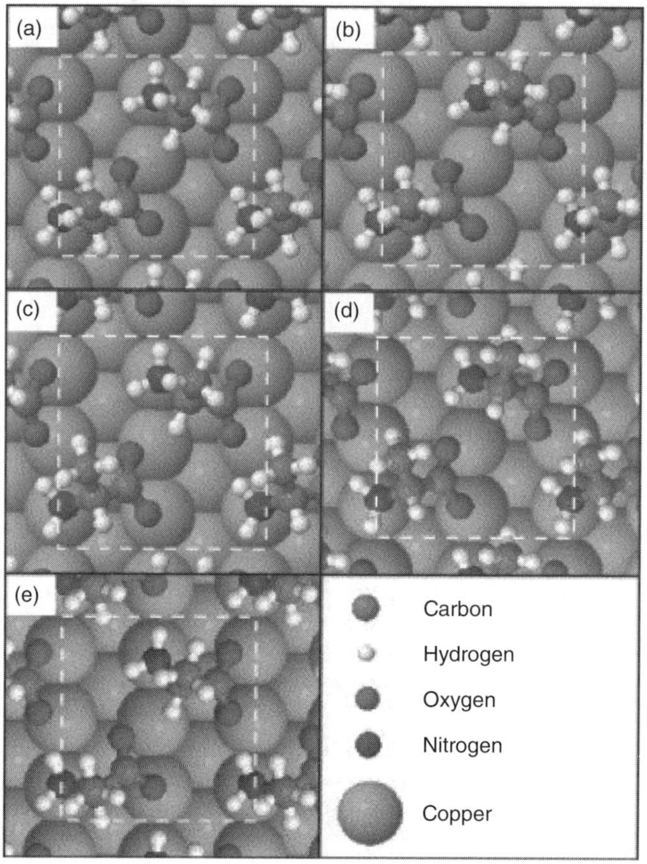

Figure 4.16 DFT-calculated local energy minima for alaninate on Cu{110}. Structures (c) and (d) are calculated to have the joint-highest adsorption energies, but structure (a) is only 0.02 eV less stable, with (b) and (e) no more than 0.05 eV less stable again. Reproduced with permission from Ref. [38]. Copyright (2006) Elsevier.

The sharp (2, 1; 1, 2) LEED pattern that arises upon adsorption of enantiopure alanine at room temperature [46] is associated with a strong RAIRS peak at 1408 cm^{-1} corresponding predominantly to the $—CO_2$ symmetric stretch mode. Absent from the spectra are any peaks identifiable as deriving from the $—CO_2$ asymmetric stretch, indicating that once again the lower coverage phase is dominated by adsorbates in the μ_3 configuration. In a sequence of experiments at higher and higher coverages, the RAIR spectra gradually evolve an additional strong absorption peak at 1618 cm^{-1}, which would typically be interpreted as deriving from the $—CO_2$ asymmetric stretch mode (see Fig. 4.17). In fact, however, DFT calculations [46] for an isolated adsorbed μ_2 alaninate moiety on this surface suggest that the $—CO_2$ group is somewhat distorted towards a geometry in which a C—O single bond points down towards the substrate and a —C=O double bond points away; a vibrational mode corresponding to the —C=O stretch and calculated at 1657 cm^{-1} lies plausibly close to the observed

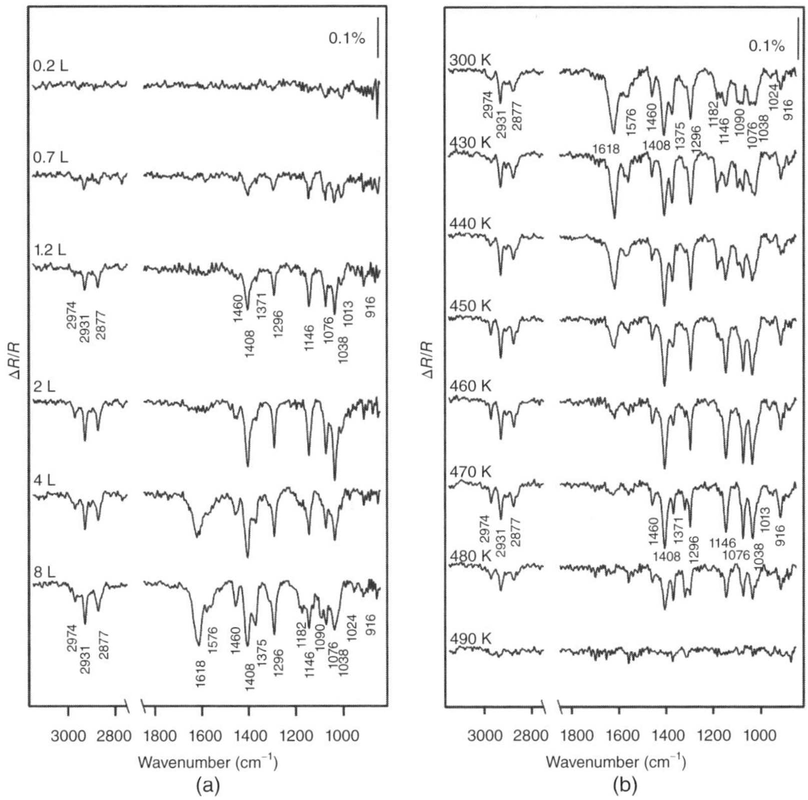

Figure 4.17 RAIR spectra obtained (a) upon 300 K exposure of Cu{311} to increasing doses of L-alanine, and (b) upon annealing the saturated overlayer at temperatures up to 490 K. Reproduced with permission from Ref. [46]. Copyright (2014) American Chemical Society.

mode at 1618 cm^{-1}. Whether one interprets this absorption peak as being due to a —C=O stretch or to the —CO_2 asymmetric stretch, it undoubtedly signals adsorption in a μ_2 configuration. Some distortion at the amine end of the molecule is also indicated, as an absorption peak emerging at 1576 cm^{-1} is interpreted as being due to the —NH_2 scissor mode; DFT calculations for the isolated μ_2 adsorbate place this at 1516 cm^{-1} [46]. These changes in the spectra with increasing coverage correlate with the emergence of streaks in the LEED pattern, consistent with a long-range disordered mixture of domains with (1, 2; 6, 4) and (1, 2; 7, 4) local structure (these particular experiments being performed with L-alanine). Upon sequentially annealing the highest coverage surface to temperatures in the range 430–460 K, these streaks are gradually lost, the pattern reverting to (2, 1; 1, 2), and the absorption peaks at 1576 and 1618 cm^{-1} disappear; annealing above 460 K, the LEED pattern gradually reverts to (1×1) and the RAIR spectra become featureless, indicating either desorption or dissociation of the adsorbates (or both).

STM images obtained whilst following similar sequences of deposition and annealing shed a great deal of light on the various transformations. At coverages

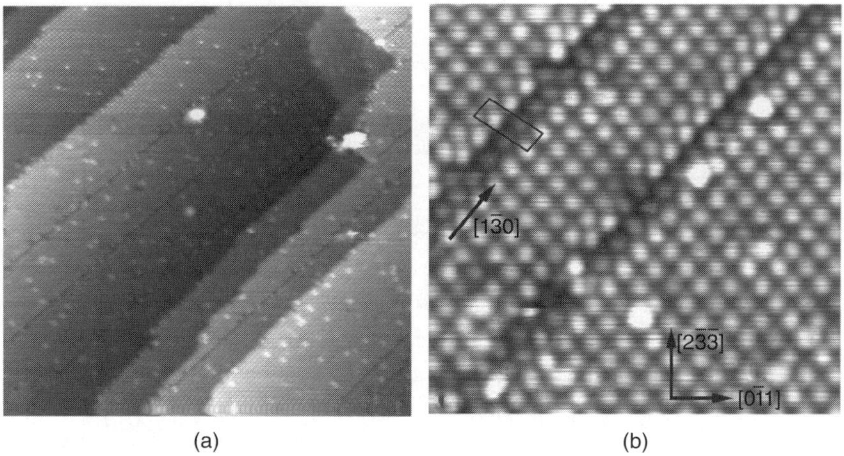

Figure 4.18 STM images of the (2, 1; 1, 2) phase observed on Cu{311} after exposure to L-alanine, at scan areas of (a) 500 × 500 Å^2, and (b) 100 × 100 Å^2. Reproduced with permission from Ref. [46]. Copyright (2014) American Chemical Society.

up to 1/3 ML, the dominant surface structure does indeed conform to the expected (2, 1; 1, 2) periodicity, with a single bright feature within each unit cell (Fig. 4.18). Occasionally, however, the long-range order is disrupted by domain boundaries, which in the case of L-alanine run exclusively in the ⟨1$\bar{3}$0⟩ direction and hence break the symmetry of the two-dimensional lattice; a disordered spacing between consecutive domain boundaries would be expected to result in precisely the form of streaking observed in the LEED experiments, and indeed the occurrence of these boundaries in the STM images does increase with increasing coverage. Careful examination of the domain boundaries reveals that they are, in fact, regions of slightly higher adsorbate density than is found in the domains that they separate; the structure of each boundary involves bright features interpreted as μ_3 species displaced somewhat from the regular (2, 1; 1, 2) positions, together with dim features interpreted as μ_2 species (*cf*. the zig–zag lines apparent upon adsorption of glycine, Fig. 4.11). As the coverage is increased into the regime where RAIRS indicates a significant μ_2 population, STM confirms that not only does the overall ratio of dim to bright features increase, but also the domain boundaries approach so close to one another that they start to occur in an ordered array. At a coverage estimated at 0.375 ML, STM reveals (1, 2; 6, 4) ordering with a μ_2:μ_3 ratio of 1:2 (see Fig. 4.19). Small regions of local (1, 2; 7, 4) structure, 0.4 ML local coverage, and 1:1 local μ_2:μ_3 ratio, are also observed, and these become more prevalent with increasing coverage (albeit they never saturate the surface).

In many respects, the adsorption of racemic alanine on Cu{311} follows a similar pattern to that described above for the enantiopure systems. At coverages below about 1/3 ML, a (2, 1; 1, 2) LEED pattern was again observed, while at higher coverages streaking occurred in two distinct azimuthal directions [29]. STM images in the high-coverage regime revealed domains separated by ⟨1$\bar{3}$0⟩-oriented boundaries, grouped into superdomains within which all the

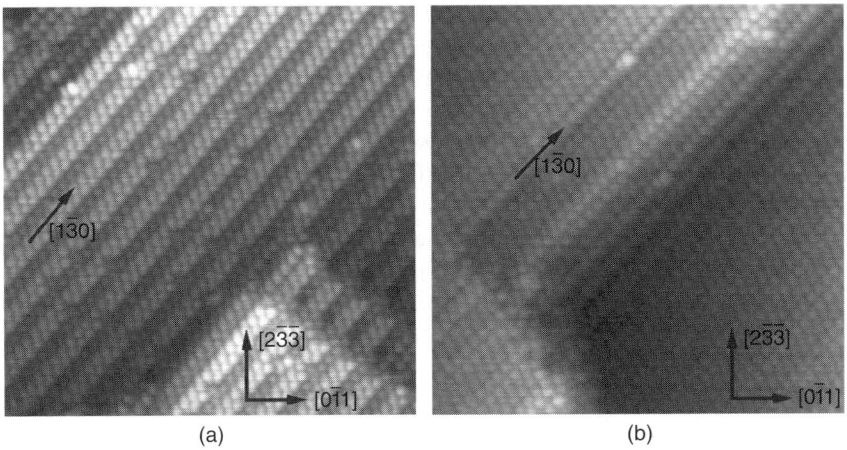

Figure 4.19 STM images of (a) the (1, 2; 6, 4) phase, and (b) the (1, 2; 7, 4) phase observed on Cu{311} after saturation exposure to L-alanine (scan areas are 200 × 200 Å^2). Reproduced with permission from Ref. [46]. Copyright (2014) American Chemical Society.

Figure 4.20 STM images (200 × 200 Å^2) obtained upon 300 K exposure of Cu{311} to (a) enantiopure L-alanine, (b) racemic alanine, and (c) glycine, followed by annealing in the range 410–450 K. Reproduced with permission from Ref. [29]. Copyright (2014) Elsevier.

boundaries are parallel but between which the boundaries change direction (Fig. 4.20). It is tempting to associate the superdomains with complete segregation, so that one will contain solely D-alaninate and the other solely L-alaninate, but this cannot yet be confirmed; it is alternatively possible, for example, that a small enantiomeric excess at the boundary of a superdomain could skew the structure of an otherwise racemic mixture of adsorbates within the body of the superdomain. In comparison with the boundaries noted upon the adsorption of glycine on Cu{311} these are much longer and relatively well ordered, but they are considerably shorter than the equivalent boundaries found upon adsorption of enantiopure alanine.

Amongst all this detailed work, there is one key point of similarity between the adsorption of alanine on Cu{110} and Cu{311} surfaces and one key point of difference. The main similarity is that on both surfaces adsorption at low coverages results in molecules with purely μ_3 configuration and no chirality in their long-range ordering, whilst at higher coverages the advent of μ_2

adsorbates coincides with evidence of supramolecular chirality. This similarity presumably arises because both surfaces preferentially bind the adsorbates into triangular footprints when possible, but resort to two-point binding when the surface becomes densely packed; evidently, intermolecular interactions between adsorbates in the three-point geometry is relatively insensitive to chirality compared to the case when some molecules are bound at only two points. The main difference, on the other hand, is that well-defined long-range order arises on the Cu{311} surface upon deposition at room temperature, whereas on the Cu{110} surface this requires annealing to at least 430 K; one may speculate that the relative ease of formation observed for long-range adsorbate networks on Cu{311} may relate to the absence of footprint chirality, which may impede the self-assembly of chiral molecules on Cu{110}.

4.3.2 Proline on Cu{110} and Cu{311}

Uniquely, amongst the naturally occurring amino acids, the only amine functionality of proline is incorporated into a ring[24] (*see Plate XI*). Immediately, this confers considerable stiffness upon the molecule, compared with acyclic amino acids, and one consequently expects an impact upon the ability of adsorbed proline to conform to the footprints available on any given surface. Moreover, the inclusion of the amine functionality within a ring implies that proline is, in fact, a secondary amine, and this in turn reduces the number of hydrogen bonds that it can donate from two to one. Accordingly, we should not be surprised if the expression of chirality upon the adsorption of proline on Cu{110} and Cu{311} were to be rather different from the behaviours observed for alanine (and indeed glycine). A good starting point for discussion, however, will be the points of similarity, most evident in vibrational spectroscopy.

Upon low-coverage 300 K adsorption of enantiopure proline, on either Cu{110} or Cu{311}, RAIRS experiments reveal the existence of an absorption band corresponding to the symmetric stretch mode of a $—CO_2$ group (1414 cm^{-1} on the former surface [47], 1410 cm^{-1} on the latter [48]) confirming deprotonation of the acid group (see Fig. 4.21). No feature is observed that could be attributed in either case to the asymmetric $—CO_2$ stretch, leading to the conclusion that the carboxylate group binds symmetrically to the surface, in close analogy to the μ_3 configuration adopted by both glycine and alanine at low coverages on both of these surfaces. Furthermore, the absence of any absorption feature around 1550 cm^{-1} argues against the existence of a secondary ammonium cation ($=NH_2^+$) suggesting that the molecule adsorbs in anionic (rather than zwitterionic) form [47, 48]. In summary, therefore, the adsorption geometry adopted by prolinate in this regime is as similar to the previously discussed acyclic amino acids as could reasonably be expected. A very significant difference emerges, however, upon increasing towards saturation coverage: on the Cu{311} surface, a strong absorption band appears at 1616 cm^{-1}, indicating the adoption of a μ_2 configuration similar to that seen for glycine and alanine at

24 Tryptophan and histidine both also feature amine-containing rings, but *additional* to a primary amine group attached to the α carbon; in proline, the α carbon itself forms part of the amine-containing ring.

high coverage [23], but on the Cu{110} surface such a band remains stubbornly almost entirely absent [47]. As for the orientation of the ring, spectra obtained on Cu{110} were originally interpreted as indicating a surface-parallel geometry [47], but subsequent DFT calculations [49] have shown that an equal mixture of surface-parallel and surface-perpendicular rings matches the experimental data very well; the most important observation, however, may simply be that the ring-related modes vary little between low and high coverage on this surface. In contrast, the ring-related modes of prolinate do vary with coverage on the Cu{311} surface [48], suggesting that the observed shift from a μ_3 to a μ_2 bonding configuration is accompanied by a corresponding re-orientation of the ring.

On Cu{110}, the adsorption of enantiopure proline at room temperature results initially in a weak (4×2) LEED pattern, which sharpens as the coverage is increased towards saturation [47], and STM observations demonstrate that the unit cell contains two adsorbed entities [49]. It is natural, given the RAIRS results discussed above, to interpret these as prolinate moieties in a μ_3 configuration, so that saturation coverage is determined to be 1/4 ML rather than the 1/3 ML that could be achieved by both glycine and alanine; presumably, the structural rigidity of the cyclic amino acid is responsible for the inability to pack more densely.

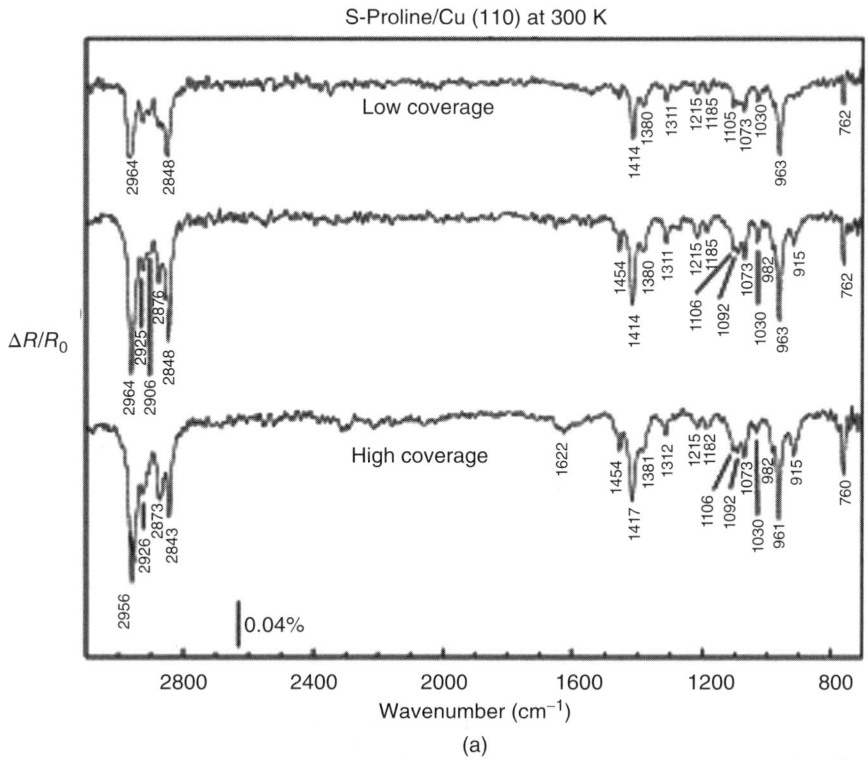

Figure 4.21 RAIR spectra obtained upon 300 K exposure of (a) Cu{110}, and (b) Cu{311}, to increasing doses of L-proline. Reproduced with permission from Ref. [47] (Copyright (2002) Elsevier) and under Creative Commons Attribution 4.0 from Ref. [48].

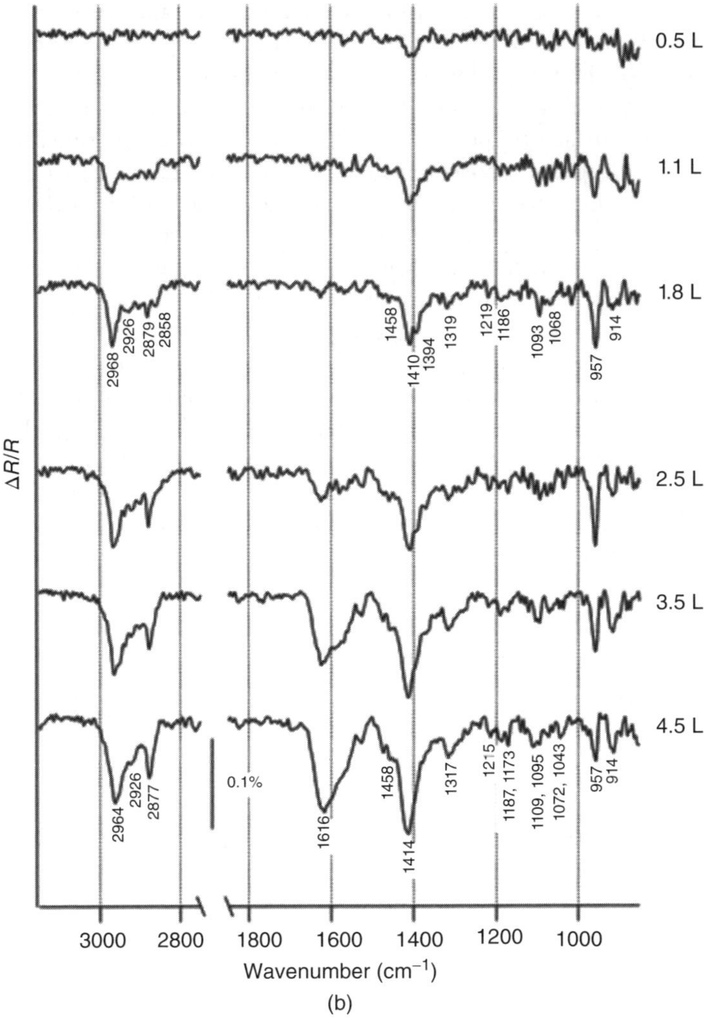

Figure 4.21 (*Continued*)

Note that this restriction to a relatively low-coverage regime prevents the formation of a long-range chiral network upon adsorption of prolinate. Interestingly, the STM images (see Fig. 4.22) reveal a strict alternation between bright and dim features, which DFT simulations suggest correspond to prolinate adsorbed with rings oriented perpendicular and parallel to the surface, respectively [49]. When racemic proline is substituted, however, the overall disposition of the adsorbed molecules remains essentially identical to the enantiopure case but the strict alternation is replaced with a random alternation between perpendicular and parallel ring orientations [50]. In other words, although the supramolecular ordering of adsorbed prolinate on Cu{110} never yields a chiral LEED pattern, it does nevertheless distinguish between enantiopure and racemic cases through the orientational disorder evident in STM experiments.

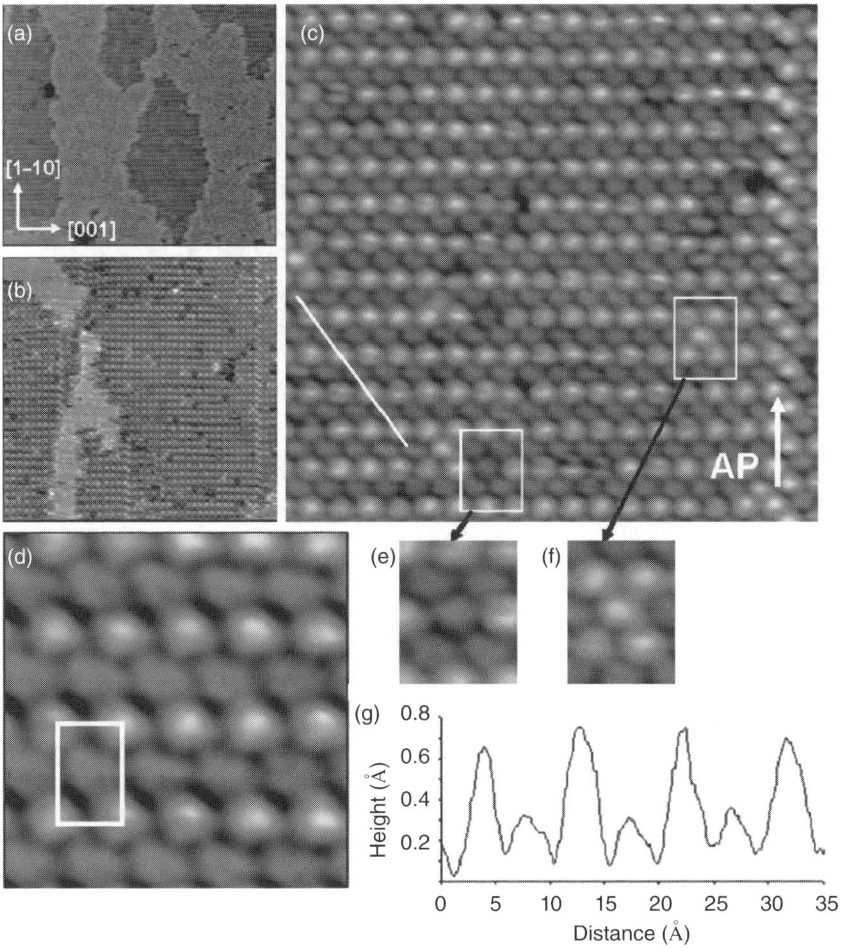

Figure 4.22 STM images of the (4×2) phase obtained upon 300 K exposure of Cu{110} to proline. Panels (a) and (b) show island size increasing with coverage, and formation of antiphase boundaries, over a 451 × 420 Å^2 scan area. In (c) and (d) the scan areas are 132 × 128 Å^2 and 37 × 39 Å^2. Panels (e) and (f) provide close-ups of defects, while (g) is a line scan crossing features of two distinct brightnesses. Reproduced with permission from Ref. [49]. Copyright (2009) American Chemical Society.

In contrast, the picture on Cu{311} is a *little* more familiar. Room-temperature adsorption of proline at coverages up to 1/3 ML yields a LEED pattern reminiscent of the (2, 1; 1, 2) phase found upon the adsorption of both glycine and alanine, but with a clear asymmetric splitting of the diffraction spots indicative of some chiral distortion to the two-dimensional lattice; corresponding STM images reveal a largely (2, 1; 1, 2) overlayer modulated subtly by imperfections in the order that betray an asymmetric bias [48]. As the coverage is increased towards saturation, the LEED data resolves ever more clearly into a true (2, 1; 1, 2) pattern, albeit with some residual asymmetric elongation of the spots; this perturbation is revealed by STM to be due to occasional boundaries similar to those found at

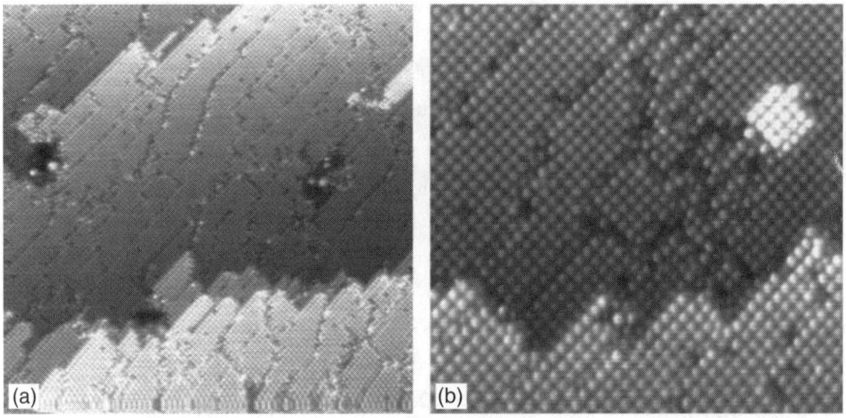

Figure 4.23 STM images of the (2, 1; 1, 2) phase obtained after saturation exposure of Cu{311} to proline at 300 K, followed by annealing at 350 K. Panels (a) and (b) show 500×500 Å^2 and 100×100 Å^2 scan areas, respectively. Reproduced under Creative Commons Attribution 4.0 from Ref. [48].

high coverage upon the adsorption of either glycine or alanine, but lacking the regularity seen in those cases (see Fig. 4.23). Unexpectedly, therefore, the chirality observed in LEED experiments on prolinate overlayers is most evident at low coverages, when only the μ_3 species is present, and rather less evident (though never absent) at high coverages, when both μ_3 and μ_2 species are present, in direct opposition to the behaviour of the acyclic amino acids.

Further interesting observations have been reported upon annealing the prolinate overlayers described above. On the Cu{110} surface, heating above 450 K was found to result in the evolution of hydrogen, surmised to be associated with aromatisation of the ring, whereas complete decomposition was observed only at around 535 K [47]. More detail of a similar process has been obtained on Cu{311}, where hydrogen is again evolved at temperatures in excess of 450 K. In this case, RAIRS reveals a gradual weakening of the 1616 cm^{-1} absorption band at successively higher annealing temperatures in the range 300–420 K, but this trend is reversed for annealing temperatures in the range 420–450 K; this key signature of μ_2 adsorption only decisively vanishes upon annealing above 470 K, by which stage most of the other absorption bands have likewise disappeared [48]. The absorption band at 1410 cm^{-1}, by way of contrast, decreases only a little in intensity over the entire range 300–480 K, disappearing only upon annealing up to 500 K; in the range 420–460 K this feature becomes noticeably more broad, but it recovers its original sharpness once all other absorption bands have vanished above 470 K. The overlayer structures obtained after annealing at these temperatures are rather complex (see Fig. 4.24) but have been tentatively interpreted as consistent with a dimerised arrangement of pyrrole-2-carboxylate moieties (*see Plate XVIII*) formed by the dehydrogenative aromatisation of the prolinate species [48]. Although pyrrole-2-carboxylate is prochiral in isolation, adsorption with its ring parallel to the surface results in a chiral complex, and this remains true even after the formation of dimers. The situation is, therefore, most

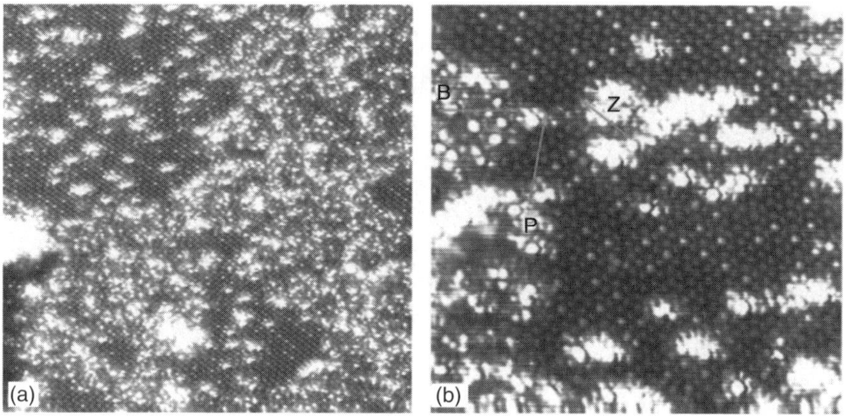

Figure 4.24 STM images of the (5, 3; 4, 6) and (3, 5; 6, 4) phases obtained after saturation exposure of Cu{311} to proline at 300 K, followed by annealing at 490 K. Panels (a) and (b) show 500 × 500 Å^2 and 100 × 100 Å^2 scan areas respectively. Reproduced under Creative Commons Attribution 4.0 from Ref. [48].

similar to that of guanine, described in Section 4.2.1, and results in the formation of (presumably homochiral) domains of (5, 3; 4, 6) and (3, 5; 6, 4) periodicity[25] in equal proportion [48].

4.3.3 Serine and Lysine on Cu{110}

The amino acids described above (glycine, alanine and proline) display a common propensity to bind to the Cu{110} and Cu{311} surfaces in either three-point (μ_3) or two-point (μ_2) fashion – variations between specific systems have proved to be explicable in terms of chirality (or its absence) and differing degrees of molecular flexibility. A very interesting line of enquiry might, therefore, be to investigate how the presence of additional functional groups may alter the behaviour more drastically, and this question has spawned a number of studies on the amino acids serine and lysine. The structure of serine is very similar to that of alanine, except that one hydrogen atom from the methyl group is replaced by a hydroxyl group, while in lysine the substituent is the propylammonium cation (*see Plate XI*). In both cases, it may be expected that adsorbates will bind to the surface and/or each other not only by the means already considered in the simpler amino acids, but also potentially through the additional hydroxyl or propylammonium groups.

The most comprehensive study of serine adsorption on Cu{110} was reported in joint work from the groups of Held and Linderoth, whose room-temperature LEED experiments on the enantiopure species at low coverage (up to an estimated 0.25 ML) revealed an intrinsically chiral (1, 2; 4, 0) pattern for L-serine and a corresponding (1, $\bar{2}$; 4, 0) pattern for D-serine[26] [51] in agreement with

25 Originally described as (4, 6; 5, 3) and (6, 4; 3, 5).

26 Originally described as ($\bar{1}$, $\bar{2}$; 4, 0) and ($\bar{1}$, 2; 4, 0).

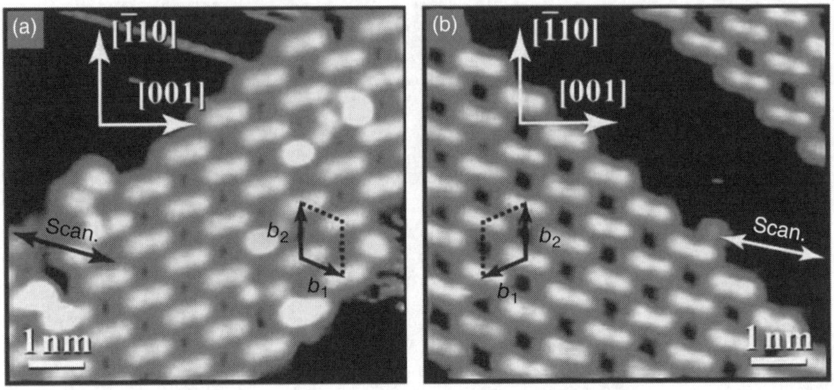

Figure 4.25 STM images obtained after 300 K exposure of Cu{110} to (a) L-serine, and (b) D-serine, followed by annealing at 370 K. Reproduced with permission from Ref. [51]. Copyright (2010) American Chemical Society.

earlier work reported by Barlow and Raval [52]. At higher coverages, however, no LEED pattern was observed at all until after annealing at 420 K; the patterns thus obtained were more intense than those found at lower coverage, but were of identical symmetry. Further experiments on racemic serine yielded a superimposition of the chiral LEED patterns, and were interpreted as indicating segregation into homochiral domains [51]. Regarding the chemical nature of the adsorbates, XPS experiments confirmed (via a peak at binding energy 399.4 eV in the N 1s region of the spectra) that the amine group exists in the $—NH_2$ form, while the acid group was shown (via a peak at binding energy 531.4 eV in the O 1s region) to be deprotonated; the additional hydroxyl group, however, remains intact [51].

Images of these serinate overlayers have been obtained via STM (see Fig. 4.25), which reveal the existence of rather elongated features, aligned at about 73° from the close-packed surface direction, and line scans along these features suggest that they are actually dimers [51]. Furthermore, O 1s NEXAFS data support a model in which each adsorbate adopts a μ_3 binding geometry and forms two hydrogen bonds with a single neighbour [51]; one molecule in each pair acts as a double donor (through its amine and hydroxyl groups) and the other acts as a double acceptor (through the two oxygen atoms of its carboxylate group). In the cases of enantiopure serine, the dimers are all rotated from the close-packed direction in the same sense (clockwise for L-serine, anticlockwise for D-serine) allowing them to pack neatly into an extended network with a single well-defined chirality. When racemic serine is used, however, dimers exhibiting clockwise orientation cannot pack neatly with those exhibiting anticlockwise rotation, leading to spontaneous resolution into the homochiral domains inferred from the LEED patterns.

Turning to lysine, room-temperature RAIRS experiments from the group of Pradier (see Fig. 4.26) revealed an absorption band at 1416 cm^{-1} interpreted as being due to the symmetric stretch mode of $—CO_2$, hence indicating

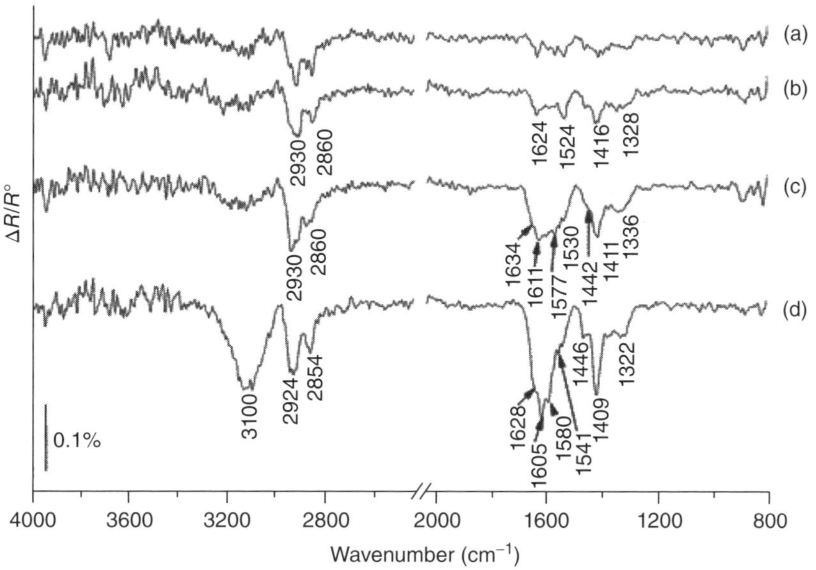

Figure 4.26 RAIR spectra obtained upon 300 K exposure of Cu{110} to increasing doses of L-lysine. Reproduced with permission from Ref. [53]. Copyright (2006) American Chemical Society.

deprotonation of the acid group [53]. Additional bands at 1524 cm^{-1} and 1624 cm^{-1} observed at low coverage were attributed, respectively, to the symmetric and antisymmetric deformation modes of a primary ammonium cation (i.e. $—NH_3^+$). Note that there is some ambiguity concerning the asserted absence of a low-coverage $—CO_2$ asymmetric stretch mode, since results for other amino acids (see above) suggest that its frequency would lie in the range 1610–1630 cm^{-1}, making it hard to distinguish from the ammonium band that the authors assert they observe; perhaps the strongest inherent argument in favour of the proffered interpretation is that the asymmetric stretch would be expected to give rise to a very strong absorption peak, whereas the data shows only rather weak absorption. Much clearer evidence, however, is provided by XPS experiments carried out by Held and co-workers [54] that demonstrate clearly the existence of two distinct chemical environments for the nitrogen atoms of the adsorbate; a peak in the N 1s region of the spectrum at a binding energy of 400 eV is attributed to nitrogen in an amine environment ($—NH_2$) whilst another at 402 eV is consistent with the primary ammonium cation ($—NH_3^+$). A single binding energy of 532 eV is observed in the O 1s region, confirming that the acid group is deprotonated and that both oxygen atoms experience identical chemical environments. At room temperature, therefore, adsorbed lysinate most likely binds to the surface in μ_3 fashion at low coverage, in close analogy with other amino acids, while the ammonium-terminated alkyl chain dangles without attachment. At higher coverages, the emergence of a strong infrared absorption band at 1611 cm^{-1} [53] indicates the likely adoption of a μ_2 binding configuration[27] with no obvious change in orientation of the side chain.

In the low-coverage regime (estimated 1/6 ML) STM images from Pradier's group reveal the growth of domains both of $(1, 2; 5, \bar{2})$ and $(1, \bar{2}; 5, 2)$ periodicities (see Fig. 4.27) [55]. Despite the use of enantiopure lysine, two opposing long-range chiralities are thus apparent in roughly equal proportion, suggesting only relatively weak interactions between neighbouring molecules. At higher coverage, however, a more conventionally chiral overlayer emerges: after prolonged exposure to L-lysine, Humblot *et al.* [55] report an extrinsically chiral $(\bar{3}, 3; 5, \bar{1})$ LEED pattern,[28] with two elongated features per unit cell identifiable in STM images; in similar fashion to the case of serine described above, these features are interpreted as dimers (see Fig. 4.28) giving a nominal coverage of 1/3 ML.

Upon heating the high-coverage phase, Humblot *et al.* [55] report an interesting reversion to achiral long-range order. Annealing to 425 K results in a (6×4) LEED pattern, while further annealing at 450 K generates a (3×2) LEED pattern. Whilst the latter is reminiscent of the low-coverage phases reported for glycine and alanine, STM images from the same authors show that each unit cell here contains only a single bright feature, suggesting a coverage of just 1/6 ML; in the lower temperature structure, each unit cell contains four bright features, suggesting precisely the same coverage, but grouped into two dimers (see Fig. 4.29). Structural models suggested for these phases by Humblot *et al.* [55] are consistent with both STM and LEED, but deviate from the familiar μ_3 binding arrangement by a 90° rotation of each molecule about the surface normal and an assumption that all nitrogen atoms exist within ammonium cations ($—NH_3^+$). Subsequent O 1s NEXAFS data from Held's group, however, provides strong evidence for the more usual orientation. Moreover, their XPS results show clearly that all nitrogen atoms exist within amine groups ($—NH_2$) and are most likely bound to the surface; based on these observations, they conclude that a μ_4 binding geometry is adopted under these conditions [54].

4.3.4 Cysteine on Cu{110} and Au{110}

Progressing still further (in chemical space) away from the simplest amino acids, we now turn to cysteine, where the methyl group of alanine is substituted not by either a hydroxyl group or a propylammonium cation (as in serine and lysine) but rather by a thiol group (*see Plate XI*). The well-known affinity of sulfur compounds for adsorption on gold has understandably induced many researchers to turn away from the familiar copper surfaces extensively studied in relation to the other amino acids, and efforts with cysteine have focused as much on Au{110} surfaces as on Cu{110}. We shall discuss both here, as the copper surface provides a nice point of contact with the systems described above. The clean Au{110}

27 Pradier and co-workers argue for a μ_1 binding configuration, on the assumption that *both* nitrogen atoms exist in primary ammonium cations. Evidence for this from RAIRS is weak, and unfortunately the XPS data from Held's group does not extend to this coverage regime.

28 Originally described as $(\bar{2}, \bar{2}; 5, \bar{1})$, which fails to emphasise the rhombic symmetry of the overlayer lattice.

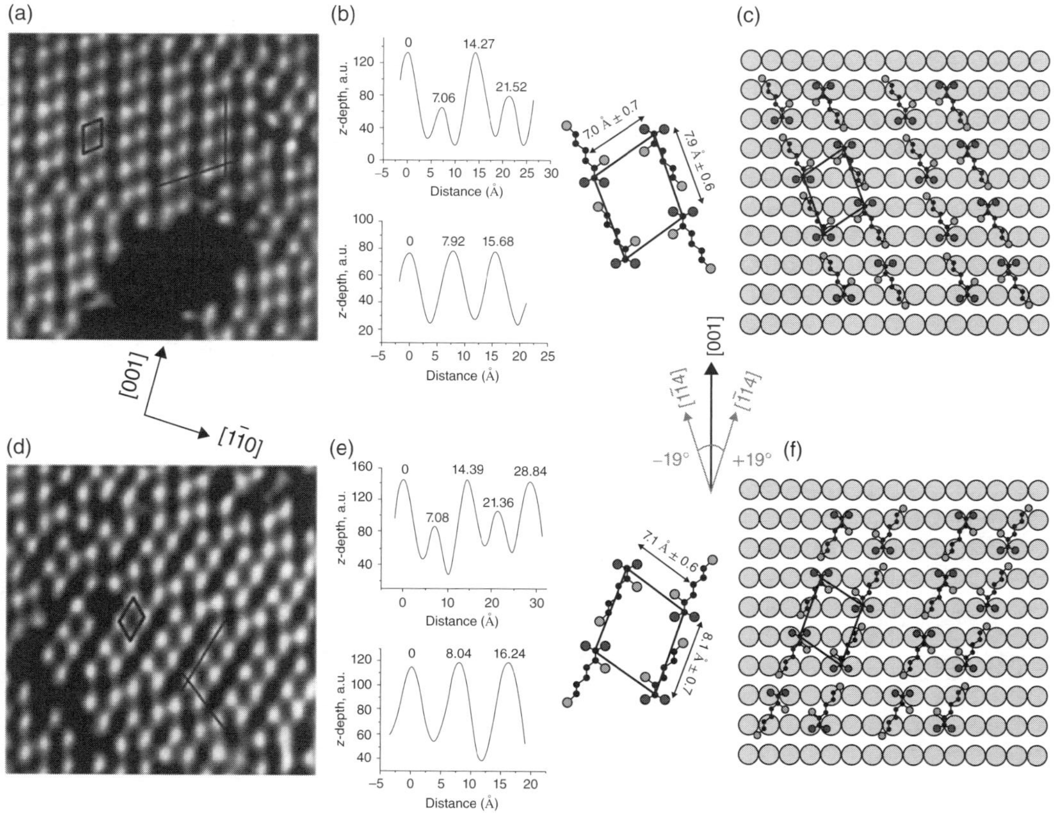

Figure 4.27 Low-coverage phase obtained after 300 K exposure of Cu{110} to L-lysine. Panels (a), (b) and (c) show, respectively, a 110 × 110 Å^2 STM image, selected STM line scans and a proposed model for the (1, $\bar{2}$; 5, 2) regions; panels (d)–(f) the same for the (1, 2; 5, $\bar{2}$) regions. Reproduced with permission from Ref. [55]. Copyright (2007) Elsevier.

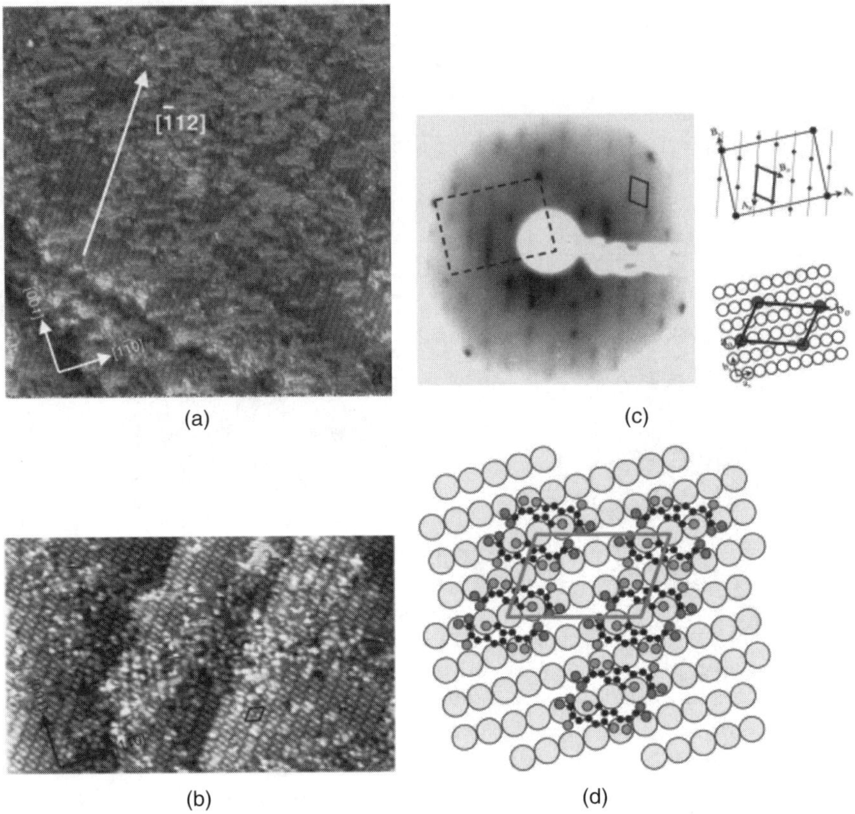

Figure 4.28 High-coverage phase obtained after 300 K exposure of Cu{110} to L-lysine. Panels (a) and (b) show 800 × 800 Å^2 and 400 × 250 Å^2 STM images, and panel (c) the corresponding LEED pattern, of the ($\bar{3}$, 3; 5, $\bar{1}$) structure. Panel (d) shows the proposed model for this phase. Note that the unit cells highlighted do not emphasise the full rhombic symmetry of the overlayer lattice. Reproduced with permission from Ref. [55]. Copyright (2007) Elsevier.

surface, it should be noted, adopts a (1×2) missing-row reconstruction, adding a structural, as well as a chemical, aspect to its difference from Cu{110}.

On the Cu{110} surface, Pradier's group reported the results of RAIRS experiments conducted after room-temperature deposition of cysteine [56]. At saturation, the absence of any absorption peak around 2550 cm^{-1}, attributable to the –SH stretch mode, was taken to indicate partial dissociation of cysteine to its thiolate form. The further absence of any peak in the 1700–1750 cm^{-1} range that might correspond to a —C=O stretch provided evidence of deprotonation of the acid group, while a strong absorption peak at 1652 cm^{-1} was interpreted as being due to the asymmetric deformation mode of a primary ammonium cation (—NH_3^+) and a weaker peak at 1510 cm^{-1} was identified with the symmetric deformation mode of the same group. An absorption peak at 1396 cm^{-1} was furthermore attributed to the symmetric —CO_2 stretch mode, but the absence of any corresponding feature due to the asymmetric —CO_2 stretch, expected in the range 1610–1630 cm^{-1}, implied that both oxygen atoms are at the same

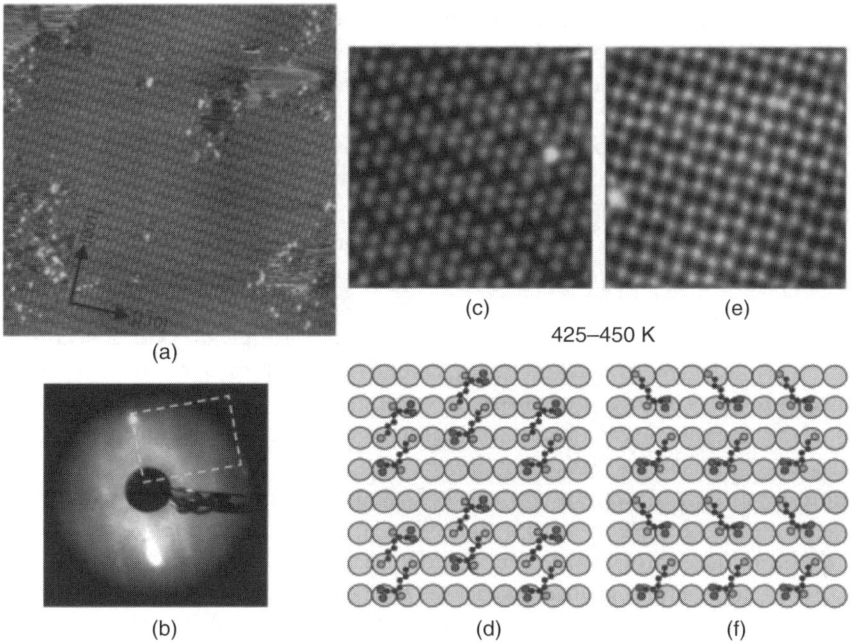

Figure 4.29 High-coverage phases obtained by heat treatment after 300 K exposure of Cu{110} to L-lysine. Panel (a) shows a 500 × 500 Å^2 STM image of the (6×4) phase obtained following annealing at 425 K, while panel (b) shows the corresponding LEED pattern. Panels (c) and (e) respectively show 100 × 100 Å^2 STM images of the (6×4) and (3×2) phases obtained following annealing at 425 and 450 K, while panels (d) and (f) show corresponding structural models for these phases. Reproduced with permission from Ref. [55]. Copyright (2007) Elsevier.

distance from the surface. The adsorbed molecules thus appear to be zwitterionic, as regards their amino acid functionality, with the carboxylate group tilted somewhat from the horizontal plane but not twisted. A model has been proposed in which the molecule adopts three-point binding (i.e. a μ_3 configuration) but with the thiolate attaching to the surface in place of the amine group [56].

The coverage dependence of cysteine adsorption on Cu{110} has been addressed by means of XPS experiments reported by Kim *et al.* [57, 58]. After room-temperature deposition, the S 2p region of the spectra obtained at low coverages (up to around a quarter of saturation coverage) provides evidence for only a single thiolate species ($2p_{3/2}$ binding energy 161.31 eV).[29] At intermediate coverages (between around one quarter and three quarters of saturation coverage) this species is joined by a second distinct thiolate species ($2p_{3/2}$ binding energy 162.33 eV). Finally, at coverages approaching saturation, the thiolate species are joined by a third species whose binding energy is consistent with a non-dissociated thiol ($2p_{3/2}$ binding energy 164.34 eV). The lowest coverage phase is associated with a weak (2×1) LEED pattern, while a similarly faint c(2×2) pattern emerges in the intermediate regime; all fractional order spots are lost as

29 Each doublet consists of a $2p_{3/2}$ peak at the energy quoted, accompanied by a $2p_{1/2}$ peak of half the intensity shifted to higher binding energy by approximately 1.2 eV.

the coverage approaches saturation, implying increasingly disordered adsorption. No hint of chiral ordering is, therefore, evident at any coverage. Analysis of the N 1s region of the XPS spectra concurs with the earlier RAIRS results that the amine group is protonated to form a primary ammonium cation (binding energy 401.7 eV) at saturation coverage, but below about half of saturation coverage there is evidence only of the amine group (binding energy 399.6 eV). A single binding energy (531.5 eV) is recorded in the O 1s region across all coverages up to saturation, indicating ubiquitous deprotonation of the acid group and equivalent local environments for the oxygen atoms. In light of these findings, the authors proposed a four-point binding model (i.e. a μ_4 configuration) at low coverage, involving the thiolate group, the carboxylate group and the amine group, which then converts to the μ_3 configuration (previously identified by RAIRS) at higher coverage, as the amine group is protonated and detaches from the surface; the intact thiol found at saturation coverage is probably associated with incipient multilayer formation [58].

Upon deposition of cysteine on Au{110} at 120 K, followed by annealing to 270 K, Kühnle *et al.* [59] reported the formation of highly monodisperse clusters, each estimated to comprise eight individual molecules. Their STM images revealed a clearly chiral internal structure to these clusters, with adsorption of L- or D-cysteine leading to mirror-image results (see Fig. 4.30). The clusters were found to diffuse readily along the close-packed surface direction when nudged by the STM tip at 260 K (although not at 120 K) and the lack of disruption to

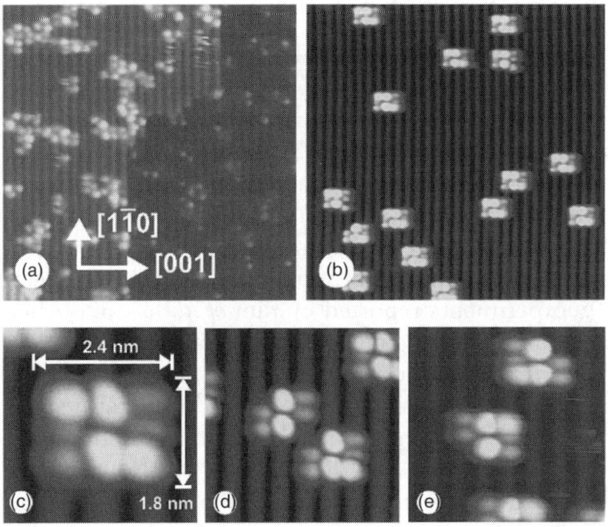

Figure 4.30 STM images obtained after 120 K exposure of Au{110} to cysteine. Panel (a) shows a 245 × 245 Å^2 image of irregular clusters formed upon initial adsorption of L-cysteine without annealing; panel (b) the same after annealing at 270 K. Panel (c) shows, in a 65 × 65 Å^2 image, the structure of a complete eight-molecule cluster deriving from the L enantiomer. Panels (d) and (e) show, at the same resolution, imperfect clusters deriving from the L and D enantiomers, respectively. Reproduced with permission from Ref. [59]. Copyright (2003) American Chemical Society.

the underlying gold surface after doing so was taken to indicate adsorption of cysteine in its thiol form. The precise nature of binding to the surface was, however, not determined, and neither was the internal structure of the clusters. Much more effort has, however, been devoted to higher-temperature phases.

In other studies by Kühnle *et al.* [60], a low coverage of cysteine was deposited on Au{110} at room temperature, then annealed at 340–380 K and imaged by STM. The surface was found to feature double-peaked structures interpreted as cysteine dimers, bound strongly to gold by thiolate groups. When using L-cysteine, the axis of each dimer was found to be rotated by 20° clockwise relative to the close-packed surface direction, whereas using D-cysteine the dimers were rotated by the same angle in the anticlockwise sense; use of racemic cysteine, on the other hand, resulted in a mixture of both clockwise and anticlockwise features, attributed to LL dimers and DD dimers respectively, but no new features were observed that might have been attributed to dimers of mixed chirality (see Fig. 4.31). Based upon STM images and DFT simulations, a model for the dimers was proposed, in which the acid groups of neutral thiolates face one another, permitting the formation of two OH–O hydrogen bonds per dimer; the substrate immediately beneath each dimer is reconstructed by the removal of four gold atoms, which relocate into islands elsewhere on the surface [60].

We note here in passing that, some years later, an intriguing follow-up study from the same group made use of a modified experimental protocol whereby racemic cysteine was initially deposited at coverages beyond a monolayer, but annealed at 420 K to achieve a low final coverage by means of both desorption and decomposition [61]. In this work, substantial reconstruction of the substrate was observed, with the formation of large rectangular terraces, and regions of c(2×4) periodicity attributed to residual adsorbed sulfur atoms lingering after cysteine dissociation. Curiously, however, the remaining cysteine dimers exhibited a distinctly asymmetric distribution, with 92% identifiable as the LL form and only 8% as the DD form. The authors described extensive control experiments aimed at clarifying the reason for this puzzling asymmetry, but ultimately this remains a mystery [61]. Clearly, there must be *some* unidentified source of asymmetry inherent in the experimental conditions, despite the best efforts of a group of highly experienced researchers, so this result must stand as a salutary reminder of the difficulties in unambiguously pinpointing true chiral effects.

Returning to our discussion of results obtained via the 340–380 K annealing of sub-monolayer cysteine-covered surfaces, we turn to the observation, again by Kühnle *et al.* [63], that cysteine dimers can be found not only on the pristine terraces of the Au{110} surface but also at kink sites where the terraces terminate. Since the surface is inherently achiral, left- and right-handed versions of these kink sites must exist in equal proportion, but individual sites are chiral and can be labelled S or R according to the microfacet convention described in Sections 1.4.2 and 2.4.1. For LL dimers adsorbed at an S kink (or for DD dimers adsorbed at an R kink) the dimer axis was found to be rotated by 30° from the close-packed surface direction (i.e. 10° further than on a pristine terrace) while for DD dimers adsorbed at an S kink (or for LL dimers adsorbed at an R kink) the rotation angle is 10° (i.e. 10° less than on the pristine terrace). Furthermore, small homochiral islands of densely packed dimers were observed to nucleate at these kink sites,

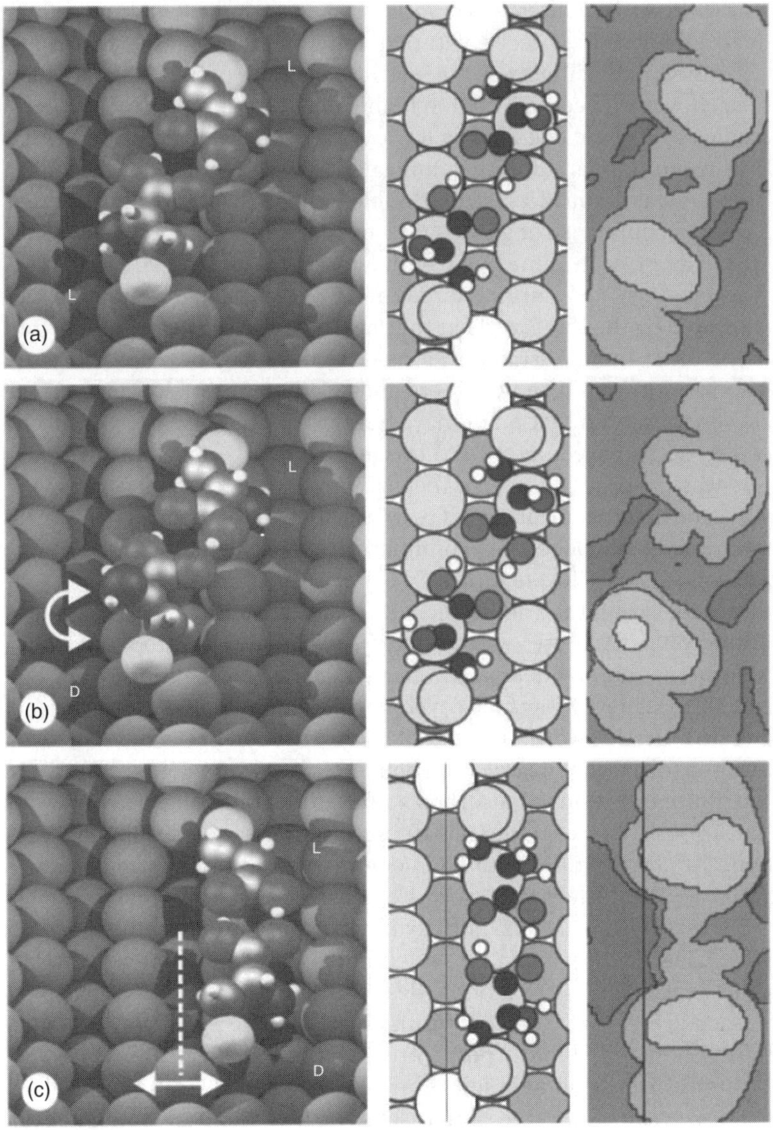

Figure 4.31 Models and simulated STM images for cysteine dimers on Au{110}, assuming molecules in their thiolate form bound at a cluster of four gold vacancies. Panel (a) shows the LL dimer structure, observed upon adsorption of either enantiopure L-cysteine or racemic cysteine; the mirror-image DD dimer (not shown) is observed upon adsorption of D-cysteine or racemic cysteine. Panels (b) and (c) depict hypothetical mixed dimer structures, but these are not observed upon adsorption of racemic cysteine. Reproduced with permission from Ref. [60]. Copyright (2002) Nature Publishing Group.

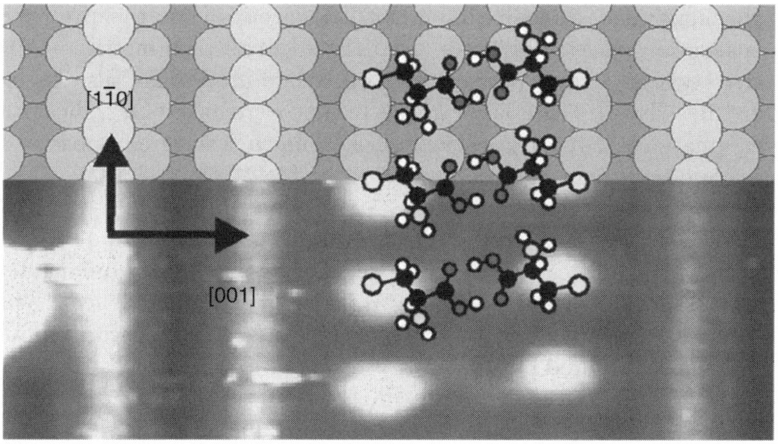

Figure 4.32 DFT-calculated structure and STM image for linear features observed upon room-temperature exposure of Au{110} to D-cysteine, followed by annealing at 380 K. Reproduced with permission from Ref. [62]. Copyright (2004) American Physical Society.

with islands prefentially growing from S kinks upon deposition of D-cysteine, or from R kinks upon the deposition of L-cysteine [63].

In addition to the isolated dimers and small homochiral kink-nucleated islands, extended linear structures were also observed and interpreted as comprising a chain of dimers stacked along the close-packed direction of the surface (see Fig. 4.32) [62]. Subsequent XPS studies reported by Gonella *et al.* [64], however, whilst confirming the ubiquity of thiolate groups (S 2p binding energy 161.95 eV) indicated that the low-coverage regime (presumably dominated by isolated dimers) consisted mainly of neutral molecules while the higher coverage sub-monolayer regime (presumably dominated by the linear features) comprised mainly zwitterionic species. It seems, therefore, that the precise nature of the linear features remains obscure.

4.3.5 Tartaric Acid on Cu{110}

It will be recalled that adsorption of *meso*-tartaric acid on Cu{110} has been discussed above (Section 4.2.5) as an example of a prochiral molecule with two acid groups (see also the discussion of succinic and fumaric acids in Section 4.2.4). Now we return to consider the adsorption of the chiral forms of the molecule, namely D- and L-tartaric acid (alternatively *S,S*- and *R,R*-tartaric acid respectively). Infrared experiments on L-tartaric acid, reported by Raval's group [65], show that chiral tartaric acid remains intact upon adsorption at low temperature (83 K) with very little difference observable between the surface spectra and that of the molecular solid. Moreover, no long-range order was detected by LEED, and the evidence therefore points towards formation of a disordered multilayer.

Of greater interest is the result of L-tartaric acid adsorption at room temperature, where the behaviour depends somewhat upon the coverage. At low coverage (up to an estimated 0.25 ML) the presence of an absorption peak at 1711 cm^{-1} was attributed by Raval's team to the carbonyl stretch mode, indicating the continued existence of intact acid groups, but the presence also of an absorption peak at 1437 cm^{-1}, attributed to the symmetric $—CO_2$ stretch mode, suggests that deprotonated carboxylate groups are also to be found within the adsorbed overlayer; these results were taken to indicate a monotartrate phase [65]. A clear c(4×6) LEED pattern was briefly observed under these conditions, but this converted to a c(4×2) pattern after a few seconds of exposure to the electron beam (or upon gentle annealing).[30] In STM experiments, the authors reported that the c(4×6) phase could only be imaged occasionally, and a second scan of the same surface region would subsequently reveal the c(4×2) phase. Evidently, the c(4×6) phase is extremely delicate and susceptible to restructuring either thermally or due to electronic perturbation. Raval and co-workers pointed out that it is possible to explain both structures with models in which the molecules occupy the *same* positions, differing only in the arrangement of the intermolecular hydrogen bonds [65]. Relatively recently, PhD experiments reported by Woodruff and co-workers [66] have been analysed to provide structural details of a similar monotartrate phase, revealing that the carboxylate group binds in μ_2 fashion to the substrate, with the intact acid group dangling some distance from the surface.

Annealing the c(4×2) phase at 350 K was found to result in slow conversion to a $(1, 2; 8, \bar{2})$ phase for L-tartaric acid [65] (or the $(1, \bar{2}; 8, 2)$ phase for D-tartaric acid [68]) which could also be formed immediately by adsorption at 405 K. Neither of these periodicities implies an intrinsically chiral lattice, since both are rectangular, but because they do not align with the high-symmetry directions of the substrate they nevertheless imply extrinsic lattice chirality.[31] In this state, the absorption peak at 1711 cm^{-1} was absent from the RAIR spectra, indicating that no intact acid groups remain in the overlayer at these temperatures, and additional subtle changes to the spectra further strengthen the identification of a pure bitartrate phase. Not unreasonably, Raval and co-workers attributed the extrinsically chiral lattice arrangement to intermolecular hydrogen bonding [68] but subsequent DFT calculations by Barbosa and Sautet [69] suggested that intramolecular hydrogen bonding is in fact dominant, with lattice chirality conferred by virtue of repulsive interactions between hydroxyl groups on adjacent molecules; analysis of PhD data by Woodruff and co-workers [66] yielded a structural model consistent with the calculated geometry, lending such conclusions additional credibility. Later STM images obtained by Raval's group (see Fig. 4.33) revealed that the gradual conversion at 350 K from c(4×2) to $(1, 2; 8, \bar{2})$ takes place (with an activation energy of 73 kJ mol^{-1}) at the edge of c(4×2) islands, with the extrinsically lattice-chiral structure spreading out onto adjacent regions of bare copper; if the coverage is such that the c(4×2) structure covers the entire surface before heating to 350 K, then conversion to the $(1, 2; 8, \bar{2})$ phase is entirely suppressed [67].

Upon room-temperature adsorption of L-tartaric acid at coverages beyond the completion of the c(4×2) phase, it was found by Raval and co-workers that an

30 These phases were originally described as (4, 0; 1, 3) and (4, 0; 1, 2) phases, respectively, thus obscuring their rhombic symmetry.

31 The original designations of $(1, 2; \bar{9}, 0)$ and $(9, 0; \bar{1}, 2)$ entirely fail to emphasise the rectangular nature of these lattices.

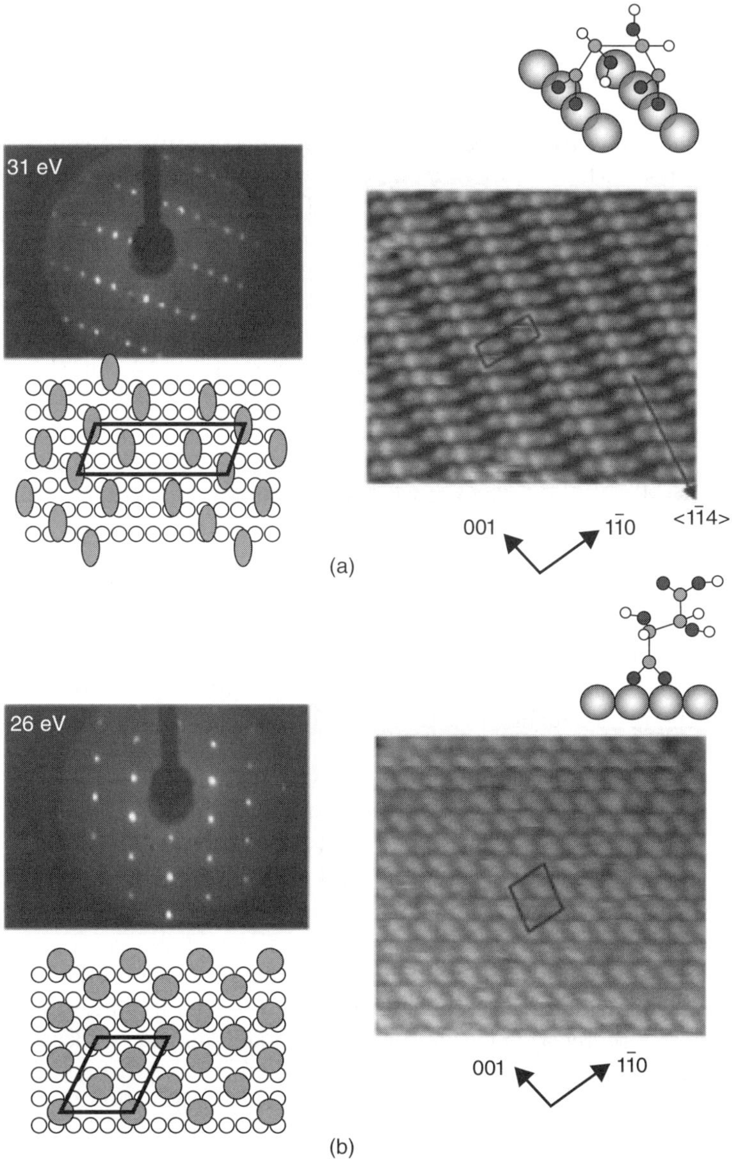

Figure 4.33 LEED patterns and STM images obtained after exposure of Cu{110} to L-tartaric acid. Panel (a) shows, over an 80 × 75 Å^2 scan area, the (1, 2; 8, $\bar{2}$) phase observed at around 350 K, while panel (b) shows, over an 80 × 70 Å^2 scan area, the c(4×2) phase observed at 300 K. Reproduced with permission from Ref. [67]. Copyright (2001) Elsevier.

intrinsically lattice-chiral $(\bar{2}, 2; 4, 1)$[32] phase could be generated [65]. Infrared data revealed that the 1711 cm^{-1} peak present in the lower-coverage phase was now split into peaks at 1674 and 1759 cm^{-1}. Although still interpreted as deriving from intact acid groups, these were taken to indicate that two different types of acid group were present at the surface. Specifically, the strongly redshifted frequency is typical of structures in which two acid groups face each other and make a pair of hydrogen bonds. The STM images obtained under these conditions show clearly two distinct types of features, one of which is consistent with a dimer. The authors therefore proposed a plausible model (coverage 0.30 ML) in which monotartrate dimers and monomers alternate in a 1:1 ratio; the dimers are held together by hydrogen bonding between their acid groups, while the monomers form hydrogen bonds with the dimers through their hydroxyl groups [65]. Annealing this phase at 405 K (or adsorbing a sufficiently high coverage at 405 K) led to the formation of a $(4, 1; \bar{2}, 4)$[33] LEED pattern with similar coverage, for which a structural model, with 1:3 dimer:monomer ratio and 0.28 ML coverage, was later proposed by Ernst and co-workers [70]. Once again, this periodicity does not, in itself, display intrinsic lattice chirality (being rectangular in nature) but it does display extrinsic lattice chirality by virtue of the fact that it does not align with the high-symmetry directions of the substrate.

Adsorption of racemic tartaric acid at 405 K was also studied by Ernst's group, who reported the formation, at low coverages, of a LEED pattern consistent with superposition of $(1, 2; 8, \bar{2})$ and $(1, \bar{2}; 8, 2)$ patterns, indicating the existence of extrinsically lattice-chiral domains equivalent to the structures formed in isolation by enantiopure tartaric acid [70]. At saturation coverage, a c(4×2) LEED pattern was observed, with coverage estimated via XPS, on the basis of the C 1s region of the spectrum, to be 0.25 ML. This contrasts with the case of enantiopure tartaric acid, where saturation corresponded to 0.30 ML coverage, with the extrinsically lattice-chiral $(4, 1; \bar{2}, 4)$ LEED pattern mentioned above. Importantly, this argues strongly against the possibility that the racemate segregates on the surface into c(4×2) regions containing only L-tartaric acid and others containing only D-tartaric acid; instead, we must assume that all domains of c(4×2) structure contain roughly equal numbers of both enantiomers. It is interesting to note, therefore, that segregation into homochiral domains appears to occur in the low-coverage regime, but not in the high.

Finally, we note an interesting example of chiral imprinting in this system, whereby the chirality of the adsorbed overlayer may be semi-permanently imparted to the underlying substrate.[34] Preparing the $(4, 1; \bar{2}, 4)$ phase by deposition of L-tartaric acid at room temperature and annealing at 405 K, Lawton *et al.* [71] proceeded to anneal further at 445 K, ultimately obtaining[35] an intrinsically lattice-chiral $(2, \bar{1}; 6, 7)$ phase characterised (in STM) by distinct bright features running diagonally across the surface (see Fig. 4.34).

32 Originally described as (4, 1; 2, 3).
33 Originally described as (4, 1; 2, 5).
34 Somewhat similar results, which we shall not discuss in detail, have been reported by Karageorgaki and Ernst [72] upon the adsorption of malic acid.
35 Via an intermediate c(4×2) phase.

Figure 4.34 STM images (with 5 nm scale bars in lower left) obtained after 300 K exposure of Cu{110} to L-tartaric acid, followed by annealing at 445 K. Panel (a) shows a region with regular $(2, \bar{1}; 6, 7)$ order, with an inset linescan demonstrating the height of the bright lines to be consistent with a monoatomic copper row; panel (b) shows a Fourier transform of this image, highlighting the periodicity of these features, and panel (c) provides a real-space schematic. Panel (d) shows an isolated added row, adjacent to trenches of single-atom depth. Reproduced with permission from Ref. [71]. Copyright (2013) American Chemical Society.

These features appear to underlie the tartaric acid overlayer, and their height and width is consistent with the addition of a single row of copper atoms; trenches of comparable depth appear transiently, running parallel to the raised features, suggesting that atoms are removed from the one to form the other (with the trenches eventually subsumed at terrace-edges). The periodicity of this phase is corroborated by a sharp LEED pattern obtained after annealing at 483 K, and the chirality of both this and the STM images could be reversed simply by repeating the entire experiment with D-tartaric acid [71]. Rather, more extensive instances of chiral imprinting, amounting to complete facetting of the substrate, will be discussed later in this chapter.

4.3.6 Glutamic Acid on Ag{110} and Ag{100}

Another interesting molecule containing two acids groups, but this time only a single chiral centre, is glutamic acid; this is an amino acid that may be obtained by substituting one hydrogen atom from alanine's methyl group by ethanoic acid (*see Plate XI*). Studies concerning the adsorption of this molecule on a variety of silver surfaces have revealed not only several striking similarities with the adsorption of amino acids on copper, but also several equally striking differences.

To focus upon similarities first, we turn initially to the adsorption of L-glutamic acid on Ag{110}, as reported by Jones *et al.* [73]. Here, RAIRS has been employed to ascertain the chemical nature of the surface species upon adsorption at 300 K, whereupon the existence of only a single peak ascribable to carbonyl (at 1651 cm^{-1} in the low-coverage regime) implies that one of the two acid groups becomes deprotonated at the surface. Consistent with this interpretation, the observed peak at 1387 cm^{-1} has been ascribed to the $—CO_2$ symmetric stretch mode of the carboxylate group, while the absence of any corresponding asymmetric mode was taken as evidence of this group's bidentate binding to the surface. Together with a weak band at 1574 cm^{-1} attributed to the $—NH_2$ scissor mode of the amine group, the evidence points towards a μ_3 configuration at the amino acid end of the molecule, highly reminiscent of the binding seen for other amino acids on Cu{110} and Cu{311}. In contrast, the acid group at the distal end of the molecule remains intact, and may interact either weakly with the surface or strongly with neighbouring adsorbed molecules.

At low coverages on Ag{110}, STM reveals significant ridge-and-furrow facetting of the surface, although the exact facet orientations have not been identified; the ridges run along the $\langle\bar{1}10\rangle$ surface direction, however, implying that the facets themselves are necessarily achiral. Other regions of the surface retained the original {110} orientation, and it was only upon these that a molecular overlayer could be discerned [73]. This implies that the facetted regions either host very few adsorbed molecules, or that molecules bound to the facets are highly mobile and thus effectively invisible to STM. In the non-facetted regions, an intrinsically chiral lattice, describable in matrix notation as $(4, 2; \bar{5}, 1)$, was observed both via STM and LEED (becoming increasingly ordered upon annealing at 375 K). Each cell in the STM images contained six distinct moieties, suggesting a local coverage of 3/11 ML (see Fig. 4.35).

Figure 4.35 Low-coverage STM images obtained after 300 K exposure of Ag{110} to L-glutamic acid. Panel (a) shows the initial coexistence of ridge-and-furrow facets with the (4, 2; $\bar{5}$, 1) phase, while panel (b) reveals the effect of annealing at 375 K. Panels (c) and (d) show close-ups of the (4, 2; $\bar{5}$, 1) phase, and panel (e) provides a schematic model. Reproduced with permission from Ref. [73]. Copyright (2005) American Chemical Society.

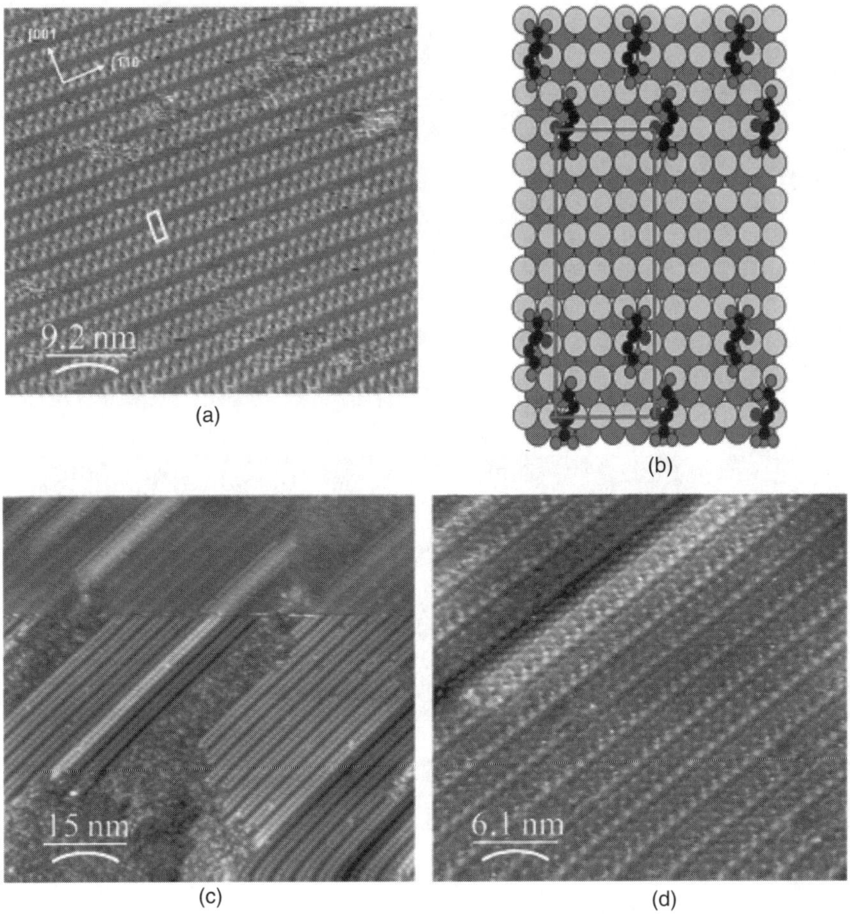

Figure 4.36 High-coverage STM images obtained after 300 K exposure of Ag{110} to L-glutamic acid. Panel (a) shows the initial (4×8) phase, while panel (b) provides a schematic model. Panels (c) and (d) show predominantly the c(4×12) phase observed after annealing at 423 K. Reproduced with permission from Ref. [73]. Copyright (2005) American Chemical Society.

Upon further exposure to L-glutamic acid, the ridge-and-furrow facets were lost, the structure of the molecular overlayer changed, and the carbonyl frequency associated with the intact acid group shifted up in frequency. This last mentioned occurrence was interpreted as signalling the formation of adsorbed dimers, bound together by hydrogen bonding between dangling acid groups [73]. Indeed, STM images reveal a (4×8) overlayer (consistent with the observed LEED pattern) with two moieties per unit cell (see Fig. 4.36) suggestive of dimer formation; the implied local coverage of 1/16 ML is actually *lower* than that found in the intrinsically chiral lattice mentioned above, but the absence of ridge-and-furrow facets (of unknown coverage) means that it is entirely plausible for the global coverage now to be higher. Upon annealing still higher coverages in the range 375–475 K, the observed LEED pattern indicated c(4×12) periodicity, while the STM images featured regions of c(4×12) order interspersed with regions showing (4×6) order (see Fig. 4.36). Counting moieties in the STM images suggested a local coverage of 1/4 ML while further changes in the RAIR spectra indicated continued intermolecular interactions and even the possibility of an incipient second layer.

Finally, exposure to L-glutamic acid at 375 K (as opposed to exposure at 300 K followed by annealing) led to the formation of a phase entirely unrelated to those just described [73]. In this case, the LEED pattern reveals $(3, 2; 3, \overline{3})$[36] periodicity, while STM confirms two distinct moieties per unit cell (implying a local coverage of 2/15 ML). Notwithstanding the apparent long-range chiral order associated with this two-dimensional lattice, a super-order of high-density domain boundaries was also observed in STM, which the authors speculated was linked to the similar strength of adsorbate–adsorbate and adsorbate–substrate binding in this system.

While the results described above for adsorption of glutamic acid on Ag{110} are reasonably comparable with those for the adsorption of similar molecules on Cu{110} and Cu{311} surfaces, adsorption of the same molecule on Ag{100} leads to a very different range of behaviours, as described by Vattuone and co-workers [74, 75]. Upon deposition of L-glutamic acid at 250 K, STM reveals the formation of one-dimensional 'row-like' structures that are extended in a direction rotated approximately 40° clockwise from the nearest ⟨011⟩ direction (and hence display one-dimensional extrinsic lattice chirality). Each row appears to consist of individual horizontally oriented adsorbed moieties stacked side-by-side, but these rows are also often found in pairs (see Fig. 4.37a). Deposition at 300 K leads, in contrast, to the formation of so-called 'comb-like' structures, aligned more or less precisely along one or other of the ⟨011⟩ directions (see Fig. 4.37b). Although the orientation of these quasi-one-dimensional combs is hence achiral, the 'teeth' break the symmetry and render the overall disposition of molecules chiral. Upon deposition at 325 K, the comb structure coexists with a minority two-dimensional phase described as 'flower-like' (see Fig. 4.37c) in which groups of four particularly bright features in STM (interpreted as tetramers comprising vertically oriented molecules) are separated by pairs of less bright features (interpreted as dimers comprising horizontally oriented molecules). Finally, upon deposition at 350 K, the comb-like structures are absent, but the flower-like structure coexists with a 'square-like' structure (see Fig. 4.37d) in which individual horizontally oriented adsorbed moieties define the edges of square pores. Although the latter two structures possess square two-dimensional lattices, the flowers of the former and the vertices of the latter break all improper symmetries and render the arrangement chiral. Furthermore, the sides of the unit cell in the flower-like structure are rotated approximately 5° anticlockwise relative to their nearest ⟨011⟩ direction, while those of the unit cell in the square-like structure are rotated some 10° anticlockwise relative to theirs (which would imply extrinsic lattice chirality, regardless of the internal details of each unit cell).

36 Originally described as $(3, 2; \overline{3}, 3)$.

Figure 4.37 STM images obtained after exposure of Ag{111} to L-glutamic acid. Panels show (a) the row-like structure formed at 250 K, (b) the comb-like structure formed at 300 K, (c) the flower-like structure formed at 350 K, and (d) the square-like structure, also formed at 350 K. All scan areas are 80 × 80 Å^2. Reproduced with permission from Ref. [74]. Copyright (2011) American Chemical Society.

A careful analysis of the square-like structure, supported by first-principles DFT, conclusively demonstrated that it involves neutral glutamic acid molecules binding to one another via a ring of hydrogen bonds and only loosely bound to the substrate by Van der Waals interactions [75]. The main role of the substrate in this case is, therefore, to tip the balance of chemistry away from the zwitterionic form favoured in bulk glutamic acid. Nevertheless, sufficient site-specificity remains in the weak adsorbate-substrate interaction to define a preferred orientation of the overlayer lattice. As for the chemistry of the row-like, comb-like and flower-like structures, RAIRS and XPS are insufficient to determine the ratio between neutral and anionic forms, although both are certainly present in variable quantities; the zwitterionic form is ruled out by the presence of only a single N 1s peak in XPS at 399.4–400.2 eV for all but the multilayer regime [74].

Figure 4.38 STM image and model (panels (a) and (b)) for the (21, 19; $\overline{10}$, 13) phase obtained after exposure of Au{111} to *R*-2-butanol. Reproduced with permission from Ref. [76]. Copyright (2016) American Institute of Physics.

4.3.7 2-Butanol on Au{111}

Tetrameric structures are also evident upon the adsorption of 2-butanol on the Au{111} surface. This rather simple chiral molecule (*see Plate XV*) has very recently been studied experimentally by the group of Sykes, with supporting DFT calculations provided by Carrasco and Michaelides [76]. Strikingly, the adsorbed molecules are seen in STM to organise into an essentially square lattice, with a tetramer located at each vertex, despite the triangular symmetry of the underlying substrate. Indeed, the adsorbate–substrate interaction appears to be rather weak, in that the herringbone reconstruction of Au{111} is not lifted upon adsorption, but not altogether so weak as to preclude any influence on geometry; the overlayer lattice maintains a well-defined orientation with respect to that of the substrate, with *S*-2-butanol producing domains displaying square unit cells rotated 25° clockwise from one of the three equivalent $\langle\bar{2}11\rangle$ surface directions, and *R*-2-butanol producing domains rotated by an equal angle in the opposite sense. The authors present a model (see Fig. 4.38) for the latter lattice based on a 5×5 grid of overlayer squares coincident with a (21, 19; $\overline{10}$, 13) periodicity[37] of the substrate lattice; the tetramers are believed to comprise four intact *R*-2-butanol molecules, bound to the surface by a weak dative covalent bond and to each other by two hydrogen bonds each (one as donor, one as acceptor). The pores of the overlayer lattice expose chiral ensembles of gold atoms, which could hold potential as sites for asymmetric adsorption and/or reaction of guest species.

37 The internal angles of the 'squares' thus created are actually 90.6° and 89.4°, with side lengths differing by 0.5%; it may be that the actual overlayer is truly incommensurate, but the proffered model is clearly at least a very good approximation.

4.3.8 Tartaric Acid on Ni{111}

It will not have escaped the reader's attention that all the case studies of prochiral and chiral adsorbates discussed thus far have involved adsorption on coinage metals – predominantly copper – and indeed this bias is reflective of the weight of published studies in the literature. These moderately reactive metals provide an ideal backdrop for studies in which one hopes to encounter interesting chiral effects, being neither so assertive as to entirely dominate overlayer structure (leaving little scope for the role of interadsorbate interactions) nor so passive as to leave the adsorbates entirely to their own devices (as if they were free-floating molecular rafts). Nevertheless, a few cases clearly warrant inclusion here, of which the first that we shall address is the adsorption of tartaric acid on Ni{111}.

This system has been studied in depth by Jones and Baddeley [77] through a combination of RAIRS and STM. Investigation by LEED was also attempted, but the overlayer was found to be highly sensitive to electron beam damage, in contrast to the situation on Cu{110} described above (Section 4.3.5) where only relatively mild beam sensitivity was reported. Also contrasting with adsorption on Cu{110} is the fact that results here appear to be influenced not only by internal factors such as surface temperature, but also by external factors such as the molecular flux. At low flux, with a surface temperature of 300 K, absorption bands appear in RAIRS at 1428 and 1656 cm^{-1}, attributed respectively to the symmetric $—CO_2$ stretch mode of a carboxylate group and to the —C=O stretch mode either of an intact acid group or of a carboxylate group bound to the surface in monodentate fashion; the latter is present only at rather low coverage, however, suggesting that whilst a low-coverage monotartrate species may be one possible interpretation (as per the Cu{110} case) one might alternatively (as do the authors) prefer to take it as evidence of a monodentate bitartrate species[38] that reprotonates at high coverage to form monotartrate dimers with carbonyl groups lying parallel to the surface and hence invisible to RAIRS. At high flux, with the same surface temperature, the carbonyl band is entirely absent, which the authors take as evidence that dimers are formed rapidly and then kinetically preclude the monodentate bitartrate species altogether, despite the latter being thermodynamically preferred at low coverage [77].

After deposition of L-tartaric acid (i.e. *R*,*R*-tartaric acid) in high flux conditions at 300 K, STM images revealed ordered adsorption in islands covering approximately half the surface (with disordered regions between); within these islands, the dominant phase displayed an intrinsically chiral lattice with $(3, 2; \bar{2}, 2)$ periodicity, while a minority phase displayed an intrinsically chiral lattice with (3, 1; 1, 4) periodicity.[39] In both cases, two moieties were visible within each primitive unit cell (see Fig. 4.39) suggesting coverages of 1/5 ML and 2/11 ML, respectively.

At a slightly higher surface temperature of 350 K, the picture is a little different. Under these conditions, a RAIRS absorption band at 1651 cm^{-1} is observed at low coverage even at high flux. The authors interpret this as indicating that

38 That is, a bitartrate species in which each carboxylate group binds to the surface through only one of its two oxygen atoms.

39 Originally described as (5, 0; 1, 2) and $(4, \bar{1}; \bar{1}, 3)$.

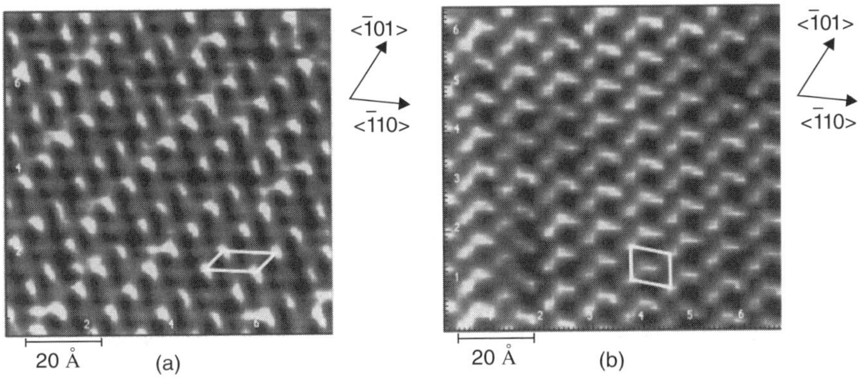

Figure 4.39 STM images obtained after 300 K exposure of Ni{111} to L-tartaric acid. Panel (a) shows the majority (3, 2; $\bar{2}$, 2) phase and panel (b) the minority (3, 1; 1, 4) phase. Reproduced with permission from Ref. [77]. Copyright (2002) Elsevier.

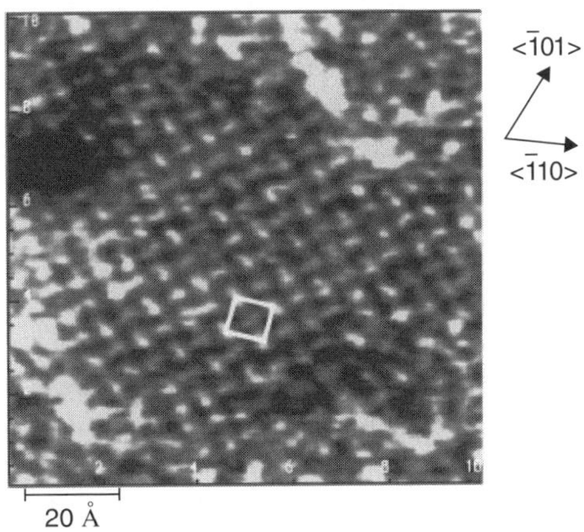

Figure 4.40 STM image obtained after 350 K exposure of Ni{111} to L-tartaric acid, showing a (3, 1; 1, 4) periodicity but with detail differing distinctly from the minority phase of Fig. 4.39. Reproduced with permission from Ref. [77]. Copyright (2002) Elsevier.

monotartrate dimers are no longer kinetically stable, due to thermal energy in the overlayer [77]. At low coverage, therefore, the authors favour a monodentate bitartrate species in rationalising their RAIRS data. At higher coverages, the band at 1651 cm^{-1} becomes less prominent, suggesting that a more conventional bidentate binding of carboxylate groups in the bitartrate is favoured. Whilst STM once again reveals islands with (3, 1; 1, 4) periodicity (see Fig. 4.40) the internal details of each primitive unit cell differ from the images obtained at lower temperature (albeit there still appear to be two moieties per primitive unit cell, and hence the same local coverage). There is, however, some evidence to suggest that very small regions similar to the 350 K overlayer are already present in the 300 K overlayer, implying that the latter might gradually convert into the former over a sufficiently long time [77].

4.3.9 Alanine on Pd{111}

Another interesting system involving adsorption of chiral molecules onto a non-coinage metal is that of alanine on Pd{111}. Here, Tysoe and co-workers reported XPS studies some years ago [78], together with similar data on glycine [79] and proline [80] adsorption, but more recently have supplemented this with RAIRS and intriguing STM imaging showing rather unique chiral behaviour [81].

In the early XPS work, on alanine deposited at 100 K, Gao *et al.* [78] reported sub-monolayer peaks in the N 1s region of their spectra at binding energies of 398.8–399.4 and 401.3–401.6 eV, respectively attributed to a neutral amine group (—NH_2) and a primary ammonium cation (—NH_3^+). The relative intensities of these features suggest that ~65% of adsorbed moieties were in the zwitterionic form, with the remaining ~35% adopting the anionic form; although no data was provided for the O 1s region, a peak in the C 1s region at 288.0–288.5 eV was interpreted as more likely due to carboxylate than to an intact acid group. Subsequently obtained RAIRS data, recorded after deposition at 300 K, confirmed this assessment [81], on the basis of a peak at 1402 cm^{-1} interpreted as arising from the symmetric —CO_2 stretch mode of the carboxylate group and the absence of any mode attributable to an intact acid group; a band deriving from the asymmetric —CO_2 stretch mode was also evident at 1606 cm^{-1}, from the lowest coverages, indicating that the binding geometry is not straightforwardly comparable with that typically adopted on copper surfaces. Additional XPS data, also obtained after deposition at 300 K, make it clear that binding of alanine in the anionic form is strongly preferred at the lowest coverages, with the zwitterionic form increasing to approximate parity only as the coverage approaches saturation of the monolayer [81].

STM images obtained at 120 K, subsequent to alanine deposition at 290 K, revealed two main structural motifs [81]. The more prevalent of these are dimers, conjoined in short chains aligned along the close-packed $\langle 1\bar{1}0 \rangle$ crystallographic directions of the surface. Also present, however, are discrete tetramers, similarly aligned with the symmetry of the underlying substrate (see Fig. 4.41). While earlier DFT calculations [82] had indeed indicated that isolated monomers are most

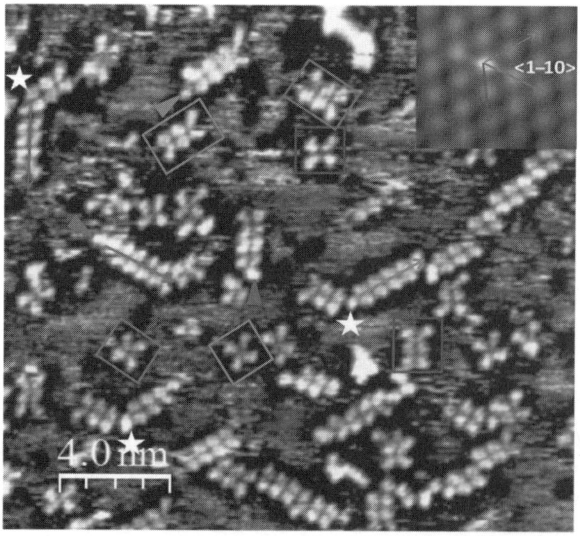

Figure 4.41 STM image obtained after 290 K exposure of Pd{111} to L-alanine, showing short chains of dimers and isolated tetramers. Reproduced with permission from Ref. [81]. Copyright (2014) American Chemical Society.

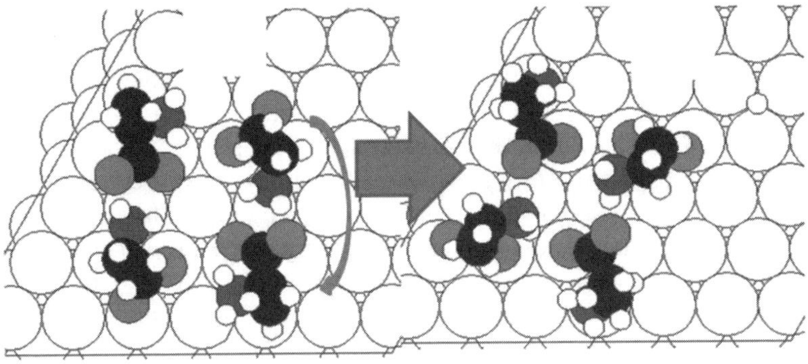

Figure 4.42 DFT-calculated models for the tetramers formed upon exposure of Pd{111} to alanine. Panel (a) shows the initial-guess structure, based on antiparallel anions and zwitterions. Panel (b) shows the energy-minimised structure, in which the rotation of the anions is the major change. Reproduced with permission from Ref. [81]. Copyright (2014) American Chemical Society.

stable in the anionic form, it turned out on further investigation that dimers containing two anions are unstable (relative to separation of their component monomers) while those containing two zwitterions, or one anion and a zwitterion, are not; the result is that dimers containing both species are more stable than those containing only one or the other [81]. Perhaps unsurprisingly, it is the interaction between the cationic end of the zwitterion and the carboxylate end of the anion that is predominantly responsible for this stability, the —NH_3^+ group eschewing contact with the surface in favour of intimacy with its neighbour. Moreover, it is the carboxylate group of the anion that adopts a monodentate configuration with respect to the surface, giving rise to the 1606 cm^{-1} band in the RAIRS data, while that of the zwitterion retains a more conventional bidentate binding arrangement. Similarly, a highly stable tetramer structure may be constructed in which the cationic ends of two zwitterions converge with the carboxylate ends of two anions, with the resulting simulated STM images in excellent agreement with experimental observations (see Fig. 4.42). Once again, the carboxylate groups of the anions adopt monodentate binding configurations, while those of the zwitterions remain in bidentate geometries.

Whilst this system lacks true long-range order, it is nevertheless significant that chirality is expressed at an organisational level higher than that of individual molecules. The footprint occupied by an enantiopure tetramer is chiral, although that occupied by similarly enantiopure dimers is not. This fact will turn out to be of some importance regarding the effectiveness of amino acids as chiral modifiers on this surface, as we shall see in the following chapter.

4.4 Chiral Facetting: Remodelling the Surface

Just as we have previously noted our bias towards adsorption on coinage metals, it is equally apparent that (aside from the two immediately preceding sections) we have focussed almost entirely upon the medium-symmetry {110} and {311} surfaces of these metals. In part, this again reflects the preponderance of available literature, but it must certainly be admitted that we have essentially neglected a fair number of studies conducted on the higher symmetry {111} and {100} surfaces.[40] Our best justification in doing so is that adsorption on these substrates is often disordered, and does not usually show clear evidence of chiral lattice organisation. Since our purpose is selectively to illustrate chiral surface phenomena, rather than exhaustively to catalogue the surface chemistry of organic molecules, the omission is probably forgiveable. There are, however, some notable exceptions, and chief amongst these are those cases where the adsorbate actually reshapes a high-symmetry surface to expose low-symmetry facets more amenable to binding a particular enantiomer. It is to such systems that we now turn our attention.

4.4.1 Glycine, Alanine and Lysine on Cu{100}

Upon room-temperature adsorption of glycine onto the Cu{100} surface, Zhao and co-workers [83] reported that native monoatomic terrace edges[41] showed evidence in STM of bunching and reorientation towards a variety of ⟨013⟩ alignments. Moreover (especially upon annealing to 430 K) bunching was so pronounced that distinct facetting of the surface could be said to have occurred. The majority facets retained {100} orientation, but the morphology was best characterised not with reference to occasional monoatomic terrace edges but instead with reference to distinct facets of {3 1 17} orientation (see Fig. 4.43). A clear c(2×4) overlayer was visible on the {100} facets, explicable in terms of a simple model in which glycinate moieties in the μ_3 configuration occupied three

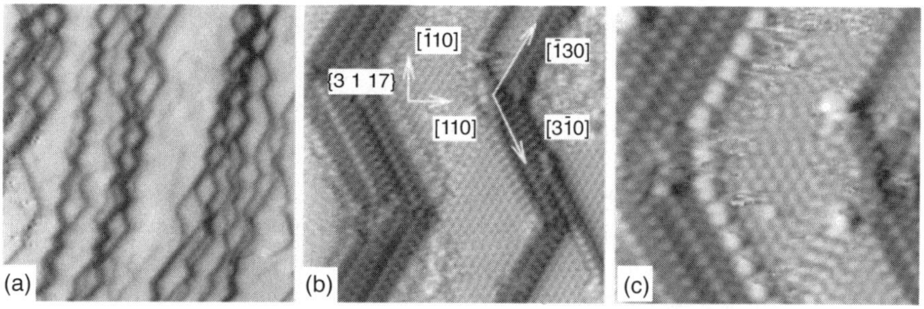

Figure 4.43 STM images obtained after room-temperature exposure of Cu{100} to glycine. Panel (a) shows a 1450 × 1450 Å^2 scan area, (b) a 175 × 175 Å^2 scan area, and (c) a 110 × 110 Å^2 scan area. Note the symmetric distribution of two enantiomorphically related {3 1 17} facet orientations, and the c(2×4) periodicity on the {100} facets. Reproduced with permission from Ref. [83]. Copyright (1999) Elsevier.

40 Some of these will crop up in later chapters, when the presence of chiral modifiers breaks symmetry and makes these often less interesting systems very interesting indeed.
41 Randomly distributed terrace edges are, of course, inevitable features of any surface, no matter how ideal its preparation. These may arise due to a slight miscut of the crystal, in which case their alignment will be statistically skewed in one particular direction, or due to subsequent polishing, in which case no statistically preferred alignment will be apparent.

top-layer copper atoms out of every four. Such a model necessarily implies an asymmetric contribution to the surface stress, and since the highest symmetry primitive unit cell of this model is spanned by lattice vectors of ⟨013⟩ alignment it is perhaps unsurprising that such directions prove key to the facetting behaviour. It might also be noted that the ⟨013⟩ directions are notable at the level of tertiary structure (see Chapter 2) and this also could predispose the surface to facet along the corresponding zone; the {3 1 17} poles each lie only a short distance (within the stereographic projection) along one of the ⟨013⟩ zones from their nearest {100} pole. Within each {3 1 17} facet, STM images revealed molecules arranged in a simple (1×1) pattern with respect to the primitive unit cell of that facet. Although each {3 1 17} facet is, in itself, intrinsically chiral (and must, therefore, induce some chiral distortion of any glycine molecules adsorbed upon it) the global symmetry of the {100} surface, coupled with the achirality of glycine, ensures that all eight possible instances of {3 1 17} orientation[42] were found at the surface (as made clear in a subsequent publication by Zhao [84]). That is to say both {3 1 17}-D and {3 1 17}-L were found with essentially equal probability.

Repeating the same experiment with enantiopure and racemic alanine, Zhao *et al.* [85] once again found a propensity towards the formation of {3 1 17}-oriented facets. As before, STM revealed a simple (1×1) disposition of molecules on the chiral facets, contrasting with a c(2×4) overlayer on the {100} facets. Interestingly, all eight possible {3 1 17} orientations were *again* observed, even when enantiopure molecules were used, implying that there was no strong coupling between the chirality of the adsorbate and that of the substrate facets. A quite different result was obtained, however, when enantiopure lysine was dosed instead; in this case, adsorption and annealing of L-lysine promoted exclusively the formation of {3 1 17}-D facets, while {3 1 17}-L facets were never observed [84]. It may be that the possibility of μ_4 binding between lysine and the surface is of some importance; certainly, the detailed structure within each {3 1 17} facet was distinctly different from that of either glycine or alanine, displaying a fourfold increase in repeat distance parallel to the ⟨013⟩ facet edge. On the {100} terraces, intrinsically chiral lattices were observed both at room temperature (where (4, 1; $\bar{3}$, 4) and (4, 1; $\bar{2}$, 4) overlayers were found to coexist) and after annealing to 430 K (where only the latter was evident) [86]; this would appear to contrast with the achiral c(2×4) overlayer seen on these terraces after adsorption and annealing of either glycine or alanine, but it ought to be noted that a later paper by Cheong and Gellman [87] reports only the presence of the achiral phase for lysine.

4.5 Chiral Metallorganic Frameworks: Into the Second Dimension

In the examples discussed above, it has typically been found that overlayers of chiral molecules remain somewhat aloof from the substrate that hosts them.

42 For a specific instance of the {100} surface type, only 8 of the 16 {3 1 17} surfaces may occur without creating overhangs.

That is, the molecular network sits *above* the metal surface, rather than being in any sense incorporated within it (or, indeed, *vice versa*). Nevertheless, the induction of chiral facets on Cu{100} does suggest that it might be possible, under certain circumstances, for the underlying metal to play a rather more active role than simply providing an inert template for the growth of purely molecular networks. Indeed, a number of recent experiments highlight the intriguing prospect of two-dimensionally chiral metallorganic frameworks, spontaneously formed by the adsorption of chiral molecules upon suitably chosen bimetallic surfaces. Such networks might combine the high catalytic activity of single transition metal atoms with enantioselectivity conferred by their chiral molecular component.

4.5.1 Glutamic Acid on Ni/Au{111}

It has long been known that nanoclusters of transition metals can nucleate at the 'elbows' of the Au{111} herringbone reconstruction; these are localised sites where the top-layer of the reconstructed surface exhibits dislocations and hence a locally high surface energy. Amongst the metals known to nucleate in this way, nickel deposited at 300 K forms relatively monodisperse two-dimensional islands [88] and the adsorption of L-glutamic acid upon these has been studied by Trant *et al.* [89] using both RAIRS and STM. In the absence of nickel, no adsorption is detected upon exposure of the Au{111} surface to glutamic acid at 300 K, but when 0.1 ML Ni is present the story changes dramatically. Under these circumstances, RAIR spectra are reminiscent of those obtained by the same group on Ni{111} surfaces [90] with peaks at 1421 and 1516 cm^{-1} attributable, respectively, to the $—CO_2$ symmetric stretch of a carboxylate group and to the $—NH_3^+$ symmetric deformation (umbrella mode) of a primary ammonium cation; these features are interpreted as indicating that the amino acid end of the molecule adopts zwitterionic form. At the distal end, an intact acid group is thought to be responsible for an intense peak that shifts from 1647 cm^{-1} at low coverage to around 1750 cm^{-1} at higher coverage, reflecting a significant change in its role as the source of intermolecular binding. In STM, meanwhile, the highest coverages of glutamic acid result in the apparent removal of nickel islands, but since it is known that molecular adsorption on the clean gold surface is negligible, one can conclude that much of the transition metal must somehow remain. Upon annealing to 360 K, the observable overlayer restructures into a network of extended linear features that must presumably incorporate not only molecules but also transition metal atoms.

At a higher pre-coverage of Ni (0.25 ML) the adsorption of L-glutamic acid yields relatively similar RAIR spectra, indicating the same chemical species, and whilst annealing to 350 K still yields linear chains as before, these now run between apparently persistent nickel islands distributed according to the dictates of the original herringbone reconstruction (see Fig. 4.44) [89]. Favoured orientations of the chains reflect the chirality of the adsorbed molecular species, but the spacing of features observed in STM is not consistent with the simplest interpretation that the chains constitute nickel glutamate. Instead, the authors surmise that the adsorbed moieties undergo internal cyclisation, yielding chains

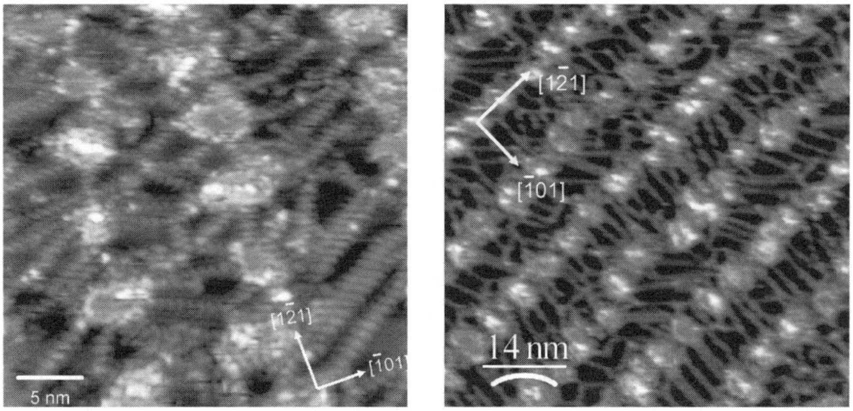

(a) (b)

Figure 4.44 STM images obtained after exposure of Au{111} at 300 K to nickel and then glutamic acid, followed by annealing at 360 K. Scan areas are (a) 300 × 300 Å^2, and (b) 700 × 700 Å^2. Reproduced with permission from Ref. [89]. Copyright (2007) American Chemical Society.

of nickel pyroglutamate. Two different types of chain are distinguished, each thought to contain nickel atoms and pyroglutamate moieties in a 1:2 ratio, with the type lying closest to the $\langle 1\bar{1}0 \rangle$ surface directions predominantly held together by hydrogen bonds, and the type lying closest to the $\langle \bar{2}11 \rangle$ surface directions bonded by predominantly ionic interactions.

In a second study, conducted at even higher Ni pre-coverage (4–8 ML) Trant *et al.* [91] investigated the effect of annealing *prior* to glutamic acid exposure. In this way, it was possible to prepare surfaces that were either Ni-rich or Au-rich in the outermost layers. With low annealing temperatures nickel remained at the surface, but at higher annealing temperatures the lower surface energy of gold caused nickel to be buried. When the surface layers were Ni-rich, a combination of RAIRS, temperature programmed desorption (TPD) and medium-energy ion scattering (MEIS) indicated the immediate formation of nickel pyroglutamate at 300 K, whereas on the Au-rich surfaces the evidence pointed to the presence of zwitterionic moieties. The fact that this latter species could be converted to nickel pyroglutamate upon subsequent heating was taken to indicate segregation of nickel through the gold capping layers. Although no STM experiments were reported in this particular study, it is not unreasonable to imagine similar morphologies to those obtained at lower nickel coverage, albeit the likely absence of any herringbone reconstruction would presumably modify the details.

4.5.2 Lysine on Ni/Au{111}

In contrast to glutamic acid, Wilson *et al.* [92] have demonstrated that lysine *does* adsorb onto the clean Au{111} surface at 300 K. Substantial changes in morphology result, particularly at terrace edges, where 'nanofingers' grow out from the upper onto the lower terraces; using L-lysine, these features extend in the $\langle 14\bar{5}\rangle$ directions lying c. 11° anticlockwise from the nearest $\langle 1\bar{1}0\rangle$ axis and thus exhibiting a distinct one-dimensional extrinsic lattice-chiral aspect. On the terraces themselves, L-lysine packs imperfectly into an overlayer approximately corresponding to (3, 0; 1, 5) periodicity, modulated by the persistent herringbone reconstruction of the underlying gold, and again displays intrinsic lattice chirality.

In the presence of 0.5 ML pre-adsorbed nickel islands, however, the picture becomes quite different [93]. Upon limited exposure to lysine at 300 K, RAIRS experiments on the bimetallic surface reveal significantly stronger absorption bands than on gold alone (perhaps 50 times more intense for similar exposure) indicative of substantially increased uptake. Peaks at 1414 and 1578 cm^{-1} are identified as arising from $—CO_2$ symmetric and asymmetric stretch modes, respectively, suggesting not only the deprotonation of the acid group[43] but also the adoption of a twisted geometry relative to the surface plane. At the same time, a peak observed at 1643 cm^{-1} is attributed to the $—NH_3^+$ asymmetric deformation mode of the primary ammonium cation; the absence of a corresponding symmetric deformation (umbrella mode) suggests that this group lies with its axis of symmetry almost parallel to the plane of the surface. An additional peak at 1616 cm^{-1}, consistent with a possible $—NH_2$ scissor mode, implies that anionic and zwitterionic species may coexist.

Concurrent with the emergence of these vibrational modes, STM images reveal the gradual erosion of distinct nickel islands, but both the observed morphology and infrared spectra are sufficient to imply that at least some of the 'lost' atoms must somehow remain at the surface. As was the case for glutamic acid (see above) the authors proposed that chain-like metallorganic structures are formed – in this instance, nickel lysinate, with an ideal ratio of nickel atoms to lysinate moieties of 1:2 precisely. Unlike the previous system, however, the chains here are quite tightly packed, to the extent that their individual identity is somewhat obscured. This, in turn, promotes a high degree of order, such that a well-defined $(2, \bar{1}; 2, 7)$[44] periodicity is observed. A minority phase exhibiting $(4, \bar{2}; 2, 6)$[45] periodicity is also present upon adsorption at 300 K, but becomes considerably more prevalent after annealing at 350 K. The local coverage of nickel atoms in the metastable phase is believed to be 1/16 ML, while in the stable post-annealing phase it is 1/14 ML.

4.5.3 Proline on Ni/Au{111}

For the final case study of this chapter, we turn to the adsorption of L-proline on 0.15 ML nickel-covered Au{111} reported by Seljämae-Green *et al.* [94]. To some extent, one would be forgiven for expecting a result similar to that obtained upon the adsorption of L-glutamic acid on the same surface – not because the two molecules are inherently similar in structure (they are not) but because proline bears a striking similarity to the pyroglutamic acid molecule into which glutamic acid is thought to transform upon adsorption (see above). However, although the erosion of nickel islands occurs as before, the ubiquitous chain-like

43 The tell-tale carbonyl stretch at 1700–1750 cm^{-1} is also conspicuously absent.

44 Originally described as $(7, \bar{2}; 1, 2)$.

45 Originally described as $(6, \bar{2}; 2, 4)$.

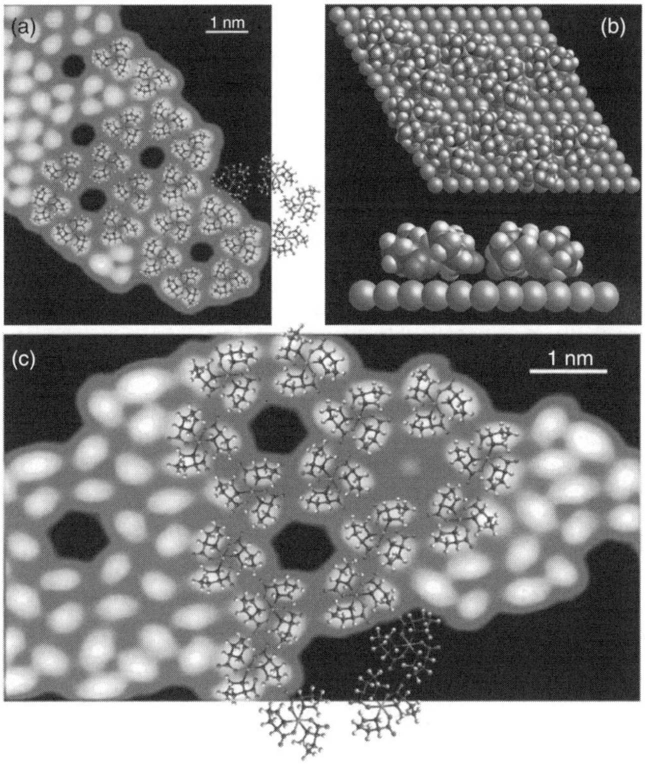

Figure 4.45 STM images obtained after exposure of Au{111} at 300 K to nickel and then proline, followed by annealing at around 400 K. Overlaid models are based upon prolinate trimers centred upon individual nickel adatoms. Reproduced with permission from Ref. [94]. Copyright (2015) American Chemical Society.

features seen with glutamic acid (and indeed lysine) are not the only prolific mode of adsorption on this occasion (after annealing in the range 340–470 K). Instead, islands of a hexagonal porous structure are also commonly observed (see Fig. 4.45) and attributed to a conglomeration of trimers comprising some combination of metallic and molecular components. Those chains that do occur appear similar in structure to the previously described exemplars. The hexagonal phase, meanwhile, orients in domains that respect the threefold rotational symmetry of the underlying substrate but break its mirror symmetry.

In addition to the STM images, the authors also report XPS data that suggests adsorption at 300 K in a mixture of zwitterionic, anionic and neutral forms [94]. The N 1s region of the spectrum at this temperature shows peaks of roughly equal integrated intensity at binding energies of 399.8 and 401.7 eV, consistent with the existence of a neutral amine group and a secondary ammonium cation, respectively. Similarly, the O 1s region reveals not only a majority peak at a binding energy of 531.5 eV, attributed to a deprotonated carboxylate group, but also two minority peaks at binding energies of 533.7 and 532.5 eV, respectively interpreted as being due to the hydroxyl and carbonyl components of an intact acid group. It thus appears that the zwitterionic and anionic forms of the molecule dominate under these conditions, while the neutral form is much less prevalent. Upon annealing, the peak attributed to the secondary ammonium cation decreases sharply in its intensity, disappearing completely when the anneal goes beyond 370 K. The authors note that adsorption of proline in the absence of nickel yields zwitterionic and anionic species that decompose at 370 K [95], suggesting that this may account for these features in the experiments conducted with pre-dosed nickel. Above 370 K, the STM images reveal considerably better-defined long-range structure, coincident with the temperature at which XPS data becomes consistent only with the anionic species.

Importantly, the XPS results also confirm the interpretation that runs throughout all the work described above on Ni/Au{111}: that nickel coordinates with molecular species and remains on the surface even when its original islands are fully eroded. Prior to adsorption of proline, the Ni $2p_{3/2}$ peak is observed at a binding energy of 852.1 eV. Adsorption of proline at 300 K leads to a slight shift in binding energy for this peak to 852.5 eV, coupled with a drop in its integrated intensity, and to the appearance of an entirely new peak at a binding energy of 855.4 eV; both features persist upon annealing up to 450 K [94]. The peak at the lower binding energy is interpreted as being due to metallic nickel, while that at higher binding energy is suggested to correspond to the Ni(III) oxidation state. Such a conclusion strongly suggests that the trimers seen in the STM experiments each consist of a central trivalent Ni cation, surrounded by three (monovalent) prolinate anions; first-principles DFT calculations also support this view.

4.6 Executive Summary

As will be clear from the foregoing discussion, rather a large body of work has been published in recent years on the expression of chirality at intrinsically achiral surfaces. The danger is that one fails to 'see the wood for the trees,' so it will be worthwhile, at this stage, to take stock of the phenomenological range that has thus far been uncovered. Let us, therefore, revisit the systems described above, from a thematic perspective:

Hetereochiral Domains

Arguably, the least promising scenario for the observation of chirality at an intrinsically achiral surface would appear to be the case where the adsorbates are themselves achiral and possess zero (or very little) internal conformational flexibility. Nevertheless, as the example of the nitrogen-induced clock reconstruction of Ni{111} demonstrates, it is entirely possible for the substrate to reconstruct in a manner that breaks mirror symmetry at a highly local level, albeit retaining glide symmetry (and hence achirality) at the level of the surface unit cell. The example of CO adsorption on Pt{110} reveals similar highly local symmetry breaking due to the orientation of the adsorbates, despite minimal modification of the internal adsorbate structure. In either case, each domain on the surface contains adsorbate-substrate complexes that break the symmetry in enantiomerically related senses, so that the resulting structure may most sensibly be described as heterochiral (or racemic). With more flexible, though still achiral, molecules, the asymmetry of the adsorbate–substrate binding footprint may result in substantial deformation of the adsorbate. A case in point, described in detail above, is that of glycine adsorbed on Cu{110}, where the molecular backbone is twisted in one sense or the other in order to conform to mirror-image right-angled triangular footprints for three-point binding. Each surface unit cell contains two molecules subject to mirror-image distortion, resulting in glide symmetry and hence racemic domains. The surface unit cell, in all such cases, must necessarily be of achiral shape.

Homochiral Domains

An alternative to the formation of heterochiral domains is, of course, the formation of homochiral domains with opposing chirality. That this behaviour can arise even for entirely rigid achiral adsorbates is demonstrated by the system of benzene on Ni{111} and has been generalised through Monte Carlo simulations of rigid tripod and cruciform adsorbates on generalised substrates. Despite the rigidity of the adsorbates and substrates in these cases, local chirality emerges due to their asymmetric orientation relative to one another. Alternatively, it is of course possible for homochiral domains to form upon adsorption of more flexible molecules, with asymmetric deformation accounting for local chirality within each domain. Here, the case of glycine adsorbed on Cu{311} provides an exemplary instance. The footprint for three-point binding on *fcc* {311} surfaces is an isosceles triangle, so there is no tendency towards forming heterochiral domains. In the low-coverage regime, there is no evidence for chirality, but once two-point binding becomes important, close to saturation coverage, domains of opposing chirality are formed. Note that the chirality of a domain may be expressed purely in the contents of a surface unit cell of achiral shape, as in the benzene example, or it may *also* be expressed in the existence of an irreducibly oblique surface unit cell, as in the glycine example.

Chiral Phases

When enantiopure adsorbates are used, the resulting surface structure is necessarily chiral (unless racemisation takes place upon adsorption) but this may not be readily apparent in typical experiments. Where the chirality is expressed by the formation of a chiral two-dimensional lattice (i.e. where the surface unit cell is irreducibly oblique) this will be evident from examination of LEED patterns or STM images, as many of the examples presented in this chapter attest. It is entirely possible, however, for the surface unit cell to take an achiral shape, even though its contents may be chiral, and in such circumstances the situation will be less clear; LEED patterns will be symmetric in the spatial disposition of spots (if not the spot intensities) and STM images may show asymmetry only in a manner that invites the invocation of tip artefacts and scanner miscalibration as alternative explanations. To cite just a couple of examples, the low-coverage adsorption of alanine on either Cu{110} or Cu{311} gives rise to LEED patterns and STM images that betray no indication of the chiral structure of the adsorbates; only at higher coverages does an unambiguous chiral two-dimensional lattice arise.

Role of Adsorbate Flexibility

The role of adsorbate flexibility has already been alluded to, but it bears underlining that the ability to form extended hydrogen-bonded networks relies upon the facility with which adsorbed molecules can twist to orient their functional groups in favourable directions. The importance of this fact is demonstrated particularly clearly by the adsorption on proline on Cu{110}, where the high-coverage reorientation from μ_3 to μ_2 binding, observed in the cases of glycine and alanine, simply does not occur; the most natural explanation for this difference lies in the additional stiffness conferred upon the molecular backbone of proline by its cyclic nature. The μ_2 species *is* seen upon adsorption of proline on Cu{311}, but its appearance in the dosing/annealing sequence differs from the cases of glycine and alanine, and the long-range order of the system is rather less strict.

Two-, Three- and Four-Point Binding

We have seen that amino acids are typically capable of forming long-range networks, stabilised by intermolecular hydrogen bonds, when adsorbed at moderate coverage in a μ_3 configuration. The resulting two-dimensional network is often achiral in nature, or else domains of opposing lattice chirality are formed. The chirality, or otherwise, of the adsorbate molecules seems to be of secondary importance in these situations. In contrast, when the coverage is increased beyond the point at which μ_3 molecules would saturate the surface atoms, it becomes necessary for some fraction of the adsorbed molecules to adopt a μ_2 configuration, in order to 'free up' additional sites. It is apparent that new opportunities for hydrogen bonding arise as a side effect, and these may stabilise chains of μ_2 molecules extending over many nanometres in directions that break the symmetry of the surface. It is thus the μ_2 molecules that are critical in linking the highly localised chirality of individual molecules to the long-range chirality of the two-dimensional network. When the adsorbates feature an additional potential binding functionality, such as an alcohol, propylammonium or thiol group, the possibility of μ_4 binding arises, but only at low coverage and typically (in those few cases where it has been reported) associated with an achiral two-dimensional lattice.

Formation of Oligomers

Apart from raising the possibility of μ_4 binding, another consequence of an amino acid having more than three potential loci for hydrogen bonding is that the formation of distinct oligomers may be favourable. In the adsorption of serine on Cu{110}, for example, serinate dimers are formed, with each molecule bound to the surface in a μ_3 configuration and to one neighbour as *either* a double hydrogen bond donor *or* a double acceptor; dimers have also been reported upon the adsorption of lysine on the same surface. In contrast, well-defined octomers have been identified in the case of cysteinate on the Au{110} surface, in all probability involving binding to the surface through a thiolate group while the carboxylate and amine groups are responsible for the intermolecular interactions. Dimers and tetramers comprising both anionic and zwitterionic μ_2 species are formed by alanine on Pd{111}. Looking beyond the amino acids, dimers have been reported upon adsorption of either succinic or tartaric acid on Cu{110}, whilst tetramers of guanine have been reported on the

Au{111} surface. Clearly, the formation of localised oligomers may, in itself, reflect the chirality or otherwise of the constituent monomers, but may equally limit the transmission of molecular chirality to the long-range chirality of the two-dimensional network. As ever, the devil lies in the detail.

Formation of Facets

In summarising the issues surrounding the facetting of achiral surfaces under the influence of adsorbed molecules, we begin by noting that even achiral adsorbates are capable of creating chiral facets; the crucial difference when chiral adsorbates are involved is merely the possibility of creating facets of one handedness in preference to their mirror images. Indeed, in the examples presented within this chapter, it is likely that the orientation of the facets has as much to do with the underlying 'fault lines' of the surface (i.e. the primary, secondary and tertiary zones) as it has to do with the details of the adsorbate. Considering just three simple amino acids on a single surface (i.e. Cu{100}) we have outlined a range of behaviour including: a racemic mixture of chiral facets due to adsorption of prochiral molecules (glycine); a racemic mixture of chiral facets despite the adsorption of chiral molecules (serine); and an enantiopure set of chiral facets due to adsorption of chiral molecules (lysine).

Formation of Metallorganic Frameworks

The majority of studies discussed above relate to rather unreactive substrates, to some degree because more active metals tend to dissociate the relatively fragile molecules in whose chiral properties we are interested. Exceptions include platinum and nickel surfaces, but extensive discussion of these has been deliberately deferred until Chapter 6, since they sit more comfortably in the context of asymmetric catalysis rather than chiral structure per se. Nevertheless, in this chapter we have reviewed a variety of structures formed when both reactive and unreactive metals are present together at the surface. We have seen that when nickel is available at a gold surface the formation of nickel pyroglutamate, nickel lysinate or nickel prolinate is really quite facile. Chain-like structures are common, manifesting chirality through their asymmetric orientation with respect to the high-symmetry substrate axes. In the case of nickel prolinate, however, a hexagonal porous structure is also evident and, although the overlayer lattice is clearly achiral in isolation, its registry with the substrate again reveals an overall chirality. The combination of highly reactive single-atom cationic sites surrounded by chiral (or at least chirally disposed) anionic ligands holds great promise for possible utility in chiral catalysis.

References

1 J.H. Onuferko, D.P. Woodruff and B.W. Holland, *Surf. Sci.* **87**, 357 (1979).
2 W. Daum, S. Lehwald and H. Ibach, *Surf. Sci.* **178**, 528 (1986).
3 S.M. Driver, R.L. Toomes and D.P. Woodruff, *Surf. Sci.* **646**, 114 (2016).
4 S. Karakatsani, Q. Ge, M.J. Gladys, G. Held and D.A. King, *Surf. Sci.* **606**, 383 (2012).

5 S. Yamagishi, S.J. Jenkins and D.A. King, *J. Chem. Phys.* **114**, 5765 (2001).
6 T. Zhang, Z.-P. Liu, S.M. Driver, S.J. Pratt, S.J. Jenkins and D.A. King, *Phys. Rev. Lett.* **95**, 266102 (2005).
7 T. Zhang, D.A. King and S.M. Driver, *J. Chem. Phys.* **137**, 074703 (2012).
8 A.J. Gellman, *ACS Nano* **4**, 5 (2010).
9 P. Szabelski and A. Kasperski, *Top. Catal.* **54**, 1368 (2011).
10 A. Kasperski and P. Szabelski, *Adsorption* **19**, 283 (2014).
11 A. Kasperski and P. Szabelski, *Surf. Sci.* **629**, 57 (2014).
12 M. Parschau, D. Passerone, K.-H. Rieder, H.J. Hug and K.-H. Ernst, *Angew. Chem. Int. eEd.* **48**, 4065 (2009).
13 O. Stetsovych, M. Švec, J. Vacek, J. Vacek Chocholoušvová, A. Jančařík, J. Rybáček, K. Kosmider, I.G. Stará, P. Jelínek and I. Starý, *Nature Nat. Chem.* **9**, 213 (2017).
14 K.-H. Ernst, *Nature Nat. Chem.* **9**, 195 (2017).
15 W. Xu, R.E.A. Kelly, H. Gersen, E. Lægsgaard, I. Stensgaard, L.N. Kantorovich and F. Besenbacher, *Small* **5**, 1952 (2009).
16 M. Lukas, R.E.A. Kelly, L.N. Kantorovich, R. Otero, W. Xu, E. Lægsgaard, I. Stensgaard and F. Besenbacher, *J. Chem. Phys.* **130**, 024705 (2009).
17 M. Böhringer, W.-D. Schneider and R. Berndt, *Angew. Chem. Int. Ed.* **39**, 792 (2000).
18 T.G. Gopakumar, F. Matino, B. Schwager, A. Bannwarth, F. Tuczek, J. Kröger and R. Berndt, *J. Phys. Chem. C* **114**, 18247 (2010).
19 B. Yang, Y. Yang, H. Cun, S. Du, M. Xu, Y. Wang, K.-H. Ernst and H.-J. Gao, *J. Am. Chem. Soc.* **132**, 10440 (2010).
20 R. Cortés, A. Mascaraque, P. Schmidt-Weber, H. Dil, T.U. Kampen and K. Horn, *Nano Lett.* **8**, 4162 (2008).
21 F. Vidal, E. Delvigne, S. Stepanow, N. Lin, J.V. Barth and K. Kern, *J. Am. Chem. Soc.* **127**, 10101 (2005).
22 S.M. Barlow, K.J. Kitching, S. Haq and N.V. Richardson, *Surf. Sci.* **401**, 322 (1998).
23 D.C. Madden, I. Temprano, M. Sacchi and S.J. Jenkins, *J. Phys. Chem. C* **119**, 13041 (2015).
24 R.B. Rankin and D. Sholl, *Surf. Sci.* **548**, 301 (2004).
25 A. Shavorskiy, F. Aksoy, M.E. Grass, Z. Liu, H. Bluhm and G. Held, *J. Am. Chem. Soc.* **133**, 6659 (2011).
26 Q. Chen, D.J. Frankel and N.V. Richardson, *Surf. Sci.* **497**, 37 (2002).
27 R.L. Toomes, J.-H. Kang, D.P. Woodruff, M. Polcik, M. Kittel and J.-T. Hoeft, *Surf. Sci.* **522**, L9 (2003).
28 Z.V. Zheleva, T. Eralp and G. Held, *J. Phys. Chem. C* **116**, 618 (2012).
29 D.C. Madden, M.L. Bentley, S.J. Jenkins and S.M. Driver, *Surf. Sci.* **629**, 81 (2014).
30 V. Humblot, M.O. Ortega Lorenzo, C.J. Baddeley, S. Haq and R. Raval, *J. Am. Chem. Soc.* **126**, 6460 (2004).
31 M. Parschau, S. Romer and K.-H. Ernst, *J. Am. Chem. Soc.* **126**, 15398 (2004).
32 V. Humblot and R. Raval, *Appl. Surf. Sci.* **241**, 150 (2005).
33 C. Karageorgaki, D. Passerone and K.-H. Ernst, *Surf. Sci.* **629**, 75 (2014).
34 M. Parschau, T. Kampen, and K.-H. Ernst, *Chem. Phys. Lett.* **407**, 433 (2005).

35 M. Parschau, B. Behzadi, S. Romer and K.-H. Ernst, *Surf. Interface Anal.* **38**, 1607 (2006).
36 S.M. Barlow, S. Louafi, D. Le Roux, J. Williams, C. Muryn, S. Haq and R. Raval, *Surf. Sci.* **590**, 243 (2005).
37 S.M. Barlow, S. Louafi, D. Le Roux, J. Williams, C. Muryn, S. Haq and R. Raval, *Langmuir* **20**, 7171 (2004).
38 G. Jones, L.B. Jones, F. Thibault-Starzyk, E.A. Seddon, R. Raval, S.J. Jenkins and G. Held, *Surf. Sci.* **600**, 1924 (2006).
39 S. Haq, A. Massey, N. Moslemzadeh, A. Robin, S.M. Barlow and R. Raval, *Langmuir* **23**, 10694 (2007).
40 J. Seifert, M. Busch, E. Meyer and H. Winter, *Phys. Rev. Lett.* **111**, 137601 (2013).
41 J. Seifert, M. Busch, E. Meyer and H. Winter, *Phys. Rev. B* **89**, 075404 (2014).
42 R.B. Rankin and D.S. Sholl, *Surf. Sci.* **574**, L1 (2005).
43 D.I. Sayago, M. Polcik, G. Nisbet, C.L.A. Lamont and D.P. Woodruff, *Surf. Sci.* **590**, 76 (2005).
44 G. Jones, S.J. Jenkins and D.A. King, *Surf. Sci.* **600**, L224 (2006).
45 M.L. Clegg, L. Morales de la Garza, S. Karakatsani, D.A. King and S.M. Driver, *Top. Catal.* **54**, 1429 (2011).
46 D.C. Madden, I. Temprano, M. Sacchi, M. Blanco-Rey, S.J. Jenkins and S.M. Driver, *J. Phys. Chem. C* **118**, 18589 (2014).
47 E. Mateo-Marti, S.M. Barlow, S. Haq and R. Raval, *Surf. Sci.* **501**, 191 (2002).
48 D.C. Madden, I.T. Temprano, S.J. Jenkins and S.M. Driver, *Top. Catal.* **58**, 591 (2015).
49 M. Forster, M.S. Dyer, M. Persson and R. Raval, *J. Am. Chem. Soc.* **131**, 10173 (2009).
50 M. Forster, M.S. Dyer, M. Persson and R. Raval, *Angew. Chem. Int. Ed.* **49**, 2344 (2009).
51 T. Eralp, A. Shavorskiy, Z.V. Zheleva, G. Held, N. Kalashnyk, Y. Nang and T.R. Linderoth, *Langmuir* **26**, 18841 (2010).
52 S.M. Barlow and R. Raval, *Surf. Sci. Rep.* **50**, 201 (2003).
53 V. Humblot, C. Méthivier and C.-M. Pradier, *Langmuir*, **22**, 3089 (2006).
54 T. Eralp, A. Shavorskiy and G. Held, *Surf. Sci.* **605**, 468 (2011).
55 V. Humblot, C. Méthivier, R. Raval and C.-M. Pradier, *Surf. Sci.* **601**, 4189 (2007).
56 E. Mateo-Marti, C. Methivier and C.-M. Pradier, *Langmuir* **20**, 10223 (2004).
57 J.W. Kim, H.-N. Hwang and C.-C. Hwang, *J. Phys. Chem. C* **111**, 13192 (2007).
58 J.W. Kim, Y.M. Lee, S.M. Lee, M.J. Son, H. Kang and Y. Park, *Langmuir* **26**, 5632 (2010).
59 A. Kühnle, T.R. Linderoth and F. Besenbacher, *J. Am. Chem. Soc.* **125**, 14680 (2003).
60 A. Kühnle, T.R. Linderoth, B. Hammer and F. Besenbacher, *Nature* **415**, 891 (2002).
61 A. Kühnle, T.R. Linderoth and F. Besenbacher, *Top. Catal.* **54**, 1384 (2011).
62 A. Kühnle, L.M. Molina, T.R. Linderoth, B. Hammer and F. Besenbacher, *Phys. Rev. Lett.* **93**, 086101 (2004).

63 A. Kühnle, T.R. Linderoth and F. Besenbacher, *J. Am. Chem. Soc.* **128**, 1076 (2006).
64 G. Gonella, S. Terreni, D. Cvetko, A. Cossaro, L. Mattera, O. Cavalleri, R. Rolandi, A. Morgante, L. Floreano and M. Canepa, *J. Phys. Chem. B* **109**, 18003 (2005).
65 M.O. Ortega Lorenzo, S. Haq, T. Bertrams, P. Murray, R. Raval and C.J. Baddeley, *J. Phys. Chem. B* **103**, 10661 (1999).
66 D.A. Duncan, W. Unterberger, D.C. Jackson, M.K. Knight, E.A. Kröger, K.A. Hogan, C.L.A. Lamont, T.J. Lerotholi and D.P. Woodruff, *Surf. Sci.* **606**, 1435 (2012).
67 M.O. Ortega Lorenzo, V. Humblot, P. Murray, C.J. Baddeley, S. Haq and R. Raval, *J. Catal.* **205**, 123 (2001).
68 M.O. Ortega Lorenzo, C.J. Baddeley, C. Muryn and R. Raval, *Nature* **404**, 376 (2000).
69 L.A.M.M. Barbosa and P. Sautet, *J. Am. Chem. Soc.* **123**, 6639 (2001).
70 S. Romer, B. Behzadi, R. Fasel and K.-H. Ernst, *Chem. Eur. J.* **11**, 4149 (2005).
71 T.J. Lawton, V. Pushkarev, D. Wei, F.R. Lucci, D.S. Sholl, A.J. Gellman and E.C.H. Sykes, *J. Phys. Chem. C* **117**, 22290 (2013).
72 C. Karageorgaki and K.-H. Ernst, *Chem. Commun.* **50**, 1814 (2014).
73 T.E. Jones, C.J. Baddeley, A. Gerbi, L. Savio, M. Rocca and L. Vattuone, *Langmuir* **21**, 9468 (2005).
74 M. Smerieri, L. Vattuone, T. Kravchuk, D. Costa, and L. Savio, *Langmuir* **27**, 2393 (2011).
75 M. Smerieri, L. Vattuone, D. Costa, F. Tielens and L. Savio, *Langmuir* **26**, 7208 (2010).
76 M.L. Liriano, J. Carrasco, E.A. Lewis, C.J. Murphy, T.J. Lawton, M.D. Marcinkowski, A.J. Therrien, A. Michaelides and E.C.H. Sykes, *J. Chem. Phys.* **144**, 094703 (2016).
77 T.E. Jones and C.J. Baddeley, *Surf. Sci.* **513**, 453 (2002).
78 F. Gao, Z. Li, Y. Wang, L. Burkholder and W.T. Tysoe, *Surf. Sci.* **601**, 3276 (2007).
79 F. Gao, Z. Li, Y. Wang, L. Burkholder and W.T. Tysoe, *J. Phys. Chem. C* **111**, 9981 (2007).
80 F. Gao, Y. Wang, L. Burkholder and W.T. Tysoe, *Surf. Sci.* **601**, 3579 (2007).
81 M. Mahapatra, L. Burkholder, Y. Bai, M. Garvey, J.A. Boscoboinik, C. Hirschmugl and W.T. Tysoe, *J. Phys. Chem. C* **118**, 6856 (2014).
82 J.N. James and D.S. Sholl, *J. Mol. Catal. A: Chem.* **281**, 44 (2008).
83 X. Zhao, Z. Gai, R.G. Zhao, W.S. Yang and T. Sakuri, *Surf. Sci.* **424**, L347 (1999).
84 X. Zhao, *J. Am. Chem. Soc.* **122**, 12584 (2000).
85 X. Zhao, R.G. Zhao and W.S. Yang, *Surf. Sci.* **442**, L995 (1999).
86 X. Zhao, R.G. Zhao and W.S. Yang, *Langmuir* **16**, 9812 (2000).
87 W.Y. Cheong and A.J. Gellman, *Langmuir* **28**, 15251 (2012).
88 D.D. Chambliss, R.J. Wilson and S. Chiang, *Phys. Rev. Lett.* **66**, 1721 (1991).
89 A.G. Trant, T.E. Jones and C.J. Baddeley, *J. Phys. Chem. C* **111**, 10534 (2007).
90 T.E. Jones, M.E. Urquhart and C.J. Baddeley, *Surf. Sci.* **69**, 587 (2005).

91 A.G. Trant, T.E. Jones, T.C.Q. Noakes, P. Bailey and C.J. Baddeley, *Surf. Sci.* **604**, 300 (2010).

92 K.E. Wilson, H.A. Früchtl, F. Grillo and C.J. Baddeley, *Chem. Commun.* **47**, 10365 (2011).

93 K.E. Wilson and C.J. Baddeley, *Surf. Sci.* **629**, 102 (2014).

94 R.T. Seljamäe-Green, G.J. Simpson, F. Grillo, J. Greenwood, S.M. Francis, R. Schaub, J.E. Gano, H.A. Früchtl, P. Lacovig and C.J. Baddeley, *Langmuir* **31**, 262 (2015).

95 R.T. Seljamäe-Green, G.J. Simpson, F. Grillo, J. Greenwood, S.M. Francis, R. Schaub, P. Lacovig and C.J. Baddeley, *Langmuir* **30**, 3495 (2014).

5

Asymmetric Adsorption on Chiral Substrates

The road of excess leads to the palace of wisdom.
The Marriage of Heaven and Hell, William Blake, 1790

The preceding chapter examined in depth myriad instances in which chirality can arise upon adsorption onto otherwise achiral substrates. These instances include racemic manifestation of local chirality, when the adsorbates are themselves achiral or prochiral, in addition to global chirality induced by the adsorption of enantiopure chiral molecules. Now we must turn to the adsorption of molecules on substrates that are already chiral. In doing so, this chapter will focus first upon intrinsically chiral surface orientations, of the kind introduced in Chapter 2, before turning to a brief consideration of chirally modified substrates; cases falling into the latter category but more appropriately considered under the headings of chiral amplification or chiral catalysis will, however, be deferred for the following two chapters.

5.1 Achiral Adsorbates on Intrinsically Chiral Substrates: Fault-Lines and Facets

Clearly, it ought not to be expected that interesting diastereomeric effects would arise upon the adsorption of achiral species onto chiral surfaces. Nevertheless, it is certainly the case that adsorption can alter the surface energies of different facets, giving rise to the possibility of facetting. This phenomenon, in turn, can modify the specific manifestation of chirality at the surface, potentially transforming the two-dimensional chirality of the ideal surface into the one-dimensional chirality of ridge-and-furrow facetted surfaces, or the linear and point-like chirality of pyramid-and-pit facetted surfaces (as per the discussion in Section 3.6).

5.1.1 Oxygen on Cu{531}

The adsorption of oxygen on high-index copper surfaces has been studied relatively intensively, albeit rarely in chiral situations (see, e.g. the work of Knight *et al.* [1, 2]). A common result reported in several studies is the formation

Chirality at Solid Surfaces, First Edition. Stephen J. Jenkins.

of facets, with the {410} surface planes being particularly favoured. This presumably indicates a strong thermodynamic preference for the adsorption of oxygen onto these favoured facets, but the exact adsorption site has yet to be convincingly determined. As ever, the lowering of specific surface energy associated with facet formation must be balanced against a concomitant increase in surface area, as described in Section 3.6, so the tendency to form {410} facets is likely to be less pronounced the further removed the macroscopic surface normal is from the corresponding poles. The Cu{531} surface, for instance, seems unlikely to form {410} facets, and indeed this expectation appears to be borne out.

In the course of preparing Cu{531} samples for low-energy electron diffraction (LEED) studies of its structure, Jones *et al.* [3] found that the removal of carbon and sulphur contaminants was best effected through annealing at 350–500 K in 10^{-7} mbar O_2. They cautioned, however, that this treatment must be avoided in the final stage of cleaning, to avoid facetting of the surface due to adsorbed oxygen. The precise facet orientations were not identified, but first-principles density functional theory (DFT) calculations, reported in the same work, indicated that two possible scenarios of ridge-and-furrow facetting ought actually to be favourable (from an enthalpy point of view) even for the clean surface. These scenarios involve the formation either of {110} and {311} facets, or of {111} and {210} facets; the former is calculated to produce a lowering of the specific surface energy by 0.0006 eV $Å^{-2}$ (0.010 J m^{-2}) and the latter by half that. The most favourable pyramid-and-pit facet combination (involving {100}, {111} and {110} facets) was associated with an increase in specific surface energy by as much as the most stable ridge-and-furrow combination was favoured.

Since the clean Cu{531} surface is *not* observed to facet (see also Clegg *et al.* [4]) we are faced with three possible alternatives: (i) that DFT is simply insufficiently accurate to faithfully capture the relative energy ordering in this rather subtle balancing act; (ii) that the entropy of atomic-scale roughening via adatom–vacancy pair creation is sufficient to dominate the enthalpy when considering the free energy of the system; or (iii) that kinetic considerations hinder facet formation over the duration of an experiment. Nevertheless, it is reasonable to suppose that the theoretically preferred facet combinations of the clean surface may well represent preferred modes of facetting that can be 'activated' by adsorption. It is worth noting that unpublished scanning tunnelling microscopy (STM) data relating to the action of oxygen on the Pt{531} surface does indeed indicate a preference for ridge-and-furrow facetting of the {110}/{311} variety [5]. Clearly, there is danger in extrapolating between such dissimilar metals, but there are certainly grounds for speculation. In this eventuality, the {110} and {311} facets would be angled at 17.0° and 22.0° to the macroscopic surface plane, respectively. Applying Eqs 3.63 and 3.64, we find that the {110} facets should ideally extend to an area equivalent to just over 46% of the total pre-facetting area, while the {311} facets should extend to just under 60% of the same.[1] The shared surface direction of these facets, corresponding to the line of the ridge tops and furrow bottoms,

---

1 Note that these facet areas sum to more than 100% because they are expressed relative to the pre-facetting surface area. Expressed in terms of the total post-facetting surface area, the {110} and {311} extend to just under 44% and just over 56% respectively.

is of $\langle\bar{1}12\rangle$ type and therefore corresponds to a secondary structural zone in the classification scheme of Chapter 2. In the hierarchy of surfaces, the {531} type is picked out at the intersection of two secondary zones, whereas the {110} and {311} surfaces are both found at intersections of a single primary structural zone with multiple secondary zones; the expression of facets would thus satisfy the conjecture of Jenkins and Pratt [6] that stability would be favoured when tracking along structural zones towards poles located at more major intersections than that of the non-facetted surface.

5.1.2 Cyclohexanone on Cu{643}

Cyclohexanone is an achiral cyclic ketone, structurally similar to cyclohexane but with the hydrogen atoms attached to one of its carbon atoms replaced by a double-bonded oxygen atom (*see Plate XIV*). Upon room-temperature adsorption onto the Cu{643}-D surface, Zhao and Perry [7] observed ridge-and-furrow facetting in their STM images (see Fig. 5.1) with a shared surface direction of $\langle 1\bar{3}2\rangle$ type (i.e. a tertiary structural zone, according to the scheme outlined in Chapter 2). Measurement of lattice parameters led the authors to identify the facets as having {11 7 5} and {532} orientation; of these, the former makes an angle of just under 1.9° with the macroscopic surface plane, while the latter is slightly more steeply angled at just under 4.4°. Applying Eqs 3.63 and 3.64, we can deduce that the {11 7 5} facets should extend to roughly 70% of the total pre-facetting surface area, whilst the {532} facets should extend to roughly 30%.[2] Once again, the facet combination accords with the facetting conjecture of Jenkins and Pratt [6], in that the {643} surface, which in the hierarchy of surfaces is not found even as an intersection of tertiary zones, breaks up along a tertiary zone into two poles that do.

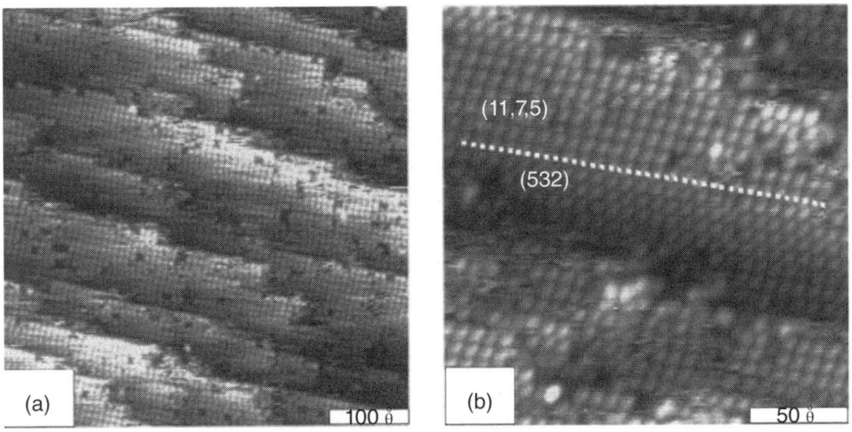

Figure 5.1 STM images obtained after room-temperature exposure of Cu{643}-D to cyclohexanone, over scan areas of (a) 500×500 Å^2, and (b) 200×200 Å^2. Reproduced with permission from [7]. Copyright (2004) Elsevier.

2 On this occasion, the observed facets lie at such shallow angles with respect to the non-facetted surface that it matters little whether the pre- or post-facetting area is used for reference.

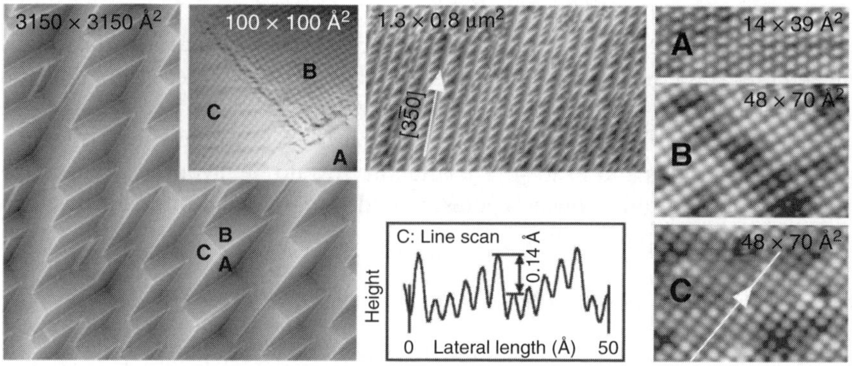

Figure 5.2 STM images obtained after exposure of Cu{532}-L to NaCl at 600 K (various scan areas). Facets marked B and C correspond to {311} and {531} orientations, respectively, and are covered by {001}-oriented films of NaCl; the facets marked A are of {111} orientation, and remain bare. The line scan reveals that occasional terrace edges exist within the {531} facets. Reproduced with permission from [8]. Copyright (2002) American Physical Society.

5.1.3 NaCl on Cu{532}

An unusual choice of adsorbate was made by Rieder and co-workers [8], who deposited NaCl onto a Cu{532} surface. Their STM images (see Fig. 5.2) revealed the formation of particularly crisply defined pyramid-and-pit facets of {111}, {311} and {531} orientation (respectively at angles of 20.5°, 11.9° and 9.2° to the macroscopic surface plane). Here, two of the newly created facet orientations are entirely achiral within themselves, although the boundaries between these facets must possess a chiral structure (i.e. linear substrate chirality). The third facet orientation does, of course, display planar substrate chirality within itself, but in addition the *mesoscopic* character of the surface has gained a chiral aspect, as the pyramids and pits of the facetted surface are asymmetric on a length-scale of many nanometres. Applying Eqs 3.70–3.72, we find that the {111}, {311} and {531} facets should extend to roughly 21%, 36% and 46% of the total pre-facetting surface area, respectively.[3] In terms of the facetting conjecture proposed by Jenkins and Pratt [6], the facet combination achieved here may best be thought of as facetting along a $\langle 1\bar{3}2 \rangle$ tertiary zone to produce {111} and {421} facets, followed (conceptually, if not temporally) by sub-facetting of the latter along a $\langle 1\bar{1}\bar{2} \rangle$ secondary zone to produce {311} and {531} facets.

5.2 Prochiral Adsorbates on Intrinsically Chiral Substrates: Familiar and Strange

On achiral surfaces, the adsorption of prochiral molecules was not sufficient to lead to global chirality, but it is certainly true that local chirality can readily

3 Once again, these facet areas sum to more than 100% because they are expressed relative to the pre-facetting surface area. Expressed in terms of the total post-facetting surface area, the {111}, {311} and {531} facets extend to roughly 20%, 35% and 45% respectively.

occur, albeit limited to finite domains occurring in both chiral forms with equal probability. On intrinsically chiral surfaces, however, the inherent asymmetry of the substrate is likely to skew the probability towards one chiral adsorption configuration in preference to its mirror image, leading in turn to the likelihood of chiral long-range order. As we shall see, the local binding of by now familiar probe molecules, such as amino acids, can be readily explained in terms of the bonding modes seen previously on achiral surfaces.

5.2.1 Glycine on Cu{531}

The adsorption of glycine on Cu{531} has been studied thoroughly by Held and co-workers [9], making use of several surface science techniques, notably LEED, near-edge X-ray absorption fine structure (NEXAFS) and X-ray photoemission spectroscopy (XPS). Upon adsorption at temperatures in the range 300–430 K, XPS reveals a single O 1s peak (c.531.3 eV) indicating that all oxygen atoms experience near-identical chemical environments and hence suggesting that the adsorbed molecules are deprotonated at the acid group. Similarly, the N 1s region generally shows a single peak (399.6–400.2 eV) consistent with the $—NH_2$ functional group, so the most reasonable interpretation is that adsorption occurs in the form of anionic glycinate moieties, just like on the achiral copper surfaces described in Section 4.2.3. The only complication is a small subsidiary peak at 397.7 eV in the N 1s region of the spectrum, found only at 300 K, and even then only in the low-coverage regime; the authors attribute this to decomposition products, but note that it is absent at higher temperatures and coverages.

Turning to NEXAFS experiments, these were performed with polarised X-rays impinging at normal incidence, rotating the sample about its surface normal to vary the azimuthal polarisation relative to the surface crystallography. When focusing upon the excitation from O 1s orbitals, the size of the NEXAFS signal is determined by the symmetry of only those unoccupied orbitals that have significant weight on the carboxylate group, and the lowest energy of these is known to have π symmetry with respect to the —COO plane. If all adsorbed molecules share the same orientation, it follows that the NEXAFS resonance corresponding to this state should vary as the squared cosine of the azimuthal polarisation angle, reducing to zero when the polarisation vector lies parallel to the vector linking the two oxygen atoms within the group. In fact, however, this NEXAFS resonance never falls below about half of its maximum value, demonstrating very clearly that the adsorbed molecules must adopt *at least two* distinctly different geometries. Modelling the adsorbates with *precisely two* adsorption geometries, the authors obtain an excellent fit to their experimental data when half the molecules adopt an orientation consistent with μ_3 adsorption onto {110} microfacets and the other half adopt an orientation consistent with μ_3 adsorption onto {311} microfacets; these results are broadly independent of both coverage and temperature [9].

As regards long-range order, Held and co-workers report evidence for a superstructure only when high-coverage overlayers are annealed to 370 K, and even then the resulting LEED pattern is streaky. Under these circumstances, the observed periodicity (on the D surface) may be described in matrix notation

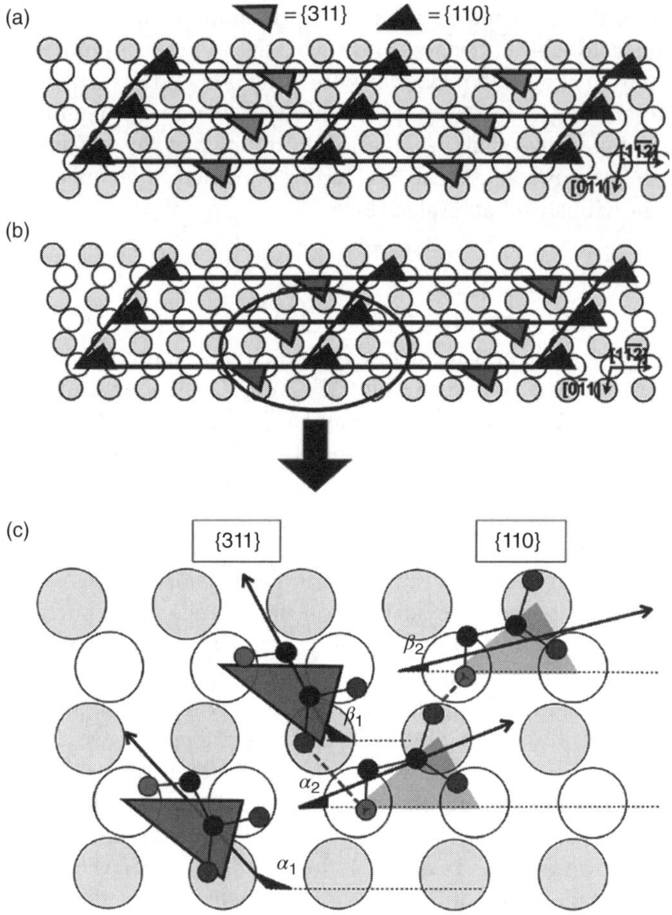

Figure 5.3 Schematic models of glycinate adsorbed on Cu{531}; only the uppermost two layers of copper atoms are shown, for clarity. Glycinate moieties occupy {311} and {110} microfacets equally, at either (a) the maximum or (b) the minimum spacing consistent with the observed surface periodicity. In (c) the favoured model, stabilised by hydrogen bonding, is shown in detail. Reproduced with permission from [9]. Copyright (2010) American Chemical Society.

as (1, 1; 0, $\bar{6}$).[4] Evidently, such a unit cell contains only six top-layer Cu atoms, and since adsorption onto either the {110} or the {311} microfacets involves two top-layer Cu atoms per molecule, it follows that there must be precisely one, two or three molecules per unit cell in the overlayer. Given that the NEXAFS data imply an equal occupancy of the two microfacet types, only the option of two molecules per unit cell can be consistent, giving a nominal coverage for the (1, 1; 0, $\bar{6}$) overlayer of 1/3 ML. Figure 5.3 shows schematically how such an overlayer would be organised, and highlights the stabilisation achieved through hydrogen bonding within two-molecule-wide stripes oriented along a ⟨0$\bar{1}$3⟩ surface direction.

4 Originally described as p(1×6).

5.3 Chiral Adsorbates on Intrinsically Chiral Substrates: Diastereomeric Effects I

Thus far, we have considered cases where the adsorbate is chiral but the substrate is not (in the preceding chapter) and where the substrate is chiral but the adsorbate is not (preceding sections within the present chapter). Now we turn to the scenario where substrate and adsorbate are *both* chiral, raising the possibility of diastereomeric effects. That is, for a surface of specific chirality, we might anticipate finding preferential adsorption of one molecular enantiomer over its mirror image, or at least different adsorption geometries for the different enantiomers. As we shall shortly see, the mechanism of **chiral recognition**[5] responsible for this varies from case to case, but generally (at least on face-centred cubic metals) depends upon multiple points of contact between adsorbate molecules and surface kink sites. Once again, most published work relates to the adsorption of amino acids on copper, so this is where we shall begin.

5.3.1 Alanine on Cu{531}

The adsorption of alanine on Cu{531} has been studied by a variety of techniques, several of which overlap with those deployed in the study of glycine adsorption onto the same surface. Starting with these, then, Held and co-workers [10] showed that XPS spectra obtained upon adsorption of either D- or L-alanine on Cu{531}-D[6] were very similar.[7] Moreover, the O 1s region revealed only a single peak (c.531.3 eV) indicating deprotonation of the acid group, while the single peak in the N 1s region (c.399.6 eV) was consistent with the $—NH_2$ functional group. In common with the adsorption of glycine onto Cu{531}, the evidence points firmly towards adsorption in the anionic form (in this case, alaninate). Differences emerge, however, when considering the long-range ordering of adsorbates on the surface. Here, LEED reveals a $(\bar{1}, \bar{1}; 4, 0)$[8] pattern in the high-coverage regime at temperatures in the range 320–400 K, in contrast to the $(1, 1; 0, \bar{6})$ pattern observed upon annealing glycinate overlayers to 370 K. Clearly, the presence of the methyl functionality in alaninate has a significant effect upon the intermolecular interactions, presumably either through the formation of additional (very weak) hydrogen bonds involving the methyl group as a donor, or through steric hindrance of the other hydrogen bonds. Interestingly, the LEED patterns obtained with D- and L-alanine are described as 'similar in principle' but with distinct broadening of the spots in the former case (the enantiospecific difference being reversed for experiments performed on the Cu{531}-L surface).

5 A few words of clarification may be useful here, to distinguish between three closely related but subtly different terms. Any situation in which the two enantiomers of a chiral species behave differently may be described as 'enantiospecific' whereas the consequent pattern of degeneracy and non-degeneracy is described as 'diastereomeric'. We reserve the term 'chiral recognition' for the detailed mechanism that leads to the expression of diastereomeric enantiospecificity in a particular system.

6 Equivalent to $Cu\{531\}^R$.

7 Minor differences between the enantiomers in relative XPS peak intensities were attributed to diffraction effects.

8 Originally described as p(1×4).

Once again, NEXAFS (on this occasion based on excitation of C 1s orbitals[9]) clearly rules out models in which a single adsorption geometry dominates, but there are some differences depending upon which enantiomer is adsorbed. For L-alanine, the data clearly favour a model in which both {110} and {311} microfacets are equally occupied by μ_3 alaninate, which is very similar to the situation upon adsorption of glycine; as before, a model in which there are two molecules per unit cell must be favoured, although in this case the nominal coverage would be 1/2 ML, not 1/3 ML. Results for D-alanine are a little more equivocal, with a model in which occupancy of {110} sites is almost double that of {311} matching the data just as well as the equal occupancy model. Given the size of the unit cell, however, it is hard to imagine how three molecules could be accommodated, but it should be borne in mind that it *is* the D-alaninate that gives rise to a much broadened LEED pattern, indicative of poor long-range order. Obviously, the behaviours upon adsorption of D- and L-alanine should again be understood to be reversed when using a Cu{531}-L substrate.

A subsequent study, by Driver and co-workers [4], explored the effect upon alaninate overlayers on Cu{531}-L of slightly higher annealing temperature – 420 K, as opposed to the maximum of 400 K used in the work described above. Remarkably, this small difference leads to radically altered behaviour, characterised by substantial facetting at the mesoscale. Intriguingly, STM images reveal that the preferred facet orientations are enantiospecific, with *different* facets induced by the adsorption of L- and D-alanine (see Fig. 5.4). In the case of D-alanine, there were two facet orientations relatively close to {531}, one of which was tentatively identified as {762}; that being the case, the other would have to be close to {831}. Precise identification of facet orientations is difficult,

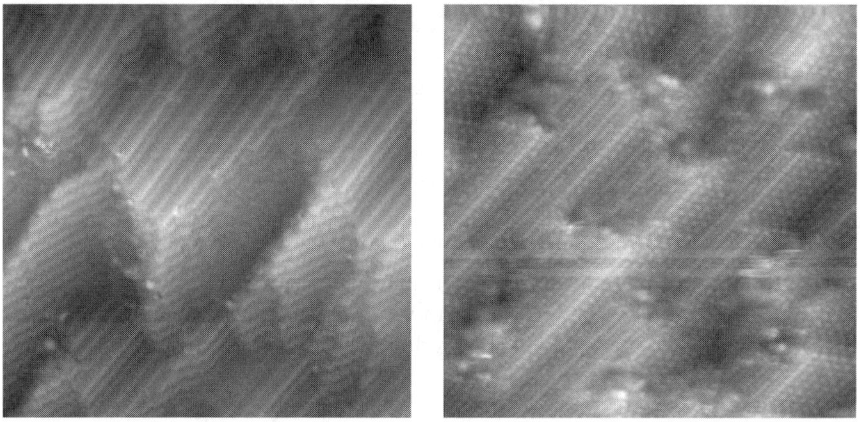

(a) (b)

Figure 5.4 STM images obtained after 300–375 K exposure of Cu{531}-L to (a) L-alanine, or (b) D-alanine, followed by annealing at 420 K. The scan areas are 500×500 Å^2. Reproduced with permission from [4]. Copyright (2011) Springer.

9 The unoccupied π orbital probed in NEXAFS by excitation of O 1s orbitals may be probed equally well by excitation of the C 1s orbitals.

however, owing to uncertainties in the calibration of STM scanners, not to mention thermal drift and Smoluchowski smoothing. In the case of L-alanine, there were again two major facet orientations relatively close to {531}, although neither were identified by the authors.

5.3.2 Serine on Cu{531}

It ought to be recalled from Section 4.3.3 that the adsorption of serine is complicated, in comparison with that of alanine, by virtue of the substitution of one hydrogen from alanine's methyl group by a hydroxyl group. In addition to the usual μ_2 and μ_3 adsorption geometries, therefore, the possibility of μ_4 binding must also be considered. In the case of serine adsorption on Cu{110}, however, only μ_3 species were observed [11].

On the Cu{531} surface, adsorption of serine has been studied in work combining experimental observations from the group of Held with computational modelling from the group of the present author [12]. Upon adsorption to saturation coverage (estimated 1/2 ML) at room temperature, followed by annealing at 370 K, enantiospecific LEED patterns were observed: $(\bar{1}, \bar{1}; 0, \bar{8})$[10] for L-serine and $(\bar{1}, \bar{1}; 0, \bar{4})$[11] for D-serine, both on Cu{531}-L. At lower coverage (c.1/3 ML) the LEED patterns are $(\bar{1}, \bar{2}; 2, 1)$ and $(\bar{1}, \bar{2}; 3, 0)$ for D- and L-serine respectively[12] on the same surface. The patterns obtained for the two enantiomers would, of course, be swapped if the Cu{531}-D surface were studied instead.

In XPS conducted at low coverage, the main O 1s peak, at a binding energy of 531.5 eV (attributed to the deprotonated acid group) was accompanied by a shoulder at 530.8–530.9 eV, consistent with a deprotonated hydroxyl group bound to the surface, implying a μ_4 configuration (a N 1s peak at 400.0 eV, consistent with an intact surface-bound amine group, was also present [13]). At saturation, on the other hand, this shoulder was absent from the O 1s peak, but a new feature was observed at 532.8–532.9 eV, interpreted as indicating the adoption of a μ_3 configuration with an intact hydroxyl group positioned away from the surface. There was little discernible difference between the two enantiomers in terms of the coverage at which this transition between μ_4 and μ_3 binding occurred, and at saturation coverage NEXAFS revealed very little difference in adsorption geometry; both {110} and {311} microfacet sites were found to be occupied equally when only μ_3 species were present (similar to the situation described for glycine and alanine adsorption above). At lower coverage, however, when XPS indicated the presence of μ_4 species, C 1s NEXAFS showed very clear differences between the enantiomers; for D-serine on Cu{531}-L, the spectral variation as a function of polarisation angle was consistent with occupation of the {110} microfacets, while for L-serine on the same surface the {311} microfacets were favoured. First-principles DFT calculations concurred that these geometry assignments were indeed highly reasonable, and that deprotonation of the hydroxyl group was indeed favourable

10 Originally described as p(1×8).
11 Originally described as p(1×4).
12 Originally described as $(\bar{1}, 1; \bar{1}, \bar{2})$ and $(\bar{1}, 1; \bar{2}, \bar{4})$.

in the μ_4 configuration but not in the μ_3 configuration. Overall, these results showed that observed diastereomeric effects in μ_3 species were limited to the long-range order reflected in LEED patterns, whereas for the μ_4 species there were also strong diastereomeric effects at the level of the local binding geometry.

5.3.3 Cysteine on Cu{531} and Au{17 11 9}

In the preceding chapter (Section 4.3.4) we saw that cysteine provided an interesting counterpoint to the relatively consistent modes of adsorption exhibited by glycine, alanine, serine and lysine. In particular, the role of its thiol group was found to be crucial – at low coverages on Cu{110} the evidence pointed towards a μ_4 configuration in which thiolate, amine and carboxylate groups all bind the molecule to the surface, while at high coverage an unusual μ_3 configuration (involving the carboxylate and thiol groups but *not* the amine group) was favoured. After adsorption of L-cysteine at 300 K on the Cu{531}-L surface, however, XPS experiments indicate adsorption in a μ_4 configuration independent of coverage [14]. In this case, the O 1s region displays a single peak at 531.48 eV, consistent with the presence of surface-bound carboxylate, while the single N 1s peak at 399.93 eV is indicative of an intact surface-bound amine group. The S 2p peaks are harder to interpret, due to their doublet structure, but deconvolution suggests the presence of two distinct sulphur species – one with a $2p_{3/2}$ binding energy of c.160.7 eV, attributed to isolated sulphur adatoms, and another with a $2p_{3/2}$ binding energy of c.161.7 eV, attributed to surface-bound thiolate.[13] It is thus clear that at least *some* of the adsorbed molecules decompose to produce the adatoms, and this reaction is found to be enhanced at higher temperatures. Upon gradual annealing up to 480 K, the authors demonstrate essentially complete decomposition by around 400 K, whereupon the S 2p region of the spectrum shows only the fingerprint of adsorbed atomic sulphur, whilst the C, N and O 1s peaks match closely those obtained upon the adsorption of alanine; the obvious inference is compelling – that cysteine has decomposed to form alanine. It should be noted, however, that C 1s NEXAFS data [14] for the annealed overlayer shows much less azimuthal variation than was found upon adsorption of alanine alone, suggesting that the presence of co-adsorbed sulphur adatoms must influence the detailed binding geometry in some manner. Prior to annealing, the overlayer prepared at 300 K provides clear evidence for the occupation of both {311} and {110} microfacets, consistent with the supposed μ_4 binding, but the precise location of the thiolate group remains the subject of speculation. Given the apparently strong diastereomeric effects seen in μ_4 species upon the adsorption of serine on this surface, it is regrettable that no experiments involving D-cysteine have been reported (or, indeed, involving L-cysteine on the Cu{531}-D surface).

In contrast, experiments involving both enantiomers of cysteine *have* been reported for the Au{17 11 9} surface; this rather esoteric choice of Miller indices yields a surface structure displaying rather wide {111} microfacets and rather isolated kink sites. Upon room-temperature deposition of either enantiomer on this surface, followed by annealing to 400 K, Greber *et al.* [15] report XPS results

13 Each doublet consists of a $2p_{3/2}$ peak at the energy quoted, accompanied by a $2p_{1/2}$ peak of half the intensity shifted to higher binding energy by approximately 1.2 eV.

including a S $2p_{3/2}$ peak at a binding energy of 162 eV, indicating (as above) the presence of a surface-bound thiolate species; no evidence of isolated sulphur adatoms is reported. The N 1s peak found at a binding energy of 399.7 eV is diagnostic of a surface-bound intact amine group, again in line with the behaviour observed on copper, but the presence of two distinct O 1s peaks, at binding energies of 531 eV and 533 eV, strongly indicates that the acid group remains intact and essentially decoupled from the surface. This state of affairs recalls that observed upon adsorption of cysteine on Cu{110} (see Section 4.3.4) and implies that each molecule binds to the surface in a μ_2 mode (via amine and thiolate groups). Photoelectron diffraction intensities, reported in the same work, show strong forward-scattering in N 1s emission, interpreted as indicating the orientation of the C—N bond of cysteine (see Fig. 5.5). The polar angle of this bond is c.75° off normal for both enantiomers, but the azimuthal angle

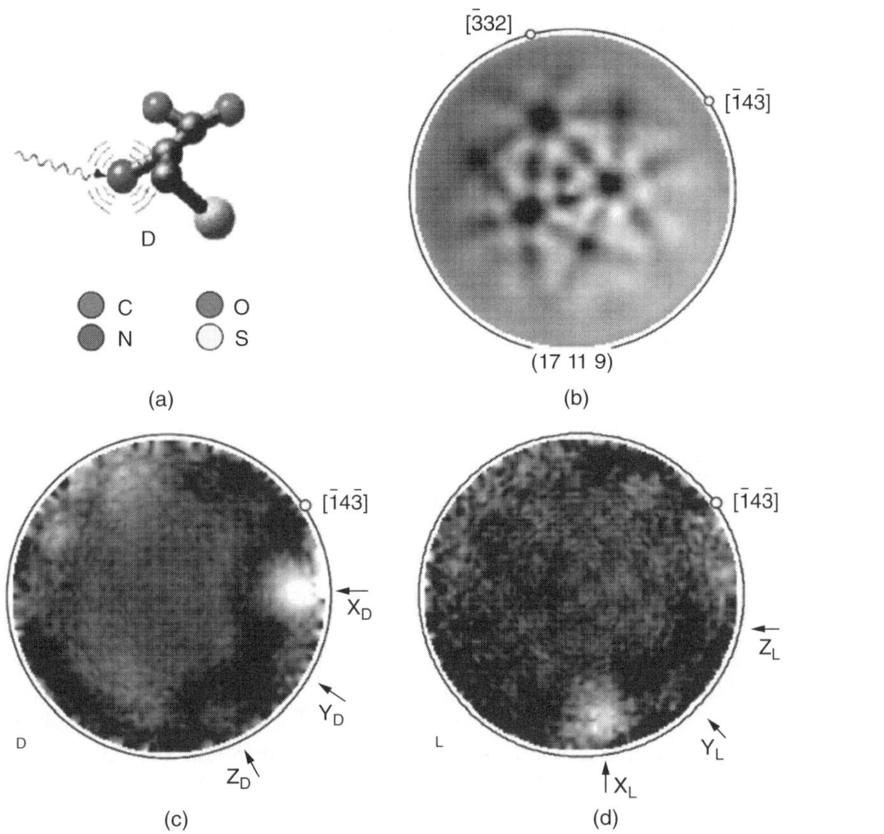

Figure 5.5 Angle-resolved photoemission from Au{17 11 9} after exposure to cysteine. In (a) is shown a schematic of the molecule (*sans* hydrogen atoms) highlighting the nitrogen atom. With a photon energy of 786 eV, diffraction patterns were recorded for (b) secondary electrons at a kinetic energy of 486 eV, and for primary electrons emitted from the N 1s state in the cases of (c) D-cysteine and (d) L-cysteine exposure. Note the strong forward-scattering at three o'clock and six o'clock positions in the latter panels. Reproduced with permission from [15]. Copyright (2006) American Physical Society.

differs by 81° between the two; first-principles DFT models of the adsorbate structures agree remarkably closely with these experimental findings. This highly diastereomeric result seemingly conflicts with the long-standing view that chiral recognition requires at least three points-of-contact between a molecule and its binding site (i.e. at least a μ_3 configuration) but the authors note that non-bonding steric effects can provide a third constraint (albeit not a third adsorbate–substrate bond) that reconciles the apparent contradiction [15].

5.3.4 Tartaric Acid on Cu{531}

An emerging theme in the discussion thus far is that amino acids are typically capable of binding, with almost equal propensity, to both the {311} and the {110} microfacets of Cu{531} surfaces.[14] Looking beyond the amino acids, the case of tartaric acid is particularly interesting in this regard, since the molecule contains two acid groups capable of binding to the surface. Upon adsorption of either D- or L-tartaric acid on Cu{531}-L at around 380–400 K, Baldanza *et al.* [16] report XPS results showing O 1s peaks at 531.4–531.6 eV, corresponding to surface-bound carboxylate; at 532.7–533.0, corresponding to dangling hydroxyl groups; and at 530.2–530.7 eV, corresponding to surface-bound deprotonated hydroxyl. Peaks in the C 1s region of the spectra at 287.6–287.9, 286.2–286.5 and 285.7–285.8 eV are associated with the same three species, respectively. For either enantiomer, the diagnosis of bitartrate adsorption is clear. Moreover, in both cases, the signature of surface-bound deprotonated hydroxyl is observed only at low coverages, while that of dangling intact hydroxyl is only observed when the former is absent; this suggests an inherent preference for a μ_6 configuration, bound through carboxylate and deprotonated hydroxyl, but crowding of the surface ultimately favours a μ_4 configuration at saturation, bound to the surface only through the carboxylate groups. Quantitative analysis of both O 1s and C 1s XPS data suggests that saturation coverage of L-tartaric acid (i.e. *R,R*-tartaric acid) amounts to only about 4/5 the saturation coverage of D-tartaric acid (i.e. *S,S*-tartaric acid) but aside from this there are few signs of any significant enantiospecific difference from this technique.

As ever, it is the NEXAFS technique that provides real insight into the precise adsorption geometry in this system, especially in the high coverage limit. Here, Baldanza *et al.* [16] report data from excitation of the O 1s orbitals into the π orbitals associated with the carboxylate groups of the adsorbed bitartrate, revealing variation with azimuthal polarisation that can only sensibly be explained by means of a model in which the molecule is twisted such that the carboxylate groups at either end lie at right angles to one another. Multiple adsorption geometries may, however, satisfy this basic requirement (see Fig. 5.6).

5.3.5 Propylene Oxide and 3-Methylcyclohexanone on Cu{643}

Whilst the studies described above have primarily focused upon structural aspects of enantiospecific adsorption, an alternative approach turns attention towards differences in adsorption energy, as revealed by the kinetics of

14 This fact alone justifies the community's focus upon the Cu{110} and Cu{311} surfaces, as evident in the preceding chapter.

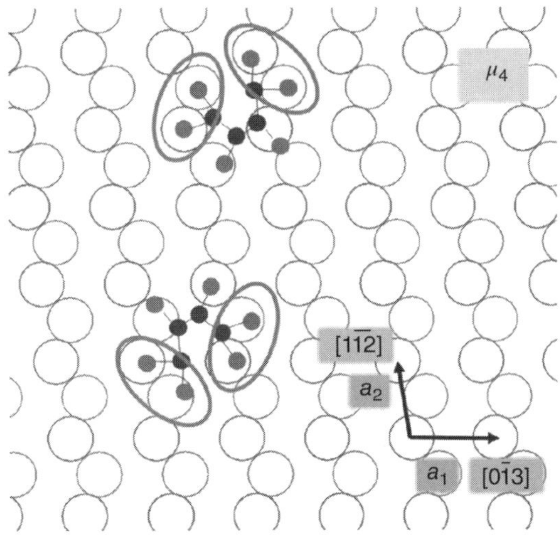

Figure 5.6 Schematic models for the adsorption geometry of tartaric acid (hydrogen atoms omitted for clarity) on Cu{531}-L. Reproduced with permission from [16]. Copyright (2016) Elsevier.

desorption. Early work along these lines, reported by Gellman and co-workers, concerned the desorption of propylene oxide (PO) and 3-methylcyclohexanone (3-MCHO) from the Cu{643} surface. As we have previously noted (in Section 5.1.2), this substrate is unstable against facetting under the influence of adsorbed cyclohexanone, but the same appears not to be true for either of the two adsorbates considered here.

Propylene oxide (more properly known as epoxypropane) is a chiral derivative of propylene (i.e. propene) readily available in racemic or enantiopure forms (*see Plate XV*). Following low-temperature adsorption of PO on Cu{643}, Horvath and Gellman [17, 18] conducted temperature programmed desorption (TPD) experiments as a function of exposure (see Fig. 5.7). At the highest exposures, three peaks were observed, at c.118, c.154 and c.222 K. The low-temperature peak was found not to saturate with increasing exposure, indicating clearly that it corresponds to a high-coverage physisorbed multilayer. Indeed, control experiments conducted on the Cu{111} surface (see Fig. 5.8) also show such a peak at much the same temperature (c.121 K)[15] [17]. In addition, a peak in the {111} data at c.145 K suggests that the mid-temperature peak in the {643} data derives from adsorbates bound to the {111} microfacets of that surface. The high-temperature peak has no analogue in the {111} data, however, and so is most likely due to binding at step and/or kink sites.

Crucially, the precise temperature of the middle peak in the {643} data depends upon the chiral identity of both the adsorbate and the substrate. The diastereomeric difference in peak temperature amounts to 1.0±0.3 K, corresponding to an estimated difference in adsorption energy of 0.27±0.06 kJ mol^{-1} with

15 It would have been surprising had multilayer desorption been sensitive to the underlying surface orientation, since one presumes that the physical and chemical properties of the multilayer ought simply to approximate those of bulk PO (subject to kinetic limitations during its formation). Had any differences been observed, one would be forced to conclude that the interface between monolayer and multilayer somehow induced the latter to adopt a metastable structure in at least one of the cases.

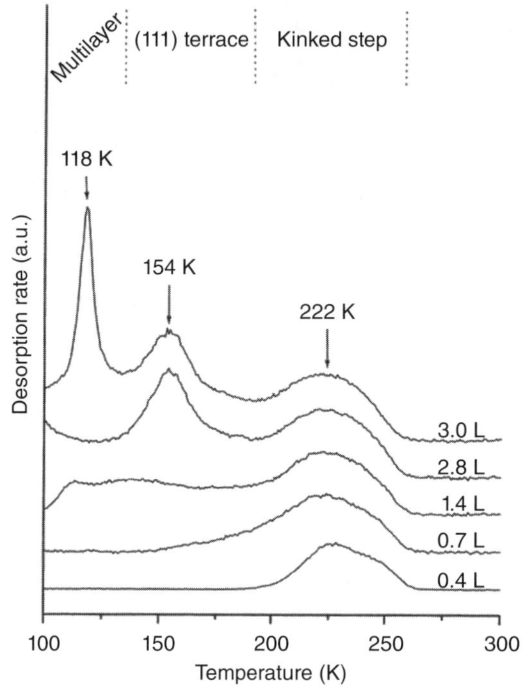

Figure 5.7 Temperature-programmed desorption results obtained at a heating rate of 1 K s^{-1} after low-temperature exposure of Cu{643} to propylene oxide. Reproduced with permission from [17]. Copyright (2002) American Chemical Society.

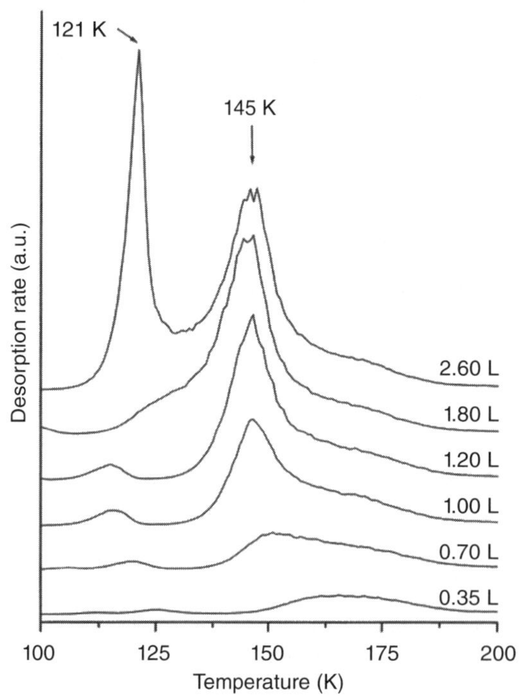

Figure 5.8 Temperature-programmed desorption results obtained at a heating rate of 1 K s^{-1} after low-temperature exposure of Cu{111} to propylene oxide. Reproduced with permission from [17]. Copyright (2002) American Chemical Society.

R-PO binding more strongly to Cu{643}-D, and *S*-PO binding more strongly to Cu{643}-L. The hypothetical enantiomeric excess in the rate constants at 300 K may be estimated as just 5%; moreover, the breadth of the TPD peaks suggests that even this is likely not achievable under practical conditions. It should also be noted, however, that all such estimates are subject to inherent *caveats* concerning the interpretation of TPD data.[16] The high-temperature peak, although associated with the step and/or kink sites (where chiral effects might be expected to be at their strongest) is rather too broad to allow a similar analysis. The quantifiable enantiospecificity in adsorption/desorption for this system is undeniably rather small, but it does serve to demonstrate that such effects are, in principle, of measurable magnitude; on the {111} surface, in contrast, differences in desorption temperature between the molecular enantiomers are (as required by symmetry) found to be statistically negligible.

Another particularly early demonstration of enantiospecific desorption may be found in the case of 3-MCHO (*see Plate XV*) on Cu{643}. Once again, the TPD experiments reveal multiple peaks, this time at c.168, c.222, c.346 and c.385 K (see Fig. 5.9) [17, 20]. The low-temperature peak lies close to a similar peak obtained after adsorption on the Cu{111} surface (c.178 K) and, since neither

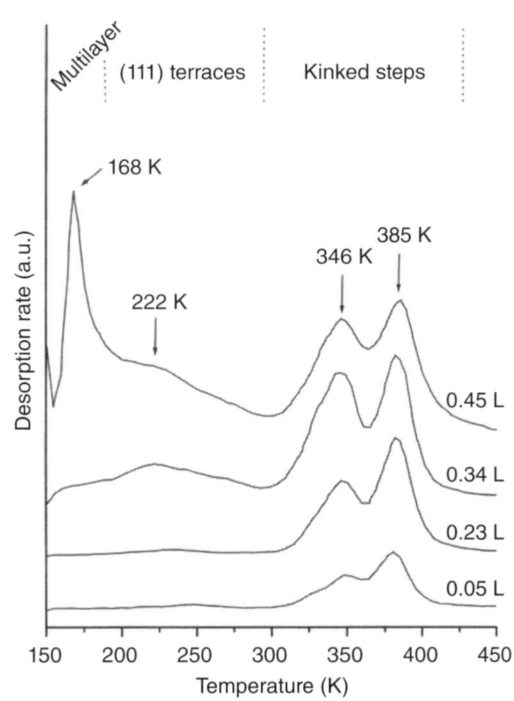

Figure 5.9 Temperature-programmed desorption results obtained at a heating rate of 1 K s^{-1} after low-temperature exposure of Cu{643}-L to 3-methylcyclohexanone. Reproduced with permission from [17]. Copyright (2002) American Chemical Society.

16 In analysing TPD data obtained with a single heating rate, it is necessary to *assume* a value for the pre-exponential factor in the desorption rate. Although it is standard to take 10^{13} s^{-1} for this factor, there exists clear evidence that realistic values could range over several orders of magnitude on either side [19]. Thankfully, the estimated heat of adsorption is only weakly dependent upon the choice of pre-exponential factor, but nevertheless a systematic error by a factor of two in either direction would not be out of the question.

saturates with increasing exposure, both are attributed to a high-coverage physisorbed multilayer. The peak at c.222 K is very broad, preventing any possibility of observing enantiospecific effects, but does again correspond nicely with a similar (albeit somewhat sharper) feature seen in the Cu{111} data at c.230 K and so is interpreted as due to adsorption on the {111} microfacets of the kinked surface [17]. The remaining two peaks may therefore be attributed to adsorption at either the kink sites or the short segments of straight step-edge that lie between them. Complementary TPD experiments conducted on stepped Cu{221} and Cu{553} surfaces, which have no kink sites at all, reveal closely similar peaks at c.345 K and a complete absence of any structure at c.385 K [21]. This strongly suggests that the high-temperature peak on Cu{643} corresponds to adsorption at kink sites, while the next highest corresponds to adsorption at the intervening straight step-edges.[17] This attribution is confirmed by pre-adsorbing iodine, adatoms of which are expected to selectively block the kink sites; needless to say, increasing iodine pre-coverage is found to result in selective removal of the high-temperature peak while having comparatively little influence on the next highest [21, 22].

As for enantiospecific effects, the peaks at c.222 and c.345 K showed no discernable disastereomeric differences, but the position of the high-temperature peak (at c.385 K) was 3.5±0.8 K higher for *R*-3-MCHO on the Cu{643}-L surface than for the same molecule on the Cu{643}-D surface[18] (see Fig. 5.10). From this, the diastereomeric difference in adsorption heat was estimated as 0.92±0.21 kJ mol^{-1}, with *R*-3-MCHO bound more strongly on the Cu{643}-L surface and *S*-3-MCHO bound more strongly on the Cu{643}-D surface. The rather larger asymmetry means that the hypothetical enantiomeric excess in the rate constants at 300 K would be 41%, although once again it should be stressed that this may not be achievable in practice due to the breadth of the TPD peaks.

5.3.6 3-Methylcyclohexanone on Cu{531}, Cu{651} and Cu{13 9 1}

Subsequent to the observation of diastereomeric differences in the adsorption heat of 3-MCHO on Cu{643}, several further studies were reported on other intrinsically chiral copper surfaces. Since these made use of very similar experimental techniques, and indeed embody much the same principles, we do not describe them in detail, but simply list the bare bones of the results for completeness.

On the Cu{531} surface, desorption from step and kink sites occurs at peak temperatures of c.345 and c.375 K respectively (together with a multilayer peak at

17 Interestingly, similar TPD experiments performed on the Cu{653} surface show much reduced contribution from adsorption at the supposed kink sites. This may be explained by bunching of the kink sites so that their density is lowered in favour of longer stretches of straight step-edge. Such a reconstruction is always possible, but is known *not* to happen when 3-MCHO is adsorbed on the Cu{643} surface; whether it in fact occurs on the Cu{653} surface remains speculative but seems likely.

18 No results were reported for L-3-MCHO, as it was not commercially available, but a racemic mixture yielded a desorption peak at roughly the temperature expected if the correct diastereomeric behaviour were to be observed.

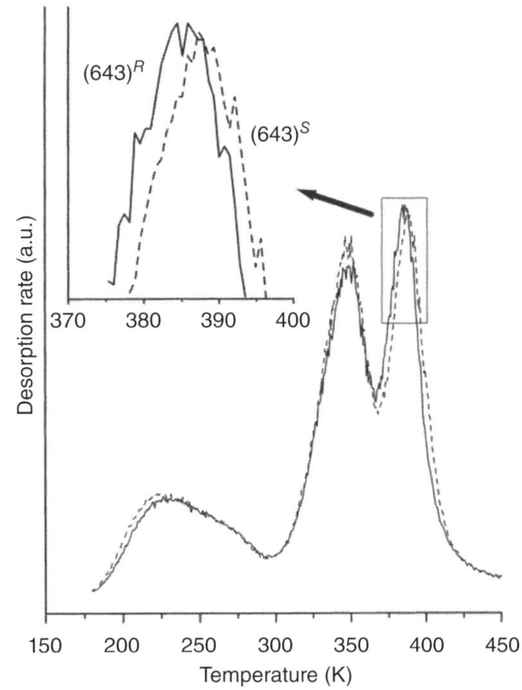

Figure 5.10 Temperature-programmed desorption results obtained at a heating rate of 1 K s^{-1} after low-temperature exposure of Cu{643}-D (solid curve) or Cu{643}-L (dashed curve) to 3-methylcyclohexanone. Reproduced with permission from [17]. Copyright (2002) American Chemical Society.

c.165 K). Desorption from the kink sites is sensitive to chirality, with *R*-3-MCHO desorbing from the Cu{531}-L surface at a peak temperature about 2.2±0.6 K higher than when desorbing from the Cu{531}-D surface; this translates to a difference in adsorption heat of 0.5±0.2 kJ mol^{-1}, favouring *R*-3-MCHO on the Cu{531}-D surface and *S*-3-MCHO on the Cu{531}-L surface [23]. This would imply a hypothetical enantiomeric excess in the rate constants at 300 K of 10% in this case.

Results taken on the Cu{651} surface again show evidence for desorption from step and kink sites, the latter giving rise to a TPD peak at c.366 K. This feature occurs, however, some 2.4±0.7 K higher in temperature for *R*-3-MCHO desorbing from the Cu{651}-L surface than when desorbing from the Cu{651}-D surface; the adsorption heat is consequently estimated to be 0.7±0.2 kJ mol^{-1} higher for *R*-3-MCHO on the Cu{651}-L surface and for *S*-3-MCHO on the Cu{651}-D surface [24]. The hypothetical enantiomeric excess in the rate constants at 300 K would then be 14%.

Finally, on the Cu{13 9 1} surface, step and kink sites are again in evidence within the TPD spectrum, with the latter producing a peak at c.385 K. This peak is shifted by 2.4±0.8 K to higher temperature for *R*-3-MCHO adsorbed on the Cu{13 9 1}-D surface relative to the Cu{13 9 1}-L surface. Such a temperature difference corresponds to an estimated difference in adsorption heat of 0.7±0.2 kJ mol^{-1} – coincidentally the same as for the Cu{651} surface, but this time favouring adsorption on the D surface rather than the L surface [24].

5.3.7 Alanine, Serine, Lysine, Phenylalanine and Aspartic Acid on Cu{3 1 17}

Returning to amino acids, a rather insightful study by Cheong and Gellman [25] took its inspiration from the etching reported upon lysine adsorption on the Cu{100} surface. It will be recalled, from the preceding chapter, that this resulted in the formation of well-defined {3 1 17} facets. Moreover, of the eight such facets compatible with a given instance of the {100} surface type, only four were observed when using each molecular enantiomer; specifically, the {3 1 17}-D facets were favoured when dosing L-lysine, while the {3 1 17}-L facets were favoured when dosing D-lysine [26].

Using a copper sample specially cut to reveal {3 1 17} orientations (D on one side, L on the other) Cheong and Gellman [25] performed TPD experiments subsequent to high-coverage lysine adsorption at 300 K (see Fig. 5.11). In both cases, multilayer desorption was observed with a rate peaking at c.360 K, which was identical to that observed in similar experiments conducted on Cu{100}. The monolayer desorption peak, however, occurred at c.470 K for L-lysine on the D surface (and for D-lysine on the L surface) but at c.460 K for L-lysine on the L surface (and for D-lysine on the D surface). The precise difference in desorption temperature was measured to be 9.5±0.8 K, while the two different enantiomers, adsorbed separately on the Cu{100} surface as a control, again yield statistically indistinguishable results. The authors estimate that the diastereomeric difference in desorption temperature on the {3 1 17} surfaces reflects a difference of adsorption heat amounting to 2.6±0.3 kJ mol^{-1}. Clearly, this is a much stronger effect than seen for adsorption on the Cu{643}, Cu{643}, Cu{651} and Cu{13 9 1} surfaces, presumably reflecting the fact that the orientation in this case has been

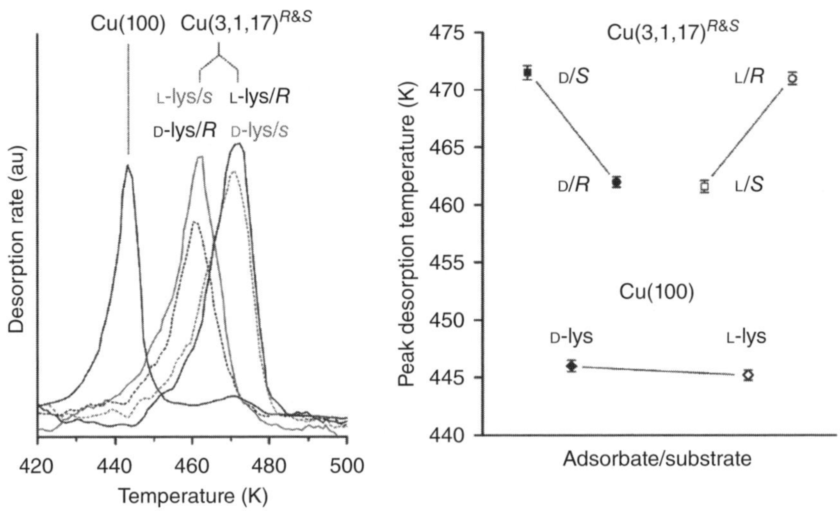

Figure 5.11 Temperature-programmed desorption results obtained at a heating rate of 1 K s^{-1} after low-temperature exposure of Cu{3 1 17} (both enantiomeric forms, $R \equiv D$ and $S \equiv L$) or Cu{100} to lysine. Reproduced with permission from [25]. Copyright (2011) American Chemical Society.

specially chosen in light of a known preference of the adsorbate for certain facets. Indeed, one can estimate the hypothetical enantiomeric excess in the rate constants at 300 K to be 48%.

It is worth noting, in passing, that whilst the TPD trace from Cu{100} shows its main monolayer peak at c.445 K, a small subsidiary peak is also visible at c.470 K. Although well-formed facets of {3 1 17} orientation only appeared in the work of Zhao [26] upon annealing, the results of Cheong and Gellman [25] strongly suggest, therefore, that preferential binding to terrace edges not only persists throughout the TPD temperature ramp, but may also herald incipient facetting.

An alternative method of determining diastereomeric adsorption energies has been employed by Yun and Gellman [27] for their studies of aspartic acid on Cu{3 1 17}. In this work, the authors first saturated the L surface with isotopically labelled L-aspartic acid at 400 K, then annealed at 460 K whilst exposing to unlabelled D-aspartic acid for a variable period of time. Aspartic acid is an amino acid that may be conceptualised (*see Plate XI*) as succinic acid with one of its alkyl hydrogen atoms replaced by an amine group; the molecule thus contains two acid groups (one adjacent to the amine group, the other more distant) and the labelling in this instance sees the carbon atom of each having an atomic mass number of 13 rather than the usual 12. Subsequently, the surface was heated (without further exposure to either form of aspartic acid) to induce decomposition of the adsorbed molecules, and the ratio of $^{13}CO_2$ to $^{12}CO_2$ was taken to indicate the ratio of L- to D-aspartate achieved at the penultimate step.[19] The entire experiment was then repeated with the order of adsorption reversed (i.e. unlabelled D-aspartic acid now followed by labelled L-aspartic acid). Plotting the relative coverages of the enantiomers as a function of exposure to the late-arriving species it was possible to extract a rate constant of 1.05 ($\pm$0.04) $\times$ 10^{-3} s^{-1} for displacement of L-aspartate by D-aspartate, and of 0.37 ($\pm$0.01) $\times$ 10^{-3} s^{-1} for the reverse process. Although these measurements were not obtained at equilibrium, they nevertheless allow us to determine a pseudo-equilibrium constant of 2.84 $\pm$ 0.15 from the ratio of the two displacement rate constants.

A true equilibrium constant was also obtained by the same authors by preparing the surface at a chosen enantiomeric ratio (following the procedure just described) and then exposing this surface to a racemic mixture of aspartic acid [27]. Regardless of the initial conditions, the L surface always tended eventually to a ratio (of D to L-aspartate) of 2.30$\pm$0.06, while the D surface always tended to a ratio of 0.46$\pm$0.02; since these asymptotic ratios are indeed equilibrium measurements, they can be properly considered to yield a true equilibrium constant of 2.30 for the displacement of L-aspartate by D-aspartate on the L surface, and of 2.18 for the displacement of D-aspartate by L-aspartate on the D surface. These two equilibrium constants ought, of course, to be equal by symmetry, so their difference reflects the inherent accuracy of the method. Additional experiments performed with exposure to unequal mixtures of L- and D-aspartate were added

19 It is assumed that prolonged exposure of copper to aspartic acid leads to a high-coverage phase comprising anionic monoaspartate moieties, by analogy with the adsorption of similar molecules such as succinic acid or tartaric acid. The precise identification of the adsorbed species is not germane to the analysis, so long as their acid or carboxylate groups behave similarly regardless of which enantiomer of aspartic acid gave rise to them.

to the analysis, allowing the authors to determine an overall best estimate of 2.29±0.17 for the equilibrium constant of displacement – not startlingly dissimilar from the pseudo-equilibrium constant derived from kinetic measurements.[20] Assuming Langmuirian adsorption, the same data provided an estimate of 3.15±0.29 kJ mol^{-1} for the diastereomeric difference in adsorption heat (D-aspartic acid yielding the larger adsorption heat on the L surface; L-aspartic acid doing the same on the D surface).[21] The hypothetical enantiomeric excess in the rate constants at 300 K would then be estimated as 56%.

Armed with this rather elegant methodology, Gellman and co-workers have extended their studies on the Cu{3 1 17} surface to include the adsorption of alanine, serine and phenylalanine [28, 29]. In the former case, they find no discernible deviation from racemic molecular coverage upon exposure to racemic alanine; indeed, when exposed to alanine prepared at enantiomeric excesses of ±33% (i.e. enantiomeric ratios of 2:1 or 1:2) the surface enantiomeric excess matched this perfectly, thus indicating negligible diastereomeric difference in adsorption heat[22] [28]. In studies of serine and phenylalanine adsorption on the same surface, very similar null results were also obtained, the authors speculating that the ability of lysine and aspartic acid to adopt μ_4 and μ_5 configurations at the surface, respectively, may be the key feature in their enantiospecific behaviour [29].

5.4 Chiral Adsorbates on Chirally Modified Substrates: Diastereomeric Effects II

In contrast to studies on intrinsically chiral surfaces, which have focused almost exclusively on copper, those aimed at chirally modified surfaces have generally made use of rather more reactive metals, such as palladium and platinum. This surely reflects,[23] in part at least, a desire to connect with practical heterogeneous catalysis, where these metals may be of use in hydrogenation

20 If the overall mechanism contains only a single step, or is dominated by a single rate-determining step, then one ought to expect the ratio of forward and reverse rates to approximate the equilibrium constant reasonably well, even quite far from equilibrium. The same need not be true if the mechanism involves multiple steps that might be considered rate determining under different conditions. The results discussed here suggest that the actual interchange of chemisorbed enantiomers is rate determining for displacement, rather than, for example, precursor-mediated adsorption or desorption of physisorbed intermediates.
21 Note that this methodology does not rely upon an estimate for the pre-exponential factor in the desorption rate, being based solely on the ratio of desorption rates rather than their absolute values. The assumption of Langmuirian adsorption may, however, be questionable.
22 In the same work, Yun *et al.* [28] also report the results of DFT calculations suggesting a diastereomeric difference in adsorption heat of 0.11 eV per molecule (10.6 kJ mol^{-1}) in favour of L-alanine on Cu{3 1 17}-L. This would, however, be large enough to be measured easily by the experimental technique applied, so the fact that it is *not* suggests that DFT is not yet capable of resolving such subtle phenomena with reliable fidelity.
23 It may also be worth noting that early efforts to demonstrate enantiospecificity in adsorption on chirally modified copper surface proved largely fruitless.

reactions; indeed, in Chapter 7, we shall devote considerable attention ourselves to just such real-world applications, in the hydrogenation of α- and β-ketoesters. The mechanism of chiral recognition is usually discussed in terms of either the **template** model or the **one-to-one** model. In the template model, chiral modifiers self-assemble into extended porous arrays, allowing the adsorbate access to the surface only via asymmetrically shaped two-dimensional pockets; there is no requirement, within this model, that the interaction with the modifier should necessarily stabilise the adsorbate relative to its binding on the bare surface, since the modifier leaves the adsorbate no alternate locations in which to adsorb. The one-to-one model, in contrast, posits that there is an attractive and localised interaction between each adsorbate and a single modifier molecule, requiring no long-range collusion; here the modifier *must* stabilise the adsorbate, presumably in some particular three-dimensionally asymmetric orientation, else there would be no particular tendency for them to interact at all. As we shall see below, however, the available data does not always clearly support one or the other model, and there may even be cases that evidently fall between the two extremes.

5.4.1 Propylene Oxide on 2-Butanol-Modified Pd{111} and Pt{111}

Following monolayer adsorption of 2-butanol on Pd{111} at 80 K, Tysoe and co-workers [30] performed a TPD experiment, noting a peak at 170 K, consistent with desorption of the intact molecule, and another at 260 K, consistent with the co-desorption of 2-butanol and 2-butanone.[24] In reflection absorption infrared spectroscopy (RAIRS) experiments conducted on a similarly prepared surface, absorption bands attributed to the C—O stretch mode of 2-butanol were initially observed at 1123 and 897 cm^{-1} (the latter being coupled with a methyl rocking mode). On heating to 150 K, these modes shifted to 1090 and 885 cm^{-1} (a change attributed to dehydrogenation of the hydroxyl group) while an entirely new mode appeared at 1670 cm^{-1} (attributed to the formation of a carbonyl). Heating further to 200 K caused the disappearance of the dehydrogenated hydroxyl feature, but considerable strengthening of the carbonyl absorption band. Taken together, these results strongly suggest adsorption at 80 K in the form of 2-butanol, with dehydrogenation to form 2-butoxy at temperatures no higher than 150 K, and complete conversion of the latter into 2-butanone below 200 K; desorption of 2-butanol at 170 K presumably involves recombination of

24 Identification of desorbing species requires comparison of the observed peaks in mass spectrometry with the expected cracking pattern, which in the case of 2-butanol is dominated by mass 45, with a significant secondary peak at mass 59; in the case of 2-butanone, the dominant fragment occurs at mass 43, with a significant secondary feature at mass 72; the secondary peaks at masses 59 and 72 are absolutely diagnostic between these two species. In the original paper, desorption at 260 K is suggested to involve 2-butanone and ethylene, but the former was identified not only by a peak in mass 72 (correctly identified with the ketone) but also by a peak in mass 59 (more properly associated with the alcohol) and the latter on the basis of peaks in masses 26 and 28 (potentially indicative of ethylene, but equally explicable by the cracking patterns of either the alcohol or the ketone). On balance, it seems that the observed spectra are most consistent with a mixture of 2-butanol and 2-butanone desorbing at 260 K, and indeed at least one later paper from the same group seems implicitly to concur with this view [31].

2-butoxy with coadsorbed hydrogen provided by the conversion of 2-butoxy to 2-butanone. At 260 K, it seems that desorption of 2-butanone is competitive with dissociation, and that hydrogen liberated in the latter process may recombine with other 2-butanone molecules to produce 2-butanol. Subsequent TPD experiments reported by Gao *et al.* [32] showed that at much lower initial coverage of 2-butanol (below ∽1/9 ML) the formation of 2-butanone is pre-empted by complete dissociation of 2-butoxy into carbon monoxide and hydrogen.

In contrast to 2-butanol, TPD experiments following the adsorption of PO onto clean Pd{111} at 80 K strongly suggest that it desorbs intact, with no dissociation whatsoever [30]. Its desorption peak does, however, shift significantly from ∽230 K at low coverage down to ∽190 K at the highest coverage before second-layer growth begins; a combination of DFT and Monte Carlo (MC) calculations attributes this shift to variations in adsorbate orientation due to the influence of neighbouring molecules, and the corresponding range of adsorption energy is estimated at 46–62 kJ mol^{-1} (DFT) or 39–57 kJ mol^{-1} (MC) [33]. These results suggest that PO is an ideal probe molecule, adsorbing reversibly on the surface while exhibiting marked sensitivity to its environment. Moreover, a strong infrared absorption band at 810 cm^{-1}, associated with deformation of the epoxide group, is sufficiently far in frequency from any modes of 2-butanol (and its surface products) that identification of the adsorbed species in RAIRS is rather straightforward [30].

In preparation for experiments aimed at demonstrating diastereomeric adsorption, Tysoe and colleagues [30] prepared chiral overlayers by the adsorption of 2-butanol on Pd{111} at 80 K, followed by annealing to 150 K. Although a small amount of achiral ketone was thus produced, the surface was (at the time) supposed to be populated mainly by chiral 2-butoxy; later work has shown that a non-negligible fraction of intact 2-butanol molecules also remain, and are indeed of considerable importance to enantiospecificity [34]. Having thus modified the surface, it was cooled back to 80 K and adsorption of PO was then studied by TPD, as a function of the modifier coverage; integrating areas under TPD desorption peaks, the ratio of uptake for *R*-PO on *R*-2-butanol-modified Pd{111} versus *S*-2-butanol-modified Pd{111} could readily be determined. Converting the ratios into enantiomeric excess values, these rose from 0% on the clean surface to around 30% when the modifier coverage was around a quarter of its saturation limit; increasing the modifier coverage beyond this point saw the enantiomeric excess fall, dropping back to 0% when the modifier coverage reached half of its saturation limit. All four chirality combinations of 2-butanol and PO were then tested in RAIRS experiments (where uptake of the latter species was monitored by integrating its characteristic absorption band at 810 cm^{-1}) with entirely consistent diastereomeric results [30].

In subsequent work, Tysoe's group revisited the same system, systematically varying the protocol for preparation of the chirally modified surface in order to elucidate the active species [34]. Crucially, they tested adsorption of PO onto 2-butanol overlayers that were never heated beyond 110 K, for which essentially no dissociation of the modifier was observed, and still obtained distinctly enantiospecific results (maximum enantiomeric excess ~25%). On the other hand, using pre-dosed oxygen to ensure formation of an almost pure 2-butoxy overlayer upon exposure to 2-butanol at 150 K, it was found that enantiospecificity in the subsequent uptake of PO could be almost entirely suppressed. In contrast to the original interpretation that chiral discrimination in this system involved primarily the alkoxide species [30] these results strongly implied that only the intact alcohol was actually responsible; moreover, the authors argued that hydrogen bonding between modifier and probe molecules must explain the observed enantiospecific effect [34].

Results obtained by Lee and Zaera [35] in similar experiments on the Pt{111} surface paint much the same picture. Monolayer adsorption of 2-butanol at 100 K gave rise to a RAIRS band at 1088 cm^{-1}, identified with an intact hydroxyl group, which shifted slightly to 1084 cm^{-1} upon heating above 170 K; together with other small redshifts in the frequencies of various $=CH_2$ and $—CH_3$ modes, this was taken to indicate conversion of 2-butanol into 2-butoxy. In addition to 2-butanol desorption at around 170 K, which may be either multilayer or monolayer derived, their TPD spectra provide evidence for desorption peaking at around 260–270 K that seems to include both 2-butanol and 2-butanone species. These findings are broadly consistent with the pattern seen on Pd{111} (described above). Propylene oxide, meanwhile, was found to adsorb intact on the clean Pt{111} surface at 100 K, its presence established in RAIRS by the epoxide deformation mode at 822 cm^{-1} and in TPD by desorption in the range between ~220 K (low coverage) and ~195 K (monolayer coverage).

Upon deposition, at 100 K, of enantiopure PO onto a Pt{111} surface modified by the adsorption at 170 K of 2-butanol, clear evidence for enantiospecificity was obtained via TPD, with an enantiomeric excess of 15% in favour of *S*-PO adsorption onto the surface modified by *S*-2-butanol rather than onto that modified by *R*-2-butanol [35]. The effect of varying modifier coverage was not explicitly addressed, but this degree and direction of enantiospecificity is broadly in line with results found previously by Stacchiola *et al.* [30] on Pd{111}. Interestingly, the enantiospecific uptake of *S*-PO on chirally modified Pt{111} also then influenced the evolution of hydrogen upon heating to higher temperatures (prompting dissociation of adsorbed species) in that considerably more (enantiomeric excess −23%) was released from the surface prepared with *R*-2-butanol than from the surface prepared with *S*-2-butanol [35].

In RAIRS, the epoxide deformation band at 822 cm^{-1} was monitored upon deposition of enantiopure PO onto 2-butanol-modified Pt{111}, with enantiomeric excess measured at 66% in favour of *R*-PO on the surface modified by *R*-2-butanol versus the same species on the surface modified by *S*-2-butanol. The fact that this is much larger than recorded by TPD indicates the likelihood of enantiospecific differences in adsorption geometry, which will affect band intensities in RAIRS but not necessarily desorption peak intensities in TPD; in this respect only, the results on Pt{111} differ markedly from those obtained previously on Pd{111}, where no such difference in enantiomeric ratios between RAIRS and TPD experiments was noted.

5.4.2 Propylene Oxide on 2-Methylbutanoic Acid-Modified Pd{111} and Pt{111}

Following up their previous work on 2-butanol modification of Pd{111} (see above) the Tysoe group next considered switching from an alcohol to an acid modifier [36]. Depositing 2-methylbutanoic acid (2-MBA, *see Plate XXII*) at 100 K, their TPD spectra featured a sharp peak at ∽220 K, associated with desorption of the intact molecule, and no further peaks below 300 K. In RAIRS experiments conducted after deposition at 300 K, a prominent absorption band at 1390 cm^{-1} was interpreted as being due to the symmetric —CO_2 stretch mode of a carboxylate group, while the absence of any feature attributable to a corresponding asymmetric mode implies that this group stands symmetrically on the surface. In the light of early views on the 2-butanol system – in which the 2-butoxy species was held (again, see above) to be responsible for chiral discrimination – this bonding arrangement was expected to yield similarly enantiospecific results upon subsequent adsorption of PO. In fact, the enantiomeric excess upon separate adsorption of enantiopure PO onto 2-MBA-modified Pd{111} (all four combinations) was found to be zero, independent of modifier coverage [36]. In retrospect, these results lend further credence to the later view that it was 2-butanol itself, not its dehydrogenated product, that gave rise to enantiospecificity in the previously studied system [34] although the fact that the chiral centre lies further from the surface in the acid case may also play a role.

In subsequent work on the chiral modification of Pt{111} by 2-MBA, Lee and Zaera [37] discovered a significant enantiospecific effect upon the adsorption of PO, in marked contrast to the work just described. Here, adsorption at 100 K followed by annealing at 230 K was found to result in a distinct RAIRS peak at 1394 cm^{-1} attributed to the symmetric —CO_2 stretch mode of a carboxylate group, and again no corresponding asymmetric mode was observed. The adsorption geometry and chemical state of the surface species thus appears to be rather similar to that found on Pd{111}. Upon adsorption at 100 K onto a surface modified by *S*-2-MBA, TPD revealed an enantiomeric excess of 11% in favour of *S*-PO adsorption versus *R*-PO adsorption, but only at moderate modifier coverage – too low or too high a modifier dose led to little or no enantiospecificity, as previously observed in several of the systems already described. In RAIRS experiments, however, enantiospecificity was observed not only at moderate modifier coverage (enantiomeric excess 22%) but also, with the reverse sign, at low modifier coverage (enantiomeric excess −13%). Clearly, there are significant chiral effects upon adsorption geometry even when uptake is chirally insensitive. Also, the peak desorption rate in TPD was found to occur at 195 K for *S*-PO but at 188 K for *R*-PO, regardless of the modifier coverage.

As a final side note on this system, it is worth mentioning in passing the recent work of Schauermann and co-workers [38] who made use of single-crystal adsorption calorimetry (SCAC) to study PO adsorption on Pt{111} modified by 2-MBA. In the SCAC technique, ultrathin single-crystal samples are exposed to a pulsed molecular beam and both sticking probability and adsorption heat are measured directly [39, 40]. In this instance, the presence of 2-MBA was found to have no significant influence on either aspect of PO adsorption, and hence no enantiospecific effects were demonstrated.[25] The authors rationalise this apparent discrepancy with earlier TPD results on the basis that their low-temperature data (obtained at 110 K) probed only the actual act of sticking and the immediate accommodation of PO with the surface, and not processes that may occur slowly or only at higher temperatures [38].

5.4.3 Propylene Oxide on Amino Acid-Modified Pd{111}

Whilst investigating the efficacy (or otherwise) of 2-MBA as a chiral modifier on Pd{111}, Tysoe and co-workers [30] briefly described experiments involving the amino acid homoalanine, which showed that replacing the methyl group with an amine group (*see Plate XI*) somehow restored the chiral discrimination that had previously been lost by substituting an acid group in place of an alcohol group. In view of this fact, the question inevitably arises as to whether other amino acids might be similarly effective as chiral modifiers.

A recent study from the Tysoe group [41] indicates that this is indeed the case, although it highlights a rather interesting dependence upon a particular adsorption motif. It will be recalled, from Section 4.3.9, that alanine, when adsorbed on Pd{111}, adopts a predominantly tetrameric (occasionally dimeric) geometry featuring both anionic and zwitterionic moieties. In fact, it turns out that very similar structures are seen in STM images obtained after the adsorption of leucine (only dimers), valine (only dimers), isoleucine (mainly dimers), norvaline (mainly tetramers) and homoalanine (only tetramers).

Remarkably, when tested with enantiopure PO, uptake (as measured using the King and Wells technique [42]) was seen to lead to enantiospecificity in rough proportion to the population of tetramers, strongly suggesting that the two-dimensional chiral pocket at the centre of each tetramer is crucial to the observed asymmetry.[26] In the cases of leucine and valine, no enantiomeric excess was detected; for isoleucine, norvaline, homoalanine and alanine, the maximum enantiomeric excess values (always obtained when the modifier coverage was approximately half of its saturation limit) were 13%, 23%, 26% and 33% respectively (always in favour of *R*-PO with the D-acid, or equivalently *S*-PO with the L acid) [41]. DFT calculations by the same authors suggest that recognition, in this case, does not involve the most stable tetramer conformation, whose central pocket is too small to accommodate PO, but instead a looser transient arrangement that exposes a favourable Pd binding site at its centre.

---

25 Further experiments using 1-(1-naphthyl)ethylamine as the modifier also failed to demonstrate enantiospecificity, although here at least there was some influence on the adsorption of PO (albeit chirally insensitive).

26 Note that this system conforms neither to the classic template model, in which the chiral pockets would cover the whole surface in an extended array, nor to the classic one-to-one model, in which only a single modifier molecule would be responsible for chiral recognition. Instead, the system combines elements of both models – the two-dimensional multi-molecular pocket from the template model, alongside the attractive and localised interaction from the one-to-one model – and clearly constitutes a distinct alternative, equidistant between the two.

Rather than a static **lock-and-key** process of chiral recognition, therefore, this system seems to involve a dynamic **induced-fit** process [41].

5.4.4 Glycidol on Tartaric Acid-Modified Pd{111}

It will be recalled (Section 4.3.5) that tartaric acid adsorption on Cu{110} surfaces typically results in either monotartrate species (at high coverage) or bitartrate (at low coverage). Much the same picture pertains to adsorption of the same molecule on Pd{111} surfaces[27] [43, 44]. Low-coverage deposition of tartaric acid at 300 K resulted in a species exhibiting RAIRS bands at 1414 and 1370 cm^{-1} attributed to the symmetric —CO_2 stretch of the carboxylate groups and the OH stretch of hydroxyl groups; at high coverage, such bands are still present, although weaker, and an additional band at 1700 cm^{-1} is also observed, attributed to the presence of carbonyl in the intact acid group [43]. Cooling the low-coverage phase to 120 K, for the purposes of STM imaging, resulted in a surface characterised by isolated asperities of somewhat oval appearance, believed to correspond to the bitartrate species [44]. In TPD experiments, tartaric acid is stable on the surface up to ∽390 K, at which point it dissociates yielding CO_2 and H_2 [43].

Testing the tartaric acid-modified Pd{111} surface with enantiopure PO failed to reveal any enantiospecific adsorption, prompting Mahapatra and Tysoe [45] to try glycidol instead. Glycidol is very closely related to PO, differing only in that the distal carbon (relative to the epoxide group) is decorated with a hydroxyl group (*see Plate XV*). In consequence, the molecule enjoys the possibility of being a hydrogen bond donor, in addition to a hydrogen bond acceptor (PO can play only the latter role). On the clean Pd{111} surface, glycidol is found to adsorb intact at 90 K, with only a small fraction of molecules decomposing (via deprotonation and subsequent ring-opening) below 200 K. The majority of adsorbed glycidol desorbs intact in TPD, either at 191 K in the case of multilayers or at around 250 K from the monolayer [46].

Upon separate deposition of enantiopure *R*-glycidol and *S*-glycidol onto a surface pre-prepared with D-tartaric acid (i.e. *S,S*-tartaric acid) it was found (through subsequent TPD experiments) that a maximum enantiomeric excess in uptake of 13% (favouring the *R* enantiomer) was achieved only over a relatively narrow range of modifier coverage within the regime dominated by the bitartate species [45]. Since STM images in this regime indicate neither long-range lattice order nor short-range structural motifs, it seems likely that chiral discrimination must take place via a one-to-one interaction between individual molecules, as opposed to conforming to a model involving several modifier molecules (cf. the case described in the preceding subsection). In DFT calculations presented by the same authors, it is found that both glycidol enantiomers bind most strongly in effectively μ_3 conformations where the epoxide attaches directly to the metal surface and the hydroxyl group acts as both a hydrogen bond donor (towards a tartaric acid carboxylate group) and an acceptor (from a tartaric acid hydroxyl group). The interaction energy of *R*-glycidol with D-tartaric acid is, however,

27 The rather more complicated flux-dependent behaviour observed on Ni{111}, described in Section 4.3.8, appears to be specific to that more reactive surface.

calculated to be about 6 kJ mol^{-1} greater than that of *S*-glycidol, rationalising the experimentally observed enantiospecific uptake [45].

5.4.5 Propylene Oxide on Lysine-Modified Cu{100}

For our final example of adsorption on a chirally modified surface, we return to our old favourite copper, via a study by Cheong and Gellman [47] involving the deposition of lysine on the {100} surface. It will be recalled from Section 4.4.1 that high coverages of lysine are capable of restructuring the underlying Cu{100} substrate after annealing at 430 K. Here, however, the surface was prepared by exposure to enantiopure lysine at 300 K and not subsequently annealed, so it is believed that it remains mesoscopically planar. Exposure to PO was then carried out with the surface cooled to 95 K, and TPD performed to assess the binding strength. On the clean surface, it was established that PO adsorbs intact at this temperature, desorbing molecularly in the range between 180 K (low coverage) and 152 K (high coverage) with no sign of dissociation.

In the presence of the lysine modifier, the desorption peak was found to vary diastereomerically, with that for *S*-PO and L-lysine (or *R*-PO and D-lysine) occurring as much as 5.5±0.5 K higher in temperature than that for *S*-PO and D-lysine (or *R*-PO and L-lysine); the corresponding diastereomeric difference in adsorption heat is estimated to be approximately 1.5 kJ mol^{-1} [47]. Interestingly, the results show similar sensitivity to the modifier coverage as revealed in earlier work on Pd{111}; the maximum enantiospecificity was achieved when the empty site density of the surface corresponded to 0.25 ML,[28] dropping sharply to zero on either side. In contrast to the Pd{111} and Pt{111} systems, however, enantiospecificity appears to be a rarity on copper surfaces; the work referred to here was the first and, to date, only reported example of this phenomenon on that metal, following unsuccessful attempts to observe it on Cu{100}, Cu{110} and Cu{111} surfaces using various combinations of alaninol, 2-butanol, 3-MCHO and PO [48].

5.5 Executive Summary

Once again, let us conclude the present chapter with a thematic recap of the major points illustrated by the foregoing discussion. Four key issues would appear to be worth highlighting:

Long-Range Order

It would be fair to say that evidence for long-range order upon adsorption onto intrinsically chiral surfaces is much less clear-cut than in the case of achiral surfaces. Only a handful of ordered phases have been reported (i.e. for glycine, alanine and serine on Cu{531}) and the corresponding LEED patterns are typically diffuse and/or streaky; the implied degree of disorder is, perhaps, unsurprising since kinked surfaces are expected to exhibit significant

28 This does not necessarily equate to a lysine coverage of 0.75 ML, since in this system it seems that multilayer growth begins before completion of the first layer.

atomic-scale roughness (see Chapter 2). Where specific LEED patterns have been reported, all have thus far corresponded to oblique two-dimensional lattices, but there is no obvious reason why this should necessarily have to be the case.

Diastereomeric Facetting

In the preceding chapter, we saw that even very stable achiral surfaces, such as Cu{100}, could be prone to facetting under the influence of adsorbed chiral molecules. In the present chapter, we began by examining some cases where relatively unstable chiral surfaces, such as Cu{531}, could be induced to facet by the adsorption of achiral molecules. In both these situations, the facetting is best described as enantiospecific, in the sense that the precise facets obtained depend upon the chirality either of the adsorbate (in the former situation) or of the substrate (in the latter). Truly diastereomeric facetting, however, occurs only when both the adsorbate *and* the substrate are independently chiral. Here, we have seen just a single example (i.e. alanine on Cu{531}) where altogether different facets are stabilised depending upon whether the chirality symbols of adsorbate and substrate match or differ. Doubtless this should be a general phenomenon, and one worth seeking out in future experiments.

Two-, Three-, Four- and Five-Point Binding

Just as amino acids were most often found to adopt μ_3 binding configurations on achiral metal surfaces, this remains the most common mode of adsorption observed on intrinsically chiral surfaces. In many cases, binding occurs through the amine and carboxylate groups, although in at least one case it is thiolate and carboxylate groups that are responsible. However, whereas μ_2 species, binding through amine and intact acid groups, were important in stabilising two-dimensionally chiral molecular networks on achiral surfaces, on intrinsically chiral surfaces they seem to be disfavoured; the only μ_2 species reported thus far seems to be that bound by amine and thiolate groups in the case of cysteine on gold. Whilst it is hardly surprising that μ_3 species show some quite strong diastereomeric effects in their local binding geometry, this latter case of μ_2 binding reminds us that chiral recognition can involve steric contributions in addition to the existence of formal bonds. Nevertheless, it seems that the strongest diastereomeric effects occur when μ_4 binding is evident, as has been seen in the adsorption of serine on Cu{531}, or even μ_5, as has been inferred in the case of aspartic acid on Cu{3 1 17}.

Diastereomeric Adsorption and Desorption

Reported diastereomeric differences in the adsorption heats of chiral molecules (measured by TPD) on a variety of intrinsically chiral copper surfaces do not generally exceed 1 kJ mol^{-1} and may be much smaller. Exceptions to this rule occur for the desorption of lysine or aspartic acid from the Cu{3 1 17} surface, which was chosen specifically because of its affinity with the former molecule in the facetting of Cu{100}. In these situations, the diastereomeric difference comfortably exceeds 2 kJ mol^{-1} and may be rather larger. The specific adsorption site responsible for chiral discrimination may vary from case to case – in the examples described above, terrace, step and kink sites all have roles to play, albeit the width of peaks in the TPD spectra may serve to obscure these. Configuration of the adsorbed molecules is also undoubtedly important, with multiple points of contact favouring high enantiospecificity. On chirally modified surfaces, on the other hand, the focus has predominantly (albeit not exclusively) been on diastereomeric differences not in adsorption heat per se but rather in uptake, and here, the role of hydrogen bonding between probe and modifier molecules has been emphasised.

Models and Processes

In discussing the means by which enantiospecific adsorption occurs, we have introduced two separate dichotomies, namely template/one-to-one models and lock-and-key/induced-fit processes.[29] In most of the cases described above, the lack of real-space images hampers efforts to distinguish between template and one-to-one models, although one fairly clear example of the latter is to be found in the interaction between glycidol and tartaric acid-modified Pd{111}. Enantiospecific binding of PO on amino acid-modified Pd{111}, on the other hand, exemplifies a 'third way' between the two extreme models. Regardless of which model best applies, however, the process by which the modifier exerts its influence may be described as of lock-and-key type when either the individual modifier molecule or template chiral pocket presents an essentially rigid environment into which the adsorbate must insert. Alternatively, if the molecule or pocket adapts its shape in the presence of the adsorbate, so that flexibility rather than rigidity promotes enantiospecificity, the process of asymmetric adsorption falls into the induced-fit category.

References

1 P. Knight, S.M. Driver and D.P. Woodruff, *J. Phys. Condens. Matter* **9**, 21 (1997).

2 P. Knight, S.M. Driver and D.P. Woodruff, *Surf. Sci.* **376**, 274 (1997).

3 G. Jones, M.J. Gladys, J. Ottal, S.J. Jenkins and G. Held, *Phys. Rev. B* **79**, 165420 (2009).

4 M.L. Clegg, L. Morales de la Garza, S. Karakatsani, D.A. King and S.M. Driver, *Top. Catal.* **54**, 1429 (2011).

5 S.M. Driver, private communication.

6 S.J. Jenkins and S.J. Pratt, *Surf. Sci. Rep.* **62**, 373 (2007).

7 X.Y. Zhao and S.S. Perry, *J. Mol. Catal. A: Chem.* **216**, 257 (2004).

8 S. Fölsch, A. Riemann, J. Repp, G. Meyer and K.H. Rieder, *Phys. Rev. B* **66**, 161409 (2002).

9 T. Eralp, A. Shavorskiy, Z.V. Zheleva, V.R. Dhanak and G. Held, *Langmuir* **26**, 10918 (2010).

10 M.J. Gladys, A.V. Stevens, N.R. Scott, G. Jones, D. Batchelor and G. Held, *J. Phys. Chem. C* **111**, 8333 (2007).

29 The use of the different words 'model' and 'process' is deliberate here, in order to emphasise that these terms describe two essentially orthogonal axes of categorisation for the chiral recognition mechanism, so that all possible combinations of model and process may, in principle, exist in different systems.

11 T. Eralp, A. Shavorskiy, Z.V. Zheleva, G. Held, N. Kalashnyk, Y. Nang and T.R. Linderoth, *Langmuir* **26**, 18841 (2010).
12 T. Eralp, A. Ievins, A. Shavorskiy, S.J. Jenkins and G. Held, *J. Am. Chem. Soc.* **134**, 9615 (2012).
13 T. Eralp, A. Cornish, A. Shavorskiy and G. Held, *Top. Catal.* **54**, 1414 (2011).
14 L. Thomsen, M.T. Wharmby, D.P. Riley, G. Held and M.J. Gladys, *Surf. Sci.* **603**, 1253 (2009).
15 T. Greber, Ž. Šljivančanin, R. Schillinger, J. Wider and B. Hammer, *Phys. Rev. Lett.* **96**, 056103 (2006).
16 S. Baldanza, J. Ardini, A. Giglia and G. Held, *Surf. Sci.* **643**, 108 (2016).
17 J.D. Horvath and A.J. Gellman, *J. Am. Chem. Soc.* **124**, 2384 (2002).
18 J.D. Horvath and A.J. Gellman, *J. Am. Chem. Soc.* **123**, 7953 (2001).
19 V. Fiorin, D. Borthwick and D.A. King, in *Model Systems in Catalysis: Single-Crystals to Supported Enzyme Mimics*, ed. R. Rioux (Springer, 2009).
20 A.J. Gellman, J.D. Horvath and M.T. Buelow, *J. Mol. Catal. A: Chem.* **167**, 3 (2001).
21 J.D. Horvath, A. Koritnik, P. Kamakoti, D. Sholl and A.J. Gellman, *J. Am. Chem. Soc.* **126**, 14988 (2004).
22 P. Kamakoti, J. Horvath, A.J. Gellman and D.S. Sholl, *Surf. Sci.* **563**, 206 (2004).
23 Y. Huang and A.J. Gellman, *Catal. Lett.* **125**, 177 (2008).
24 Y. Huang and A.J. Gellman, *Top. Catal.* **54**, 1403 (2011).
25 W.Y. Cheong and A.J. Gellman, *J. Phys. Chem. C* **115**, 1031 (2011).
26 X. Zhao, *J. Am. Chem. Soc.* **122**, 12584 (2000).
27 Y. Yun and A.J. Gellman, *Angew. Chem. Int. Ed.* **52**, 3394 (2013).
28 Y. Yun, D. Wei, D.S. Sholl and A.J. Gellman, *J. Phys. Chem. C* **118**, 14957 (2014).
29 Y. Yun and A.J. Gellman, *Langmuir* **31**, 6055 (2015).
30 D. Stacchiola, L. Burkholder and W.T. Tysoe, *J. Am. Chem. Soc.* **124**, 8984 (2002).
31 F. Gao, Y. Wang and W.T. Tysoe, *J. Phys. Chem. C* **112**, 6145 (2008).
32 F. Gao, Y. Wang, L. Burkholder, D.K. Saldin, H.C. Poon, D. Sholl, J. James and W.T. Tysoe, *Surf. Sci.* **602**, 2264 (2008).
33 V. Bustos, D. Linares, A. Gil Rabaza, W.T. Tysoe, D. Stacchiola, L. Burkholder and G. Zgrablich, *J. Phys. Chem. C* **113**, 3254 (2009).
34 F. Gao, Y. Wang, L. Burkholder and W.T. Tysoe, *J. Am. Chem. Soc.* **129**, 15240 (2007).
35 I. Lee and F. Zaera, *J. Phys. Chem. B* **109**, 12920 (2005).
36 D. Stacchiola, L. Burkholder, T. Zheng, M. Weinert and W.T. Tysoe, *J. Phys. Chem. B* **109**, 851 (2005).
37 I. Lee and F. Zaera, *J. Am. Chem. Soc.* **128**, 8890 (2006).
38 P. Dementyev, M. Peter, S. Adamovsky and S. Schauerman, *Phys. Chem. Chem. Phys.* **17**, 22726 (2015).
39 C.E. Borroni-Bird and D.A. King, *Rev. Sci. Instrum.* **62**, 2177 (1991).
40 J.H. Fischer-Wolfarth, J. Hartmann, J.A. Farmer, J.M. Flores-camacho, C.T. Campbell, S. Schauerman and H.-J. Freund, *Rev. Sci. Instrum.* **82**, 024102 (2011).

41 M. Mahapatra, L. Burkholder, S.P. Devarajan, A. Boscoboinik, M. Garvey, Y. Bai and W.T. Tysoe, *J. Phys. Chem. C* **119**, 3556 (2015).
42 D.A. King and M.G. Wells, *Proc. R. Soc. London, Ser. A* **339**, 245 (1974).
43 M. Mahapatra and W.T. Tysoe, *Surf. Sci.* **629**, 132 (2014).
44 M. Mahapatra and W.T. Tysoe, *J. Phys. Chem. C* **120**, 2309 (2016).
45 M. Mahapatra and W.T. Tysoe, *Phys. Chem. Chem. Phys.* **17**, 5450 (2015).
46 M. Mahapatra and W.T. Tysoe, *Catal. Sci. Technol.* **5**, 738 (2015).
47 W.Y. Cheong and A.J. Gellman, *Langmuir* **28**, 15251 (2012).
48 W.Y. Cheong, Y. Huang, N. Dangarla and A.J. Gellman, *Langmuir* **26**, 16412 (2010).

6

Chiral Amplification

The moon was white on the barricks,
The road was white an' wide,
An' the Orderly Orf'cer took it all,
An' the ten-foot ditch beside.
An' the corporal pulled an' the sergeant pushed,
An' the three they wagged along,
But I'd shut my eyes in the sentry-box,
So I didn't see nothin' wrong.

The Shut-Eye Sentry, Rudyard Kipling, 1922

Over the preceding two chapters, we have seen a wide variety of ways in which chirality may be manifest at surfaces. Valuable though such a catalogue is, however, the ultimate goal of research in this area must surely be to achieve active *control* over the expression of chirality, with a view towards asymmetric synthesis, separation or sensing (recall Section 1.8). As we shall shortly see, the realisation of this aspiration rests upon the way in which adsorbed molecules act in concert with one another, either in the formation of regimented patterns on the surface or in the disorderly evacuation of the surface upon desorption.

We shall discuss practical examples of asymmetric synthesis in some detail in the following chapter, and the various effects of surface chirality on optical, magnetic and electronic properties (for possible sensing applications) will form the focus of the final three chapters of this book. In this chapter, we shall focus upon the goal of chiral separation, either entirely on the surface or through the twin processes of adsorption and desorption. In doing so, we shall encounter the concept of **chiral amplification**, whereby a small enantiomeric excess in one aspect of a system is leveraged to create a much larger enantiomeric excess elsewhere. Broadly speaking, amplification of this kind can be achieved either by kinetic or by thermodynamic means, and it is along these lines that we shall divide our attention.

6.1 Kinetic Amplification: Surface Explosions

In the last chapter, we discussed several examples in which desorption of one enantiomer from an intrinsically chiral surface was favoured over that of its

Chirality at Solid Surfaces, First Edition. Stephen J. Jenkins.

mirror image. The diastereomeric difference in desorption temperature (during temperature programmed desorption (TPD) experiments) was, however, never in excess of 10 K, and the corresponding difference in adsorption energy was never estimated to be much above around 3 kJ mol^{-1}. Given that peaks in TPD spectra are often considerably broader than this, it must be admitted that enantiospecificity in these systems is not always very readily apparent or impressive. What would clearly be desirable, absent any obvious way to increase the actual diastereomeric differences themselves, is to find a type of system in which TPD peaks are much sharper, so that small diastereomeric differences may be amplified. Amongst the sharpest of all TPD peaks are those found in systems exhibiting so-called **explosive desorption**.

To understand the concept of explosive desorption it will be worthwhile first to recapitulate the more frequently encountered modes of zero-order, first-order and second-order desorption. These are generally represented by rate equations of the form

$$-\frac{d\theta}{dt} = k\theta^n \tag{6.1}$$

where n represents the order of the process and k is a rate constant. Ordinarily, this rate constant may in turn be written as

$$k = \nu \exp(-E_d/k_B T) \tag{6.2}$$

in which ν is the pre-exponential factor, E_d is the activation energy for desorption, k_B is the Boltzmann constant and T is the surface temperature.

The coverage dependence implicit in this rate equation produces characteristic behaviour in desorption as a function of temperature (ramped from low to high at constant rate) and these are summarised in Fig. 6.1. For zero-order desorption (typical of desorption from multilayers) a set of TPD spectra obtained with progressively higher starting coverage will display a common leading edge, together with a peak that moves to progressively higher temperatures. In the case of first-order desorption (typical of intact molecules desorbing from a monolayer) a similar series of spectra will display no commonality of either

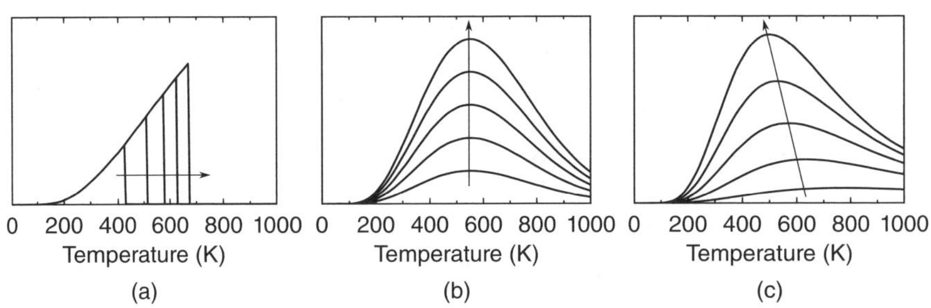

Figure 6.1 Temperature-programmed desorption profiles simulated for (a) zero-order, (b) first-order and (c) second-order processes. In each case, the same heating rates and desorption energies have been used, and the arrow indicates data obtained from a sequence of increasing initial coverages – 0.2, 0.4, 0.6, 0.8 and 1.0 relative to saturation.

leading or trailing edge, but the peak will remain at constant temperature. Finally, for second-order desorption (typical of the recombinant desorption of adsorbed species) the equivalent spectra will display a common trailing edge, while the peak shifts to lower temperatures with increasing initial coverage.

The term explosive desorption, first introduced by Madix and co-workers in the context of formic acid decomposition on nickel surfaces [1–3], refers to a rather different set of circumstances. It generally occurs when the adsorbed species possesses the potential to dissociate on the surface into moieties having much lower activation energy for desorption than the activation energy for dissociation of their parent molecule. The dissociation products will, therefore, desorb extremely rapidly upon formation. Crucially, however, in order for dissociation to take place at all, the original molecule must be adjacent to sufficient empty sites capable of (briefly) accommodating its dissociation products. Accordingly, in the simplest model (where only a single empty site is required) we expect a desorption rate of the form [3]

$$-\frac{d\theta}{dt} = k\theta(1-\theta) \tag{6.3}$$

which is first order in the coverage of molecules (θ) and first order in the coverage of vacant sites ($1-\theta$). More generally, the dependence upon vacancies may be rather more complex. For instance, Bowker and co-workers modified the rate expression to cope with cases where dissociation (and hence desorption) was limited to the perimeters of densely packed islands, obtaining

$$-\frac{d\theta}{dt} = k\theta(1-\theta)^n \tag{6.4}$$

with n equal to a half [4–6]. More recently, Mhatre *et al.* [7] have proposed adjusting the value of n to best fit experimental data; a value of 2, for example, would merely imply that two adjacent vacant sites would be required for dissociation.

The practical upshot of a rate equation of this form is that dissociation (and hence desorption) is essentially blocked at coverages approaching monolayer saturation, due to the extreme lack of any vacant sites. Only when some vacant sites start to appear, via some (often unspecified) other process, will significant desorption begin. Once it does so, however, the key point is that desorption becomes autocatalytic. The more molecules dissociate and their products desorb, the more vacant sites are created, and the faster the rate will become – at least at first. When, eventually, the reservoir of molecules on the surface has been sufficiently depleted, the rate will come crashing back down, implying not only a sudden inception but also a sudden cessation – in short, a surface explosion. None of these considerations are important, it should be stressed, when the coverage is rather low – in this case, desorption follows a more sedate path, akin to that of normal first-order desorption. If one plots a sequence of TPD spectra, again obtained with different starting coverages, the behaviour is really quite striking (see Fig. 6.2) compared with the desorption modes described above. Notably, the peak shape becomes sharper and sharper with increasing initial coverage, as a result of the positive feedback built into the rate equation.

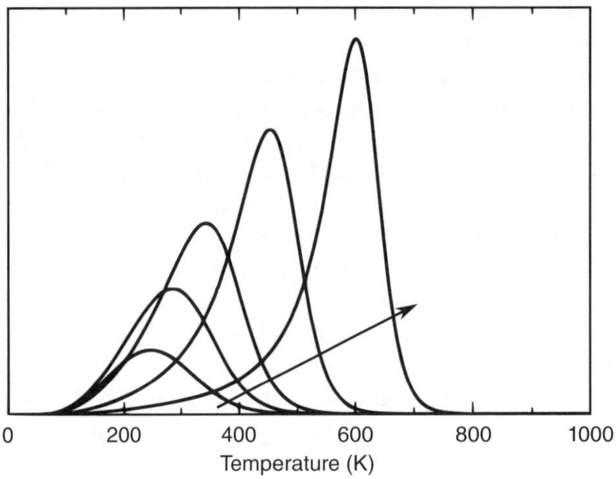

Figure 6.2 Temperature-programmed desorption profiles simulated for an explosive process, with order $n = 2$ (in Eq. (6.4)). The arrow indicates data obtained from a sequence of increasing initial coverages – 0.2, 0.4, 0.6, 0.8 and 0.9, relative to saturation.

6.1.1 Tartaric and Malic Acids on Cu{110}

The first in-depth discussion of explosive desorption in a chiral system was provided by Ernst and co-workers [8, 9] in relation to tartaric acid on Cu{110}.[1] The structural details of this system have already been described at some length in Section 4.3.5, so they need not be rehearsed here. To recap the major points, however, L-tartaric acid binds in its bitartrate form at low-to-moderate coverage (up to perhaps 90% saturation) but converts to its monotartrate form in the highest coverage regime. Racemic tartaric acid, by way of comparison, saturates at a slightly lower coverage than the enantiopure form (and in a different two-dimensional lattice) but behaves in a fairly similar manner at lower coverages.

On heating an adsorbed overlayer prepared at 405 K, CO_2, H_2O and H_2 are evolved from the surface, leaving only (according to X-ray photoemission spectroscopy (XPS)) some residual graphitic carbon [9]. In TPD experiments, only a single CO_2 peak was reported. For low initial coverages of either racemic or enantiopure tartaric acid, the peak was centred at c.450 K, but it shifted markedly to higher temperatures in experiments repeated with progressively higher initial coverages. When starting with a fully saturated overlayer, the peak was centred a little below 520 K for enantiopure tartaric acid, compared with a little below 510 K for racemic tartaric acid. The precise temperatures of the peaks were not reported, but the difference was stated to be 8 K. In later work, Mhatre *et al.* [7] conducted TPD experiments, finding the CO_2 desorption

1 Humblot *et al.* [10] published TPD spectra for tartaric acid (and for succinic acid) very slightly earlier than this work, describing desorption as 'explosive' but not discussing the implications. Since they began their TPD experiments at coverages some way below saturation, the peaks they reported were not, in any case, particularly sharp.

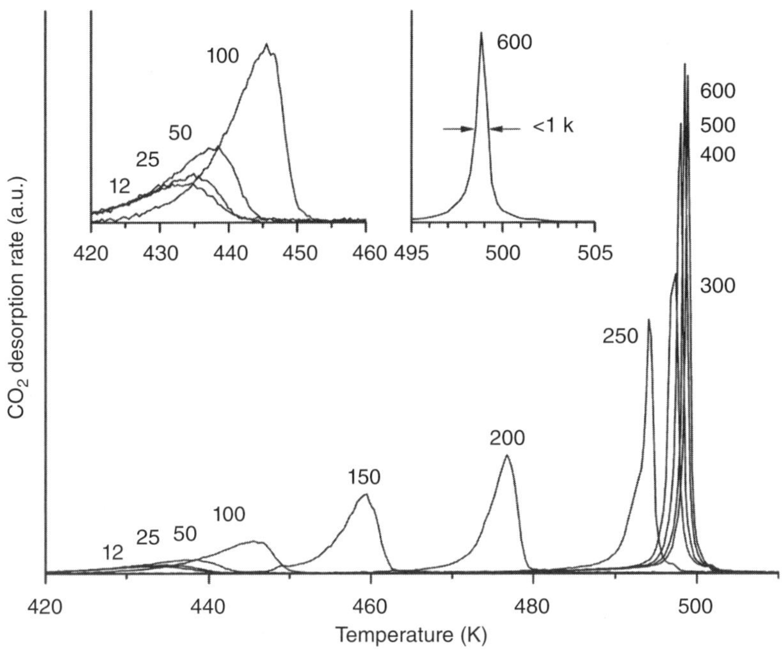

Figure 6.3 Temperature-programmed desorption profiles (obtained at a heating rate of 1 K s^{-1}) for CO_2 evolved from a Cu{110} surface following exposure to D-tartaric acid (exposure times in seconds noted alongside each trace). Insets provide more details on the lowest and highest exposure results. Reproduced with permission from [7]. Copyright (2013) American Chemical Society.

peak for enantiopure tartaric acid to be centred at 499 K when starting with a saturated overlayer (see Fig. 6.3).[2]

More importantly, the work of Mhatre *et al.* [7] was aimed at addressing the vexed question of precisely how explosive desorption from a saturated overlayer gets started in the first place. With no vacant sites at all, the kinetic models described above would indicate a rate of precisely zero, so clearly some additional process must be needed to initiate the explosion. As far back as the original work of Madix and co-workers [3], one suggestion had been that certain 'special' sites on the surface could act to initiate desorption in some way. Although the precise details of such an initiator mechanism were left vague, the idea was presumably that desorption from defect sites of some kind might create the necessary vacancies to spark an explosion. By comparing results for tartaric acid desorption from Cu{110} with those obtained on Cu{651}, however, these authors cast considerable doubt on this proposal; conducting experiments at fixed temperature (in addition to conventional TPD) they were able to demonstrate that the 'lag time' between abruptly raising the temperature and the onset of explosive desorption was similar on both surfaces, despite

2 The apparent discrepancy in peak temperature compared with the earlier value is likely due simply to the different heating rates employed. Slower heating rates in TPD tend to shift peak positions to lower temperature.

radically different densities of kink sites. Instead, Mhatre *et al.* [7] proposed that initiation was somehow intrinsic to the surface, regardless of defects, and could best be described by inserting an additional activated term into the basic rate equation. By doing so, and then fitting the kinetic model to their experimental data, they showed that desorption was second order in the coverage of vacant sites, obeying

$$-\frac{d\theta}{dt} = k_i\theta + k_e\theta(1-\theta)^2 \tag{6.5}$$

where k_i is now the rate constant for the initiation step and k_e the same for the explosion step. Within this model, the initiation step was found to fit best with an activation energy of 63 kJ mol^{-1} while the explosion step was activated with an energy of 125 kJ mol^{-1}. Intriguingly, whilst the pre-exponential factor for the explosion step took a fairly normal value of 6.1×10^{13} s^{-1}, that for the initiation step was particularly low, at 3.7×10^{3} s^{-1}. The physical nature of the initiation step remains, however, unknown at this time.

A second example of explosive desorption amongst chiral adsorbates was found in the case of malic acid, whose behaviour on Cu{110} was described by Roth and Ernst [11]. The molecule in question (*see Plate XII*) is another bicarboxylic acid, similar to tartaric acid but with one of the two hydroxyl groups of the backbone replaced by a hydrogen atom (or, indeed, similar to succinic acid with just one backbone hydrogen atom replaced by a hydroxyl group). Adsorbed at saturation coverage, XPS indicates (through both C 1s and O 1s regions of the spectra) the presence of only the monomalate form, while the bimalate form is found at lower coverages [11–13]. Upon heating, CO_2, H_2O and H_2 are evolved and the O 1s XPS peak weakens, vanishing entirely at temperatures beyond 570 K; some graphitic carbon remains, however, at temperatures up to at least 800 K.

In reporting on TPD experiments, Roth and Ernst [11] note the existence of two peaks in the CO_2 desorption rate. One of these, centred at c.570 K, is broad and does not shift either up or down in temperature upon increasing the initial coverage; this corresponds, therefore, to a first-order desorption process and need not detain us further (although it is interesting that such a feature is absent from the TPD spectra of tartaric and succinic acids). The other peak, however, shifts from c. 460 to 512 K upon increasing the initial coverage. Moreover, it sharpens considerably, achieving an eventual full-width at half maximum (FWHM) of just 2.3 K (see Fig. 6.4). This is clearly classic explosive behaviour, along the lines discussed above. Where tartaric acid showed a significant difference between desorption of its enantiopure and racemic forms, however, malic acid does not [11] despite the fact that they exhibit different sets of structural phases [12, 13].

6.1.2 Tartaric Acid on Cu{643}, Cu{17 5 1}, Cu{531} and Cu{651}

As mentioned briefly above, Mhatre *et al.* [7] conducted experiments concerning the dissociative desorption of tartaric acid from the Cu{651} surface, focusing primarily upon the fact that the lag time for onset of explosive behaviour at constant temperature was similar to that observed on Cu{110}. Their conclusion was that defect sites (i.e. kinks) could not reasonably be held responsible for the initiation of autocatalytic decomposition, pointing instead to an activated non-explosive

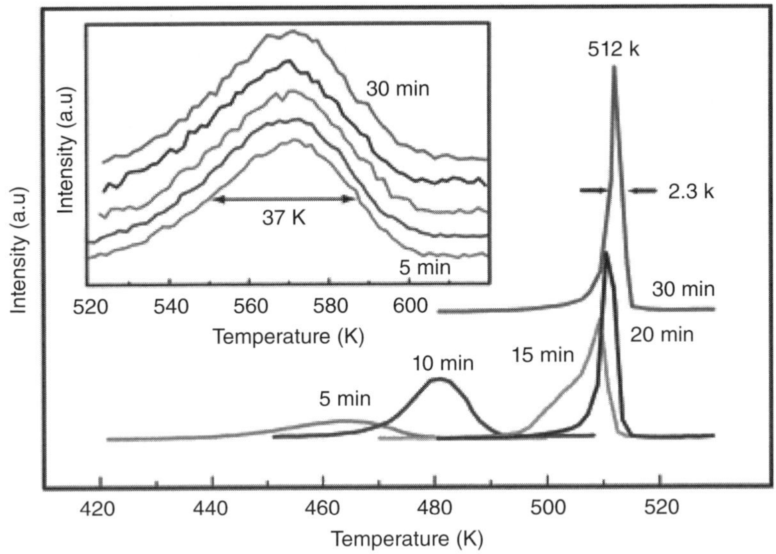

Figure 6.4 Temperature-programmed desorption profiles (obtained at a heating rate of $4\ K\,s^{-1}$) for CO_2 evolved from a Cu{110} surface following exposure to malic acid (exposure times noted alongside each trace). The inset shows a non-explosive peak that occurs after the explosive process is complete. Reproduced with permission from [11]. Copyright (2011) Springer.

process with its own barrier and pre-exponential factor distinct from that of the explosion step. In a subsequent paper, Gellman *et al.* [14] turned their attention to the now obvious question of whether explosive desorption on chiral surfaces might be notably enantiospecific, extending their survey to include not only the Cu{651} surface, but also the Cu{643}, Cu{17 5 1} and Cu{531} surfaces.

Following adsorption at 405 K of D-tartaric acid (i.e. *S*,*S*-tartaric acid) on either enantiomer of the Cu{643} surface, TPD revealed a single desorption peak shifting upwards in temperature and narrowing as the experiment was repeated with increasing dose, as expected for an explosive process. At the highest doses, the peak was centred at 486 K for the Cu{643}-L surface and at 496 K for the Cu{643}-D surface (see Fig. 6.5). Not only does this represent a particularly high diastereomeric difference in peak temperature (comparable with the highest reported in non-explosive systems) but the extremely sharp nature of the peaks (FWHM below 1 K) means that they are exceptionally well resolved. Indeed, as the temperature passes 485 K, the rate of dissociative desorption on the Cu{643}-L surface exceeds that on the Cu{643}-D surface by a factor of 50 (corresponding to an enantiomeric excess in the rates of −96%[3]),

3 Note that this (and other similar values stated in this section) is not directly comparable with the values for enantiomeric excess in the rate constants given in the preceding chapter. The enantiomeric excess in the rates, as discussed in Section 1.7, depends not only on the rate constants (which vary, in turn, only as a function of temperature) but also upon the current state of the system – specifically here, the instantaneous coverage.

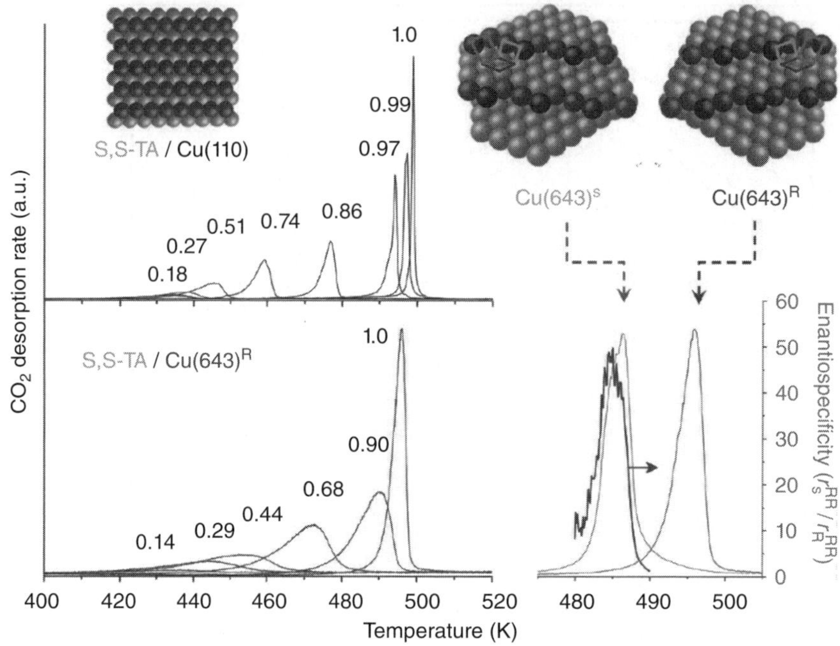

Figure 6.5 Temperature-programmed desorption profiles (obtained at a heating rate of 1 K s^{-1}) for CO_2 evolved from Cu{110} and Cu{643} surfaces following exposure to D-tartaric acid (coverages as a fraction of saturation are noted alongside each trace). The inset shows a zoomed-in view of desorption from the saturated overlayer on both enantiomeric forms of the chiral surface (R ≡ D and S ≡ L) together with the enantiomeric ratio. Reproduced with permission from [14]. Copyright (2013) American Chemical Society.

which represents a far greater degree of enantiospecificity than observed in any other surface process yet reported [14]. The same degree of enantiospecificity would, of course, be expected were D- and L-tartaric acid to have been studied on just one of the surface enantiomers.

Indeed, in similar TPD experiments on the Cu{17 5 1} surface, Gellman *et al.* [14] confirmed that the effect was truly diastereomeric, with the rates for D-tartaric acid on the D and L surfaces exactly mirroring those for L-tartaric acid on the L and D surfaces, as expected. The maximum ratio in rates for explosive desorption was found to occur at c.493 K with a value of 14 (corresponding to an enantiomeric excess in the rates of 87%). On the Cu{531} surface, the maximum ratio was 17 (corresponding to an enantiomeric excess in the rates of −89%) and occurred at a temperature of c.490 K.

Intriguingly, whilst racemic tartaric acid was found to desorb explosively at the same temperature as the *least* stable molecular enantiomer on either of the Cu{17 5 1} surfaces, it did so at the same temperature as the *most* stable molecular enantiomer on either of the Cu{531} surfaces. This strongly suggests that adsorption of racemic tartaric acid onto Cu{17 5 1} results in the formation of an intimate racemic mixture on the surface, such that the onset of desorption for either molecular enantiomer is sufficient to initiate explosive desorption of both. Clearly, the same cannot be true for the case of Cu{531}, hinting that D-tartaric acid preferentially adsorbs on the Cu{531}-D surface while L-tartaric acid does the same on the Cu{531}-L surface; such behaviour is entirely consistent, in broad terms, with experiments reported by Yun and Gellman [15] concerning the adsorption of aspartic acid on Cu{3 1 17} and discussed in the preceding chapter (Section 5.3.7). In this respect, however, we might also recall the XPS results reported by Baldanza *et al.* [16] (Section 5.3.4), which showed that the saturation coverage of D-tartaric acid on Cu{531}-L exceeds that of L-tartaric acid by about 25%, which might appear to argue for preferential adsorption in precisely the opposite sense to that implied by the work of Gellman *et al.* [14]; this matter remains to be resolved.

Returning, at length, to Cu{651}, Gellman *et al.* [14] reported the results of fixed-temperature experiments for both enantiomers of tartaric acid on the L surface, showing that there is a pronounced enantiospecific effect upon the lag time prior to initiation of explosive desorption (see Fig. 6.6). For instance, at 450 K the peak desorption rate of L-tartaric acid is achieved 400 s after the start of the experiment, while for D-tartaric acid the same point is reached some 100 s later; the ratio of rates for the two enantiomers in this experiment reaches a maximum of 12 (corresponding to an enantiomeric excess in the rates of 85%). The authors stress that this is *not* the result of a particularly large diastereomeric difference in activation energies, but instead occurs due to the non-linearity of the desorption rate equation; the same property that causes the extreme sharpness of TPD peaks also engenders a situation in which the coverage accentuates the differential in desorption rate at constant temperature.

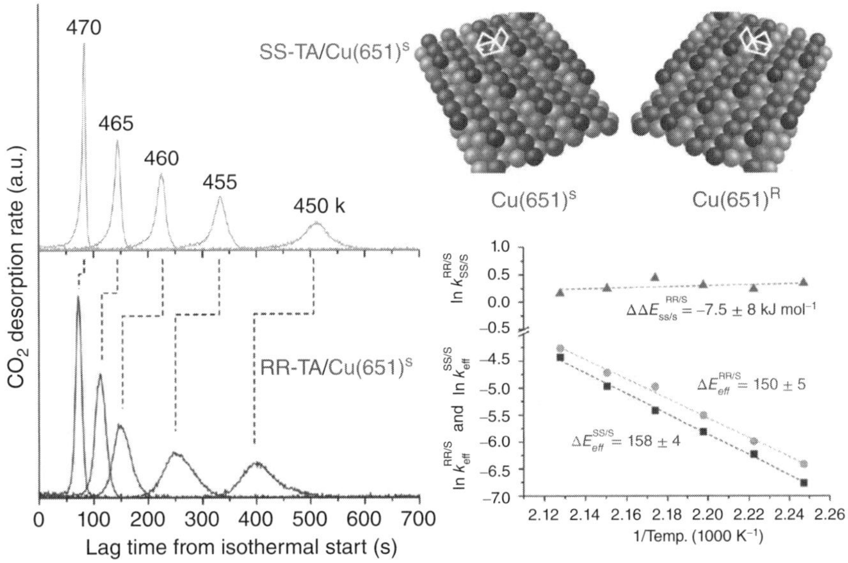

Figure 6.6 Isothermal desorption profiles (obtained at temperatures noted alongside each trace) for CO_2 evolved from Cu{643}-L following saturation exposure to D-tartaric and L-tartaric acid (D ≡ SS and L ≡ RR). The inset shows effective rate constants plotted as a function of inverse temperature (i.e. Arrhenius plots) indicating minimal diastereomeric difference in activation energy. Reproduced with permission from [14]. Copyright (2013) American Chemical Society.

6.2 Thermodynamic Amplification: Sergeants, Soldiers and Majority Rule

We have already seen a handful of examples, in Chapter 4, of systems in which adsorption of a racemic mixture on an achiral substrate appears to result in resolution into chiral domains. For both serine and tartaric acid on Cu{110}, and for alanine on Cu{311}, this interpretation was based partly on the observation that the racemic low-energy electron diffraction (LEED) patterns could be viewed as simple achiral superpositions of the enantiopure LEED patterns, and partly on scanning tunnelling microscopy (STM) images showing the existence of domains with appropriately asymmetric two-dimensional lattices. It should be stressed, however, that some caution must be exercised in identifying a particular LEED pattern or asymmetric two-dimensional lattice too closely with a specific enantiomer. One ought to recall, for instance, that we also saw chiral domains upon the adsorption of glycine on Cu{311}, even though this species is achiral, and that the local disposition of molecules within each domain appeared practically identical to that seen in the case of alanine; similar observations may be made in respect of succinic acid, in relation to the phases seen upon adsorption of tartaric acid, although here at least the periodicity is different. It is not, therefore, entirely beyond the realms of possibility that chiral domains arising from adsorption of racemic mixtures are themselves compositionally racemic (or very nearly so) and display chirality only in their long-range order.

The fact that we must separate questions of molecular chirality from the two-dimensional chirality of the lattice does, however, raise the intriguing question of whether we can controllably alter one without significantly altering the other. Specifically, it may be possible to engineer a situation in which a very small enantiomeric excess in the molecular composition of an overlayer may induce a much greater enantiomeric excess in the long-range order [17]. Such amplification would potentially be greatly beneficial in transduction between chemical and physical chirality, linking essentially localised properties of individual molecules to delocalised properties of the entire surface. In this regard, important lessons can be learnt from three-dimensional supramolecular chemistry, where two relevant effects have been recognised for some time. The first of these, the so-called **sergeants and soldiers** effect [18], occurs when large numbers of achiral sub-units (the 'soldiers') assemble preferentially in an asymmetric manner due to the influence of a small number of chiral sub-units (the 'sergeants'). The second, known as the **majority rule** effect [19], occurs when a small enantiomeric excess amongst otherwise identical chiral sub-units leads them to assemble in a disproportionately asymmetric manner. Since both effects have been observed in surface systems, it is perhaps best to explain the subtle difference between them by means of example.

6.2.1 Tartaric, Succinic and Malic Acids on Cu{110}

The first illustration of the sergeants and soldiers effect in a surface system appeared in the work of Parschau *et al.* [20] on the coadsorption of tartaric and succinic acids on Cu{110}. It will be recalled, from Section 4.2.4, that room-temperature adsorption of prochiral succinic acid leads to the formation of chiral domains characterised by a racemic superposition of extrinsically chiral $(1,\bar{1};\ 6,3)$ and $(1,1;\ 6,\bar{3})$ LEED patterns, so long as the coverage is low enough to permit the adsorbed moieties to adopt a bisuccinate configuration. Enantiopure tartaric acid, meanwhile, gives rise to distinctly different LEED patterns for each enantiomer, after annealing above 350 K in a similar coverage regime: $(1,\bar{2};\ 8,2)$ followed from the adsorption of D-tartaric acid and $(1,2;\ 8,\bar{2})$ from the adsorption of L-tartaric acid.[4]

In their coadsorption study, Parschau *et al.* [20] first deposited succinic acid, before adding a small amount of either D- or L-tartaric acid and annealing at 500 K. Remarkably, they found that just 2% adsorbed bitartrate[5] (relative to total coverage) was sufficient to result in either a purely $(1,\bar{1};\ 6,3)$ LEED pattern or a purely $(1,1;\ 6,\bar{3})$ LEED pattern, rather than a racemic superposition of the two (Fig. 6.7). Even fewer bitartrate 'sergeants' were capable of producing a marked imbalance between the LEED intensities of the bisuccinate 'soldiers', albeit not achieving complete suppression of the disfavoured periodicity. Heating sufficiently to desorb the bitartrate species but not the bisuccinate species resulted in the formation of an achiral superposition of intrinsically chiral $(1,1;\ 5,\bar{2})$ and $(1,\bar{1};\ 5,2)$ LEED patterns[6] that could also be obtained by heating pure bisuccinate overlayers; the induction of global chirality by the bitartrate species is clearly entirely reversible – as soon as the sergeants are absent, the soldiers revert to their own sense of order.

A similar manifestation of the sergeants and soldiers effect was subsequently reported upon the coadsorption of enantiopure tartaric acid with *meso*-tartaric acid [21]. The latter species is prochiral and its adsorption at intermediate coverages on Cu{110} leads to a racemic superposition of intrinsically chiral $(1,2;\ 6,\bar{1})$ and $(1,\bar{2};\ 6,1)$ LEED patterns,[7] again suggesting the formation of chiral domains (Section 4.2.5). When coadsorbed with a small amount of enantiopure tartaric acid, however, exclusive dominance of the $(1,2;\ 6,\bar{1})$ phase was induced by L-tartaric acid while the $(1,\bar{2};\ 6,1)$ phase was similarly induced by D-tartaric acid (Fig. 6.7). In comparison with the case of succinic acid, a rather greater proportion of sergeants is required to effect complete chiral control (8% in this case). In both these examples, however, it is clear that the influence of the sergeants propagates over a very long range, presumably through a combination of hydrogen bonding (and its knock-on effect via molecular distortion) and asymmetric surface stress (both induced and experienced by adsorbed molecules). It is notable, however, that the chiral order adopted in these situations is that of the chiral domains found even in the absence of sergeants, indicating that their role is to tip the balance between two equally favourable long-range periodicities, not to impose a wholly alien one. Whether the sergeants themselves are interspersed

---

4 Originally described as $(9,0;\ 1,\bar{2})$ and $(9,0;\ 1,2)$.

5 The authors note that the dissociative desorption temperature of tartaric and succinic acids remain distinct upon coadsorption, allowing the relative coverages of the adsorbed species to be measured.

6 Originally described as $(1,1;\ \bar{7},0)$ and $(7,0;\ \bar{1},1)$.

7 Originally described as $(6,\bar{1};\ \bar{1},\bar{2})$ and $(6,1;\ \bar{1},2)$.

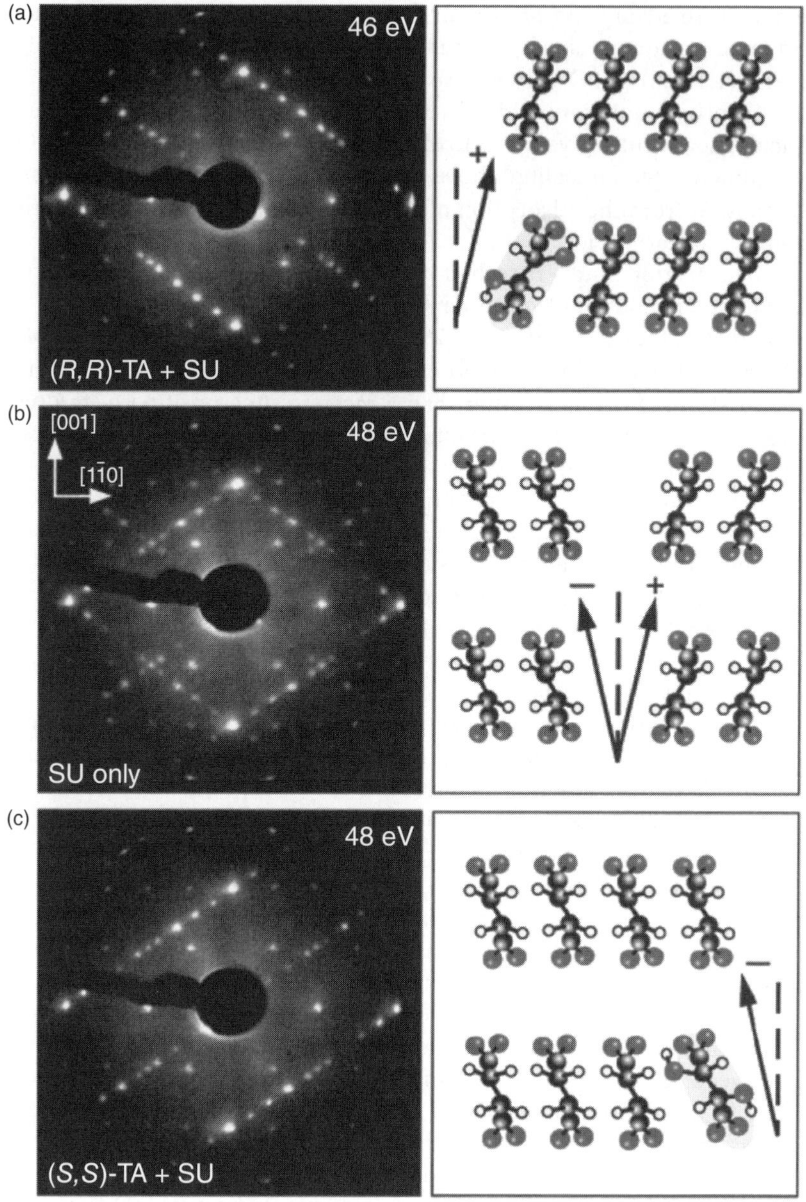

Figure 6.7 Low-energy electron diffraction patterns obtained after room-temperature exposure, followed by annealing at 500 K, of Cu{110} to succinic acid, either alone (b) or doped with a small fraction of L-tartaric acid (a) or D-tartaric acid (c). Schematic models indicate the corresponding real-space arrangement of molecules. Reproduced with permission from [20]. Copyright (2004) American Chemical Society.

intimately with the soldiers or remain aloof at domain boundaries is not, as yet, altogether clear.

A related, but subtly different, effect was first reported (in the surface context) by Haq *et al.* [22], who studied the adsorption of chiral tartaric acid on Cu{110} as a function of enantiomeric excess. As described previously in Section 4.3.5, the adsorption of racemic tartaric acid leads to a racemic superposition of $(1,2;\ 8,\bar{2})$ and $(1,\bar{2};\ 8,2)$ LEED patterns[8] presumed to be attributable to the L- and D-enantiomers (though see the *caveat* at the top of this section). By carefully calibrating against reflection absorption infrared spectroscopy (RAIRS) and TPD experiments, these authors were able to deposit both racemic and enantiopure tartaric acid in controlled amounts, in order to achieve specific values of enantiomeric excess in the adsorbed moieties. On doing so, they found that the observed LEED pattern could be switched between pure $(1,2;\ 8,\bar{2})$ and pure $(1,\bar{2};\ 8,2)$ by the imposition of enantiomeric excess values beyond only $\pm 20\%$. Furthermore, by defining an enantiomeric excess in the LEED intensities for beams associated with only one or other of the chiral patterns, they were able to quantify the relationship between long-range surface order and the chiral molecular composition of the overlayer (see Fig. 6.8). In STM images recorded for surfaces prepared with $\pm 20\%$ enantiomeric excess (in the molecular composition) it was revealed that, although the periodicity suppressed in LEED was still present to some degree at the nanoscale, it occupied a much smaller proportion of the surface than the favoured periodicity and did so in much smaller (and more disordered) domains. As in the sergeants and soldiers examples discussed above, it must evidently be the case that chiral information is being transmitted over long range, through either hydrogen-bond- or surface-mediated interactions (or some combination of both). Again, it is not clear whether the molecular chirality is uniformly distributed through the overlayer or concentrated at domain boundaries. Either way, even if it were possible to identify certain molecules as being particularly influential in dictating the chiral periodicity of this system, they would otherwise be chemically identical to those amongst the rank and file. That is, long-range order is determined by nothing more permanent than 'acting' sergeants, and all of the soldiers are equally eligible for 'promotion' to a similar position. This more democratic arrangement is, therefore, a classic example of the majority rule effect mentioned above.

Whether our final example in this section constitutes a distinct phenomenon or merely a point along a continuum between the two effects already outlined rests somewhat upon semantics. In experiments reported by Roth *et al.* [23], the addition of a small quantity of enantiopure malic acid was found to be capable of favouring one chiral domain over the other in the adsorption of racemic tartaric acid. Akin to the majority rule effect, all moieties in this case exhibit irreducible molecular chirality, but the asymmetry is driven by a subset of 'special' molecules that is chemically distinct from the common herd, as per the sergeants and soldiers effect. The authors of the original paper have coined the term **chiral conflict**, which neatly encapsulates a state of affairs governed by neither wholly authoritarian nor wholly democratic principles.

---

8 Originally described as $(1,2;\ \bar{9},0)$ and $(9,0;\ \bar{1},2)$.

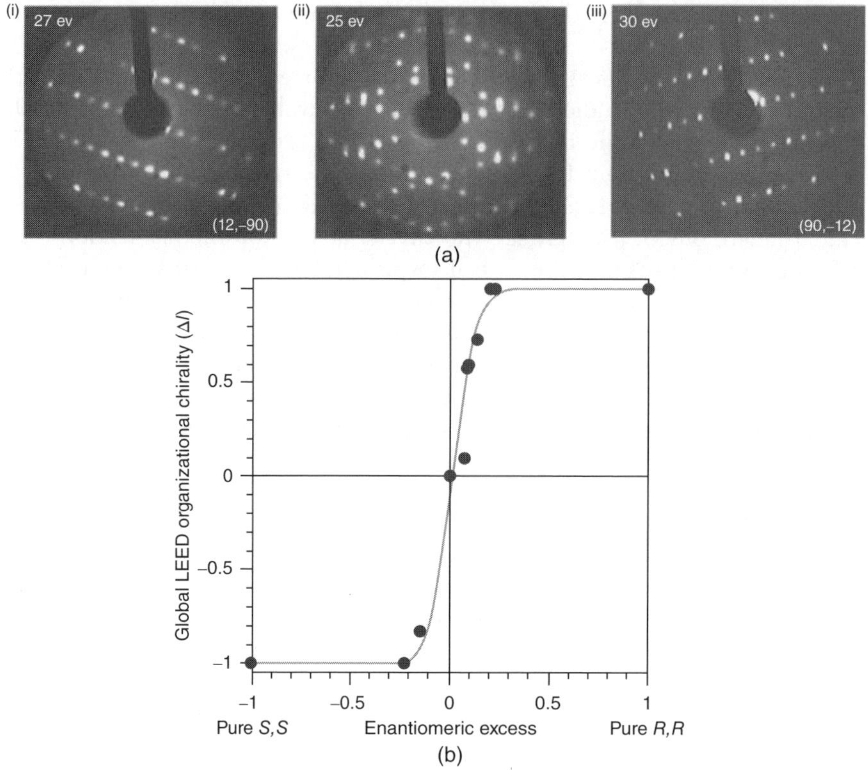

Figure 6.8 In (a) are shown low-energy electron diffraction patterns obtained after room-temperature exposure of Cu{110} to tartaric acid under conditions of (i) enantiomeric excess of 0.2 favouring the L enantiomer, (ii) enantiomeric excess of 0.0 and (iii) enantiomeric excess of 0.2 favouring the D enantiomer. In (b) the enantiomeric excess of the diffracted intensities is plotted against the enantiomeric excess of the molecular species. Reproduced with permission from [22]. Copyright (2005) Nature Publishing Group.

6.2.2 Heptahelicene on Cu{111}, Ag{111} and Au{111}

Arguably the most nuanced investigation of surface majority rule effects is to be found amongst an extensive series of studies (reported by Ernst and co-workers) concerning the adsorption of heptahelicene. This molecule can be conceived of as seven fused benzene rings, arranged in a spiral pattern (*see Plate XVIII*). Lacking a single chiral centre, the species is nevertheless most certainly chiral, and is usually labelled either P or M according to the sense of its spiral geometry: if the spiral matches that of a standard corkscrew, the label is P (standing for plus) while the mirror image would be labelled M (standing for minus). In their initial photoelectron diffraction (PhD) work, Fasel *et al.* [24] established not only that the molecule adsorbed intact on Cu{111}, but also that its spiral became distorted such that the first three rings lie essentially parallel to the surface; the adsorbate retains a distinctly chiral geometry, however, as evidenced by the observed and calculated diffraction patterns.

In a subsequent STM-based study of enantiopure heptahelicene on Cu{111}, Fasel *et al.* [25] reported the formation, at monolayer saturation coverage, of a $(2\sqrt{13}\times2\sqrt{13})$ phase, rotated by 13.9° clockwise for the M enantiomer or by the same angle anticlockwise for the P enantiomer.[9] These overlayer lattices are triangular in symmetry and hence not intrinsically chiral in and of themselves, but they do not align along high-symmetry directions of the substrate and so display extrinsic lattice chirality. Each primitive unit cell of the overlayer contains a 'cloverleaf' cluster (comprising three distinct entities, thought to be individual molecules) of rather symmetric triangular appearance and aligned with high-symmetry directions of the overlayer lattice. At slightly lower coverage (c.95% monolayer saturation) a different structure is preferred, with $(2\sqrt{39}\times2\sqrt{39})$ periodicity, rotated by 16.1° anticlockwise for the M enantiomer, or by the same angle clockwise for the P enantiomer. Once again, these represent triangular overlayer lattices, which in isolation would be intrinsically achiral; they nevertheless display extrinsic lattice chirality because they do not align with the high-symmetry directions of the underlying substrate. As for the detailed structure, the cloverleaf clusters are again in evidence (one per primitive unit cell of the overlayer) but joined now by a six-moiety 'pinwheel' cluster whose spokes are deflected in the anticlockwise sense for the M enantiomer and the clockwise sense for the P enantiomer (see Fig. 6.9). At both coverages, 180° rotational domains are observed, which is not (in this case) a demand of symmetry but, rather, a contingent feature of the energy landscape.

Upon deposition of racemic heptahelicene, the same authors [26] observed spontaneous formation of mirror-image domains organised according to intrinsically chiral lattices with periodicities $(4,5;\,5,\bar{3})$ and $(5,4;\,\bar{3},5)$;[10] in each case, the primitive unit cell of the overlayer contains two entities, arranged so as to produce a 'zig–zag' line running in a $\langle\bar{5}41\rangle$ surface direction. In the absence of any further information, one might have been tempted to speculate that these domains correspond to the segregation of P- and M-enantiomers, but here this would be hard to justify in light of the earlier results for enantiopure adsorption – not only are the lattices different (intrinsically rather than extrinsically chiral, for a start) but the zig–zag motif differs markedly from the previously observed cloverleaf and pinwheel motifs of the enantiopure case.[11]

9 The original paper does not explicitly state the overlayer periodicity, but the assignments quoted in this paragraph may be derived from close examination of the structural models presented by the authors; these models were, in turn, based upon their interpretation of the STM images.

10 Originally described as $(4,5;\,\bar{1},8)$ and $(8,\bar{1};\,5,4)$.

11 It ought, perhaps, to be mentioned that earlier LEED results reported by Ernst and co-workers *had*, in fact, been interpreted as indicating that resolution of the racemate had occurred, since the pattern obtained after deposition of the racemate *did* appear to represent a simple superposition of the pattern obtained upon adsorption of M-heptahelicene with its inferred mirror image [27]. However, the periodicities observed in those experiments – $(\sqrt{21}\times\sqrt{21})$ rotated by ±10.9° – do not *entirely* match up with any reported in the later STM-based work [25, 26] so it is not clear how much weight to attach to the apparent discrepancy. The overlayers studied by LEED were deposited at (or annealed at) slightly higher temperature (400 K) than those studied by STM (c.300 K), and while the LEED patterns were obtained at 130 K, the STM images were acquired at 50 K. It is, therefore, conceivable that these studies genuinely address slightly different structures. Alternatively, one might point to the inherent difficulty in accurately identifying LEED patterns when only a very small proportion of the expected spots are intense enough to pinpoint – only a very small error would be needed to bring matters into line.

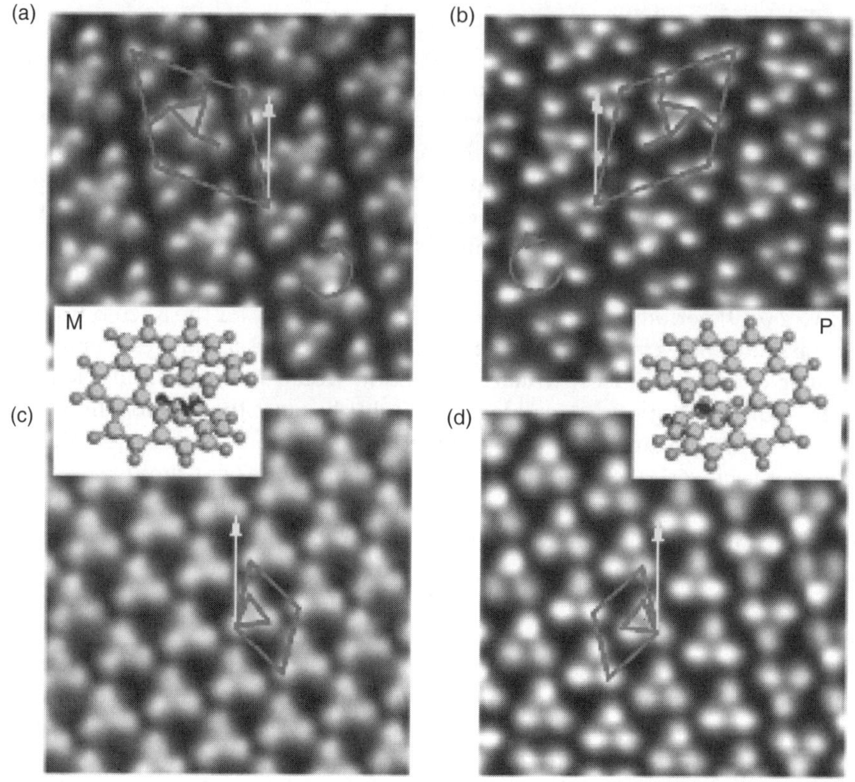

Figure 6.9 STM images obtained upon room-temperature exposure of Cu{111} to enantiopure heptahelicene. Panels (a) and (b) correspond to the M and P enantiomers, respectively, at a coverage estimated to be 95% of saturation; panels (c) and (d) are the corresponding images for saturation coverage. Reproduced with permission from [25]. Copyright (2003) Springer.

In fact, high-resolution STM images are consistent with the view that each primitive unit cell of the overlayer contains precisely one M enantiomer and one P enantiomer (see Fig. 6.10). Hence, each domain exhibits intrinsic lattice chirality whilst remaining racemic from the molecular point of view.[12] Notably, however, dosing heptahelicene prepared with a small enantiomeric excess of just 8% in favour of the M enantiomer is sufficient to exclusively favour the $(4,5;\ 5,\bar{3})$ phase over its mirror image, in another manifestation of the majority rule effect. High-resolution STM continues to confirm that the ordered structure remains racemic from the molecular perspective, so the authors conclude that the excess molecules of M-heptahelicene must reside elsewhere on the surface; small 'residual' domains of poorly ordered material accumulate in proportion to the enantiomeric excess, particularly in the region of terrace edges, and these do contain pinwheel clusters of chirality consistent with the majority

12 We note in passing, but will not discuss in detail, that resolution of molecular enantiomers associated with chiral lattice domains has been reported recently by Seibel *et al.* [28] upon deposition of racemic heptahelicene on Cu{100}.

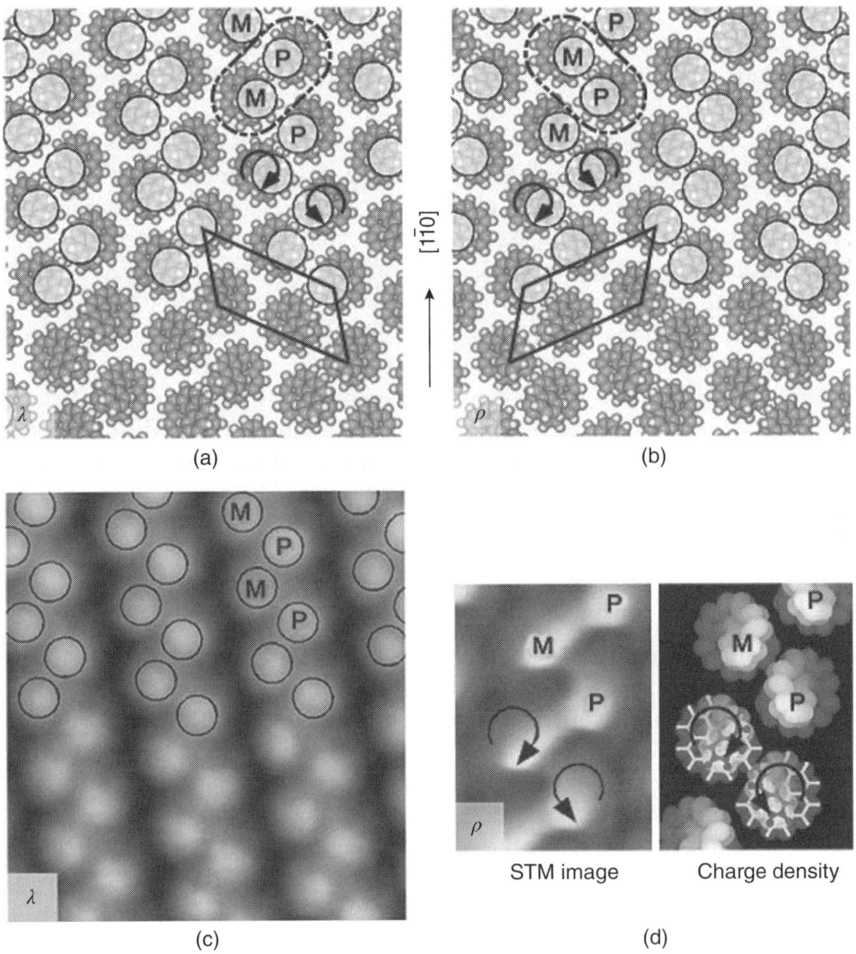

Figure 6.10 Schematic models for the chiral phases obtained upon room-temperature exposure of Cu{111} to racemic heptahelicene are presented in panels (a) and (b). Panel (c) shows an STM image (scan area 67 × 62 Å^2) corresponding to the structure shown in panel (a), while panel (d) shows a high-resolution STM image of the same structure (scan area 30 × 22 Å^2) alongside an Extended Hückel Theory simulation of the charge density associated with the lowest four unoccupied states. Reproduced with permission from [26]. Copyright (2006) Nature Publishing Group.

molecular enantiomer. The authors argue that boundaries between the lattice chiral domains are of high energy, and are hence disfavoured, so that the system requires only a very slight asymmetric perturbation to push it towards a highly asymmetric configuration; the boundary energy between lattice-chiral domains and molecularly chiral residual domains provides just such an impetus [25].

Intriguingly different behaviour has recently been observed by Seibel *et al.* [29] on the Ag{111} and Au{111} surfaces. In both cases, well-defined molecularly racemic zig–zag rows are again in evidence upon dosing with racemic heptahelicene, but running along high-symmetry directions of the substrate and with insufficient inter-row ordering to produce a well-defined two-dimensional lattice, chiral or otherwise.[13] On Ag{111} the zig–zag rows pack into domains in which all share a common $\langle 1\bar{1}0\rangle$ axis (out of the three available) but close inspection reveals a high density of internal boundaries between sub-domains with a variety of different angles for the zigs and the zags. On Au{111} there is even less order – individual zig–zag rows are each oriented along one of the $\langle 1\bar{1}0\rangle$ axes seemingly at random, with little evidence of shared-axis domains. Although the internal structure of each zig–zag row ought technically to exhibit chirality, this is quite subtle compared with the wholesale rotation of the zig–zag axis seen on Cu{111}, and the authors report no luck in identifying it within their images [29].

Returning briefly to the Cu{111} substrate, a fascinating new phenomenon has been elegantly documented by Parschau and Ernst [30] upon dosing racemic heptahelicene beyond monolayer saturation coverage. Growth apparently follows a Stranski–Krastanov type, with second-layer completion preceding the nucleation of further layers, but with fourth and subsequent layers nucleating before completion of the layer(s) beneath. Through a combination of LEED and STM, the bilayer structure was identified as displaying domains having either the intrinsically chiral $(2,5;\bar{5},\bar{1})$ lattice or its mirror image, the $(5,2;\bar{1},\bar{5})$ lattice.[14] Careful comparison with high-resolution images for a bilayer of enantiopure heptahelicene allowed the authors to convincingly argue that the first of these lattices is exclusively associated with a structure having the P enantiomer in the second layer, whilst the second lattice displays only the M enantiomer in this position. More significantly, experiments involving a small enantiomeric excess for the molecular species showed that it was possible not only to skew the second layer to display solely the minority enantiomer and its associated two-dimensional lattice, but also to determine (by simply counting molecules) that the underlying first layer must itself consist exclusively of the majority enantiomer. Indeed, even regions of the surface not yet covered to bilayer depth display (poorly ordered) cloverleaf clusters consistent with enantiomerically pure heptahelicene of the majority persuasion. It seems, therefore, that a most remarkable form of chiral resolution takes place in this system, whereby (i) the enantiomers spontaneously segregate to different layers; (ii) the preferential occupancy of different layers can be controlled by a small enantiomeric excess and (iii) molecules of the type needed for the second layer are drawn from the entirety of the first layer, including distant regions where the local coverage remains no higher than a monolayer. Indeed, the alternation of chirality is reported to continue into the third layer [30] and as yet it remains unclear at what coverage the effect of the surface will peter out; suffice to say that the bulk molecular crystal does not exhibit layered resolution of enantiomers.

---

13 Enantiopure heptahelicene does yield chiral lattices of various periodicities, populated with triangular cloverleaf clusters for both metals, with square 'four-leaved clover' structures also occurring on gold.

14 This contrasts strikingly with the behaviour of enantiopure heptahelicene on the same surface, where the second layer is found to display a small lattice mismatch with the first, leading to a chiral Moiré pattern at considerably larger length scale than the unit cells mentioned here [31].

6.2.3 Aspartic Acid on Cu{111}

Our penultimate example of chiral amplification was observed very recently by Yun and Gellman [32] whilst following up their earlier work [15, 33, 34] on the adsorption of amino acids on Cu{3 1 17}. Exposing a Cu{111} surface to aspartic acid prepared at a variety of enantiomeric excess values, the authors employed their isotopic labelling technique (as discussed in Section 5.3.7) to assess the enantiomeric excess achieved on the surface. At 400 K, they found that the surface enantiomeric excess matched precisely that of the gas-phase mixture to which the surface was exposed, in line with all their observations on the Cu{3 1 17} surface. Once saturation coverage was reached, however, they raised the surface temperature to 460 K (to permit exchange between adsorbed molecules and new arrivals from the gas phase) whereupon it was found that the surface enantiomeric excess increased dramatically, reaching c. 50% when that of the gas phase was only c.10%, and c.90% when that of the gas phase was just c.50% (see Fig. 6.11). Converting these values to enantiomeric ratios, one sees that there is a distinctly non-linear response at work – increasing the gas-phase enantiomeric ratio from 1.00 to 1.25 increased the surface ratio from 1.00 to 3.00, whereas raising the gas-phase ratio from 1.25 to 3.00 increased the surface ratio from 3.00 to 19.00.

At first blush, this seems an odd result, when even the previously studied chiral surface maintained a strict proportionality between gas-phase and surface enantiomeric ratios. In fact, we ought perhaps not to be quite so surprised. Using the expressions given in Section 1.8, we can write the surface enantiomeric excess, ee_s, in terms of the gas-phase equivalent, ee_g, in the case of proportional enantiomeric ratios, obtaining

$$ee_s = \frac{(\alpha - 1) + (\alpha + 1)ee_g}{(\alpha + 1) + (\alpha - 1)ee_g} \tag{6.6}$$

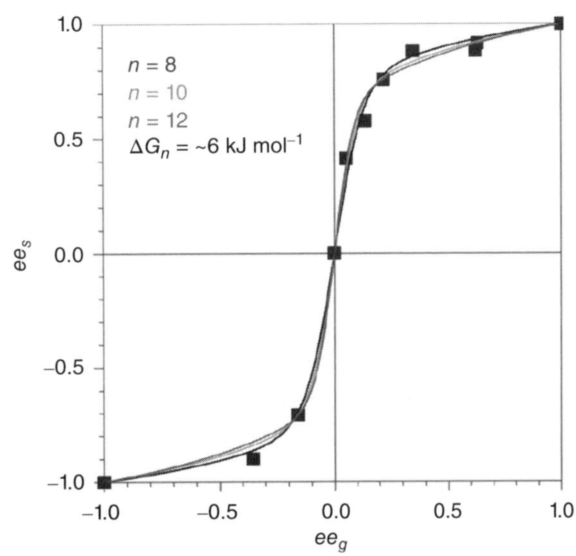

Figure 6.11 Surface enantiomeric excess versus gas-phase enantiomeric excess after annealing at 460 K a Cu{111} surface exposed to aspartic acid. Data points are experimental, while the solid curves represent fits based upon a Langmuirian model incorporating the formation of clusters containing $n = 8 - 12$ molecules. Reproduced with permission from [32]. Copyright (2015) Nature Publishing Group.

where α is just the constant of proportionality.[15] Clearly any constant value of α other than unity implies non-zero surface enantiomeric excess when the gas-phase is perfectly racemic, which is fine for the case of a chiral surface, but which would be troubling for the case of an achiral one. Evidently, we should *expect* a value of α that reduces to unity when an achiral surface is exposed to a racemic mixture of molecules – the real question is why it should ever be anything *other* than unity for an achiral surface. If we can answer that, the fact that it is not constant is a necessity, not a puzzle.

The origin of this singular effect must, of course, lie in some non-linear positive feedback mechanism related to the balance between adsorption and desorption on the surface. If the surface enantiomeric excess were constrained (in a thought experiment) to be strictly equal to that of the gas phase, the resulting surface would necessarily be chirally modified and this, in turn, would be expected to have some effect on both adsorption and desorption kinetics. If adsorption of the majority species is enhanced and/or desorption of the minority species suppressed, the result (upon removal of our imaginary constraint) ought to be a disproportionate auto-catalytic acceleration towards ever greater surface enantiomeric excess. In their analysis, Yun and Gellman [32] invoke the concept of homochiral clusters (containing 8–12 molecules and with heat of formation c.6.0 kJ mol^{-1}) as a modification to the simple Langmuirian adsorption isotherm, and show that it is capable of reproducing their experimental data extremely well (see Fig. 6.11). Of course, power laws in chemical kinetics must always be invoked cautiously when deducing mechanisms that may involve multiple competitive steps, and complex condensed phases must surely constitute a potentially pathological case.

In fact, one might equivalently conceive of this system in terms similar to the Ising models familiar from the theory of magnetism. In such a model, a lattice of sites is populated with spins that may be oriented either 'up' or 'down', and which can flip between states according to a Monte Carlo algorithm based on Boltzmann probabilities at a given temperature. With suitable energy couplings between same-spin and opposite-spin neighbours, this kind of model can readily display either paramagnetic or ferromagnetic behaviour; in the former, domains containing a single spin direction grow or diminish in response to an external forcing field, whereas in the latter the system may spontaneously collapse into a state characterised by an asymmetric distribution of domains favouring one spin direction over the other. In the present situation, we can imagine the surface as providing a lattice of sites, each populated (at saturation coverage) with either a left- or right-handed molecule. Occasional desorption frees up individual lattice sites, which are then repopulated from the gas phase by a new molecule whose handedness may either match or oppose that of the original occupant. If a molecule of the opposite sign is inserted, the site may be said to have flipped. Furthermore, any intermolecular interaction that favours neighbours of the same handedness replicates the essential coupling necessary for Ising-like behaviour.

15 That is, we have $er_s = \alpha er_g$ with er_s and er_g the surface and gas-phase enantiomeric ratios.

Indeed, the relationship between surface enantiomeric excess and gas-phase enantiomeric excess, revealed in the work of Yun and Gellman [32], is rather reminiscent of plots showing the magnetisation of a paramagnetic material versus an externally applied magnetic field. This begs the question as to whether a surface system may, in future, be found that replicates *ferromagnetic* behaviour in its chiral chemistry. Such a system would be one in which the history of exposure to enantiomeric excess must play a key role, with the single line of Fig. 6.11 splitting to create a hysteresis loop. One might then speak of **remanent chirality** and **chiral coercivity**[16] – the former referring to persistent surface enantiomeric excess in the presence of a purely racemic gas-phase species, having once exposed the surface to the enantiopure species; the latter referring to the opposite-handed enantiomeric excess that would need to be provided via the gas phase in order to force the surface enantiomeric excess back to zero from its remanent value. One might reasonably speculate that such effects should be sought in systems where chiral reconstruction of the substrate[17] carries the possibility of locking-in memory of previous chiral exposure.

6.2.4 Supramolecular Assemblies on Highly Ordered Pyrolytic Graphite

Turning aside (for the first and only time in this book) from the path of 'dry' surface science, a series of remarkable studies by De Feyter and co-workers provides insight into the expression of chirality at the liquid–solid interface. Taking highly ordered pyrolytic graphite (HOPG) as their substrate of choice, they deposit rather large organic molecules from solution, forming extended two-dimensional networks whose symmetries are influenced not only by intermolecular interactions within the overlayer but also by those between the overlayer and the supernatant solvent. In early work on two chiral oligo-(*p*-phenylenevinylene) diaminotriazine derivatives, designated *S*-OPV*n*T,[18] deposition from achiral 1-phenyloctane was found, by STM imaging, to result in the formation of adsorbed 'rosette' structures comprising hexamers of the solute molecules linked centrally by hydrogen bonds (see Fig. 6.12). Rosettes formed by *S*-OPV3T were exclusively twisted clockwise, while those formed by the slightly longer *S*-OPV4T species were exclusively twisted anticlockwise; in both cases, rosettes were arranged in (technically oblique but) very nearly triangular lattices, misaligned from the underlying graphite lattice by around 17° in the clockwise sense for *S*-OPV3T and 2° in the anticlockwise sense for *S*-OPV4T [35]. In the case of *S*-OPV5T, no ordered lattice was observed, and rosette formation was sporadic rather than ubiquitous, but where a twist could be unambiguously determined it was in the anticlockwise sense [36]. Molecular mechanics (MM) calculations by

16 The equivalent concepts in actual magnetic systems are briefly summarised in Chapter 9.

17 Such chiral reconstruction may be superficial, as in the cases of tartaric or malic acid adsorbed on Cu{110} (see Section 4.3.5), or profound, as in the cases of alanine or lysine adsorbed on Cu{100} (see Section 4.4.1). The only requirement is that it be at least metastable.

18 These rod-like molecules feature a bulky inert group at one end, a reactive triazine group at the other, and a backbone containing $n - 2$ phenylenevinylene units each bearing two (*S*)-2-methylbutoxy side groups.

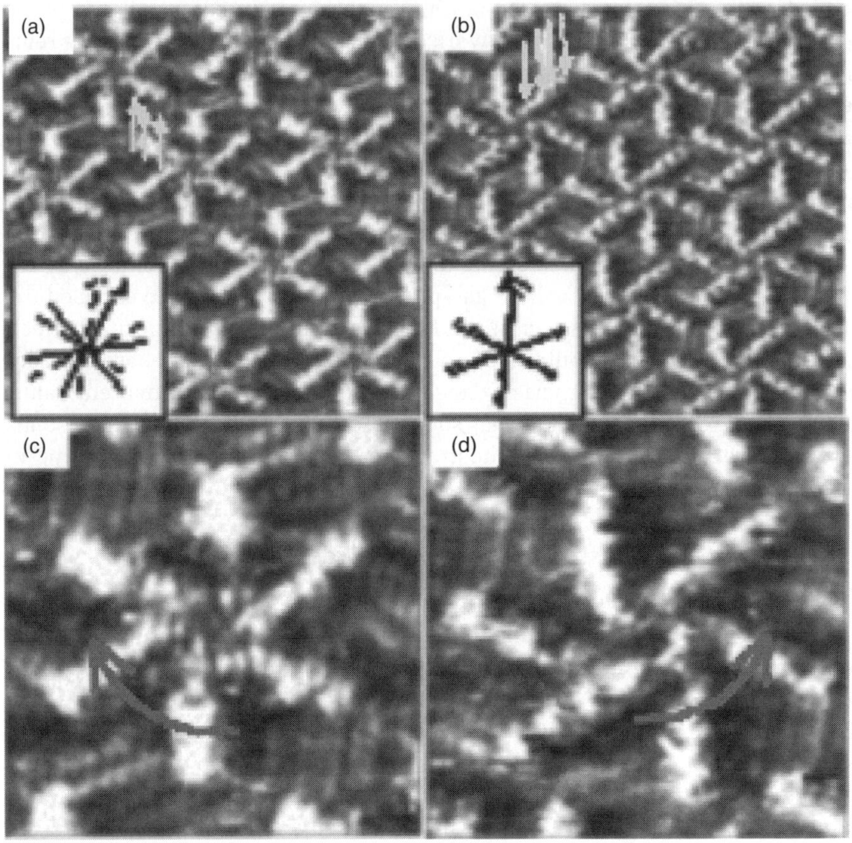

Figure 6.12 STM images obtained after exposure of HOPG to a solution containing (a) *S*-OPV3T and (b) *S*-OPV4T, with scan areas of 144 × 144 Å^2 and 184 × 184 Å^2 respectively. Panels (c) and (d) show close-up images of the constituent rosettes. Reproduced with permission from [35]. Copyright (2004) Wiley.

Minoia *et al.* [37] rationalise the *S*-OPV3T results in terms of steric hindrance between the arms of each rosette, and similar arguments are likely to apply also to *S*-OPV4T and *S*-OPV5T.

Subsequent STM investigations of an achiral analogue of the *S*-OPV4T molecule, designated A-OPV4T[19] revealed that (when deposited on HOPG from achiral 1-phenyloctane) rosettes of both clockwise and anticlockwise type were formed in essentially equal abundance. Deposition from enantiopure 1-phenyl-1-octanol, however, gave rise overwhelmingly (enantiomeric excess ∽75%) to rosettes with a clockwise (anticlockwise) twist when the *S* (*R*) form of the solvent was used [38]. The authors argue convincingly that solvent molecules are neither adsorbed beneath nor between the A-OPV4T molecules, asserting

19 This molecule is identical to *S*-OPV4T except that the chiral 2-methylbutoxy side groups are replaced with achiral butoxy side groups.

thereby that this system exhibits a sergeants and soldiers effect in which the sergeants are not themselves present on the surface. Unlike the more conventional version, where sergeants are statically incorporated into the overlayer itself, here the picture is essentially dynamic, with individual sergeants from the solvent visiting the near-surface region transiently, exerting their influence and departing to be replaced by others.

Similar solvent-induced overlayer chirality has been reported also by Destoop *et al.* [39] whilst investigating alkoxylated dehydrobenzo[12]annulene derivatives, designated DBA-OC*n*.[20] Upon deposition from achiral 1,2,4-trichlorobenzene, DBA-OC10 forms a porous honeycomb structure involving the pairwise interdigitation of alkoxy chains belonging to neighbouring molecules; although individual pores are necessarily chiral, having either a clockwise or an anticlockwise rotary sense according to an essentially arbitrary definition, no overall chirality is either expected or observed [40]. Switching the solvent to either *S*- or *R*-2-octanol, however, induces the overwhelming formation (enantiomeric excess ∽85%) of anticlockwise or clockwise pores respectively; the triangular lattice of the honeycomb itself is likewise extrinsically chiral, being rotated from orientational conformity with the underlying substrate by around 4.5° in the opposite sense to that defined for the individual pores (see Fig. 6.13). This version of the sergeants and soldiers effect is reportedly common to all DBA-OC*n* molecules with *n* up to at least 16, albeit the shorter the alkoxide chains the higher the enantiomeric excess achieved [39].

A more traditional sergeants and soldiers effect has also been reported for the DBA-OC*n* series of molecules, achieved by introducing a low concentration of chirally modified solute molecules, designated cDBA-OC*n*.[21] When *S*-cDBA-OC12 is deposited on HOPG from achiral 1-phenyloctane, STM reveals the formation of purely clockwise pores, within an essentially triangular lattice rotated ∽25° clockwise relative to the underlying substrate [41]. Premixing cDBA-OC12 with DBA-OC12 in the same solvent (in 3:7 ratio) gave rise to an enantiomeric excess of ∽75% in favour of clockwise pores (when the *S* enantiomer was used) or anticlockwise pores (when the *R* enantiomer was used) [42]. Interestingly, the minority pore type, in each case, is found often to be occupied by a molecule, presumed to be cDBA-OC12; these results therefore imply enantiospecific adsorption into a self-modified chiral environment, bringing a new level of complexity to the concept of chiral amplification.

A fascinating corollary to this finding emerges in very recent work by Fang *et al.* [43] in which variations of total concentration were probed (see Fig. 6.14). Preparing HOPG by co-deposition at 295 K of *S*-cDBA-OC12 and DBA-OC12 (in 1:9 ratio) from 1-phenyloctane at a total concentration of 5.0×10^{-7} M, the authors obtained a surface exhibiting predominantly clockwise pores, at 82% enantiomeric excess; after annealing to 353 K, the

20 These molecules are based on a triangular dehydrobenzo[12]annulene unit, decorated by six alkoxy chains of length *n*.

21 These molecules feature methyl-alkoxy chains, of well-defined chirality, in place of the alkoxy chains used in the achiral molecules.

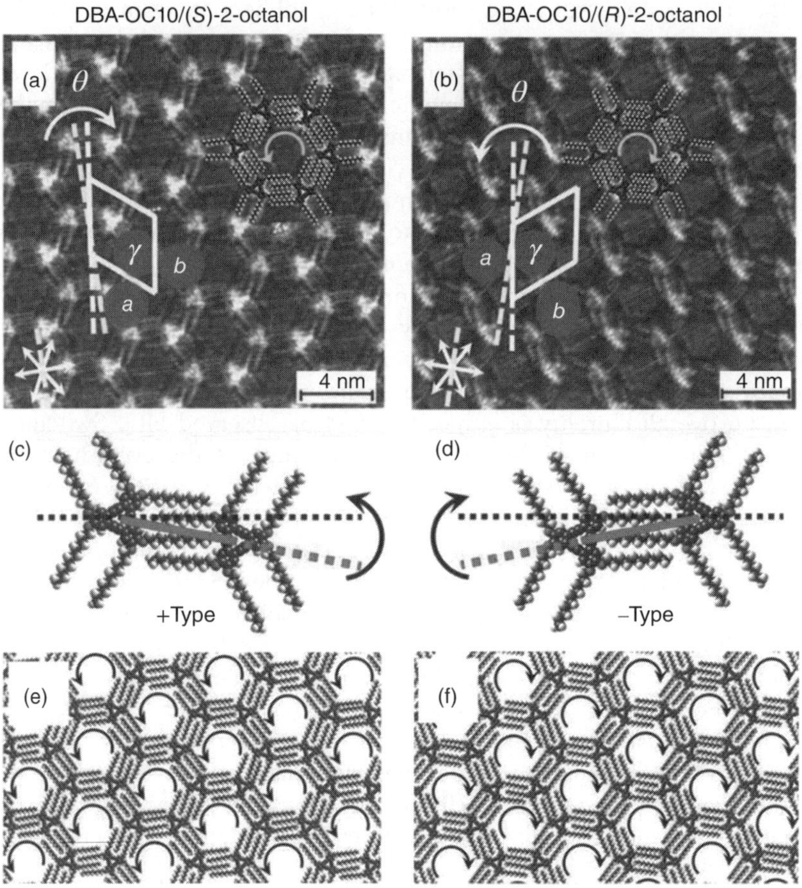

Figure 6.13 Panels (a) and (b) show STM images obtained after exposure of HOPG to a solution containing DBA-OC10 in *S*-2-octanol and *R*-2-octanol respectively. Panels (c) and (d) show schematic models of the interdigitation of alkyl chains between neighbouring molecules, while (e) and (f) show schematically the resulting chiral sense of the pore structure. Reproduced with permission from [39]. Copyright (2012) American Chemical Society.

enantiomeric excess rose to 100%, suggesting that greater mobility aids the soldiers' responsiveness to the sergeants' commands. So far, so predictable. Repeating the entire experiment with a total concentration of 9.0×10^{-7} M, however, resulted in a room-temperature enantiomeric excess of 74%, which actually *fell* to 38% upon annealing to 353 K. Even more puzzling was the observation that when the total concentration was raised to 1.5×10^{-6} M a 68% enantiomeric excess in favour of clockwise pores at 295 K was *reversed* upon annealing to an enantiomeric excess of −78% in favour of the anticlockwise variety. For total concentrations of 2.5×10^{-6} M and 4.5×10^{-5} M, annealing to 353 K yielded exclusively anticlockwise pores, despite respective enantiomeric excesses of 62% and 48% in favour of clockwise pores at 295 K. The authors suggest that at these higher concentrations, the occupation of minority pores with strongly bound cDBA-OC12 (as noted above) provides a driving force

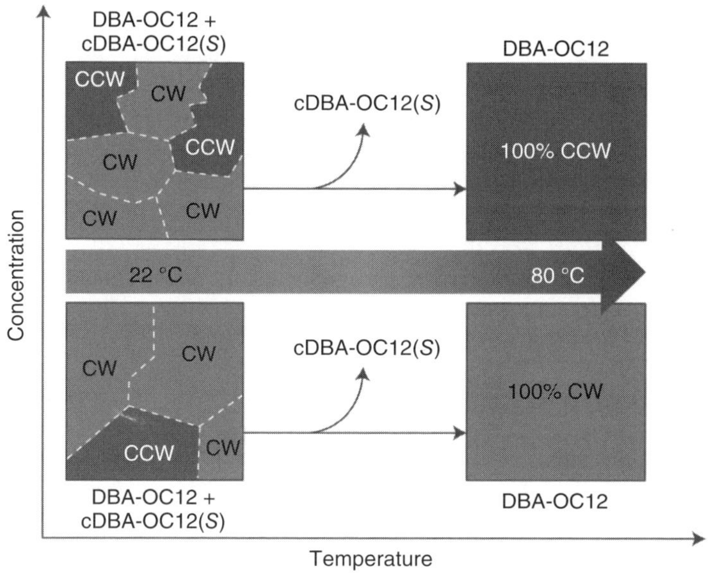

Figure 6.14 Schematic summary of the concentration dependence of chiral structures obtained upon exposure of HOPG to solutions containing both *S*-cDBA-OC12 and DBA-OC12 (in 1:9 ratio). At room temperature, clockwise (CW) pores are favoured over counterclockwise (CCW) pores, regardless of overall concentration; annealing at sufficiently high temperature, however, results in purely clockwise pores at low overall concentration but purely counterclockwise pores at high overall concentration. Reproduced with permission from [43]. Copyright (2016) Nature Publishing Group. (*See Plate V for the colour representation of this figure.*)

towards chirality reversal that awaits only the thermal desorption of more weakly bound cDBA-OC12 sergeants to assert control over the surface.

Finally, we note in passing the possibility of conflicting influences from both statically coadsorbed chiral molecules and dynamically transient solvent molecules. Such a situation has been realised in work reported by Destoop *et al.* [44] involving chiral and achiral porphyrin derivatives co-deposited on HOPG from either the achiral solvent heptanol or the chiral solvent 2-octanol. The resulting competition has been dubbed by the authors the 'sergeants and corporals' effect, whereby one entity (in this particular case, the chiral porphyrin) dictates the overall symmetry to first order, but another (in this instance, the chiral solvent) exerts a non-negligible second-order influence. Where the sergeants and the corporals disagree, the soldiers fundamentally obey the former, but tempered by respect for the latter.[22]

22 With regard to divided loyalties, we might call to mind the commentary of Niccolò Machiavelli on 'whether it is better to be loved than to be feared' (The Prince, c.1513). Noting first that one would, of course, like to be both loved *and* feared, he goes on to advise that if that is not possible, one should rather be feared, since that is more to be relied upon in time of crisis. From this, we conclude (as Macchiavelli himself inexplicably failed to do) that achiral porphyrins fear chiral porphyrins more than they love chiral solvents.

References

1 J.G. McCarty, J. Falconer and R.J. Madix, *J. Catal.* **30**, 235 (1973).
2 J.L. Falconer, J.G. McCarty and R.J. Madix, *Surf. Sci.* **42**, 329 (1974).
3 J.L. Falconer and R.J. Madix, *Surf. Sci.* **46**, 473 (1974).
4 M. Bowker and Y.X. Li, *Catal. Lett.* **10**, 249 (1991).
5 N. Aas and M. Bowker, *J. Chem. Soc., Faraday Trans.* **89**, 1249 (1993).
6 Y.X. Li and M. Bowker, *Surf. Sci.* **285**, 219 (1993).
7 B.S. Mhatre, V. Pushkarev, B. Holsclaw, T.J. Lawton, E.C.H. Sykes and A.J. Gellman, *J. Phys. Chem. C* **117**, 7577 (2013).
8 B. Behzadi, S. Romer, R. Fasel and K.-H. Ernst, *J. Am. Chem. Soc.* **126**, 9176 (2004).
9 S. Romer, B. Behzadi, R. Fasel and K.-H. Ernst, *Chem. Eur. J.* **11**, 4149 (2005).
10 V. Humblot, M. Ortega Lorenzo, C.J. Baddeley, S. Haq and R. Raval, *J. Am. Chem. Soc.* **126**, 6460 (2004).
11 C. Roth and K.-H. Ernst, *Top. Catal.* **54**, 1378 (2011).
12 C. Roth, D. Passerone, L. Merz, M. Parschau and K.-H. Ernst, *J. Chem. Phys. C* **115**, 1240 (2011).
13 C. Roth, M. Parschau and K.-H. Ernst, *ChemPhysChem* **12**, 1572 (2011).
14 A.J. Gellman, Y. Huang, X. Feng, V.V. Pushkarev, B. Holsclaw and B.S. Mhatre, *J. Am. Chem. Soc.* **135**, 19208 (2013).
15 Y. Yun and A.J. Gellman, *Angew. Chem. Int. Ed.* **52**, 1 (2013).
16 S. Baldanza, J. Ardini, A. Giglia and G. Held, *Surf. Sci.* **643**, 108 (2016).
17 K.-H. Ernst, *Curr. Opin. Colloid Interface Sci.* **13**, 54 (2007).
18 M.M. Green, M.P. Reidy, R.J. Johnson, G. Darling, D.J. O'Leary and G. Willson, *J. Am. Chem. Soc.* **111**, 6452 (1989).
19 M.M. Green, B.A. Garetz, B. Munoz, H. Chang, S. Hoke and R.G. Cooks, *J. Am. Chem. Soc.* **117**, 4182 (1995).
20 M. Parschau, S. Romer and K.-H. Ernst, *J. Am. Chem. Soc.* **126**, 15398 (2004).
21 M. Parschau, T. Kampen and K.-H. Ernst, *Chem. Phys. Lett.* **407**, 433 (2005).
22 S. Haq, N. Liu, V. Humblot, A.P.J. Jansen and R. Raval, *Nat. Chem.* **1**, 409 (2005).
23 C. Roth, D. Passerone and K.-H. Ernst, *Chem. Commun.* **46**, 8645 (2010).
24 R. Fasel, A. Cossy, K.-H. Ernst, F. Baumberger, T. Greber and J. Osterwalder, *J. Chem. Phys.* **115**, 1020 (2001).
25 R. Fasel, M. Parschau and K.-H. Ernst, *Angew. Chem. Int. Ed.* **42**, 5178 (2003).
26 R. Fasel, M. Parschau and K.-H. Ernst, *Nature* **439**, 449 (2006).
27 K.-H. Ernst, Y. Kuster, R. Fasel, M. Müller and U. Ellerbeck, *Chirality* **13**, 675 (2001).
28 J. Seibel, L. Zoppi and K.-H. Ernst, *Chem. Commun.* **50**, 8751 (2014).
29 J. Seibel, M. Parschau and K.-H. Ernst, *J. Phys. Chem. C* **118**, 29135 (2014).
30 M. Parschau and K.-H. Ernst, *Angew. Chem. Int. Ed.* **54**, 14422 (2015).
31 M. Parschau, U. Ellerbeck and K.-H. Ernst, *Colloids Surf., A* **354**, 240 (2010).
32 Y. Yun and A.J. Gellman, *Nat. Chem.* **7**, 520 (2015).
33 Y. Yun, D. Wei, D.S. Sholl and A.J. Gellman, *J. Phys. Chem. C* **118**, 14957 (2014).

34 Y. Yun and A.J. Gellman, *Langmuir* **31**, 6055 (2015).
35 P. Jonkheijm, A. Miura, M. Zdanowska, F.J.M. Hoeben, S. De Feyter, A.P.H.J. Schenning, F.C. De Schryver and E.W. Meijer, *Angew. Chem. Int. Ed.* **43**, 74 (2004).
36 A. Miura, P. Jonkheijm, S. De Feyter, A.P.H.J Schenning, E.W. Meijer and F.C. De Schryver, *Small* **1**, 131 (2005).
37 A. Minoia, Z. Guo, H. Xu, S.J. George, A.P.H.J. Schenning, S. De Feyter and R. Lazzaroni, *Chem. Commun.* **47**, 10924 (2011).
38 N. Katsonis, H. Xu, R.M. Haak, T. Kudernac, ž. Tomović, S. George, M. Van der Auweraer, A.P.H.J. Schenning, E.W. Meijer, B.L. Feringa and S. De Feyter, *Angew. Chem. Int. Ed.* **47**, 4997 (2008).
39 I. Destoop, E. Ghijsens, K. Katayama, K. Tahara, K.S. Mali, Y. Tobe and S. De Feyter, *J. Am. Chem. Soc.* **134**, 19568 (2012).
40 K. Tahara, S. Lei, J. Adisoejoso, S. De Feyter and Y. Tobe, *Chem. Commun.* **46**, 8507 (2010).
41 K. Tahara, H. Yamaga, E. Ghijsens, K. Inukai, J. Adisoejoso, M.O. Blunt, S. De Feyter and Y. Tobe, *Nat. Chem.* **3**, 714 (2011).
42 E. Ghijsens, H. Cao, A. Noguchi, O. Ivasenko, Y. Fang, K. Tahara, Y. Tobe and S. De Feyter, *Chem. Commun.* **51**, 4766 (2015).
43 Y. Fang, E. Ghijsens, O. Ivasenko, H. Cao, A. Noguchi, K.S. Mali, K. Tahara, Y. Tobe and S. De Feyter, *Nat. Chem.* **8**, 711 (2016).
44 I. Destoop, H. Xu, C. Oliveras-González, E. Ghijsens, D.B. Amabilino and S. De Feyter, *Chem. Commun.* **49**, 7477 (2013).

7

Asymmetric Heterogeneous Catalysis

Some things are hurrying into existence, and others are hurrying out of it; and of that which is coming into existence part is already extinguished. Motions and changes are continually renewing the world, just as the uninterrupted course of time is always renewing the infinite duration of ages. In this flowing stream then, on which there is no abiding, what is there of the things which hurry by on which a man would set a high price?

Meditations, Marcus Aurelius, AD 170–180 (tr. George Long, 1862)

Thus far, this book has concentrated predominantly on chirality as a state of being – a property inherent in an object, immutable as the hills.[1] As philosophers have recognised from time immemorial, however, the essential nature of the world is to be in flux – a state less of *being* than of *becoming*. The time has come, therefore, to turn our attention from the conservative expression of mere chiral adsorption to the transformative processes involved in asymmetric catalytic reactions – to ask whether it is possible to heterogeneously synthesise complex molecules in enantiopure (or near-enantiopure) form. The answer is emphatically affirmative, in principle, but practical examples remain rather rare, and innovative approaches would be most welcome. Whatever particular synthetic strategy is to be adopted, however, two key issues must be distinguished, namely chiral recognition and enantioselectivity – the former referring to the mechanism by which a chiral active site responds differentially to two different molecular enantiomers, and the latter to the ability of such a system to leverage that response into an asymmetric chemical reaction.

We have touched upon chiral recognition in Chapter 5, in the context of diastereomeric differences in adsorption and desorption, and here need only generalise the concept. It would be as well to reiterate, however, that naive notions of a lock-and-key basis for recognition have tended to give way to rather more subtle induced-fit notions as we learn more about individual systems. Furthermore, the chiral entities responsible for asymmetry in either case may act in a highly localised one-to-one manner or in a more delocalised template-like fashion. For the purposes of this chapter, however, we shall highlight one further dichotomy – between those systems in which enantioselectivity is conferred by

1 But not quite so green – *vide* Kipling.

Chirality at Solid Surfaces, First Edition. Stephen J. Jenkins.

an intrinsically chiral surface (the chiral entities being steps or kinks) and those in which it arises from the influence of an extrinsic chiral modifier (the chiral entities being adsorbed molecules).

Turning initially to the topic of catalysis at intrinsically chiral surfaces, one is immediately struck by how few studies actually tackle this interesting subject head-on. Discounting those desorption studies discussed in the previous two chapters (where the focus, even in the explosive cases, is upon how molecules arrive at and depart from the surface, rather than what they might do whilst they are there), there have only been a handful of relevant reports. One might speculate that this has to do with a perception that intrinsically chiral surfaces are inherently unsuitable from the perspective of real-world catalyst preparation, and it is certainly true that the single-crystal samples necessary for academic research in this area will never make practical catalysts. Nevertheless, real catalysts most likely do boast significant densities of chiral kink sites, and although these must necessarily occur in racemic proportion it is not entirely unrealistic to imagine selectively blocking half of them with a suitably chosen chiral species prior to use. In a sense, such a system would be a hybrid, involving both an extrinsic chiral modifier *and* intrinsically chiral surface sites, but it would fall to the latter to confer chirality upon the reaction for which the catalyst is designed. In efforts to grasp the potential of this approach, studies on intrinsically chiral single-crystal samples are certainly a necessary step, if not per se an entirely sufficient one. Here, we shall examine two such studies, concerning firstly the electro-oxidation of glucose on platinum electrodes, and secondly the electron-stimulated oxidation of methyl lactate on copper.

In contrast to asymmetric catalysis at intrinsically chiral surfaces, which appears thus far to be limited to cases of selective oxidation, examples involving chirally modified surfaces appear to be equally strongly skewed towards cases of selective hydrogenation.[2] Two reactions in particular, the hydrogenation of α-ketoesters over platinum and of β-ketoesters over nickel, have garnered intense attention for decades, and we shall focus on these below. In both cases, hydrogenation reduces a carbonyl group to an alcohol, converting the involved carbon atom into a chiral centre in the process. An alternative class of reaction, in which a carbon double bond is reduced over a palladium catalyst, has also received some interest but is comparatively underdeveloped; for this topic, the reader is referred to an excellent review by Kyriakou *et al.* [3]. Although a great deal of literature exists concerning the practical catalytic science of these chirally modified systems, we shall focus here upon the insights gained via the fundamental surface science approach, in keeping with the tenor of preceding chapters.

2 Indeed, the reactions described in detail within this chapter might readily be supplemented by at least two further examples involving hydrogenation, namely the hydrogenation of 2-butoxy either in the presence of propylene oxide on Au/Pd{111} [1] or of proline on otherwise clean Pd{111} [2]. In both these cases, however, it may be argued that enantioselectivity arises primarily due to preferential adsorption and/or desorption of reactants/products, rather than due to preferential formation of one specific enantiomer per se. Whilst this is a perfectly valid strategy for asymmetric catalysis, it is likely to be highly sensitive to conditions of temperature and pressure to an extent that genuinely enantioselective synthesis ought not.

7.1 Electro-Oxidation of Glucose on Pt{643} and Pt{321}

The first successful demonstration of catalytic enantioselectivity conferred by an intrinsically chiral surface actually arose in an electrochemical context, when Attard and co-workers [4, 5] applied the technique of cyclic voltammetry (CV) to the electro-oxidation of glucose (*see Plate XVI*) over stepped and kinked platinum electrodes. During such experiments, sweeping the potential of the electrode in the positive sense reveals peaks in the measured current attributable to stages in the oxidation of glucose to glucolactone

$$C_6H_{12}O_6 \rightarrow C_6H_{10}O_6 + 2H^+ + 2e^- \qquad (7.1)$$

although side reactions may also occur and the precise mechanism is not entirely clear [6]. Ideally, sweeping the potential in the reverse sense would reveal peaks in the current corresponding to desorption of the product; in principle, the cycle could then be repeated *ad infinitum*, but in reality one cannot discount the possibility of irreversible changes to the surface. In particular, earlier work [7] had already indicated a build-up of irreversibly adsorbed carbon monoxide (from a side-reaction) prone to blocking active sites on the surface and gradually attenuating all structure in successive voltammograms.

Using a Pt{211} electrode and a sweep rate of 50 mV s^{-1}, the voltammogram for 0.005 M D-glucose in 0.05 M sulphuric acid was found to be essentially identical to that obtained for the same concentration of L-glucose in the same acidic environment [4]. Both displayed a sharp peak at 0.22 V, interpreted as indicative of dehydrogenation/adsorption at step sites, and a rather broader peak at ~0.3 V, attributed to the same process occurring on the intervening {111} terraces of the surface. Finally, a further broad peak at ~0.5 V may be identified with the oxidation of adsorbed dehydrogenation products (Fig. 7.1a and b). Repeating the same experiment with a Pt{332} electrode, the voltammograms of the two enantiomers were again shown to be practically identical, although the peak structure differed a little from that obtained with the Pt{211} electrode [4]. Here, the sharp peak at 0.22 V was absent, but a very small feature was present at 0.07 V. Moreover, the broad peak at ~0.3 V was much stronger in this case (Fig. 7.1c and d). These differences can be rationalised by the contrasting nature of the step sites (exposing a {100} microfacet on the {211} surface, but a {110} microfacet on the {332} surface) and the relative widths of the {111} terraces (three atoms wide on the {211} surface versus five atoms wide on the {332} surface). On the Pt{111} surface, only the peaks at ~0.3 and ~0.5 V were in evidence, consistent with the complete absence of steps [5].

Whatever the precise interpretation of peaks, voltammograms obtained on Pt{643} electrodes display a combination of the features seen on Pt{111}, Pt{211} and Pt{332}. That is, step-related dehydrogenation/adsorption peaks are seen at both 0.07 and 0.22 V, together with a broad oxidation peak at ~0.5 V. A peak at 0.35 V, attributed to dehydrogenation/adsorption on bare {111} terraces, is present only when D-glucose is used with the {643}-L electrode or L-glucose with the {643}-D electrode; for the other two combinations,

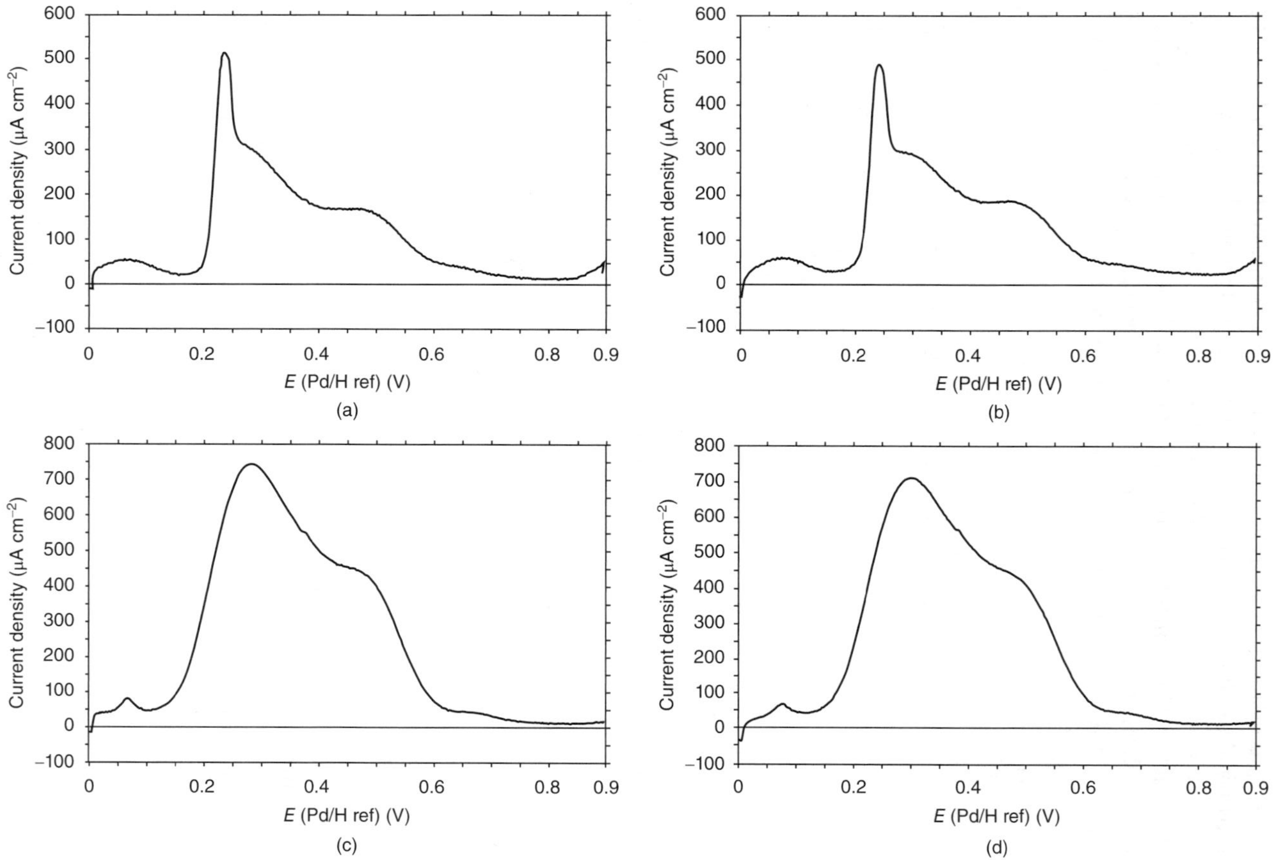

Figure 7.1 Cyclic voltammograms obtained in sulphuric acid for (a) D-glucose and (b) L-glucose on Pt{211}, and for (c) D-glucose and (d) L-glucose on Pt{332}. Reproduced with permission from [4]. Copyright (1999) American Chemical Society.

it is entirely absent (see Fig. 7.2). The authors rationalised this finding as evidence for higher reactivity in the latter pair of cases, leading to a more rapid build-up of site-blocking CO on the terraces [4]. Whilst absolute rates of electro-oxidation drop upon lowering the temperature, enantioselectivity rises, consistent with an estimated diastereomeric difference in the effective activation barrier of 0.14 kJ mol^{-1} when evaluated at a potential of 0.39 V. The estimated enantiomeric ratio in the rates would then amount to 9:5 (equivalent to an enantiomeric excess in the rates of ~36%) at 298 K [5]. Similarly enantioselective results were obtained on Pt{531}, Pt{321} and Pt{431} electrodes (with selectivity scaling in proportion with the surface density of kinks) and related sugars including mannose, arabinose and xylose (*see Plate XVI*) were also found to behave in the same manner [8]. On the other hand, Attard and co-workers [9] subsequently showed that electro-oxidation of 2-chloropropanol, 2-butanol, 1,3-butanediol and 2,3-butanediol (*see Plate XV*) was insensitive to chirality on various surfaces, including Pt{531}, Pt{321}, Pt{643}, Pt{976} and Pt{11 7 1}, concluding that the cyclic structures (pyranose rings) adopted by the sugars in solution may be in some manner crucial to enantioselectivity in this reaction.

Further insight was subsequently provided by Hazzazi *et al.* [10] who selectively blocked step sites on Pt{321} by the electro-deposition of bismuth from $Bi(NO_3)_3$ prior to the electro-oxidation of glucose. As expected, the step-related peaks were progressively suppressed as the bismuth coverage increased up to 2/3 ML (see Fig. 7.3). Notably, the enantioselectivity was also influenced: at bismuth coverages between around 1/10 and 1/2 ML, a peak at just over 0.3 V (corresponding to adsorption on terrace sites) was observed for all four combinations of substrate and adsorbate chirality, whereas at bismuth coverages below 1/10 ML it was observable only for D-glucose on the Pt{321}-L surface and for L-glucose on the Pt{321}-D surface. Diastereomeric differences in the voltammograms disappear more or less entirely once the steps are completely saturated with bismuth, with only the peak above 0.4 V remaining. Indeed, this latter peak actually grows with increasing bismuth coverage, indicating that oxidation of dehydrogenation products is actually enhanced by taking step sites out of the equation.

In contrast, prior electro-deposition of silver from $Ag(ClO_4)$ does not result in any enhancement of the oxidation rate – quite the opposite, in fact (see Fig. 7.4). Whilst the peak above 0.4 V (corresponding to oxidation of adsorbed dehydrogenation products) rapidly disappears with increasing silver coverage, those at 0.07 and 0.22 V (corresponding to dehydrogenation/adsorption at steps) remain very much in evidence. Dehydrogenation/adsorption at terrace sites (indicated by the peak at ~0.3 V) declines gradually with increasing silver coverage. Chiral discrimination is maintained at all but the highest coverages, and the authors propose that whereas bismuth blocks step sites, silver actually substitutes into terrace sites. Thus the processes thought to occur on those terraces are progressively switched off, whilst those located at the step sites persist [10]. The influence of gold, electro-deposited from $HAuCl_4$, is thus of particular interest, producing a uniform suppression of all peaks that conforms neither to the behaviour of bismuth nor to that of silver. Gold, therefore, is surmised simply to block step and terrace sites with equanimity, unlike bismuth (which selectively blocks the former) and silver (which selectively blocks the latter).

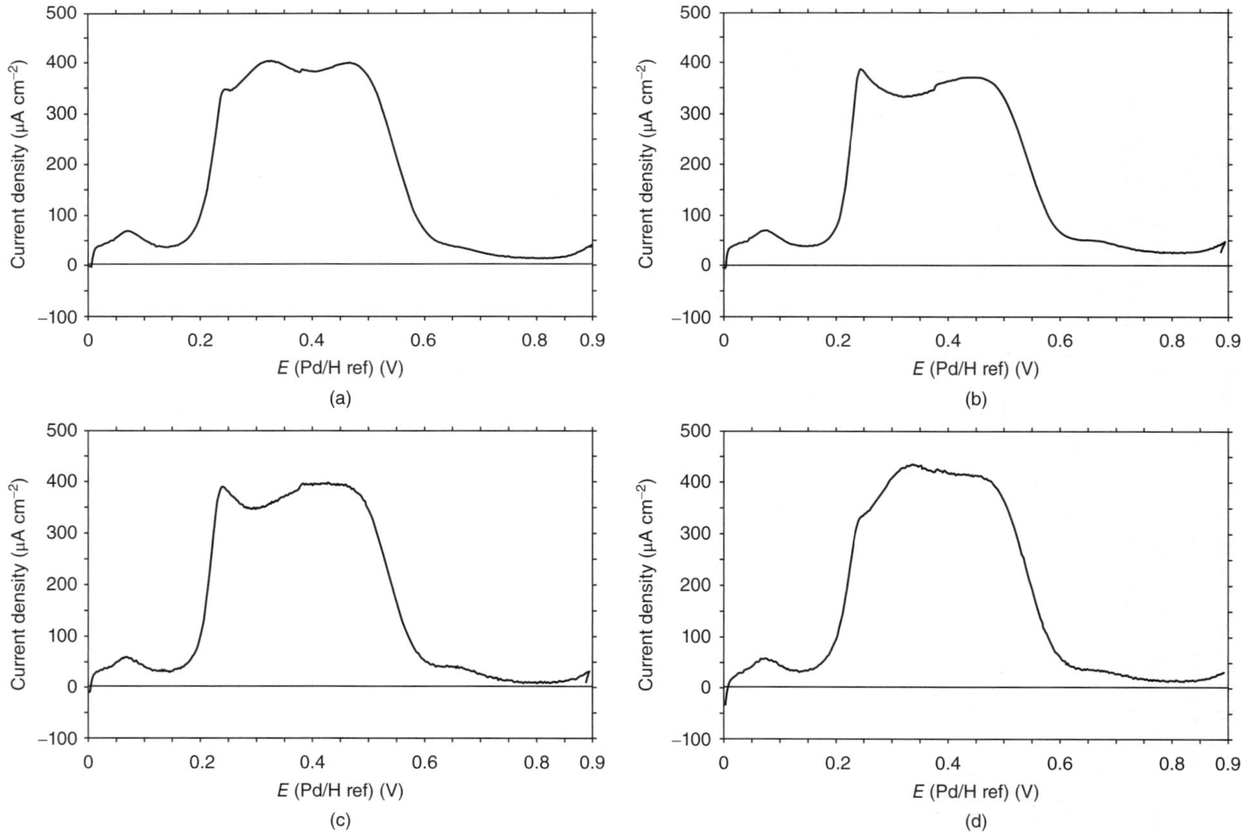

Figure 7.2 Cyclic voltammograms obtained in sulphuric acid for (a) D-glucose and (b) L-glucose on Pt{643}-L, and for (c) D-glucose and (d) L-glucose on Pt{643}-L. Reproduced with permission from [4]. Copyright (1999) American Chemical Society.

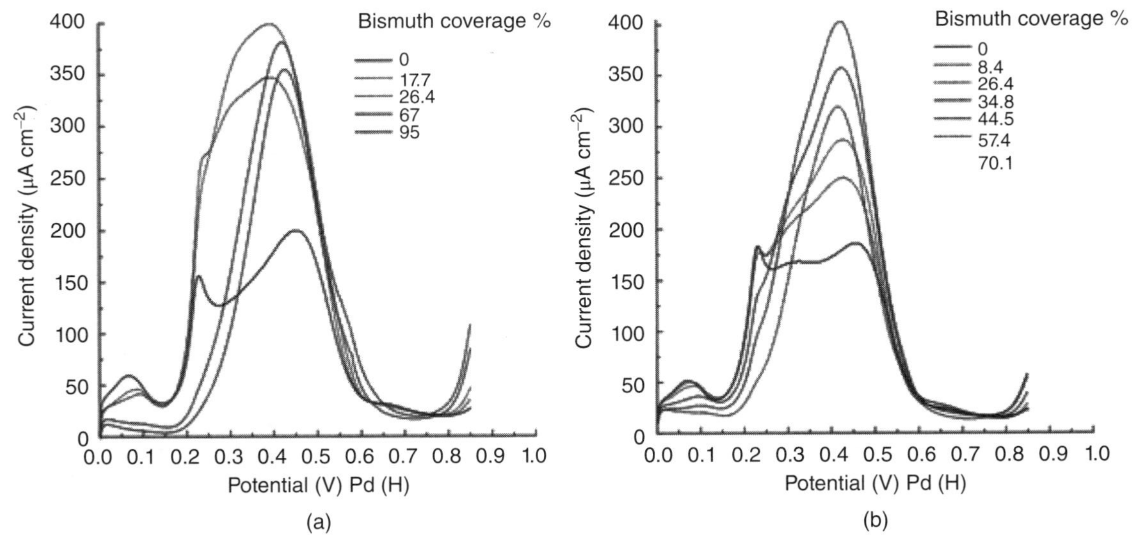

Figure 7.3 Cyclic voltammograms obtained in sulphuric acid for (a) D-glucose and (c) L-glucose on Pt{321}-D, and for (b) D-glucose and (d) L-glucose on Pt{321}-L. Pre-coverage with bismuth progressively reduces the observed peaks. Reproduced with permission from [10]. Copyright (2011) Springer.

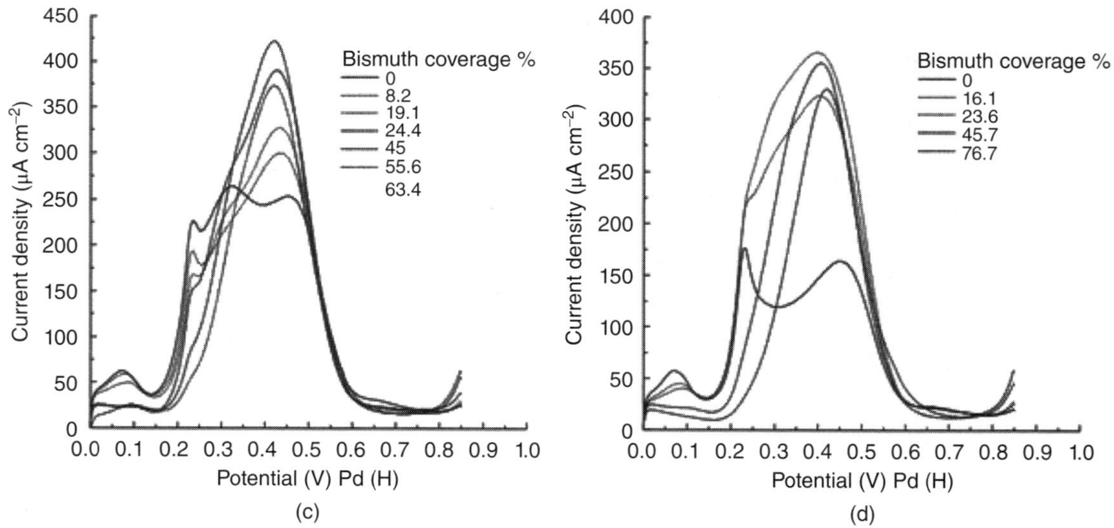

Figure 7.3 (*Continued*)

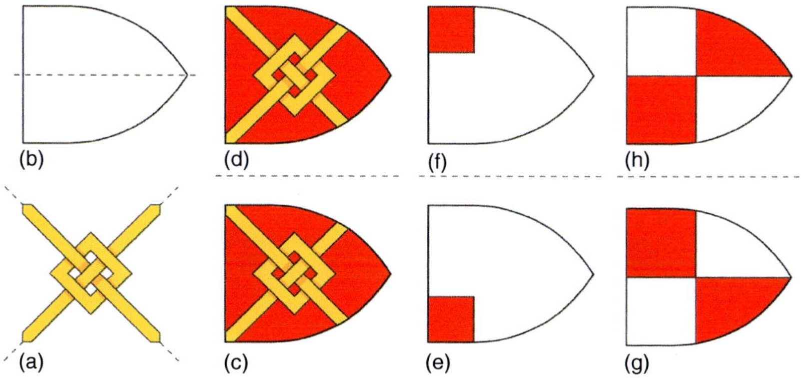

Plate I Heraldic elements including frets, cantons and quarters. See Fig. 3 for details.

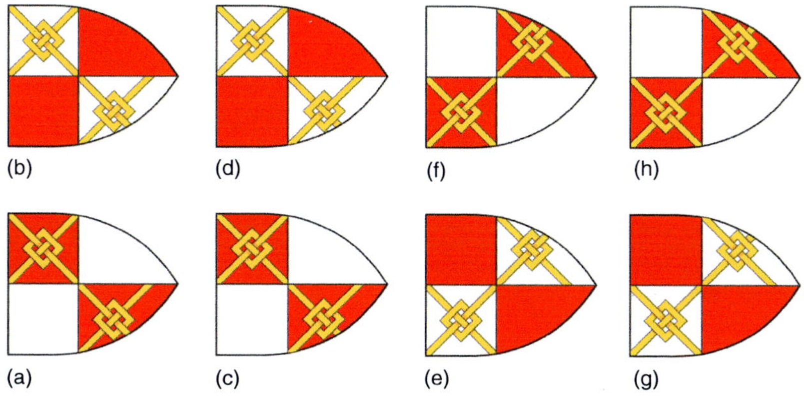

Plate II Chiral variations on an heraldic theme involving quarters and frets. See Fig. 4 for details.

Chirality at Solid Surfaces, First Edition. Stephen J. Jenkins.

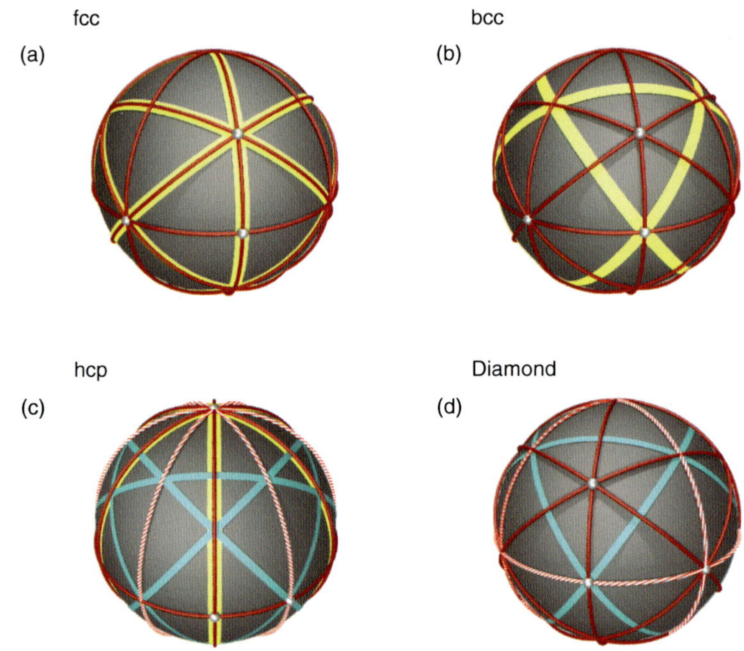

Plate III Spherical representations of surface symmetry and structure. See Figs 2.5, 2.11, 2.20 and 2.26 for details.

(a) fcc

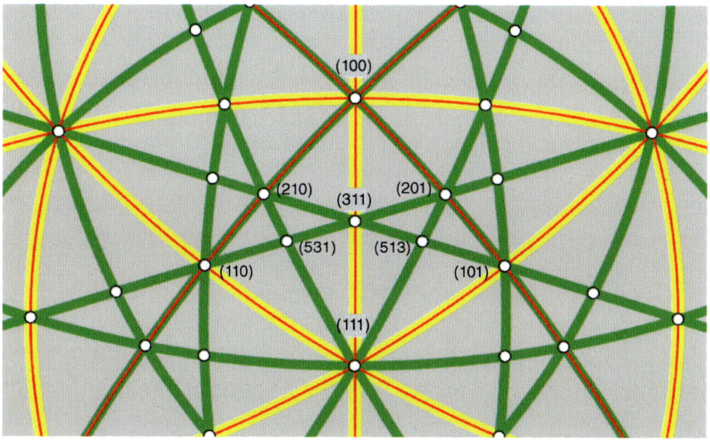

(b) bcc

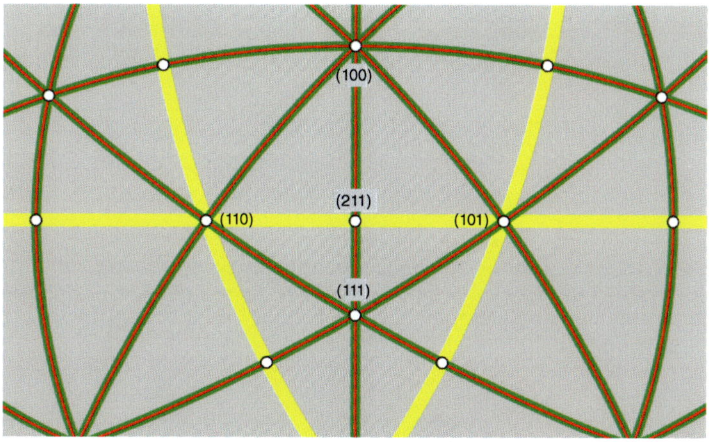

Plate IV Stereographic projections of surface symmetry and structure for face-centred and body-centred cubic materials. See Figs 2.8 and 2.13 for a full description.

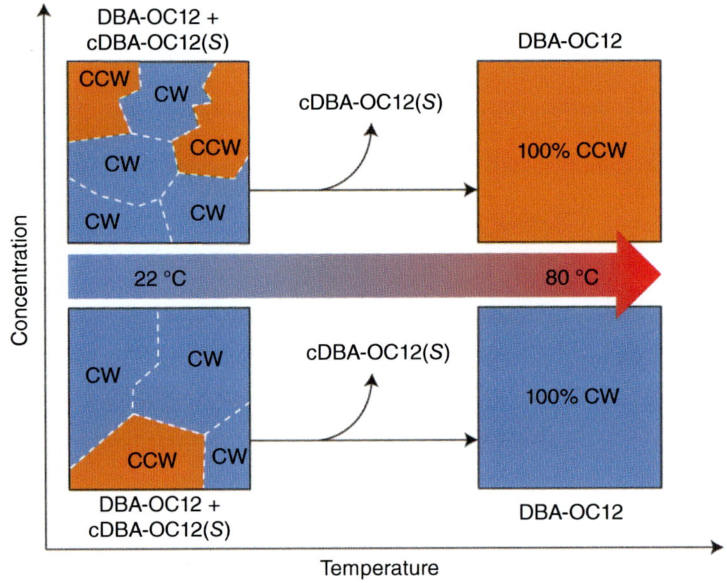

Plate V Schematic summary of the pore structure obtained upon exposure of HOPG to a 1:9 mixture of *S*-cDBA-OC12 and DBA-OC12, illustrating the relative abundances of regions displaying clockwise (CW) or counterclockwise (CCW) pores. See Fig. 6.14 for details. Reproduced with permission from Y. Fang *et al*, *Nature Chem.* **8**, 711 (2016). Copyright (2016) Nature Publishing Group.

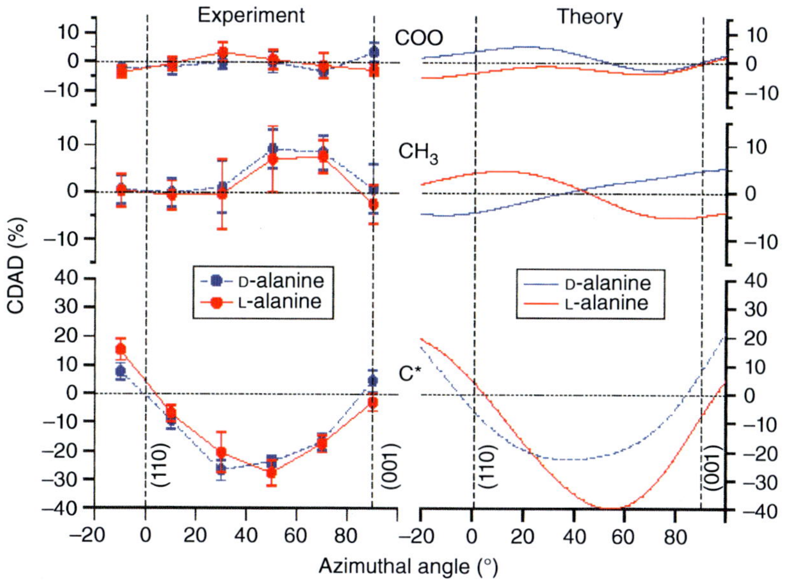

Plate VI Measured and calculated asymmetries of photoemission from the C 1s orbitals of various regions within alaninate molecules adsorbed on Cu{110}. See Fig. 8.2 for details. Reproduced with permission from F. Allegretti *et al*, *New. J. Phys.* **7**, 109 (2005).

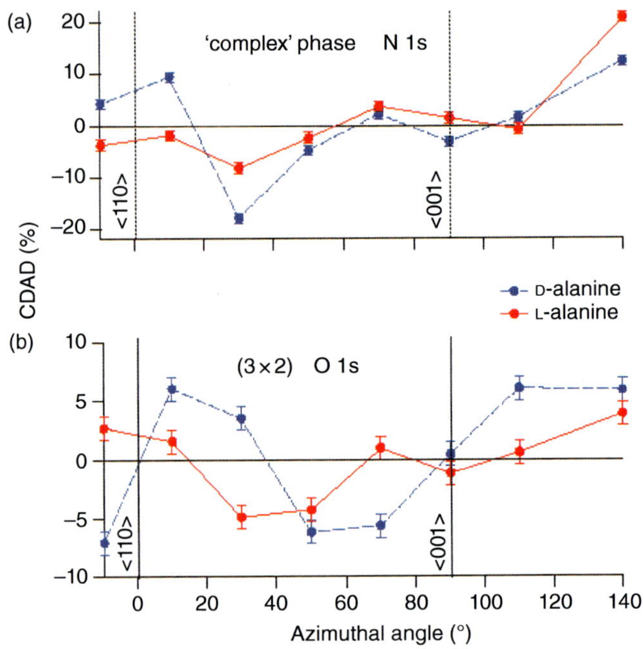

Plate VII Asymmetry of photoemission from N 1s and O 1s orbitals of alaninate adsorbed on Cu{110}. See Fig. 8.3 for details. Reproduced with permission from F. Allegretti *et al*, *New. J. Phys.* **7**, 109 (2005).

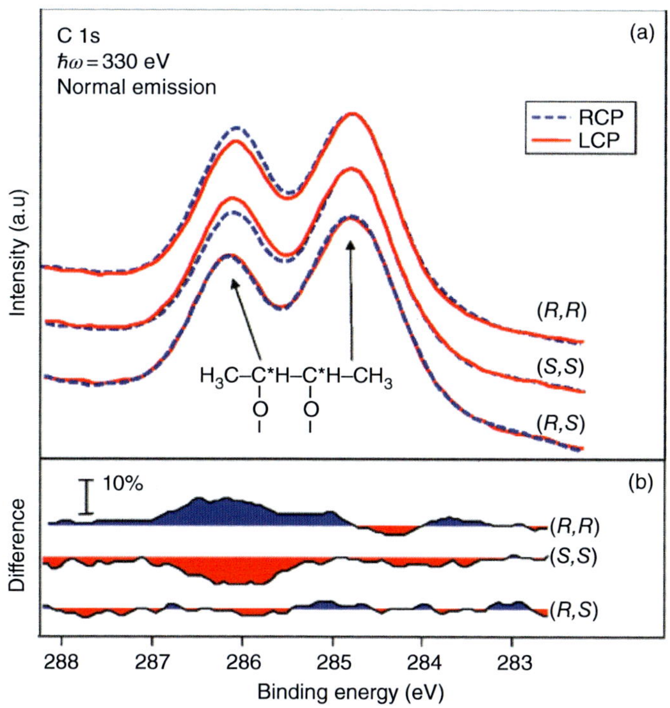

Plate VIII Asymmetry of photoemission from C 1s orbitals after adsorption of butanediol on Si{001}. See Fig. 8.4 for details. Reproduced with permission from J.W. Kim *et al*, *Phys. Rev. Lett.* **95**, 107601 (2005). Copyright (2005) American Physical Society.

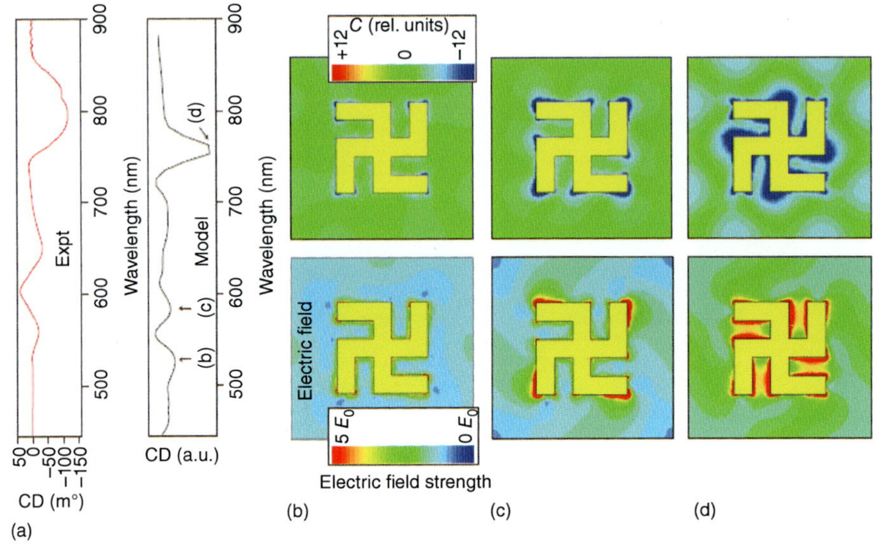

Plate IX Simulated circular dichroism and electromagnetic chirality for sub-micron gold gammadions. See Fig. 8.6 for details. Reproduced with permission from E. Hendry *et al*, *Nature Nanotech.* **5**, 783 (2010). Copyright (2010) Nature Publishing Group.

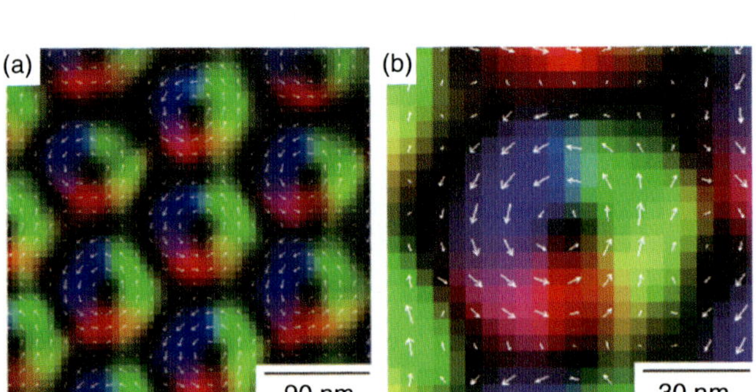

Plate X Skyrmions observed in a thin film of $Fe_{1-x}Co_xSi$. See Fig. 9.9 for details. Reproduced with permission from X.Z. Yu *et al*, *Nature* **465**, 901 (2010). Copyright (2010) Nature Publishing Group.

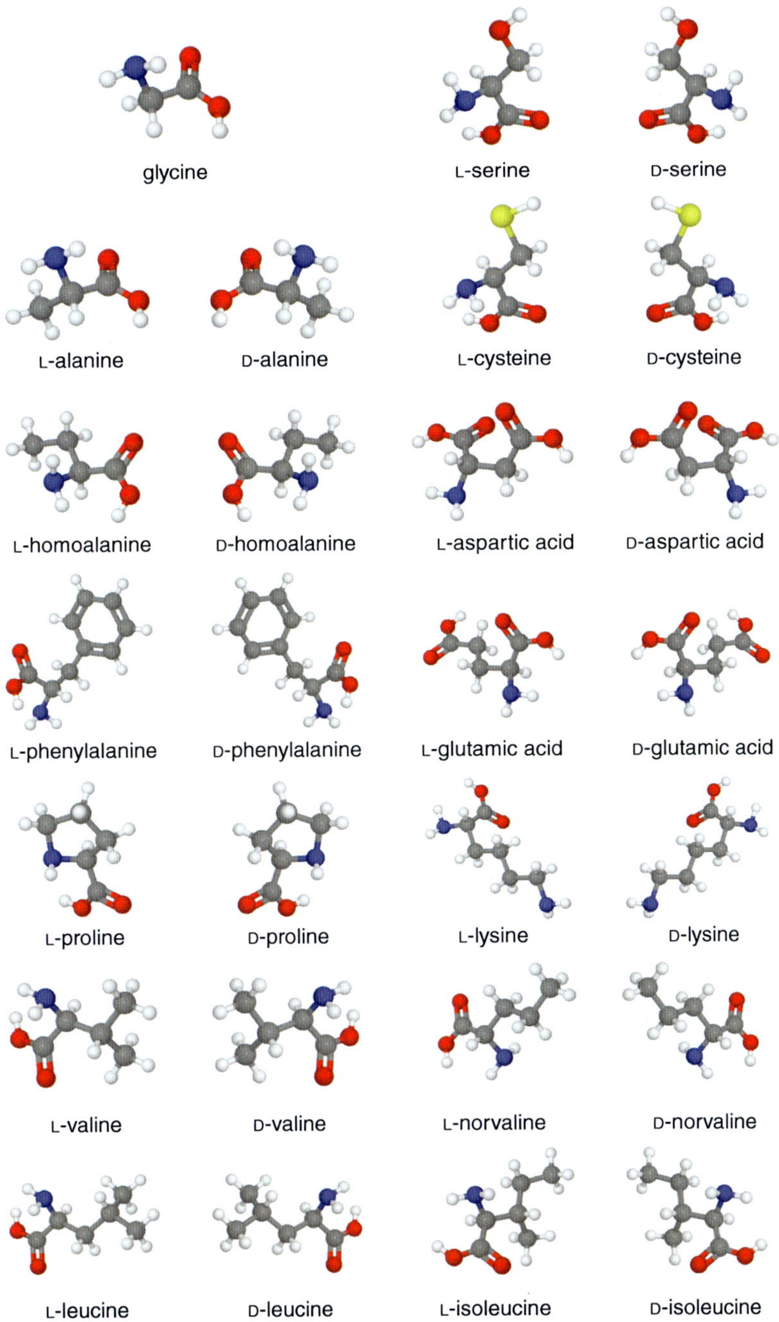

Plate XI Amino acids.

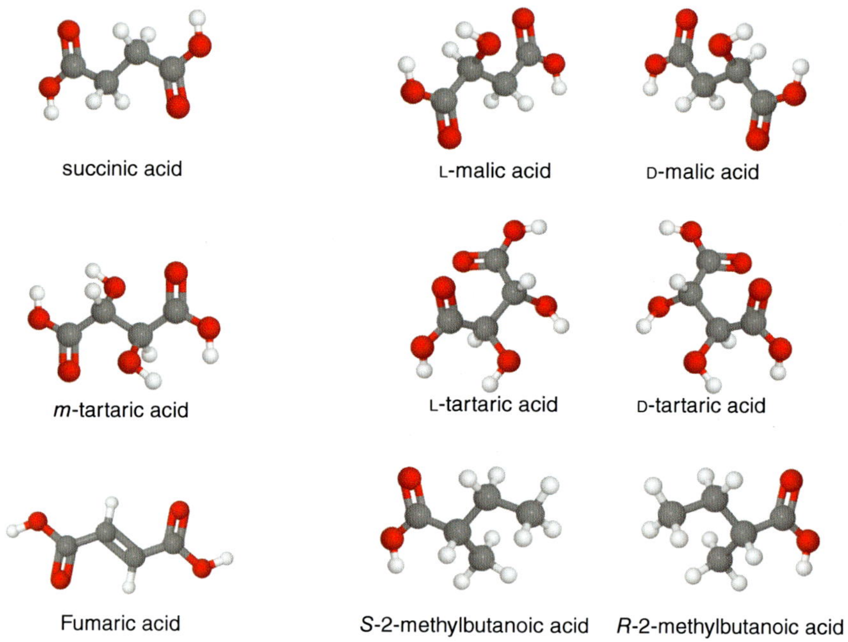

Plate XII Dicarboxylic and alkyl acids (omitting amino acids and stilbenes).

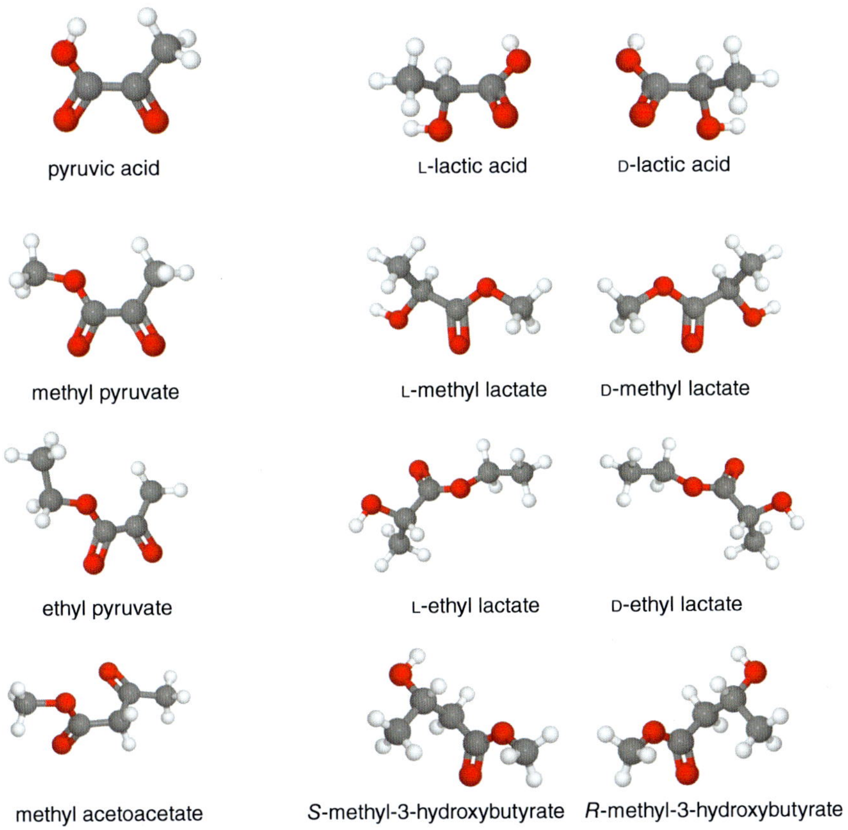

Plate XIII Keto/hydroxy-acids and keto/hydroxy-esters (omitting amino/dicarboxylic acids).

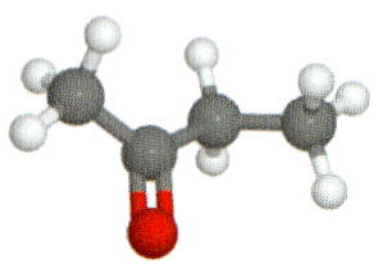

2-butanone

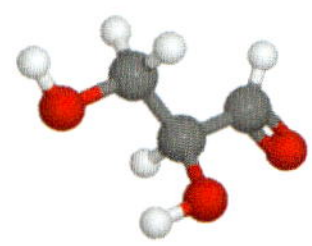

L-glyceraldehyde

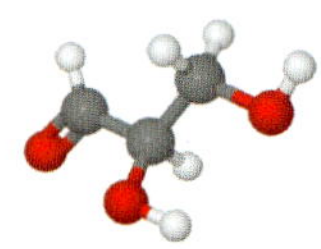

D-glyceraldehyde

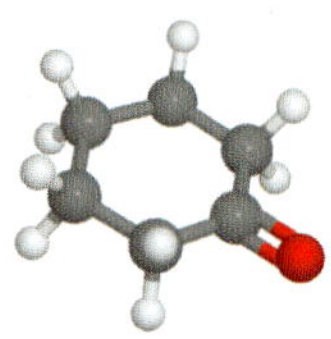

cyclohexanone

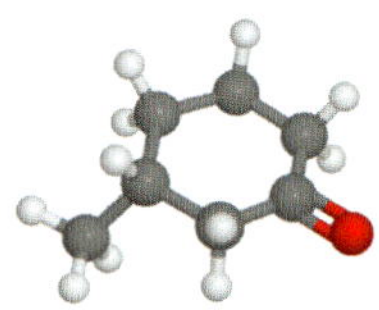

S-3-methylcyclohexanone

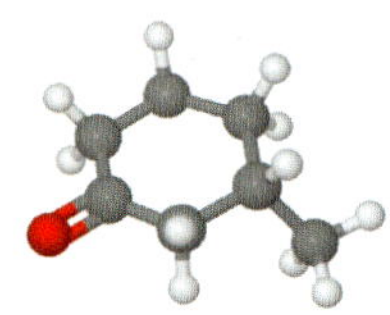

R-3-methylcyclohexanone

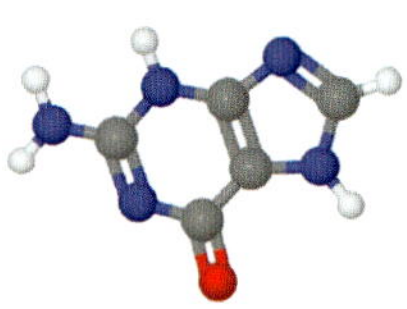

guanine

Plate XIV Ketones and aldehydes (omitting acids and esters).

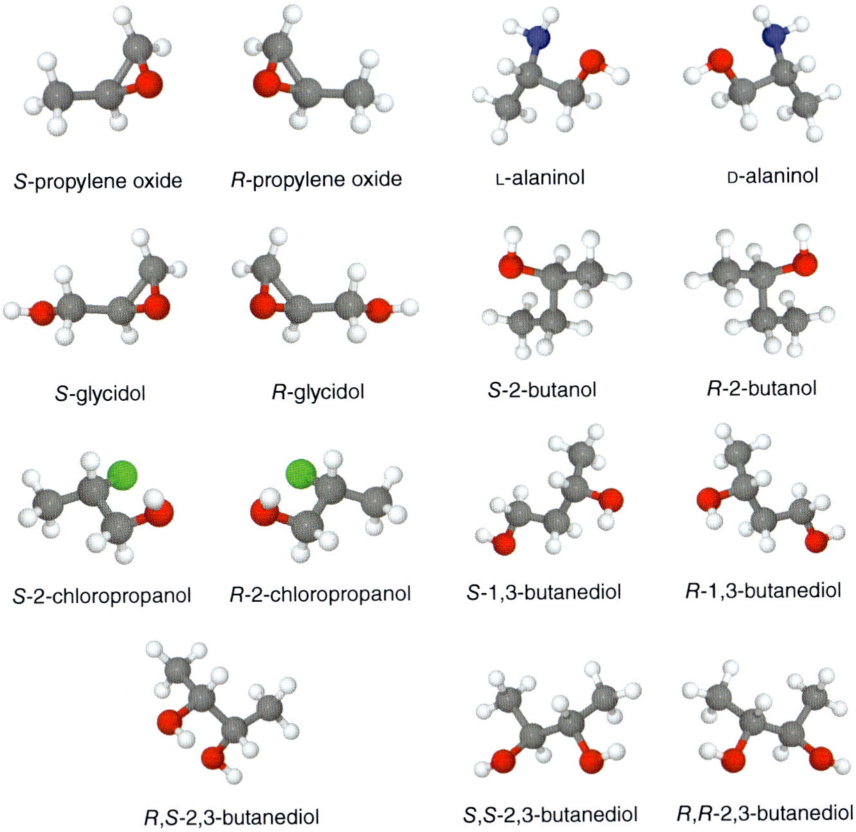

Plate XV Alcohols and epoxides (omitting acids, esters and aldehydes).

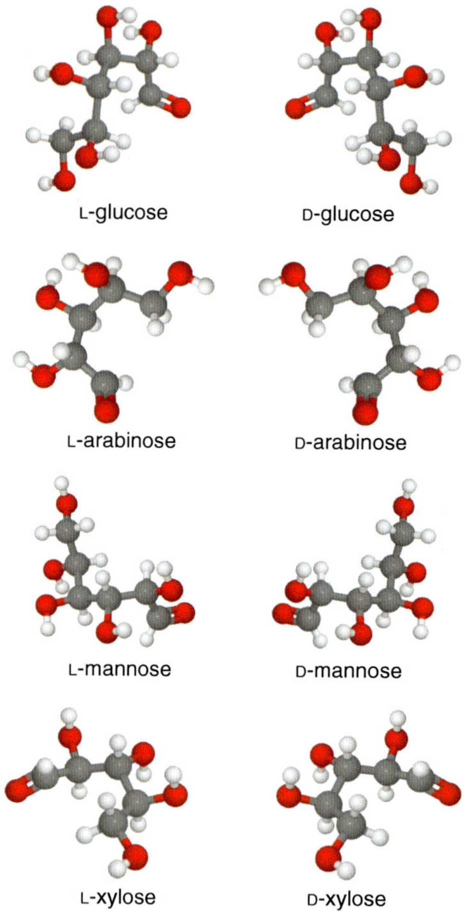

Plate XVI Sugars.

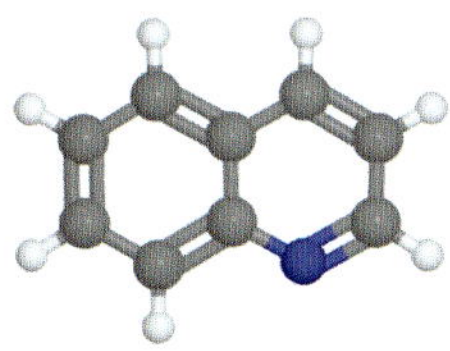

quinoline

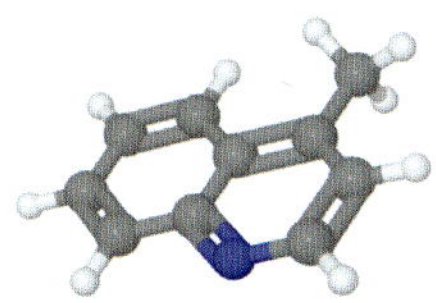

lepidine

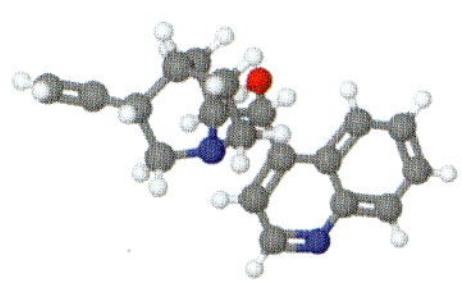

cinchonidine

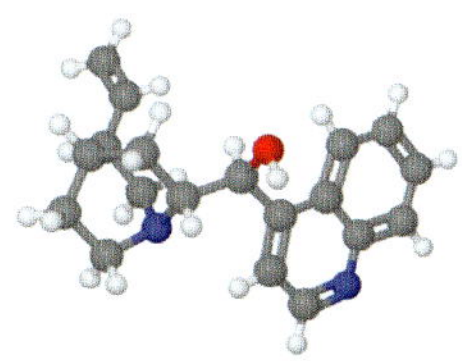

cinchonine

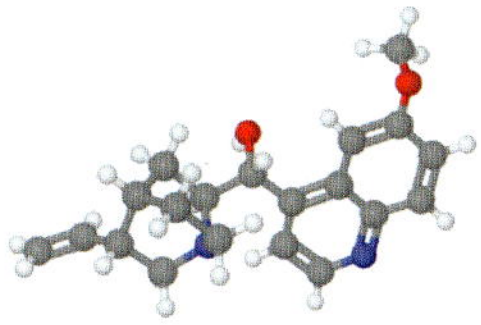

quinine

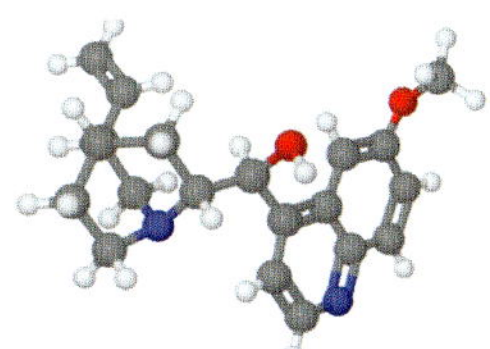

quinidine

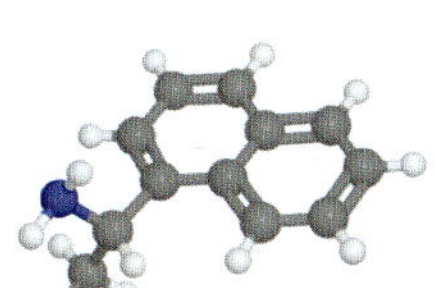

S-1-(naphthyl)-ethylamine

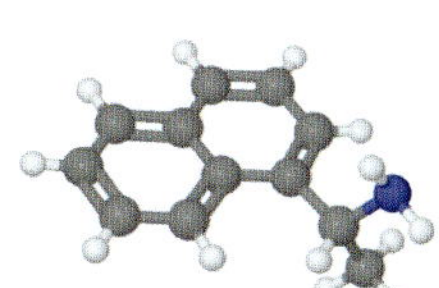

R-1-(1-naphthyl)-ethylamine

Plate XVII Alkaloids.

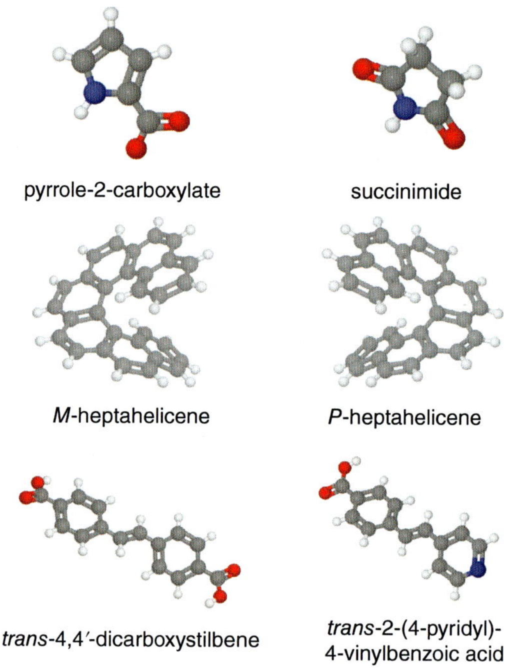

Plate XVIII Miscellaneous molecules.

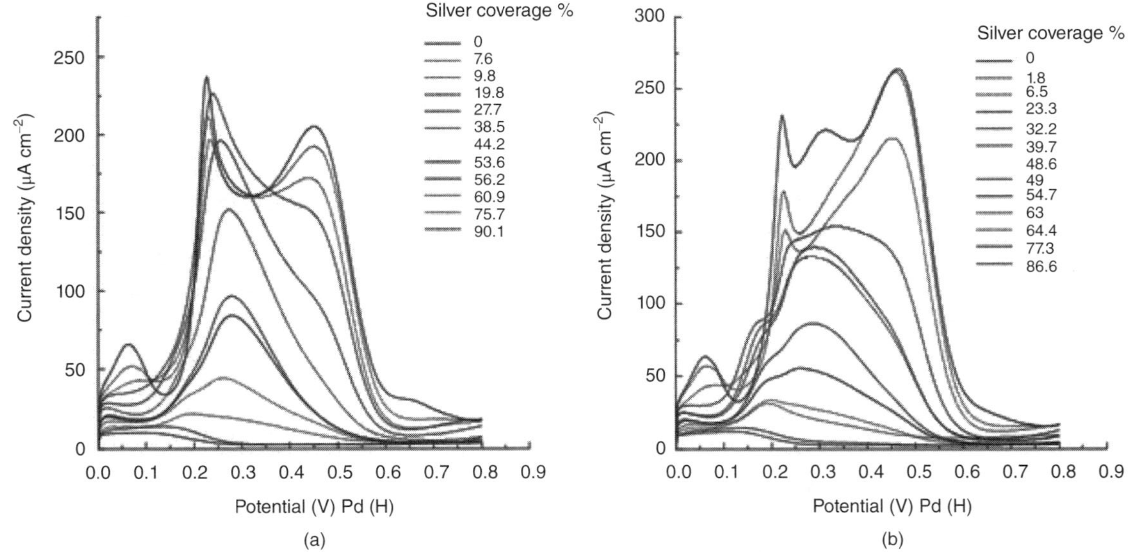

Figure 7.4 Cyclic voltammograms obtained in sulphuric acid for (a) D-glucose and (c) L-glucose on Pt{321}-D, and for (b) D-glucose and (d) L-glucose on Pt{321}-L. Pre-coverage with silver progressively reduces the observed peaks. Reproduced with permission from [10]. Copyright (2011) Springer.

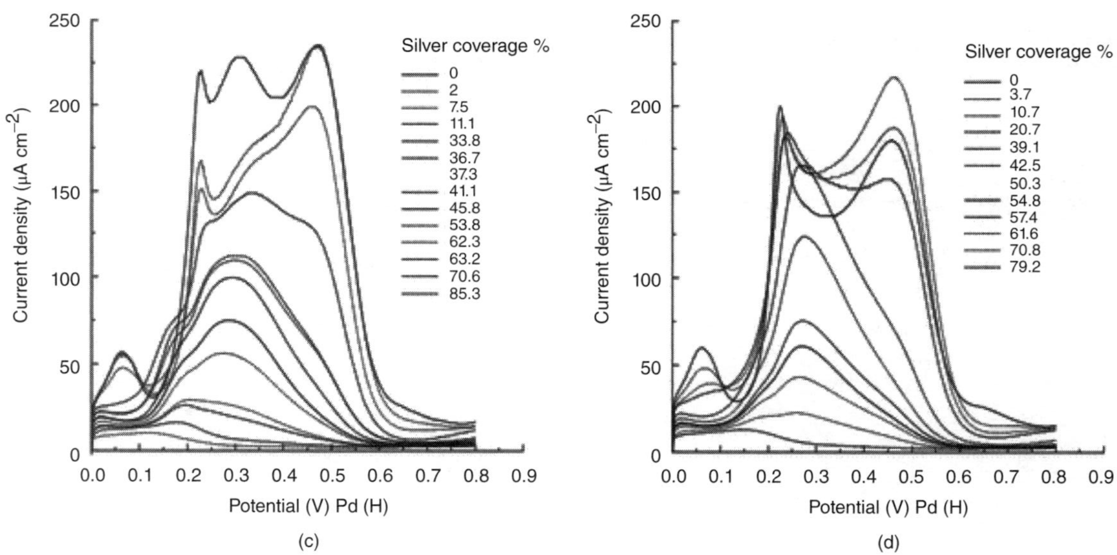

Figure 7.4 (*Continued*)

7.2 Electron-Stimulated Oxidation of Methyl Lactate on Cu{643}

Our second exemplary system comes via an intriguing study by Fleming *et al.* [11], who focused upon the desorption kinetics of methyl lactate (*see Plate XIII*) on the Cu{643}-D surface. Conducting temperature programmed desorption (TPD) measurements on overlayers prepared at 250 K, the authors reported that both enantiomers gave rise to two distinct desorption features; the first of these was moderately sharp (full-width at half maximum (FWHM) ~30 K) and clearly enantioselective, peaking at 272 K for D-methyl lactate and at 287 K for L-methyl lactate; the second was less obviously so, occurring over the same fairly broad range of temperatures (325–400 K) for both enantiomers. In fact, however, there *is* a difference even in the high-temperature peak, which is consistently less intense for L-methyl lactate. The explanation is that whilst the first TPD peak arises from direct desorption of intact methyl lactate, the second derives from the recombination of adsorbed hydrogen and methyl pyruvate (MP) (both originating from the partial dissociation of methyl lactate at lower temperature). Since methyl pyruvate (*see Plate XIII*) is prochiral rather than chiral, we should expect no enantiospecific difference in its recombinative desorption temperature, and indeed we see none, but it is not particularly surprising that the *quantity* of adsorbed methyl pyruvate should depend upon which enantiomer of chiral methyl lactate was involved in its production. Specifically, these results imply that the conversion to methyl pyruvate of D-methyl lactate is more efficient than that of L-methyl lactate on this intrinsically chiral surface; the inefficiency of the latter process is attributed to the preferential dissociation of an intermediate chiral alkoxide species [11].

The diastereomeric difference in desorption temperature for intact methyl lactate, amounting to 15 K, implies a corresponding difference in adsorption heat of 3.92 kJ mol^{-1}, which is already quite significant in the context of other systems discussed in the last two chapters.[3] Nevertheless, irradiation of the overlayer at 110 K by low-energy electrons (50 eV at 4 μA cm^{-1}) massively enhances the enantioselectivity of the system. Subsequently acquired TPD spectra reveal a progressive drop in the intensity of the low-temperature peak with increasing total electron dose, indicative of either electron-stimulated dissociation or desorption of methyl lactate [11]. A concomitant rise in the intensity of the high-temperature peak strongly argues for the former process, suggesting, as it does, an increased population of methyl pyruvate in the aftermath of irradiation. Crucially, the estimated cross-section for this electron-stimulated process is highly enantiospecific, amounting initially to $6.7 \pm 2.2 \times 10^{-16}$ cm^2 for D-methyl lactate, but just $2.9 \pm 1.0 \times 10^{-16}$ cm^2 for L-methyl lactate.[4] In consequence, it proved possible to produce an enantiomeric excess of ~20% in an initially racemic surface population of methyl lactate, simply by electron irradiation [11].

3 Once again, we should recall that adsorption heat estimates typically rely upon an assumption that the corresponding pre-exponential factor takes a value of 10^{13} s^{-1}.

4 These figures drop as the electron dose accumulates, due to poisoning by the products of side reactions.

Now, it could be argued that this phenomenon might easily have been included in the preceding chapter as just another example of chiral amplification, and indeed there is some merit to this view. Its place in the present chapter, however, is due largely to the fact that the process of electron-stimulated dissociation demonstrated in this case study is not only highly enantiospecific but also highly selective in the bonds that are broken. As the authors point out, and indeed elaborate upon in a later study of the same species on Cu{111} [12], the short lifetimes of electronically excited states on the surface are likely crucial to both of these desirable features. If replicated in other similar systems, one can foresee two possible ways in which such properties might be exploited.

Firstly, one might simply reverse the sense of the current reaction by working on a hydrogenation catalyst such as platinum. As we shall see below, this metal allows the conversion of prochiral methyl pyruvate to chiral methyl lactate, but induction of asymmetry invariably relies upon the presence of a chiral modifier; how satisfying instead to exploit electron-stimulated dissociation on an intrinsically chiral surface to continually drive the non-desired chiral product back towards the prochiral reactant! And secondly, although the conversion of lactate to pyruvate implies loss of chirality upon electron-stimulated selective partial dissociation, it is not altogether impossible to imagine a prochiral molecule whose dissociation might render it chiral. In such a case, any substantial diastereomeric difference in the cross-section for electron-stimulated desorption would directly induce similar asymmetry in the product distribution.[5]

7.3 Hydrogenation of α-Ketoesters over Platinum: The Orito Reaction

The classic example of enantioselective hydrogenation of an α-ketoester is provided by the so-called **Orito Reaction**, in which prochiral methyl pyruvate (MP) or ethyl pyruvate (EP) is converted over a platinum catalyst into chiral methyl or ethyl lactate with high enantiomeric excess (favouring the *R* enantiomer[6]) under the chirally modifying influence of the natural alkaloid cinchonidine (CD) [13]. Notably, the overall rate of reaction is enhanced by one or two orders of magnitude, alongside the enantioselective effect. By now, many variants of this canonical reaction have been tested – substituting different ketones, different metals and different alkaloid modifiers – and an excellent review of these has been provided by Blaser [14]. Here we shall restrict our focus to the core reaction and only those variants whose study sheds particularly bright light upon general mechanistic details.

5 Since dissociation typically *simplifies* molecular structure, it is not all that easy to conceive of an example where chirality is acquired in this way. 'Not easy' does not, however, necessarily equate with 'impossible'.

6 The *S* enantiomer of methyl or ethyl lactate can be favoured by use of cinchonine (a pseudo-enantiomer of cinchonidine (CD)) as the modifier; the rather similar molecules quinine and quinidine (*see Plate XVII*) may be substituted in place of cinchonidine and cinchonine respectively, but the enantiomeric excess is not so high.

The CD molecule itself (*see Plate XVII*) is a natural alkaloid derived from the bark of trees from the genus *Cinchona*. It is helpful to visualise it as comprising a quinoline moiety joined to a vinyl-substituted quinuclidine moiety through a short and flexible linking alcohol that functions as a chiral centre. The sheer size and complexity of the molecule immediately suggests that its mode of action is unlikely to be either the blocking of individual chiral kink sites or the formation of an ordered chiral lattice (both of which could be done better by a smaller and simpler molecule).[7] Instead, the consensus view is that CD forms some kind of one-to-one complex with the reactant molecule, orienting it in some specific way and hence favouring hydrogenation to a single enantiomer of the product. All the evidence points to the modifier binding to the surface via its quinoline moiety, with the quinuclidine moiety rearing up from the surface and forming a three-dimensionally chiral pocket into which the reactant must slot. Relevant questions include the precise means by which reactant and modifier bond to one another, the degree to which modifier flexibility may be important, the preferred orientation of the bound reactant and the direction from which hydrogen is supplied.

7.3.1 Adsorption Geometry of Methyl and Ethyl Pyruvate

To begin with the alkyl pyruvate species, one critical question is the conformation adopted by the C—C bond between the ketone and ester functionalities. Although this is not a double bond, spectroscopic evidence strongly suggests that planar *s-cis* and *s-trans* conformations are local energy minima. In the case of MP, the latter has long been known to be favoured exclusively in the crystalline phase (and substantially in liquid, solution or vapour phases) [17]. Rather more recently, a combination of spectroscopy and density functional theory (DFT) showed that the *s-trans* form was similarly favoured for EP, albeit again with a substantial fraction of the *s-cis* form, especially in more polar and/or alcoholic solvents [18]. On the surface, we might reasonably anticipate significant perturbation of the molecular geometry, and the possibility of a switch between *s-trans* and *s-cis* forms must be seriously entertained.

The binding geometry of MP on Pt{111} was first addressed in detail by Lambert and co-workers [19, 20] by means of X-ray photoemission spectroscopy (XPS), near-edge X-ray absorption fine structure (NEXAFS) and scanning tunnelling microscopy (STM). Upon adsorption of a monolayer at 298 K, XPS experiments revealed peaks in the C 1s region at binding energies of 286.2 and 288.6 eV, the former interpreted as deriving from the carbon atoms of the methyl groups and the latter as relating to the remaining carbon atoms.

7 Analysis of CV data obtained during deposition of cinchonidine on various kinked platinum electrodes led Attard to conclude that adsorption of the alkaloid was insensitive to the chirality of the surface; the idea that selective blocking of one or other kink type on an otherwise achiral single-crystal surface was effectively ruled out [8]. This is not quite the same thing as saying that low-coordination sites are not important, however. Jenkins *et al.* [15] (no relation) subsequently showed, using similar techniques, that enantioselectivity in the hydrogenation of EP by CD-modified platinum is greater at terrace edges than on the terraces themselves. There remains the hint of a possibility, therefore, that chiral corners at the boundaries of supported nanoparticles may be selectively blocked in the practical catalyst [16].

The O 1s region, on the other hand, featured a single broad peak at a binding energy of 533.6 eV, deriving from both ketone and ester functionalities. Compared with spectra obtained from multilayers of MP at 95 K, the ratio of carbonyl-to-methyl carbon rose from 1.0 in the multilayer to 1.8 in the monolayer, while the ratio of carbon-to-oxygen likewise rose, albeit only from 1.3 to 1.5. These differences were attributed to a change in the chemical state of the molecule when chemisorbed in the sub-monolayer regime, namely polymerisation by the elimination of methoxy. STM images obtained at room temperature show clear evidence for the existence of chain-like structures, consistent with this interpretation (see Fig. 7.5). When imaged under 2×10^{-5} mbar hydrogen overpressure, however, only isolated monomers are visible on the surface, suggesting that polymerisation is inhibited; the authors note that their preferred polymerisation mechanism (informed by their NEXAFS data) is initiated by the loss of a single hydrogen atom from the methoxy group of the monomer, which would be disfavoured in the presence of coadsorbed hydrogen. In light of these findings, the authors propose that irreversible deactivation of the catalyst under hydrogen-poor conditions – an occasional problem in the Orito reaction – may be due to the formation of persistently site-blocking polymers [19, 20]. Subsequent experiments by McBreen and co-workers [21] suggest an alternative interpretation, however.

Using MP labelled with deuterium in the methoxy group, Lavoie *et al.* [21] found in TPD studies that while H_2 was evolved at ~290 K, D_2 desorbed from the surface only above 350 K. This implies that the methoxy group remains integral to the molecule at room temperature and that loss of hydrogen from the methyl group of the pyruvate moiety is the true initiator of polymerisation. In room-temperature reflection absorption infrared spectroscopy (RAIRS) experiments, these authors observed an absorption band at 1594 cm^{-1} that

Figure 7.5 STM image of polymeric chains obtained after 298 K exposure of Pt{111} to methyl pyruvate in the absence of hydrogen. The overall scan area is 1000×1000 Å^2 and the upper portion has been filtered to enhance edge contrast. Reproduced with permission from [19]. Copyright (2000) American Chemical Society.

they identify with the —C=C— stretch mode of an enol. That is, they propose that hydrogen is transferred from the methyl group of pyruvate to the adjacent carbonyl, via a surface adatomic intermediate (in the process of which, some hydrogen recombinatively desorbs). Since the enol form of pyruvate features both hydrogen bond acceptor and donor sites, it is possible to envisage polymers based on chains of such molecules, rather than invoking the loss of methoxy inferred previously. Indeed, Lavoie *et al.* [21] report that the supramolecular structures seen in their STM images at room temperature are relatively mobile and that monomers periodically detach and reattach from polymer chains, consistent with the interpretation that they are held together only by hydrogen bonds. Quite how this picture squares with the XPS intensities mentioned above is not, however, entirely clear.

At the lower adsorption temperature of 110 K, the McBreen group reported absorption bands for sub-monolayer MP on Pt{111} at 1290 and 1543 cm^{-1}, attributed respectively to the ester bond and to a mixed mode involving C—C and C—O single bonds [22]. A further mode at 1452 cm^{-1} was not explicitly attributed, but most likely derives from the asymmetric deformation mode of one or other $—CH_3$ group. On the basis of these spectral features, the authors propose that the molecule adopts the *s-cis* configuration, binding to the surface as an enediolate; at higher coverage, close to saturation of the monolayer, an additional absorption band appears at 1742 cm^{-1}, attributed to a carbonyl group, suggesting that the *s-trans* configuration is also present (see Fig. 7.6).

The geometry of adsorbed EP, meanwhile, was studied by Baiker and co-workers [23] through the use of NEXAFS. Upon adsorption on Pt{111} at room temperature and a partial pressure of 0.065 mbar, several clear π resonances were observed at the C 1s and O 1s absorption edges, associated with transitions to unoccupied states straying onto the C=O double bonds of pyruvate. Varying the polarisation of incident X-rays, these features permitted the orientation of the molecule to be deduced, yielding a mean tilt angle for the carbonyl groups of $18\pm7°$ from the surface normal. Such a result would be perfectly consistent with either the *s-cis* or the *s-trans* conformation, so long as the plane of the molecule is relatively close to upright on the surface. In the presence of 0.5 mbar hydrogen overpressure, the tilt angle from the surface normal was found to increase, reaching $32\pm6°$, implying that the molecule tends to keel over under these circumstances.

The results of XPS and ultraviolet photoemission spectroscopy (UPS) experiments performed by the same group are consistent with the overall picture of relatively upright adsorption taking place through at least one of the carbonyl

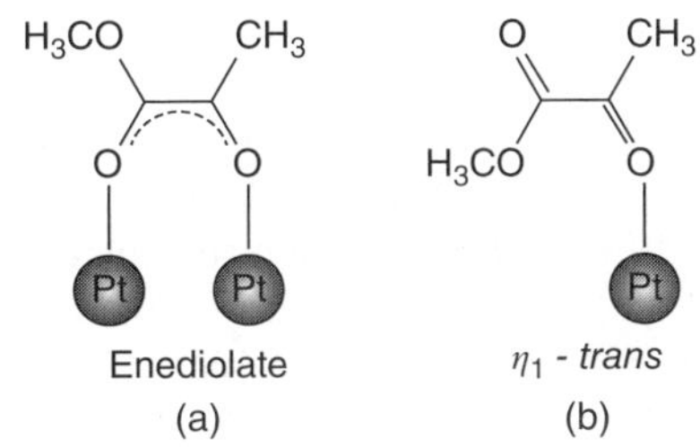

Figure 7.6 Schematic models of methyl pyruvate binding on platinum in (a) its *s-cis* (enediolate) form and (b) its *s-trans* (η_1-*trans*) form. Reproduced with permission from [22]. Copyright (2003) American Chemical Society.

groups [24]. In XPS obtained at 153 K in the monolayer regime, the C 1s region featured partially resolved peaks at binding energies of 284.8, 285.4, 286.1 and 288.1 eV, interpreted respectively as arising from the pair of methyl carbon atoms, the methylene carbon atom, the carbon atom of the ketone functionality and the carbon atom of the ester functionality. The O 1s region, on the other hand, featured a single asymmetric feature deconvoluted into three peaks at binding energies of 531.5, 532.2 and 532.7 eV, attributed to the oxygen atom of the ester carbonyl, the oxygen atom of the ketone carbonyl and the oxygen atom of the ester bridge. The coverage corresponding to saturation of the chemisorbed monolayer, based on comparison with XPS intensities for a known coverage of CO on the same surface, was estimated at one adsorbate molecule per four or five Pt atoms, supporting the notion of an upright rather than flat-lying geometry. In UPS experiments, meanwhile, a marked stabilisation (by 0.7 eV) of the highest occupied molecular orbital (HOMO) was noted [24]. Since this orbital is distributed across the carbonyl groups of the molecule, whether in the *s-cis* or the *s-trans* configuration, it is probable that one or both of these groups is involved in binding to the surface. In view of the apparent upright geometry, and in light of the RAIRS results for MP, it seems fairly likely that the *s-cis* conformation is adopted and that EP also binds through both carbonyl groups.

7.3.2 Adsorption Geometry of Cinchonidine and its Cousins

As mentioned above, it is generally believed that CD binds to the platinum surface via its quinoline moiety, with the quinuclidine moiety remaining essentially out of contact. This chemically appealing notion was essentially confirmed at an early stage by isotopic exchange experiments conducted by Bond and Wells [25] who monitored the uptake of deuterium at 293 K by 10,11-dihydrocinchonidine (HCD) on a platinum catalyst in C_2H_5OD solvent (by means of nuclear magnetic resonance (NMR)) finding that after facile exchange at the molecule's OH group, the next most rapid reaction was near-simultaneous exchange at all of the quinoline positions; no exchange was reported at the quinuclidine positions. Since the exchange reaction is catalysed by the platinum surface, the implications for adsorption geometry are obvious. Above 350 K, similar exchange experiments are reported to indicate a rather more tilted orientation for the quinoline moiety, and this has been linked with a concomitant loss of enantioselectivity and activity in the Orito reaction [26].

Some of the most convincing direct evidence concerning the conformation and orientation of adsorbed HCD was subsequently provided by the work of Evans *et al.* [27]. Performing NEXAFS experiments upon deposition onto Pt{111} at 298 K, they monitored a sharp π resonance at the N 1s edge as a function of X-ray polarisation, concluding that the adsorbed molecules adopted a geometry in which the quinoline moiety was almost perfectly flat. Repeating their experiment at 323 K, however, the appearance of a new resonance in the N 1s region was taken to indicate a degree of dissociation, leading to a surface hosting roughly the same coverage of the intact molecule as achieved previously, but now packed alongside a similar coverage of quinoline (the surface-stable dissociation product). Under these more congested conditions, the NEXAFS data supported a model in which the quinoline moiety of the larger molecule was tilted by some 60° from the surface plane, in line with the earlier isotopic exchange experiments.

Additional very strong evidence in favour of this view was provided in 2001 by two groups using specialised infra-red techniques. Firstly, we shall mention the work of Ferri *et al.* [28, 29] on the adsorption of CD (and other alkaloids) from solution onto an alumina-supported platinum catalyst. Making use of the attenuated total internal reflection (ATR) geometry, they were able to obtain vibrational spectra as a function of alkaloid concentration in dichloromethane. At low concentration (corresponding to low surface coverage) only a single absorption band in the diagnostic 1500–1650 cm^{-1} region was noted, at 1567 cm^{-1}, and this was attributed to an in-plane deformation mode of the quinoline moiety. The absence of any other modes associated with quinoline was taken (in accordance with the metal surface selection rules) as evidence for a species in which that portion of the molecule lies flat against the surface. At somewhat higher concentration, the band at 1567 cm^{-1} is joined by a second at 1530 cm^{-1}, interpreted [28] as a ring mode associated with the formation of an α-quinolyl species[8] (see Fig. 7.7). At the highest concentrations, the spectrum acquires several further bands in the diagnostic region, at 1511, 1590 and 1610 cm^{-1} (all interpreted as arising from in-plane quinoline deformation modes) while the band at 1530 cm^{-1} disappears. The spectrum thus indicates a highly tilted species very weakly bound to the surface by a dative covalent bond involving the quinoline moiety's nitrogen atom [28].

Secondly, we turn to complementary polarisation-modulation RAIRS (PM-RAIRS) experiments reported by Kubota and Zaera [30] for a saturated solution of CD in carbon tetrachloride. In comparison with a control spectrum recorded for the bulk solution, data obtained from a sample in contact with (polycrystalline) platinum foil revealed only very small changes to the in-plane deformation modes of the quinoline moiety, indicating that it remains intact on the surface. That this part of the molecule interacts strongly with the surface is inferred from the shift of a band attributed to its C—H wag modes (from 1228 cm^{-1} in the bulk liquid to 1236 cm^{-1} on the surface). Absorption bands attributed to vibrational modes of the quinuclidine moiety, on the other hand, do not shift in frequency, suggesting that it indeed remains both intact and relatively distant from the surface; changes in the relative intensity of these bands may, however, be interpreted as arising from the adoption of a well-defined orientation upon adsorption.

Further insight was gained by varying the concentration of the CD solution, whereupon the following points were noted: (i) that below 5% of the saturation concentration, no spectroscopic evidence for adsorption could be found; (ii) that between 5% and 20% of the saturation concentration, only three weak peaks, all attributable to the quinuclidine moiety, were observed and (iii) that above 20% of saturation concentration, spectra increasingly resembled the saturation spectrum. The absence of quinoline modes in the intermediate spectra was taken to

---

8 That is, a species in which a hydrogen atom is abstracted from a carbon atom adjacent to the nitrogen atom of the quinoline moiety. The preferred binding geometry of this species is strongly tilted, involving a dative covalent bond between the nitrogen atom and the surface, together with an ordinary covalent bond between the adjacent carbon atom and the surface.

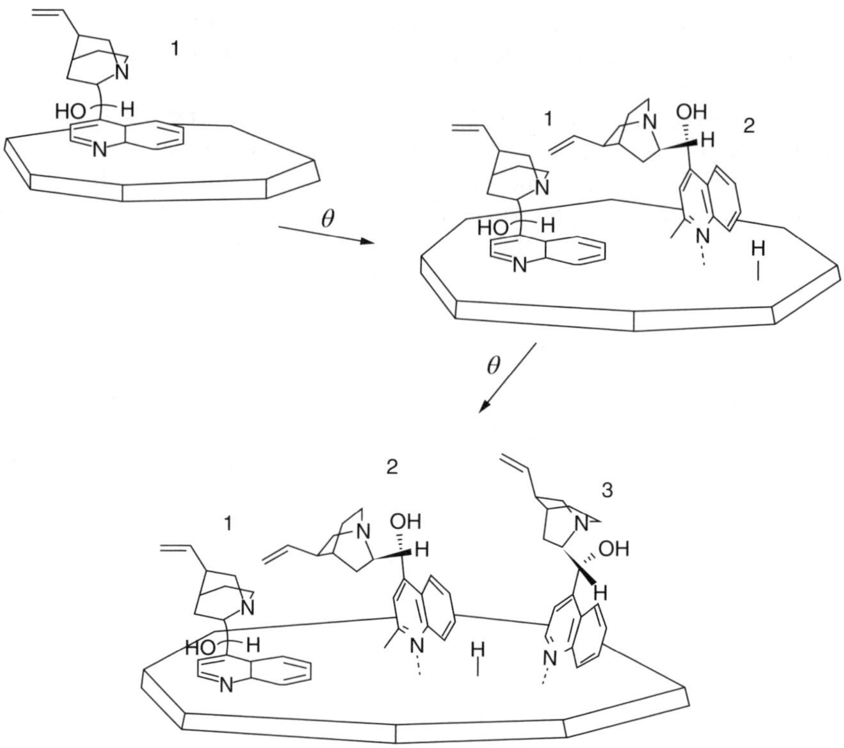

Figure 7.7 Schematic models of cinchonidine adsorption on platinum. At low coverage, binding occurs through the flat-lying quinoline moiety (1). At moderate coverage, an α-quinolyl species arises, in which a carbon atom adjacent to the quinolinic nitrogen atom is deprotonated and strongly involved in binding to the surface (2). At the highest coverage, the quinolinic nitrogen atom is solely responsible for binding (3). Reproduced with permission from [28]. Copyright (2001) American Chemical Society.

imply that this moiety lies practically flat in this regime, while their appearance in the high concentration regime suggests that a more or less pronounced tilt develops as the surface coverage increases, presumably for steric reasons. The authors note a remarkable correspondence between the evolution with concentration of bands pertaining to flat-lying/tilted quinoline and the observation of high/low enantiomeric excess in the hydrogenation of EP.[9] Earlier studies by Sun and co-workers [31] had already identified a marked decline in enantioselectivity associated with too high a modifier concentration, but the work of Kubota and Zaera [30] was crucial in convincingly rationalising this in terms of a conformational switch from flat to tilted in the adsorbed quinoline moiety. An impressively exhaustive assignment of RAIRS bands for CD was subsequently published by Zaera's group, confirming their previous conclusions to a high degree of confidence and also providing information on the orientation of the quinuclidine moiety [32]. Different solvent choices have been found by Ma and Zaera [33] to

9 The authors also note that their reactivity studies were performed in acetic acid, not carbon tetrachloride, but the coincidence remains highly suggestive.

alter greatly the adsorption/desorption equilibrium of CD (and indeed of other alkaloid modifiers) according to a volcano plot with respect to solvent polarity and/or dielectric constant, but not to affect radically the quinoline-anchored adsorption conformation inferred from RAIRS data.

Armed with these observations, it is worth reviewing briefly the adsorption of a few simpler molecules whose behaviour sheds light on aspects of the more complex system. We shall begin with the adsorption geometry of quinoline on Pt{111}, as reported by Bonello and Lambert [34]. In XPS data obtained by these authors after sub-monolayer deposition at 300 K, peaks were observed in the C 1s region at binding energies of 284.7 and 285.9 eV, the latter interpreted, based on relative intensities, as deriving from the two nitrogen-adjacent carbon atoms and the former from all seven others. The N 1s region is home to a single peak at a binding energy of 398.5 eV, and the ratio of its intensity to that of the carbon peaks matches closely that obtained from multilayer quinoline at 95 K; this suggests that the molecule remains chemically unaltered upon chemisorption, but such a conclusion is somewhat in conflict with later RAIRS experiments reported by Ma *et al.* [35] that indicated substantial dissociation. Nevertheless, so long as some fraction of quinoline remains intact and constitutes the *only* aromatic species on the surface, analysis of π resonances in NEXAFS data is reasonable, albeit coadsorption with dissociation products may alter the quinoline adsorption geometry.

In their NEXAFS experiments, conducted at 300 K, Bonello and Lambert [34] reported that the fine structure due to excitation from C 1s orbitals was broadened beyond useful interpretation, presumably due to the variety of local environments inhabited by individual carbon atoms. The N 1s edge did, however, feature several well-resolved π resonances that permitted the orientation of the molecule to be deduced by varying the polarisation of incident X-rays. The resulting apparent tilt angle of $20\pm5^\circ$, obtained at monolayer coverage, is broadly consistent with expectations of a more or less flat-lying geometry; the authors point out that even a small proportion of highly tilted molecules (perhaps at terrace edges or other defects) could skew the NEXAFS results quite significantly away from zero tilt. At a coverage approximately half that of the complete monolayer, however, analysis of N 1s NEXAFS data yielded a tilt angle of $15\pm5^\circ$, consistent with the view that steric effects favour a higher degree of molecular tilt.[10] In contrast to results for the quinoline-containing molecules discussed above, the orientation of quinoline itself appears to be largely unaffected by heating to 360 K. Meanwhile, STM images obtained at 300 K reveal quinoline molecules to be uniformly but randomly distributed across the surface, but annealing beyond 350 K led to accumulation at terrace edges and/or aggregation into polymer chains. In the absence of temperature-induced tilting, this behaviour was proposed as the cause of irreversible enantioselectivity loss upon heating, and hence as further evidence for the one-to-one model of chiral recognition [34].

10 Similar experiments conducted with lepidine – differing from quinoline by the replacement of a hydrogen atom by a methyl group in the position opposite the nitrogen atom (*see Plate XVII*) – yields a tilt angle of $32\pm5^\circ$, emphasising the role played by steric effects in determining the adsorption geometry.

Finally, before addressing the MP-CD and/or EP-CD interactions in detail, it will be advisable to introduce one further potentially significant molecule (often invoked as a simple model for the alkaloid modifier) namely 1-(1-naphthyl)ethylamine (NEA). In this species, the quinoline moiety of CD is replaced with a naphthalene moiety, and the bulky quinuclidine moiety, dangling from its flexible alcoholic linker, is replaced with a compact and relatively inflexible ethylamine moiety (*see Plate XVII*). The molecule does act as a chiral modifier in its own right,[11] but importantly it *may* also provide a degree of insight into which aspects of the CD molecule are responsible for the latter's particularly high enantioselectivity.

Early data on the adsorption geometry of NEA was provided by XPS and NEXAFS experiments performed by Lambert and co-workers [36]. In these, NEA was deposited on Pt{111} at 300 K, whereupon two XPS peaks were observed in the C 1s region, at binding energies of 284.4 and 286.2 eV, attributed respectively to the carbon atoms of the naphthalene moiety and those of the ethylamine moiety. In the N 1s region, only a single peak, at a binding energy of 399.7 eV, was observed, as expected for an intact amine group. Several π resonances were identified in the NEXAFS data at the C 1s edge, allowing the molecular orientation to be deduced; a tilt angle of $46\pm5°$ between the naphthalene plane and the plane of the surface was estimated on this basis. In comparison with the geometry of quinoline, both in isolation and as a moiety within CD, the orientation of NEA appears rather far from flat, presumably as a result of steric issues; subsequent STM imaging by Bonello *et al.* [37] revealed an apparent height for adsorbed NEA molecules on this surface that was approximately twice that of either naphthalene or quinoline. Later RAIRS experiments from the group of Zaera [38] support the contention of highly tilted NEA adsorption under vacuum on Pt{111}, even at low coverage.[12] Moreover, subsequent work by Gordon and Zaera [39] provides evidence that – when deposited from solution in carbon tetrachloride, at least – NEA may actually bind to platinum via its amine functionality, and not through its naphthalene moiety at all. Clearly, the extent to which enantioselectivity induced by CD and NEA shares a similar origin is rather open to question.

7.3.3 Binding and Reaction in the Chiral Complex

Despite its apparent complexity, the chiral chemistry of cinchonidine is dominated by just three 'sticky' sites where it can potentially attach to (and orient) coadsorbed reactants such as alkyl pyruvates: the two nitrogen atoms of its quinuclidine and quinoline moieties, and the hydroxyl group of its linking alcohol; each can act either as a hydrogen bond acceptor or as a hydrogen bond donor, depending upon the nature of the solvent. Of these sticky sites, the hydroxyl group is unlikely to be of primary importance in generating enantioselectivity, since it typically sits outside (and points away from) any chiral pocket that may be formed

11 More accurately, it is the precursor to an efficient chiral modifier that forms under reaction conditions.

12 It ought, perhaps, to be mentioned that RAIRS experiments reported some years earlier by McBreen and co-workers [22] gave qualitatively different spectra, and were interpreted as consistent with a flat-lying geometry. The reasons for this apparent discrepancy are not altogether clear.

at the surface by the overhanging quinuclidine moiety.[13] In essence, therefore, attention focuses upon which (if either) of the nitrogen atoms are involved in hydrogen bonding with the reactant, and in what way. One model, in particular, has gained significant traction.

The model in question, proposed by Baiker and co-workers [41] long before the detailed adsorption geometry of CD was experimentally known, involves only the nitrogen atom of the quinuclidine group in hydrogen bonding with an alkyl pyruvate. These authors assumed attachment of CD to the surface via its (flat-lying) quinoline moiety, but otherwise did not include platinum in their calculations; protonation of the quinuclidine nitrogen atom by a protic solvent was also assumed, as was an 'open' conformation (in which the quinuclidine moiety reaches up and away from the quinoline moiety). Using a semi-empirical approach to estimate interactions between coadsorbed species, they then found an energetically favourable orientation for MP (in its *s-trans* conformation) bound to the modifier through hydrogen bonding between the protonated quinuclidine nitrogen atom and the ketone functionality of the pyruvate.[14] In this geometry, hydrogenation of the ketone carbon atom *from the direction of the surface* would result in D-methyl lactate, in preference to L-methyl lactate (see Fig. 7.8). Bürgi and Baiker [43] subsequently showed, however, through NMR spectroscopy, that the prevalence of open and 'closed' geometries (the latter involving the quinuclidine moiety arching back and over the quinoline moiety) in solution was critically dependent on the nature of the solvent – closed forms being favoured over open as solvent polarity increased.

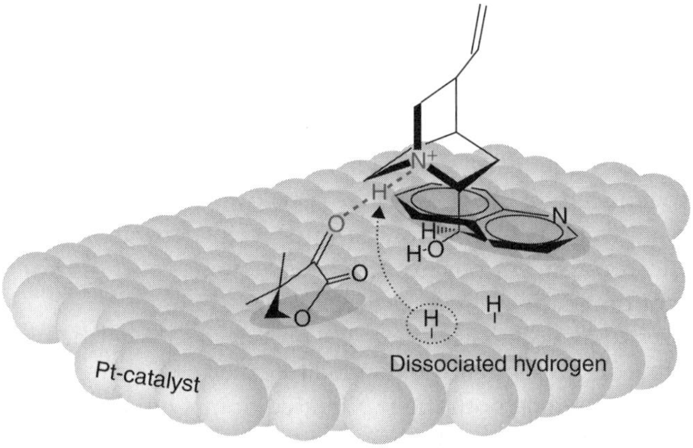

Figure 7.8 Schematic model of surface-bound cinchonidine interacting with a coadsorbed ketone, via its quinuclidine moiety. Reproduced under Creative Commons Attribution 3.0 from [44].

13 This is not to say that it must be entirely irrelevant, however, since it may be involved in the formation of supramolecular assemblages on the surface, potentially affecting the behaviour of the modifier. Moreover, bonding (or substitution) at this site could affect the geometry and chemistry of the chiral pocket, once again modulating the overall behaviour, as noted by Vargas *et al.* [40].

14 Bürgi and Baiker [42] suggest that in aprotic solvents the ketone functionality of the pyruvate moiety may be protonated, allowing it to form a hydrogen bond with the bare nitrogen atom of the quinuclidine moiety.

Some years later, extensive calculations by Vargas and Baiker [45], employing DFT on cluster models of the Pt{111} surface, fleshed out the situation considerably; systematic investigation of several possible adsorption conformations led to the conclusion that whilst closed and 'quinuclidine-bound' geometries (the quinuclidine moiety binding to the surface alongside the quinoline moiety in the latter case) are energetically favoured for CD in isolation on the surface, these are readily converted into 'open' geometries upon interaction with MP (i.e. an induced-fit process is involved). The flexibility of CD, conferred by two torsions of the linking alcohol, seems to be crucial in allowing the chiral pocket to form, although one ought to note that solvent effects were not included in the analysis.[15] A recent review from the pen of Baiker is highly recommended for its summary of these and other issues [44].

Although the Baiker scheme is pretty convincing, it would be remiss to omit mention of a plausible alternative, suggested initially by work from the McBreen group on NEA [22]. In their RAIRS experiments, they observed absorption bands upon MP/NEA coadsorption consistent with those reported for MP alone, as described earlier (Section 7.3.1). Specifically, their results suggested adsorption in the *s-cis* conformation at low coverage, evolving into the *s-trans* conformation only at higher coverage. They also attribute changes in spectral features relating to NEA as indicative of hydrogen bond formation between its amine group and one of the carbonyl groups of pyruvate. Crucially, however, they propose that the carbonyl group in question belongs to the ester functionality of the reactant, on the grounds that the ketone functionality is expected to form a strong covalent bond with the surface. In a subsequent paper, the same authors further proposed that this conformation constituted merely the precursor to an activated state in which the ketone functionality rotated towards (and formed a weak hydrogen bond with) one of the carbon atoms of the naphthalene moiety [47]. Clearly, were naphthalene to be replaced by quinoline (as in CD) this second hydrogen bond could conceivably be much stronger and the activated state correspondingly more favourable, but the nitrogen atom of the heterocycle is awkwardly placed for this to occur in practice. The activation of homocyclic aromatic rings towards hydrogen bond formation is, however, an entirely expected consequence of C—H bond polarisation upon adsorption on metal surfaces [48, 49].

Building upon these ideas, Lavoie and McBreen [48] proposed a model for alkyl pyruvate interaction with CD (subsequently extended to generic Orito reactants and alkaloid modifiers [50]) in which both the protonated quinuclidine nitrogen atom *and* the activated aromatic anchor (whether naphthalene or quinoline) play key roles. The former provides a strong hydrogen bond, capable of firmly holding the reactant in place; the latter contributes the subtle extra twist (quite literally) that imparts chiral discrimination to the complex (see Fig. 7.9) . Only when the second point of contact is established will the ketone functionality adopt the correct orientation to permit hydrogenation from the direction of the surface

15 We might reasonably expect these to be dominated by the adsorbate–substrate interaction, but there is certainly scope for doubt. Indeed, the role of solvents in facilitating fluctuations in alkaloid geometry has been emphasised by Mink *et al.* [46] on the basis of NMR data, albeit protonation of the quinuclidine nitrogen atom is found to lock the conformation to a certain extent.

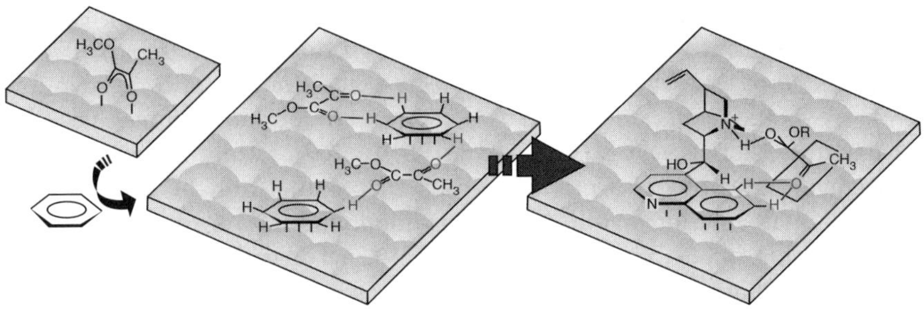

Figure 7.9 Schematic model of surface-bound cinchonidine interacting with a coadsorbed ketone, via both its quinuclidine *and* quinoline moieties . Reproduced with permission from [48]. Copyright (2005) American Chemical Society.

(yielding, e.g. D-methyl lactate in preference to L-methyl lactate from prochiral MP). Notwithstanding uncertainty regarding the orientation of NEA on platinum surfaces (see the preceding subsection) the McBreen model is nevertheless both chemically appealing and consistent with observed trends in enantioselectivity amongst various analogous reactant/modifier pairs.

7.4 Hydrogenation of β-Ketoesters over Nickel: The Izumi Reaction

In comparison with the asymmetric hydrogenation of α-ketoesters, that of β-ketoesters has been much less intensively studied by the techniques of surface science. It has, however, been the subject of a great deal of work from the viewpoint of catalytic science, much of it performed in the laboratory of Yoshiharu Izumi, who also comprehensively reviewed the topic in 1983 [51]. This work drew attention to the classic case of methyl acetoacetate (MAA, *see Plate XIII*) on Raney nickel catalysts modified by treatment with either an α-amino or α-hydroxy acid, which we might reasonably describe as the **Izumi Reaction**. Well-known modifiers for this reaction include tartaric acid, aspartic acid and glutamic acid, and we shall examine recent surface studies relevant to all three cases below. The product of this reaction is *R*-methyl-3-hydroxybutyrate (*R*-M3HB, *see Plate XIII*) when using the L enantiomer of the modifier (i.e. *R*,*R*-tartaric acid, *S*-aspartic acid or *S*-glutamic acid) at room temperature; in the case of modification by glutamic acid, enantioselectivity can be reversed by heating above 350 K.

7.4.1 Adsorption Geometry of Methyl Acetoacetate

The adsorption of MAA on Ni{111} was first studied by Jones and Baddeley [52] by means of RAIRS experiments. Upon deposition at 300 K, a rather strong absorption band observed at 1789 cm^{-1} was attributed to CO occupying threefold hollow sites, suggesting a fair degree of dissociation at this temperature.

Indeed, TPD experiments resulted in coincident CO and H_2 desorption peaks centred on 400 K but extending significantly down to around 350 K. Notwithstanding the apparent fragility of the molecule, additional RAIRS absorption bands were noted, consistent with some retention of intact molecules. Of particular interest is a very weak band at 3017 cm^{-1} (saturating at relatively low coverage) indicative of a methine group that should only be present if at least some small fraction of the molecules adopt the enol form; the authors suggest that this might be stabilised by an internal hydrogen bond between the enol functionality and the carbonyl group of the ester functionality [52]. Also noteworthy are the absorption bands at 1726 and 1301 cm^{-1}, attributed respectively to the carbonyl stretch mode and to the —CH_2 twist mode; the strength of the latter indicates that most of the adsorbed molecules retain the diketo form. Furthermore, the presence of strong out-of-plane vibrations implies a more or less flat-lying adsorption geometry [52].

7.4.2 Two-Dimensional Cocrystallisation: Tartaric/Glutamic Acid Modification

It will be recalled (Section 4.3.8) that tartaric acid forms ordered overlayers on Ni{111}, displaying majority (3,2; $\bar{2}$,2) and minority (3,1; 1,4) periodicities with the L enantiomer (i.e. *R,R*-tartaric acid) [53]. It might be expected, therefore, that enantioselectivity in this case could be due to some kind of templating effect, whereby the ordered lattice exposes chiral ensembles of metal atoms, capable of instigating the hydrogenation of MAA to one enantiomer in preference to the other. Such a model fell from favour early on amongst cognoscenti of the Orito reaction, in light of evidence for disordered modifier adsorption, but here the situation appears much more susceptible to such an explanation. In fact, however, the reality turns out to be rather more subtle, as we shall see.

Attempting deposition of MAA onto a Ni{111} surface pre-dosed with a saturated monolayer of tartaric acid, Jones and Baddeley [52] were unable to detect any change in RAIRS, implying that essentially no MAA was adsorbed under these conditions. Using a slightly lower pre-dose of tartaric acid, however, substantive changes were detected upon exposure to MAA. Evidently, the loci for hydrogen bond formation are fully engaged in the high-coverage regime of tartaric acid, leaving no scope for incorporation of new molecules into the ordered structure. In the low-coverage regime, inverse bands at 1870 and 2013 cm^{-1} were interpreted as indicating displacement of adventitious CO by MAA, suggesting that the interpolation of the former tends to disrupt the hydrogen-bonded network of tartaric acid. A conventional absorption band at 1312 cm^{-1} was attributed to the —CH_2 twist mode of MAA, seen at 1301 cm^{-1} in the absence of tartaric acid, while a further absorption band at 1589 cm^{-1} was supposed to be a strongly redshifted carbonyl stretch mode, seen previously at 1726 cm^{-1}. These modes confirm the presence of MAA on the surface, suggesting also that the molecules may be lying rather less flat than in the absence of tartaric acid, accounting for strong substrate influence upon their carbonyl groups [52].

Most interestingly, STM images revealed ordered structures upon coadsorption, but *not* the same as those reported previously for tartaric acid alone. At relatively low MAA coverage, the overlayer adopted an extrinsically chiral rhombic (4,1; 1,4) periodicity[16] with two features within each primitive unit cell (1/15 ML of each species) while at somewhat higher MAA coverage, the ordered two-dimensional lattice displayed an intrinsically chiral (2,4; 5,1) periodicity[17] with three features per cell (1/9 ML of MAA, 1/18 ML of tartaric acid). Evidently, any template that may have existed prior to deposition of MAA is entirely expunged from the surface by the formation of one or other of these cocrystalline phases; to the extent that the template model may be invoked, therefore, it must certainly operate at the extreme limit of an induced-fit process!

A highly plausible model for the lower coverage phase, suggested by the authors on the basis of high-resolution STM images, implies an adsorption geometry for MAA exclusively canted over towards one side rather than the other; such a geometry would necessarily lead to preferential hydrogenation on the face angled closer to the surface, leading to the observed enantioselectivity. In the higher coverage phase, the authors note that alternate MAA molecules tilt in opposite directions, removing any scope for an enantioselective reaction [52]. In both models, the hydroxyl groups of tartaric acid play a key role as hydrogen bond donors towards the carbonyl group of the ketone moiety within MAA. This role was further highlighted in subsequent work by Jones and Baddeley [54] in which they rationalised the beneficial effects of alkali halide promoters and acidic conditions in terms of RAIRS evidence that both factors increase the ratio of ketone to enol in the reactant.

Turning now to glutamic acid, STM studies by Baddeley and co-workers initially identified small patches of $(\sqrt{7} \times \sqrt{7})R19.1°$ [55] and $(3\sqrt{3}\times3\sqrt{3})R30°$ [56] periodicity within an otherwise disordered saturated monolayer formed at 300 K. At lower coverages, there was no indication of ordered adsorption. Upon deposition at 300 K, RAIRS data in the low-coverage regime revealed absorption bands at 1656 and 1420 cm^{-1}, the former attributed to the carbonyl stretch mode of an intact acid group, and the latter to the symmetric $—CO_2$ stretch mode of a carboxylate group; an absorption band at 1525 cm^{-1} was attributed to the symmetric $—NH_3^+$ stretch mode of a primary ammonium cation and taken, with the other bands, as evidence of zwitterionic adsorption [55]. In the high-coverage regime, after deposition at the same temperature, these absorption bands were joined by another at 1557 cm^{-1} that was attributed to the $—NH_2$ scissor mode of an amine group, suggesting that both zwitterionic *and* anionic species were present. When deposition was carried out at 350 K, on the other hand, evidence for the zwitterionic form was absent [55], underlining once again the critical importance of quite modest temperature differences in defining the surface chemistry of these systems. Indeed, subsequent work suggested the formation of thick layers of either nickel glutamate or sodium glutamate, under acidic or basic conditions respectively, when modification occurred at the higher temperature[18] [57].

16 Originally described as $(3,1; \bar{3},4)$.

17 Originally described as $(4,1; \bar{2},4)$.

18 Sodium was present because the acidity was regulated by addition of NaOH.

In line with previously discussed results for tartaric acid, surfaces dosed to monolayer saturation with glutamic acid are essentially unable to accommodate any MAA molecules to which they may subsequently be exposed. Thus the conditions that lead to ordered adsorption of the modifier are not conducive to coadsorption and the classic combination of template model with lock-and-key process once again proves inadequate. At lower modifier coverage, however, exposure to MAA results in the formation of ordered phases not seen in the absence of the reactant: in addition to a hexagonal $(\sqrt{7} \times \sqrt{7})R19.1°$ periodicity, consistent with compression of regions containing only modifier molecules, STM studies by Trant and Baddeley [58] revealed a majority phase believed to show an extrinsically chiral hexagonal $(\sqrt{21} \times \sqrt{21})R10.9°$ periodicity, coexisting with a minority phase of intrinsically chiral $(4,3;\bar{4},6)$ periodicity.[19] The majority coadsorbed phase displays two features within each primitive unit cell, and is therefore thought to involve local coverages of 1/21 ML for each species; the minority coadsorbed phase displays three features per primitive unit cell and is believed to correspond to local coverages of 1/18 ML for the modifier and 1/36 ML for the reactant. Plausible models for the two coadsorbed phases have been proposed (stabilised by hydrogen bonding between zwitterionic modifier molecules and the ketone functionality of the reactant) and once again preferential orientation of the MAA molecule canted strongly to one side (see Fig. 7.10) is invoked as the proximal cause of cocrystalline enantioselectivity [58]. Also reminiscent of the tartaric acid results, RAIRS studies as a function of decreasing modification pH demonstrated a marked tendency towards adoption by the reactant of its ketone tautomer in preference to the enol form, rationalising enhancement of room-temperature enantioselectivity under acidic conditions; moreover, raising the modification temperature beyond 350 K skewed the reactant strongly towards the enol form, providing a rationale for the reversal of enantioselectivity observed in practical catalysis [57].

7.4.3 Defect-Localised Oligomerisation: Modification by Aspartic Acid

Thus far, we have seen evidence that the classic template model can only account for the enantioselectivity of the tartaric or glutamic acid-modified Izumi reaction via an extreme induced-fit process. That is, an extended porous network of modifier molecules does indeed accommodate the adsorbed prochiral reactant, but the nature of that network is itself fundamentally altered by the presence of the reactant. Lest we be lulled into believing that all reactions of MAA to M3HB derive their enantioselectivity in similar manner, however, our final example – that of modification by aspartic acid – provides a case where matters are quite different.

Evidence for the influence of acidity during modifier deposition on Ni{111} was initially reported for this system by Wilson and Baddeley [59]. When the surface was modified with aspartic acid under low pH conditions, their RAIRS

19 The periodicities quoted here are a little uncertain, being derived ultimately from close examination of STM images, rather than low-energy electron diffraction (LEED) patterns. The matrix notation for the minority phase was not explicitly stated in the original paper, but is implicit in the model presented by the authors.

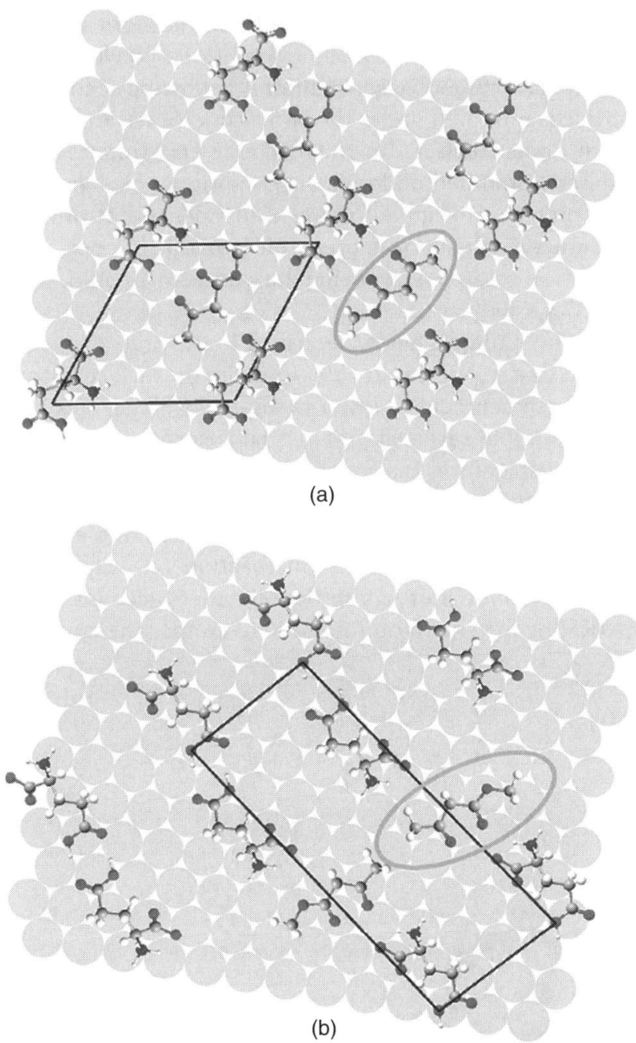

Figure 7.10 Schematic models of MAA coadsorbed with glutamic acid in (a) the $(\sqrt{21} \times \sqrt{21})R10.9°$ phase and (b) the $(4,3; \bar{4},6)$ phase. These periodicities correspond to the locations of the adsorbates, but the models show variations in the molecular orientation – the highlighted acetate molecules would preferentially be hydrogenated to the *S* form of 3-methoxybutyrate, while the others would give rise to the *R* enantiomer. Reproduced with permission from [58]. Copyright (2011) American Chemical Society.

data showed clear absorption bands at 1516 and 1656 cm^{-1}, attributed to the symmetric and antisymmetric stretch modes of a primary ammonium cation (i.e. $—NH_3^+$); absorption bands at 1712 and 1741 cm^{-1} were interpreted as arising from two distinct carbonyl groups, suggesting that the acid functionalities of the modifier remain intact. With the surface prepared at high pH values, however, the spectra revealed absorption bands at 1419 and 1631 cm^{-1}, interpreted as evidence for carboxylate groups (i.e. $—CO_2$) but the bands due to the ammonium cation were relatively diminished in their intensity. Accordingly, the authors asserted that aspartic acid adsorbs in cationic form under acidic conditions, converting via a zwitterionic state to anionic form under basic conditions. Following adsorption at 300 K of MAA onto surfaces modified under the latter conditions, an intense absorption band at 1251 cm^{-1} was attributed to the reactant's enol functionality, but when the surface had been modified under acidic conditions this band was entirely absent. Moreover, there were signs of strong dependence not only on acidity but also on temperature, with surfaces prepared at pH 5 (the optimum for enantioselectivity in practical catalysis) favouring the ketone form of MAA at 300 K but the enol form at 373 K [59].

As in the cases of tartaric and glutamic acids, described in the preceding subsection, RAIRS again showed that MAA was unable to adsorb on surfaces dosed to monolayer saturation with aspartic acid; at lower modifier coverage, however, the STM results of Wilson *et al.* [60] revealed preferential adsorption at terrace edges in proximity to localised structures interpreted as succinimide oligomers. These modifier structures were formed at 300 K in the vicinity of terrace edges, imaging as four-lobed features thought to comprise eight cyclic moieties of alternating orientation, arising from a condensation reaction (see Fig. 7.11). The absence of hydrogen bond donor sites within the oligomers accounts for a marked preference in RAIRS for the enol tautomer of MAA under the coverage

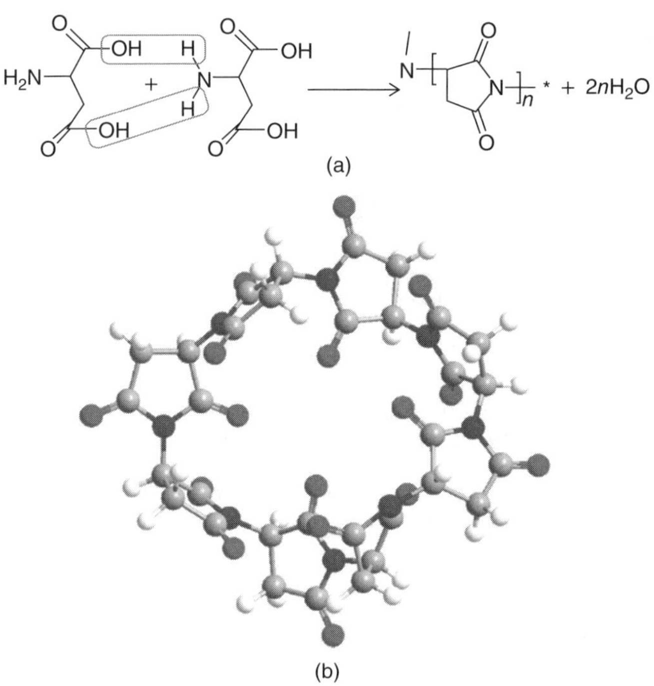

Figure 7.11 Schematic model for the four-lobed features formed upon room-temperature exposure of Ni{111} to aspartic acid. Panel (a) shows the suggested condensation reaction leading to the cyclic species that combine into the 'square' macrocycle shown in panel (b). Reproduced with permission from [60]. Copyright (2012) American Chemical Society.

conditions when intimate coadsorption occurred.[20] In contrast to the cases of tartaric and glutamic acids, therefore, room-temperature enantioselectivity induced by aspartic acid seems to rely upon a localised interaction involving the enol form of MAA, as opposed to a spatially extended interaction involving its ketone form. The situation thus corresponds to the 'third way' – intermediate between a purely one-to-one interaction and the classic template model (cf. Section 5.5). At higher modifier coverage, aspartate species accumulated on terraces, blocking the adsorption of MAA altogether [60]. Such behaviour is entirely consistent with XPS data obtained previously upon adsorption of aspartic acid on polycrystalline nickel, where deposition under pH conditions optimised for enantioselectivity led to spectroscopically undetectable modifier coverage (suggesting that a proportionately tiny number of modifier-decorated defect sites dominate the mechanism) while higher or lower pH produced clear evidence of adsorbed aspartate species [61].

References

1 F. Gao, Y. Wang, Z. Li, O. Furlong and W.T. Tysoe, *J. Phys. Chem. C* **112**, 3362 (2008).
2 F. Gao, Y. Wang and W.T. Tysoe, *J. Phys. Chem. C* **112**, 6145 (2008).
3 G. Kyriakou, S.K. Beaumont and R.M. Lambert, *Langmuir* **27**, 9687 (2011).
4 A. Ahmadi, G. Attard, J. Feliu and A. Rodes, *Langmuir* **15**, 2420 (1999).
5 G. Attard, A. Ahmadi, J. Feliu, A. Rodes, E. Herrero, S. Blais and G. Jerkiewicz, *J. Phys. Chem. B* **103**, 1381 (1999).
6 K.E. Toghill and R.G. Compton, *Int. J. Electrochem. Sci.* **5**, 1246 (2010).
7 M.J. Llorca, J.M. Feliu, A. Aldaz, J. Clavilier and A. Rodes, *J. Electroanal. Chem.* **316**, 175 (1991).
8 G. Attard, *J. Phys. Chem. B* **105**, 3158 (2001).
9 O.A. Hazzazi, G.A. Attard and P.B. Wells, *J. Mol. Catal. A: Chem.* **216**, 247 (2004).
10 O.A. Hazzazi, C.A. Harris, P.B. Wells and G.A. Attard, *Top. Catal.* **54**, 1392 (2011).
11 C. Fleming, M. King and M. Kadodwala, *J. Phys. Chem. C* **112**, 18299 (2008).
12 C. Fleming and M. Kadodwala, *Surf. Sci.* **604**, 409 (2010).
13 Y. Orito, S. Imai, S. Niwa and G.-H. Nguyen, *J. Synth. Org. Chem. Jpn.* **37**, 173 (1979).
14 H.-U. Blaser, in *Cinchona Alkaloids in Synthesis and Catalysis: Ligands, Immobilisation and Organocatalysis*, ed. C.E. Song (Wiley-VCH Verlag GmbH & Co. KGaA, Weinheim, 2009).
15 D.J. Jenkins, A.M.S. Alabdurahman, G.A. Attard, K.G. Griffin, P. Johnston and P.B. Wells, *J. Catal.* **234**, 230 (2005).

20 One should note that the STM and accompanying RAIRS experiments were performed after deposition by sublimation of aspartic acid under ultra-high vacuum conditions. They are thus not *directly* comparable with experiments performed after deposition from solution at well-defined pH values, but clearly they are of relevance and provide valuable morphological insight.

16 G.A. Attard, K.G. Griffin, D.J. Jenkins, P. Johnston and P.B. Wells, *Catal. Today* **114**, 346 (2006).
17 J.K. Wilmshurst and J.F. Horwood, *Aust. J. Chem.* **24**, 1183 (1971).
18 D. Ferri, T. Bürgi and A. Baiker, *J. Chem. Soc., Perkin Trans. 2* 221 (2000).
19 J.M. Bonello, R.M. Lambert, N. Kunzle and A. Baiker, *J. Am. Chem. Soc.* **122**, 9864 (2000).
20 J.M. Bonello, F.J. Williams, A.K. Santra and R.M. Lambert, *J. Phys. Chem. B* **104**, 9696 (2000).
21 S. Lavoie, M.-A. Laliberté, G. Mahieu, V. Demers-Carpentier and P.H. McBreen, *J. Am. Chem. Soc.* **129**, 11668 (2007).
22 S. Lavoie, M.-A. Laliberté and P.H. McBreen, *J. Am. Chem. Soc.* **125**, 15756 (2003).
23 T. Bürgi, F. Atamny, A. Knop-Gericke, M. Hävecker, T. Schedel-Niedrig, R. Schlögl and A. Baiker, *Catal. Lett.* **66**, 109 (2000).
24 T. Bürgi, F. Atamny, R. Schlögl and A. Baiker, *J. Phys. Chem. B* **104**, 5953 (2000).
25 G. Bond and P.B. Wells, *J. Catal.* **150**, 329 (1994).
26 G. Webb and P.B. Wells, *Catal. Today* **12**, 319 (1992).
27 T. Evans, A.P. Woodhead, A. Gutierrez-Soza, G. Thornton, T.J. Hall, A.A. Davis, N.A. Young, P.B. Wells, R.J. Oldman, O. Plashkevych, O. Vahtras, A. Agren and V. Carravetta, *Surf. Sci.* **436**, L691 (1999).
28 D. Ferri and T. Bürgi, *J. Am. Chem. Soc.* **123**, 12074 (2001).
29 D. Ferri, T. Bürgi and A. Baiker, *Chem. Commun.* **37**, 1172 (2001).
30 J. Kubota and F. Zaera, *J. Am. Chem. Soc.* **123**, 11115 (2001).
31 C. LeBlond, J. Wang, J. Liu, A.T. Andrews and Y.-K. Sun, *J. Am. Chem. Soc.* **121**, 4920 (1999).
32 W. Chu, R.J. LeBlanc, C.T. Williams, J. Kubota and F. Zaera, *J. Phys. Chem. B* **107**, 14365 (2003).
33 Z. Ma and F. Zaera, *J. Phys. Chem. B* **109**, 406 (2005).
34 J.M. Bonello and R.M. Lambert, *Surf. Sci.* **498**, 212 (2002).
35 Z. Ma, I. Lee, J. Kubota and F. Zaera, *J. Mol. Catal. A: Chem.* **216**, 199 (2004).
36 J.M. Bonello, E.C.H. Sykes, R. Lindsay, F.J. Williams, A.K. Santra and R.M. Lambert, *Surf. Sci.* **482**, 207 (2001).
37 J.M. Bonello, F.J. Williams and R.M. Lambert, *J. Am. Chem. Soc.* **125**, 2723 (2003).
38 I. Lee, Z. Ma, S. Kaneko and F. Zaera, *J. Am. Chem. Soc.* **130**, 14597 (2008).
39 A.D. Gordon and F. Zaera, *Angew. Chem. Int. Ed.* **52**, 3453 (2013).
40 A. Vargas, T. Bürgi and A. Baiker, *J. Catal.* **226**, 69 (2004).
41 O. Schwalm, B. Minder, J. Weber and A. Baiker, *Catal. Lett.* **23**, 271 (1994).
42 T. Bürgi and A. Baiker, *Acc. Chem. Res.* **37**, 909 (2004).
43 T. Bürgi and A. Baiker, *J. Am. Chem. Soc.* **120**, 12920 (1998).
44 A. Baiker, *Chem. Soc. Rev.* **44**, 7449 (2015).
45 A. Vargas and A. Baiker, *J. Catal.* **239**, 220 (2006).
46 L. Mink, Z. Ma, R.A. Olsen, J.N. James, D.S. Sholl, L.J. Mueller and F. Zaera, *Top. Catal.* **48**, 120 (2008).
47 S. Lavoie, M.-A. Laliberté and P.H. McBreen, *Catal. Lett.* **97**, 111 (2004).
48 S. Lavoie and P.H. McBreen, *J. Phys. Chem. B* **109**, 11986 (2005).

49 S. Lavoie, G. Mahieu and P.H. McBreen, *Angew. Chem. Int. Ed.* **45**, 7404 (2006).
50 S. Lavoie, M.-A. Laliberté, I. Temprano and P.H. McBreen, *J. Am. Chem. Soc.* **128**, 7588 (2006).
51 Y. Izumi, *Adv. Catal.* **32**, 215 (1983).
52 T.E. Jones and C.J. Baddeley, *Surf. Sci.* **519**, 237 (2002).
53 T.E. Jones and C.J. Baddeley, *Surf. Sci.* **513**, 453 (2002).
54 T.E. Jones and C.J. Baddeley, *J. Phys. Chem. C* **111**, 17558 (2007).
55 T.E. Jones, M.E. Urquhart and C.J. Baddeley, *Surf. Sci.* **587**, 69 (2005).
56 T.E. Jones and C.J. Baddeley, *Langmuir* **22**, 148 (2006).
57 T.E. Jones, A.E. Rekatas and C.J. Baddeley, *J. Phys. Chem. C* **111**, 5500 (2007).
58 A.G. Trant and C.J. Baddeley, *J. Phys. Chem. C* **115**, 1025 (2011).
59 K.E. Wilson and C.J. Baddeley, *J. Phys. Chem. C* **113**, 10706 (2009).
60 K.E. Wilson, A.G. Trant and C.J. Baddeley, *J. Phys. Chem. C* **116**, 1092 (2012).
61 K.E. Wilson and C.J. Baddeley, *J. Catal.* **278**, 41 (2011).

8

Optical Consequences of Surface Chirality

So they went down in the dusk to the doors of the house, and along the beach of the sea. And the waves beat upon the one hand, and upon the other the dead leaves ran; and the clouds raced in the sky, and the gulls flew widdershins. And when they came to that part of the beach where strange things had been done in the ancient ages, lo! there was the crone, and she was dancing widdershins.

The Song of the Morrow
Robert Louis Stevenson, 1896

The phenomenon of optical rotation, whereby the polarisation of light is reoriented upon passage through certain media, was first observed by Arago, who reported in 1811 upon the separation into colours of polarised light passing through quartz [1]. The following year (much to Arago's chagrin) Biot took up the baton and presented the first in a series of reports [2] that would eventually see optical rotation demonstrated in a wide range of materials. Since many of Biot's materials were non-crystalline (indeed, even fluid) it became clear that optical rotation must be a molecular phenomenon, rather than a property inherently reliant upon long-range order. In 1820, however, Herschel noted that the macroscopic shape of quartz crystals could be correlated with the propensity for either clockwise or anticlockwise rotation of polarisation, presenting something of a challenge to this point of view [3]. It fell to Pasteur, in 1848, to demonstrate that although synthetic tartaric acid was optically inactive, it could be resolved into active components – one form active in the same sense as biogenic tartaric acid, the other (hitherto unknown) form active in the opposite sense – by manually sorting[1] individual crystals by shape [5]; clearly, optical activity could arise from molecules, from crystallographic order, or (as in this case) the two factors could be inextricably linked. In this way was the circle squared, and the

1 Rather wonderfully, the same experimental protocol has recently been applied directly to individual molecules of heptahelicene on a Cu{111} surface; manipulating these with a suitably functionalised STM tip, Ernst *et al.* [4] were able first to separate the components of adsorbed heptahelicene dimers, and then to identify the chirality of those components. Not only was it confirmed that each dimer contained one molecule of each handedness, but it was also shown that dimers could be artificially re-formed (by further manipulation) only if two mirror-image molecules were selected; forcing two molecules of the same handedness together resulted in repulsion.

Chirality at Solid Surfaces, First Edition. Stephen J. Jenkins.

foundation for a modern understanding of optical activity laid.[2] These classic experiments, however, probe only bulk optical phenomena, and indeed strictly relate only to transmission. The purpose of the present chapter is to examine the ways in which similar behaviour may be observed at chiral surfaces, and in either reflection or absorption. In order to do so, we ought first to remind ourselves of the ways in which light itself may be considered a chiral phenomenon.

8.1 The Nature of Light

To begin with the basics, let us recall that light is a transverse electromagnetic wave, comprising oscillating electric and magnetic fields directed at right angles both to each other and to the direction of propagation. These fields oscillate not only in time but also in space, and it is the relationship between these two oscillatory behaviours that determines the speed of light, c, according to the wave equations

$$\frac{\partial^2 \mathbf{E}}{\partial t^2} = c^2 \nabla^2 \mathbf{E} \tag{8.1}$$

and

$$\frac{\partial^2 \mathbf{B}}{\partial t^2} = c^2 \nabla^2 \mathbf{B} \tag{8.2}$$

where **E** and **B** respectively correspond to the electric and magnetic fields. Note that the phase relationship between different vector components of the electric field (or between different vector components of the magnetic field) is not constrained by these fundamental equations, and it is the variety of possible solutions that gives rise to the concept of polarisation, which we shall shortly explore. Before doing so, it is worth stressing the quantum nature of light, as this has an important bearing upon nomenclature.

In common with other quantum entities, the processes of emission or absorption of light are best described by treating it as a particle-like phenomenon: that is, one speaks of the emission or absorption of individual photons, each possessed of a specific energy. Conversely, the propagation of light is best described by treating it as a wave-like phenomenon, with a wavelength linked to the energy of the corresponding photon. Crucially, the wave associated with an individual photon necessarily possesses a well-defined phase relationship between its electric and magnetic components, and hence a well-defined polarisation. The phenomenon of unpolarised light is merely the result of superposing the waves associated with multiple photons and therefore does not require a separate treatment.

8.2 Planar and Twisted Light

At any given instant, the electric (or magnetic) field associated with a polarised electromagnetic wave will be found to exhibit a series of two-dimensional

2 In my experience, people tend to have heard about Pasteur's experiment, but progressively fewer of the foregoing work by Herschel, Biot and especially Arago. A fascinating perspective on the rivalry between Arago and Biot was published on the bicentenary of the former's first report [6].

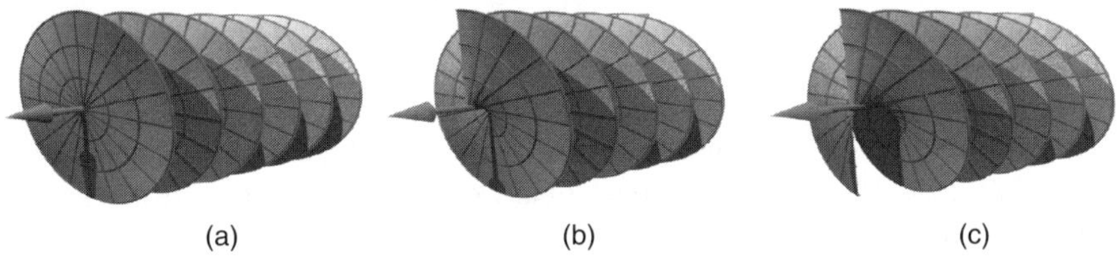

Figure 8.1 Planar and twisted light. Panel (a) depicts surfaces of constant (equal) phase for a planar light wave; panels (b) and (c) show twisted light of identical wavelength, differing in the number of 'starts' to their helices (one and two, respectively).

surfaces of constant phase, embedded into the three-dimensional space within which the wave travels. In the simplest circumstance, these surfaces are essentially flat[3] and the light may then be said to be **planar**. Within this paradigm, polarisation may be either linear or circular in nature, and both cases have been exploited as probes for the chirality of material systems. Relatively recently, however, attention has been drawn to an alternative situation where surfaces of constant phase are helical, rather than planar, and the chiral consequences of *this* scenario – which we shall describe as **twisted** – have scarcely been explored (see Fig. 8.1). Before discussing the applications of either planar *or* twisted light, however, a brief recap of some fundamental properties is probably in order.

8.2.1 Linear and Circular Polarisation

Let us begin by considering a beam of planar light travelling in the $\hat{\mathbf{z}}$ direction, polarised such that its electric field vector oscillates only in the $\hat{\mathbf{x}}$ direction

$$\mathbf{E}_x(\mathbf{r}, t) = \begin{pmatrix} 1 \\ 0 \end{pmatrix} (E_0/2) \exp[i(kz - \omega t)] + c.c. \tag{8.3}$$

and another polarised such that *its* electric field vector oscillates only in the $\hat{\mathbf{y}}$ direction

$$\mathbf{E}_y(\mathbf{r}, t) = \begin{pmatrix} 0 \\ 1 \end{pmatrix} (E_0/2) \exp[i(kz - \omega t)] + c.c. \tag{8.4}$$

where k is the wavenumber and ω the frequency of the light; the '*c.c.*' term indicates addition of the complex conjugate and ensures that the field takes a real value; the column vectors (known as **Jones vectors** [7]) should be interpreted as unit vectors in the Cartesian coordinate system. These are orthogonal examples of **linearly polarised light**, and their superposition with differing amplitude and identical phase can readily produce linearly polarised light with its electric field oscillating in any desired direction perpendicular to the $\hat{\mathbf{z}}$ direction. That is, the combination

$$\begin{aligned} \mathbf{E}_\theta(\mathbf{r}, t) &= \cos\theta \mathbf{E}_x(\mathbf{r}, t) + \sin\theta \mathbf{E}_y(\mathbf{r}, t) \\ &= \begin{pmatrix} \cos\theta \\ \sin\theta \end{pmatrix} (E_0/2) \exp[i(kz - \omega t)] + c.c. \end{aligned} \tag{8.5}$$

3 Very close to the light source, the surfaces of constant phase may not actually be perfectly planar – for instance, a point light source will generate spherical surfaces of constant phase – but they will asymptotically approach planarity as the distance from the light source increases.

produces a linearly polarised beam with its electric field oscillating within the xy plane at an angle of θ to the $\hat{\mathbf{x}}$ direction.

If instead we superpose our linearly polarised beams with identical amplitude but differing phase, we can generate **elliptically polarised light**, in which the direction of the electric field vector continually changes and its magnitude varies but never vanishes. Thus we may have

$$\mathbf{E}_{\phi}(\mathbf{r},t) = (1/\sqrt{2})[e^{-i\phi/2}\mathbf{E}_x(\mathbf{r},t) + e^{+i\phi/2}\mathbf{E}_y(\mathbf{r},t)]$$
$$= (1/\sqrt{2})\begin{pmatrix} e^{-i\phi/2} \\ e^{+i\phi/2} \end{pmatrix}(E_0/2)\exp[i(kz-\omega t)] + c.c. \tag{8.6}$$

with the phase angle ϕ controlling the shape of the ellipse traced out by the electric field vector in the xy plane.[4] In the event that $\phi = \pi/2$ we have **right circularly polarised light**

$$\mathbf{E}_{+}(r,t) = (1/2)\begin{pmatrix} 1-i \\ 1+i \end{pmatrix}(E_0/2)\exp[i(kz-\omega t)] + c.c. \tag{8.7}$$

in which the electric field vector traces a circle in a clockwise sense when viewed along the direction of travel, whereas with $\phi = -\pi/2$ we have **left circularly polarised light**

$$\mathbf{E}_{-}(r,t) = (1/2)\begin{pmatrix} 1+i \\ 1-i \end{pmatrix}(E_0/2)\exp[i(kz-\omega t)] + c.c. \tag{8.8}$$

in which the electric field vector traces the same circle in an anticlockwise sense. Furthermore, inspection of these equations reveals that our original linearly polarised light beams may themselves be viewed as superpositions of the circularly polarised beams. That is, the expressions

$$\mathbf{E}_x(\mathbf{r},t) = (1/2)[(1+i)\mathbf{E}_{+}(\mathbf{r},t) + (1-i)\mathbf{E}_{-}(\mathbf{r},t)] \tag{8.9}$$
$$\mathbf{E}_y(\mathbf{r},t) = (1/2)[(1-i)\mathbf{E}_{+}(\mathbf{r},t) + (1+i)\mathbf{E}_{-}(\mathbf{r},t)] \tag{8.10}$$

and

$$\mathbf{E}_{\theta}(\mathbf{r},t) = (1/2)[(1+i)e^{-i\theta}\mathbf{E}_{+}(\mathbf{r},t) + (1-i)e^{+i\theta}\mathbf{E}_{-}(\mathbf{r},t)] \tag{8.11}$$

can be shown to recover the forms of Eqs. (8.3)–(8.5). Indeed, we are free to consider *either* the linear polarisations *or* the circular polarisations as an orthogonal basis set, in terms of which other general polarisations may be expressed.

Since left and right circularly polarised light beams can be thought of as being enantiomerically related to one another (i.e. they are non-superimposable mirror images) and hence will behave differentially when interacting with other chiral entities, they are (arguably) the most natural choice of polarisation basis set when considering the optical properties of chiral systems. The optical rotation of linear polarisation upon passage of light through a chiral medium, for example, may be viewed as a natural consequence of differing refractive indices (and hence propagation speed) for the two different circular polarisations that together combine

4 The orientation of the ellipse *within* its plane can, of course, be varied by combining the transformations of Eqs. (8.5) and (8.6) into a single, more general, expression. We keep the two aspects separate here for purely didactic purposes.

to form the linear; as the left- and right-handed contributions progressively slip in and out of phase, so the angle θ in Eq. (8.11) monotonically varies. If, instead of differential propagation speed, the chiral medium displays differential absorbance of left and right circularly polarised light – a phenomenon known as **circular dichroism** (CD) – the effect will be gradually to convert the initial linear polarisation into elliptical polarisation as the beam traverses the sample. Measurements of CD are typically reported either as percentage difference in absorbance between left- and right-circular polarisations, or as a change in the **ellipticity angle**, defined as *half* the phase angle, ϕ, that appears in Eq. (8.6).

8.2.2 Polarisation on a Helix

In our treatment of polarisation thus far, we have restricted ourselves to cases where the surfaces of constant phase are simply planar. More generally, however, they may be of helical shape. For a wave travelling along the $\hat{\mathbf{z}}$ direction, polarised with its electric field in the $\hat{\mathbf{x}}$ direction, we may therefore write

$$\mathbf{E}_x(\mathbf{r}, t) = \begin{pmatrix} 1 \\ 0 \end{pmatrix} (E_0/2) \exp[i(kz - \omega t - n\theta\sqrt{x^2 + y^2})] + c.c. \tag{8.12}$$

where the magnitude of the integer n controls the pitch of the helix, and its sign the sense of its helicity; for positive values of n the surfaces of constant phase spiral anticlockwise around the $\hat{\mathbf{z}}$ axis, while for negative values they spiral clockwise. Importantly, the helical nature of the light is entirely independent of the direction of its linear polarisation. That is, we may equally imagine light polarised in the $\hat{\mathbf{y}}$ direction

$$\mathbf{E}_y(\mathbf{r}, t) = \begin{pmatrix} 0 \\ 1 \end{pmatrix} (E_0/2) \exp[i(kz - \omega t - n\theta\sqrt{x^2 + y^2})] + c.c. \tag{8.13}$$

or indeed any linear combination of these two orthogonal cases. By allowing the coefficients in such a linear combination to be complex, we may again (as in the preceding section) obtain elliptical polarisation, or indeed the two extremes of right circularly polarised

$$\mathbf{E}_+(\mathbf{r}, t) = (1/2) \begin{pmatrix} 1 - i \\ 1 + i \end{pmatrix} (E_0/2) \exp[i(kz - \omega t - n\theta\sqrt{x^2 + y^2})] + c.c. \tag{8.14}$$

and left circularly polarised

$$\mathbf{E}\_(\mathbf{r}, t) = (1/2) \begin{pmatrix} 1 + i \\ 1 - i \end{pmatrix} (E_0/2) \exp[i(kz - \omega t - n\theta\sqrt{x^2 + y^2})] + c.c. \tag{8.15}$$

light as desired.

We thus conclude that it is possible for light to display two forms of chirality simultaneously: one due to circular polarisation, the other due to the helicity associated with its surfaces of constant phase. From a quantum viewpoint, these different forms of chirality may be thought of as analogous to the spin and orbital angular momenta found in fermions (of which more in the following chapter).

Whilst the generation of planar light is relatively straightforward (requiring only control over the direction and phase of an electric field uniformly across the entire lateral spread of a light beam) that of helically polarised light is rather more difficult. Recent technical advances in fast-switching liquid crystal polarisers permit the necessary temporally and spatially varying control over polarisation and phase that allows helices with a desired number of 'starts' to be selected. Much of the progress in this area has been driven by potential applications in multiplex communication, whereby distinct streams of data may be encoded in helices of differing orbital angular momentum.[5] The scope for employing this technology in the study of chiral materials, let alone chiral surfaces, appears not to have been exploited to date.

8.3 Dichroic Photoemission

One of the simplest optical experiments we might imagine capable of distinguishing between adsorbed enantiomers would simply be to irradiate the sample with circularly polarised light and test for differential absorption coefficients. If right and left circularly polarised light impinges upon right- or left-handed molecules, one ought reasonably to expect diastereomeric differences to emerge (i.e. **circular dichroism**). According to calculations by Plashkevych *et al.* [8] for isolated alanine, cysteine, serine and valine molecules, however, the enantiomeric excess in absorption coefficients is expected to be no more than 0.1%, which would be very difficult to measure with any accuracy.

An alternative, rather more promising, avenue has been pioneered in recent years by a small number of workers, exploiting not the CD of absorption, but rather that of photoemission (i.e. X-ray photoemission spectroscopy (XPS)). The earliest example of this technique applied to adsorbed chiral molecules was reported by Woodruff and co-workers [9, 10] who took as their benchmark system the case of alanine on Cu{110}. The details of this system are described in Section 4.3.1 and need not be rehearsed here, but suffice to say that the sample was prepared in either the simple (3×2) phase (either enantiomer) or the complex $(2,\bar{2};\ 5,3)/(2,2;\ 5,\bar{3})$ phase (L-alanine/D-alanine). The essence of their experiment was then to conduct X-ray photoemission studies with right and left circularly polarised light, at a photon energy of 310 eV, monitoring the outgoing electron flux as a function of kinetic energy over a range of azimuthal angles.[6] By deconvoluting the raw data, three distinct peaks in the flux as a function of the photoelectron kinetic energy could be identified and associated (based on prior experience) with excitation from the C 1s orbitals of the carboxylate group, the methyl group and the backbone of the adsorbed molecules. For each

5 In plane polarisation, only two orthogonal states exist – either linear polarisation along two Cartesian axes, or circular polarisation in the clockwise or anticlockwise sense. In the case of helical polarisation, cases with differing numbers of 'starts' are orthogonal to one another, so multiple beams of the same wavelength may overlap in space and time without interference.

6 In some experiments, the X-rays were incident normally upon the sample, with electron emission measured at a polar angle of 60° relative to the surface normal; in others, the incidence and emission angles were set at 30° either side of the surface normal, within a common plane.

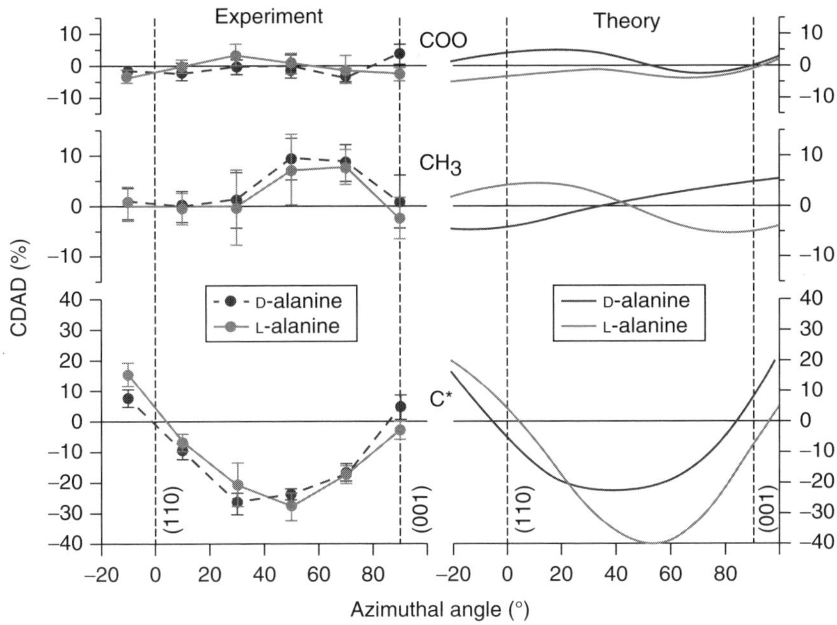

Figure 8.2 Measured and calculated asymmetries of photoelectrons emitted from the C 1s orbitals of carboxylate (top), methyl (middle) and backbone (bottom) regions of alaninate adsorbed on Cu{110}. Reported percentages are enantiomeric excess values obtained when swapping between right and left circularly polarised light. Reproduced with permission from [10]. (*See Plate VI for the colour representation of this figure.*)

measurement geometry, the enantiomeric excess in photoemission flux was then calculated and plotted (see Fig. 8.2).

Now, at first glance, the enantiomeric excess values obtained in these experiments seem large in comparison with their error bars, and indeed there is little doubt that a real effect is being measured. Unfortunately, however, the largest contribution to this effect is not something in which we are particularly interested. As the authors very carefully pointed out, one should expect enantiomeric effects between right and left circularly polarised light from *any* array of spatially oriented and/or spatially ordered molecules (regardless of whether those molecules are in themselves chiral) whenever the experimental geometry itself is chiral [9, 10]. That is, if the plane defined by the incident X-ray and the photoemitted electron does not coincide with a mirror plane of the sample studied, the combined system of sample-plus-probe will be inherently chiral *before* one even considers either the symmetry of the sample or the polarisation of the light; swapping between circular polarisation will then always lead to diastereomeric differences, commonly referred to as **circular dichroism in the angular distribution** (CDAD)[7] [11–13]. Any CDAD arising from the chirality of the adsorbed molecules constitutes a small additional signal on top of a relatively dominant

7 Occasionally, this effect is known as 'false dichroism', which is surely a misnomer as dichroism is a perfectly valid term for the phenomenon, whatever its root cause.

background arising from the geometry of the experiment and not from the chirality of the sample. Indeed, even a perfectly clean achiral substrate will exhibit CDAD in all but a subset of experimental geometries.

Naively, one might attempt to avoid the complexities of substrate-related CDAD altogether, by conducting experiments only in geometries known to be achiral in the absence of adsorbates (i.e. where the plane of incidence/emission coincides with a mirror plane of the clean surface). In this way, one would expect that *any* observed CDAD ought to arise purely from the chirality of the adsorbed molecules, or that of any two-dimensional lattice they might adopt. Two objections may be raised, however, the first being that any inaccuracy in alignment of the sample with respect to the experimental apparatus could well lead to false positives in identifying the overlayer contribution; the second is that there is no guarantee that these special geometries would happen to coincide with situations in which the overlayer contribution is maximised, or even sufficiently large to measure. It was for these reasons that Woodruff and his colleagues chose instead to systematically rotate the plane of incidence/emission: the result is a set of curves in which the difference between enantiomers may become evident at any point, and where differences visible at those particular angles corresponding to achiral experimental geometries may be judged to be robust or not. Cutting to the chase, although the CDAD effect due to overlayer chirality is small, the authors conclusion was that it was indeed large enough to be reliably measured; enantiomeric excess in the angle-resolved photoelectron flux differed between overlayers comprising L- and D-alanine by as much as twenty percentage-points at certain angles, and although the difference was typically less at those angles where substrate-related CDAD was expected to vanish, the overlayer contribution remained statistically significant for at least some functional groups (not always and not only the chiral centre) in some structural phases; indeed, effects of similar magnitude were also noted in emission from N 1s and O 1s orbitals[8] (Fig. 8.3) [9, 10].

At around the same time, Kim *et al.* [14] applied a very similar approach to a Si{001} sample exposed to *R*,*R*- *S*,*S*- or *R*,*S*-2,3-butanediol; the first two tautomers are enantiomers, while the third is an achiral meso isomer and provides for a useful null experiment – all three are believed to bind at intact silicon dimer sites via deprotonation of their hydroxyl groups. Using X-rays of energy 330 eV, incident at 54° off-normal within a mirror plane of the substrate, these authors demonstrated CDAD (in normal emission) varying as a function of binding energy in the C 1s region of the spectrum; crucially, the enantiomeric excess values obtained after dosing the two enantiomers were reversed with respect to one another, while those obtained when using the achiral tautomer were close to zero (see Fig. 8.4). It is noteworthy that in this case (in contrast to that of alanine on copper) the CDAD effect was strongest for binding energies around 286 eV, corresponding to emission from the chiral centres of the adsorbates, and much more equivocal for binding energies around 285 eV, corresponding to emission from the methyl carbon atoms. Extending their dataset to non-normal

8 These experiments employed photon energies of 450 and 570 eV, respectively, in order to maintain similar photoelectron kinetic energies to those involved in the C 1s experiments.

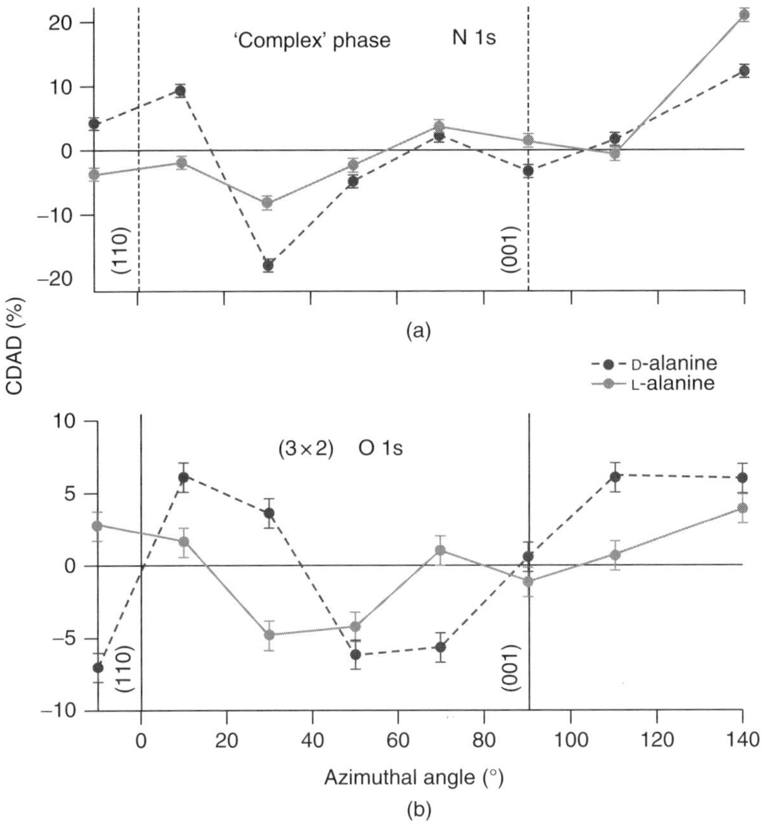

Figure 8.3 Asymmetry of photoelectrons emitted from the N 1s (a) and O 1s (b) orbitals of alaninate adsorbed on Cu{110}. Reported percentages are enantiomeric excess values obtained when swapping between right and left circularly polarised light. Reproduced with permission from [10]. (*See Plate VII for the colour representation of this figure.*)

emission angles, Kim *et al.* [14] further demonstrated that so long as incidence and emission remained coplanar (and within a mirror plane of the substrate) the observed CDAD showed (approximate) reversal upon switching between molecular enantiomers.

In the experiments described above, emission from core states was precipitated by the absorption of X-rays, and it may appear a little curious, therefore, that enantiospecific effects should arise at all. The core states of atoms are generally fairly insensitive to their environment and typically retain highly symmetric orbitals (the 1s states are, of course, essentially spherical aside from minor perturbation from the crystal field). Clearly, the reported CDAD differences between enantiomers must derive primarily from final state effects – the coupling of core electrons to outgoing plane waves may be influenced by unoccupied states in the vicinity of the vacuum level during the process of emission itself, and these virtual states may be distinctly chiral. The precise way in which this happens, and specifically how the process encodes information about chirality into the angular

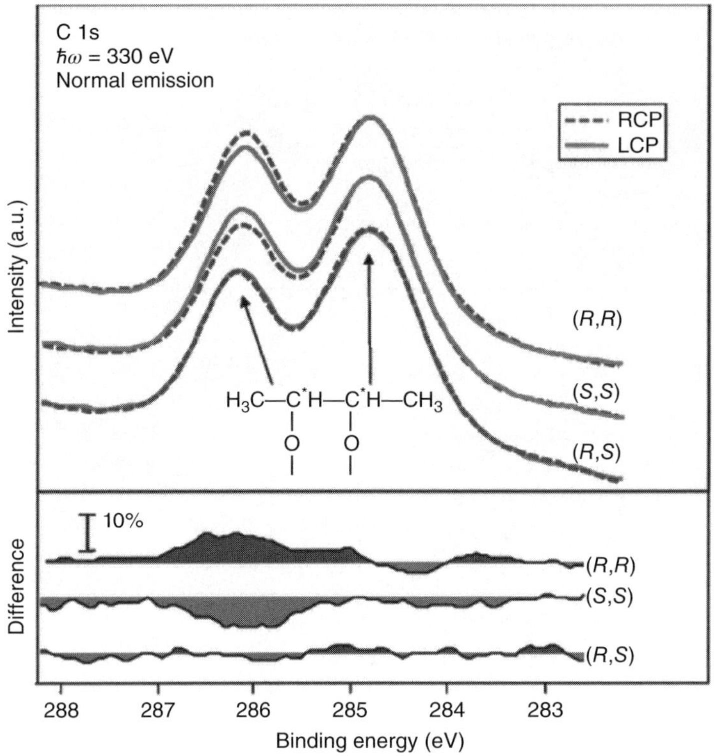

Figure 8.4 Asymmetry of photoelectrons emitted from C 1s orbitals after adsorption of butanediol on Si{001}. Reproduced with permission from [14]. Copyright (2005) American Physical Society. (*See Plate VIII for the colour representation of this figure.*)

distribution of outgoing electrons, is something that may be modelled computationally, but which is hard to grasp conceptually in predictive detail. Nevertheless, the initial focus upon emission from core states was doubtless prompted by a desire to simplify matters and to restrict complexity in the observed results. An alternative approach would clearly be to maximise the potential for CDAD by probing (chiral) valence states – conducting circularly polarised ultraviolet photoemission spectroscopy (UPS) rather than XPS experiments – at the expense of considerably more uncertainty over the exact origin of the (admittedly rather larger) observed diastereomeric differences that one may reasonably expect.

The first study to take the UPS approach was reported by Contini *et al.* [15] on a Cu{100} sample dosed with either D- or L-alaninol.[9] The two enantiomers form $(\sqrt{17} \times \sqrt{17})$ overlayers, rotated by 14.0° clockwise in the D case and by the same amount anticlockwise in the L case. The overlayers thus display both irreducible molecular chirality and extrinsic lattice chirality, both of which may contribute to any observed CDAD via the asymmetry of associated valence electronic states.

9 This aminoalcohol is both a primary amine and a primary alcohol. It is structurally similar to alanine but with the carbonyl oxygen of the acid group replaced by a hydrogen atom (*see Plate XV*).

To avoid effects due to chirality in the experimental geometry, the plane of incidence/emission was aligned with a mirror plane of the substrate (specifically, one perpendicular to a $\langle 110 \rangle$ surface direction). Using a photon energy of 30 eV, the photoelectron flux was measured as a function of binding energy, and CDAD was observed to display very consistent sign-reversal upon switching from one adsorbate enantiomer to the other. While enantiomeric excess magnitudes obtained in achiral experimental geometries did not exceed 10% in the XPS data described above, and were often much less, here the enantiomeric excess peaked at 15% [15].

In a second study of CDAD in the valence states of a chiral surface, Ferrari *et al.* [16] turned to the familiar scenario of tartaric acid on Cu{110}. Specifically, the authors studied the $c(4{\times}2)$ phase[10] reported previously for this system (Section 4.3.5). The achirality of this two-dimensional lattice allows the possibility of monitoring CDAD in a case where the only source of asymmetry is the irreducible chirality of the adsorbates themselves. Once again, enantiomeric excess magnitudes peaked well above 10%, with approximate reversal in sign upon switching between L- and D-tartaric acid.

8.4 Non-linear Optics in Chiral Systems

When light passes through any medium it will, in general, produce an electrical polarisation within it; this will, in turn, give rise to electromagnetic radiation whose superposition with the incident light modulates the original wave. To a very good approximation, this feedback may be assumed to be essentially linear, in the sense that its magnitude will be proportional to the amplitude of the incident light, and in this limit the effect of the electrical polarisation of the medium is merely to alter the beam's phase or polarisation. More generally, however, non-linear effects may be anticipated when the beam intensity is large, since the feedback is proportional to quadratic or higher order terms in the incident amplitude. The symmetry properties of such effects are of crucial importance, regardless of whether we are looking at chiral or achiral surfaces, so that is where we shall begin.

8.4.1 Symmetry Constraints on Non-linear Optical Phenomena

Let us consider the most general expression for the polarisation density (i.e. electric dipole moment per unit volume) associated with an instantaneous

10 For some reason, the authors describe this phase as having (4,0; 2,3) periodicity, despite the clear contrary evidence of their published low-energy electron diffraction (LEED) pattern. The confusion perhaps arises due to a comment in the original work by Raval's group [17] that a (4,0; 2,3) LEED pattern (equivalent to a $c(4 \times 6)$ LEED pattern) was briefly seen before gentle annealing or exposure to the electron beam caused a switch to (4,0; 2,1) periodicity (equivalent to $c(4 \times 2)$ periodicity). Scanning by STM similarly induced conversion from $c(4 \times 6)$ to $c(4 \times 2)$ on the few occasions when the former phase could be imaged at all. Consistent with the fact that both phases essentially involve the same arrangement of molecules on the surface, differing merely in the internal configuration of each molecule, it appears that the lower symmetry phase is extremely sensitive both to thermal and to electronic destabilisation. It seems rather unlikely, therefore, that it survives during UPS experiments to any significant degree.

electric field

$$P_i = \epsilon_0 \chi_{ij} E_j + \epsilon_0 \chi_{ijk} E_j E_k + \epsilon_0 \chi_{ijkl} E_j E_k E_l + \cdots \tag{8.16}$$

where χ_{ij}, χ_{ijk} and χ_{ijkl} are components of the first-, second- and third-order susceptibility tensors, and where subscripts indicate the $\hat{\mathbf{x}}$, $\hat{\mathbf{y}}$ and $\hat{\mathbf{z}}$ Cartesian directions; summation over repeated subscripts is implied.[11]

Already in the linear case (i.e. neglecting the second- and higher order terms) the tensor nature of the first-order susceptibility is critical in explaining birefringence, and even straightforward refraction, since it allows the components of a driving electric field (i.e. incident light) to mix with those of a driven electric field (i.e. refracted light) without contradiction. Inclusion of the second-order term, however, introduces further complexity.

Imagine that the electric field within a particular material may be described as the superposition of two travelling waves (associated with two beams of light) with differing polarisation, wavevector, frequency and phase:

$$\mathbf{E}_1(\mathbf{r}, t) = \frac{1}{2} \begin{pmatrix} E_{1x} \\ E_{1y} \\ E_{1z} \end{pmatrix} \exp[i(\mathbf{k}_1 \cdot \mathbf{r} - \omega_1 t - \phi/2)] + c.c. \tag{8.17}$$

and

$$\mathbf{E}_2(\mathbf{r}, t) = \frac{1}{2} \begin{pmatrix} E_{2x} \\ E_{2y} \\ E_{2z} \end{pmatrix} \exp[i(\mathbf{k}_2 \cdot \mathbf{r} - \omega_2 t + \phi/2)] + c.c. \tag{8.18}$$

where ϕ is the phase difference, ω_1 and ω_2 the frequencies, $\mathbf{k}_1$ and $\mathbf{k}_2$ the wavevectors, and the Jones vectors take care of the polarisations; linear polarisation is imposed if the vector components are all real, while circular polarisation may be obtained with suitable complex choices for these quantities.

Setting $\mathbf{E} = \mathbf{E}_1 + \mathbf{E}_2$, the first-order polarisation density may simply be written as

$$\begin{aligned} \mathbf{P}^{(1)}(\mathbf{r}, t) &= \epsilon_0 \chi_{ij} E_j \\ &= \mathbf{P}^{(1)}_{\omega_1} \exp[i(\mathbf{k}_1 \cdot \mathbf{r} - \omega_1 t - \phi/2)] + c.c. \\ &\quad + \mathbf{P}^{(2)}_{\omega_2} \exp[i(\mathbf{k}_2 \cdot \mathbf{r} - \omega_2 t + \phi/2)] + c.c. \end{aligned} \tag{8.19}$$

where

$$\mathbf{P}^{(1)}_{\omega_n} = \frac{1}{2} \epsilon_0 \underline{\underline{\chi}}^{(1)} \begin{pmatrix} E_{nx} \\ E_{ny} \\ E_{nz} \end{pmatrix} \tag{8.20}$$

with

$$\underline{\underline{\chi}}^{(1)} = \begin{pmatrix} \chi_{xx} & \chi_{xy} & \chi_{xz} \\ \chi_{yx} & \chi_{yy} & \chi_{yz} \\ \chi_{zx} & \chi_{zy} & \chi_{zz} \end{pmatrix} \tag{8.21}$$

11 We neglect, in this treatment, electric quadrupolar (and higher order) phenomena, along with phenomena associated with the magnetic field; in general, these are expected to be of much smaller magnitude than electric dipolar phenomena, except in the vicinity of plasmonic resonances.

which describes a polarisation comprising the sum of two vectors, each oscillating in sign at one or other of the two original frequencies (albeit not necessarily in the same direction as the corresponding electric fields). Note that, throughout this section, we implicitly restrict ourselves to cases where the susceptibility is independent of frequency.[12] A corollary of this restriction is that the subscripts of susceptibility tensor components may be interchanged freely, so that $\chi_{xy} = \chi_{yx}$, $\chi_{xz} = \chi_{zx}$ and $\chi_{yz} = \chi_{zy}$; the first-order susceptibility tensor thus comprises just six independent components.

Turning to the second-order polarisation density, however, this takes the form

$$\begin{aligned}\mathbf{P}^{(2)}(\mathbf{r},t) &= \epsilon_0 \chi^{(2)}_{ijk} E_j E_k \\ &= \mathbf{P}^{(2)}_0 + c.c. \\ &\quad + \mathbf{P}^{(2)}_{2\omega_1} \exp[i(2\mathbf{k}_1.\mathbf{r} - 2\omega_1 t - \phi)] + c.c. \\ &\quad + \mathbf{P}^{(2)}_{2\omega_2} \exp[i(2\mathbf{k}_2.\mathbf{r} - 2\omega_2 t + \phi)] + c.c. \\ &\quad + \mathbf{P}^{(2)}_{\omega_1+\omega_2} \exp[i((\mathbf{k}_1 + \mathbf{k}_2).\mathbf{r} - (\omega_1 + \omega_2)t)] + c.c. \\ &\quad + \mathbf{P}^{(2)}_{\omega_1-\omega_2} \exp[i((\mathbf{k}_1 - \mathbf{k}_2).\mathbf{r} - (\omega_1 - \omega_2)t)] + c.c. \end{aligned} \tag{8.22}$$

where

$$\mathbf{P}^{(2)}_0 = \frac{1}{2}\epsilon_0 \underline{\underline{\chi}}^{(2)} \begin{pmatrix} E^2_{1x} + E^2_{2x} \\ E^2_{1y} + E^2_{2y} \\ E^2_{1z} + E^2_{2z} \\ E_{1y}E_{1z} + E_{2y}E_{2z} \\ E_{1z}E_{1x} + E_{2z}E_{2x} \\ E_{1x}E_{1y} + E_{2x}E_{2y} \end{pmatrix} \tag{8.23}$$

$$\mathbf{P}^{(2)}_{2\omega_n} = \frac{1}{2}\epsilon_0 \underline{\underline{\chi}}^{(2)} \begin{pmatrix} E^2_{nx} \\ E^2_{ny} \\ E^2_{nz} \\ 2E_{ny}E_{nz} \\ 2E_{nz}E_{nx} \\ 2E_{nx}E_{ny} \end{pmatrix} \tag{8.24}$$

and

$$\mathbf{P}^{(2)}_{\omega_1+\omega_2} = \mathbf{P}^{(2)}_{\omega_1-\omega_2} = \epsilon_0 \underline{\underline{\chi}}^{(2)} \begin{pmatrix} E_{1x}E_{2x} \\ E_{1y}E_{2y} \\ E_{1z}E_{2z} \\ E_{1y}E_{2z} + E_{2y}E_{1z} \\ E_{1z}E_{2x} + E_{2z}E_{1x} \\ E_{1x}E_{2y} + E_{2x}E_{1y} \end{pmatrix} \tag{8.25}$$

12 For metallic systems, the susceptibility is approximately constant for frequencies well below the plasma frequency. Since this is typically in the ultraviolet region, our condition should be met for visible and longer wavelengths. Relaxing this condition introduces no additional physics, but would require us to keep track separately of the susceptibilities applicable to each of our two original travelling waves, leading to needless complexity in the presentation.

with

$$\underline{\underline{\chi}}^{(2)} = \begin{pmatrix} \chi_{xxx} & \chi_{xyy} & \chi_{xzz} & \chi_{xyz} & \chi_{xzx} & \chi_{xxy} \\ \chi_{yxx} & \chi_{yyy} & \chi_{yzz} & \chi_{yyz} & \chi_{yzx} & \chi_{yxy} \\ \chi_{zxx} & \chi_{zyy} & \chi_{zzz} & \chi_{zyz} & \chi_{zzx} & \chi_{zxy} \end{pmatrix} \tag{8.26}$$

in which the restriction to static (i.e. frequency-independent) susceptibility implies only 10 independent tensor components ($\chi_{xxy} = \chi_{yxx}$, etc.).

Equation 8.22 clearly indicates the existence of oscillations in the induced polarisation at frequencies equal to twice each of the original frequencies, the sum of the original frequencies and the difference of the original frequencies.[13] Since the oscillations in the polarisation must themselves *create* electromagnetic radiation, it follows that the presence of two 'driving' beams of light within a material having non-negligible second-order susceptibility necessarily implies the presence of four 'driven' beams. In practice, the intensities of the latter are typically very low unless quite intense driving beams are utilised, explaining why exploration of this aspect of optics was delayed until the advent of laser technology. In the case where one of the driving beams is absent, the phenomenon of **second harmonic generation** (SHG) still leads to a driven beam of twice the original frequency exhibited by the single driving beam. Where two driving beams are present, SHG applies to both separately, while the phenomena of **sum-frequency generation** (SFG) and **difference-frequency generation** (DFG) are also observed.

Crucially, any medium whose point group[14] contains the inversion operation will have zero second-order susceptibility, implying that SHG, SFG and DFG are absent in such cases. Amongst crystalline materials, this condition is met for 11 of the 32 crystallographic point groups (the **centrosymmetric point groups**) namely, O_h, T_h, D_{6h}, C_{6h}, D_{3d}, S_6, D_{4h}, C_{4h}, D_{2h}, C_{2h} and S_2. For bulk materials exhibiting one of these symmetries, the *only* conceivable locus for non-linear optical effects is the surface, in the vicinity of which inversion symmetry is impossible. Similarly, inversion symmetry may be considered as applying throughout any non-oriented fluid that may exist above such a surface[15] so this family of phenomena will be truly surface-specific even when the sample is immersed in either gaseous or liquid environments. This property has been exploited to great effect in studies of surface chemistry under ambient conditions [18].

Whatever the symmetry of the underlying substrate, the surface region itself can exhibit 1 of only 10 possible crystallographic point groups, namely C_6, C_4, C_3, C_2, C_1, C_{6v}, C_{4v}, C_{3v}, C_{2v} and C_{1h}; all others possess symmetry elements incompatible with the truncation of *any* crystalline material. To understand the range of non-linear optical effects arising at surfaces, therefore, it will suffice to consider each of these point groups in turn. Fortunately, several share susceptibility tensors of the same (or similar) form and can usefully be grouped together. The forms adumbrated below owe a significant debt to the treatment of Boyd [19] and once again make use of a susceptibility that is independent of frequency.

13 Interestingly, the equation also implies the induction of a constant electric field – a phenomenon known as **optical rectification**.

14 It will be recalled that we strictly mean the point group of the space group.

15 Liquid crystals would, of course, constitute a special case.

C_{6v}, C_{4v}, C_6 and C_4

The four surface point groups of highest symmetry all share precisely the same type of susceptibility tensor. Setting $\hat{\mathbf{z}}$ as the surface-normal direction, it can be shown that only five non-zero tensor components remain (and four of *these* must be identical) yielding

$$\underline{\underline{\chi}}^{(2)} = \begin{pmatrix} 0 & 0 & 0 & 0 & \chi_{xzx} & 0 \\ 0 & 0 & 0 & \chi_{xzx} & 0 & 0 \\ \chi_{xzx} & \chi_{xzx} & \chi_{zzz} & 0 & 0 & 0 \end{pmatrix} \tag{8.27}$$

where we have explicitly permuted subscripts where possible to emphasise related components. The high symmetry of these point groups ensures that the form of this tensor remains entirely unchanged upon rotation of our coordinate system about the surface normal.

C_{3v} and C_3

Reduction of the symmetry, such that the highest order rotational axis is threefold, relaxes the constraints upon the tensor components. In the case of the C_{3v} point group, we have

$$\underline{\underline{\chi}}^{(2)} = \begin{pmatrix} 0 & 0 & 0 & 0 & \chi_{xzx} & \chi_{xxy} \\ \chi_{xxy} & \overline{\chi_{xxy}} & 0 & \chi_{xzx} & 0 & 0 \\ \chi_{xzx} & \chi_{xzx} & \chi_{zzz} & 0 & 0 & 0 \end{pmatrix} \tag{8.28}$$

where one of the three mirror planes has been taken to be perpendicular to the $\hat{\mathbf{x}}$ direction. In this coordinate system, there are only three independent non-zero tensor components, out of a total of eight; rotating our coordinate system around the $\hat{\mathbf{z}}$ direction (again aligned with the surface normal) would, on this occasion, increase the number of non-zero components, but the present choice emphasises the full symmetry of the system and should be preferred. In the absence of any of the C_{3v} mirror planes, however, the C_3 point group yields

$$\underline{\underline{\chi}}^{(2)} = \begin{pmatrix} \chi_{xxx} & \overline{\chi_{xxx}} & 0 & 0 & \chi_{xzx} & \chi_{xxy} \\ \chi_{xxy} & \overline{\chi_{xxy}} & 0 & \chi_{xzx} & 0 & \overline{\chi_{xxx}} \\ \chi_{xzx} & \chi_{xzx} & \chi_{zzz} & 0 & 0 & 0 \end{pmatrix} \tag{8.29}$$

where there are now clearly 4 independent non-zero tensor components, out of a total of 11, regardless of any rotation of the coordinate system about the $\hat{\mathbf{z}}$ axis.

C_{2v} and C_2

Turning to the C_{2v} point group, appropriate for achiral surfaces whose highest order rotational symmetry is twofold, we find

$$\underline{\underline{\chi}}^{(2)} = \begin{pmatrix} 0 & 0 & 0 & 0 & \chi_{xzx} & 0 \\ 0 & 0 & 0 & \chi_{yyz} & 0 & 0 \\ \chi_{xzx} & \chi_{yyz} & \chi_{zzz} & 0 & 0 & 0 \end{pmatrix} \tag{8.30}$$

in which there are three independent non-zero tensor components, out of a total of five; once again, rotation of our coordinate system about the $\hat{\mathbf{z}}$ axis is immaterial in this assessment. Without the mirror planes of the C_{2v} case, the second-order susceptibility tensor of the C_2 point group conforms to

$$\underline{\underline{\chi}}^{(2)} = \begin{pmatrix} 0 & 0 & 0 & \chi_{xyz} & 0 & \chi_{xxy} \\ \chi_{xxy} & \chi_{yyy} & \chi_{yzz} & 0 & \chi_{xyz} & 0 \\ 0 & 0 & 0 & \chi_{yzz} & 0 & \chi_{xyz} \end{pmatrix} \tag{8.31}$$

which may be characterised by four independent non-zero tensor components, out of a total of eight, regardless of the azimuthal orientation of our coordinate system.

C_{1h} and C_1

Finally, when an achiral surface entirely lacks rotational symmetry, we arrive at the C_{1h} point group, whose second-order susceptibility tensor takes the form

$$\underline{\underline{\chi}}^{(2)} = \begin{pmatrix} \chi_{xxx} & \chi_{xyy} & \chi_{xzz} & 0 & \chi_{xzx} & 0 \\ 0 & 0 & 0 & \chi_{yyz} & 0 & \chi_{xyy} \\ \chi_{xzx} & \chi_{yyz} & \chi_{zzz} & 0 & \chi_{xzz} & 0 \end{pmatrix} \tag{8.32}$$

where the sole mirror plane has been taken to be perpendicular to the $\hat{\mathbf{y}}$ axis. In such a coordinate system, there are evidently 6 independent non-zero tensor components, out of a total of 10; rotating our coordinate system around the $\hat{\mathbf{z}}$ direction would increase the number of non-zero components, but would obscure the symmetry of the situation and should be deprecated. Removing the remaining mirror plane yields the C_1 point group, in which case we have

$$\underline{\underline{\chi}}^{(2)} = \begin{pmatrix} \chi_{xxx} & \chi_{xyy} & \chi_{xzz} & \chi_{xyz} & \chi_{xzx} & \chi_{xxy} \\ \chi_{xxy} & \chi_{yyy} & \chi_{yzz} & \chi_{yyz} & \chi_{xyz} & \chi_{xyy} \\ \chi_{xzx} & \chi_{yyz} & \chi_{zzz} & \chi_{yzz} & \chi_{xzz} & \chi_{xyz} \end{pmatrix} \tag{8.33}$$

where all 10 independent non-zero tensor components of the general case remain.

8.4.2 Implications for Chiral Surfaces

The results outlined in the preceding subsection suggest an interesting way in which SHG, SFG or DFG methods can be focused upon purely chiral aspects of a particular system. Given an achiral substrate, the corresponding point group of the surface (C_{6v}, C_{4v}, C_{3v}, C_{2v} or C_{1h}) will imply zero magnitude for several components of the second-order susceptibility tensor (either intrinsically, or subject to a particular choice of coordinate system). Upon adsorption of chiral molecules, the surface point group must reduce to a chiral sub-group: C_6, C_3, C_2 or C_1 if starting from C_{6v}; C_4, C_2 or C_1 if starting from C_{4v}; C_3 or C_1 if starting from C_{3v}; C_2 or C_1 if starting from C_{2v}; or C_1 if starting from C_{1h}. Subject to adopting the coordinate system conventions used above, this will cause irreducibly non-zero components to appear in all but the $C_{6v} \rightarrow C_6$ and $C_{4v} \rightarrow C_4$ cases.

To give a concrete example, consider the face-centred cubic {111} surface, which exhibits threefold rotational symmetry. By choosing our coordinate system such that $\hat{\mathbf{x}}$ lies perpendicular to one of the mirror planes of the surface, we ensure that the χ_{xxx}, χ_{xyy} and χ_{yxy} components vanish from the second-order susceptibility tensor (as in Eq. (8.28)). It then follows that the $\hat{\mathbf{x}}$ component of $\mathbf{P}^{(2)}_{2\omega_1}$

(to explore just one possible avenue) must be related to the electric fields of the driving beams by

$$\hat{\mathbf{x}} \cdot \mathbf{P}^{(2)}_{2\omega_1} = E_{1x}[\chi_{xzx}E_{1z} + \chi_{xxy}E_{1y}] \tag{8.34}$$

from which it is evident that one might, in principle, modulate the amplitudes E_{1x}, E_{1y} and E_{1z} in such a manner as to change all three but leave the resulting SHG signal unchanged. Upon adsorption of chiral molecules, however, the symmetry must reduce to either the C_3 point group or the C_1 point group. In the first of these cases, the $\hat{\mathbf{x}}$ component of $\mathbf{P}^{(2)}_{2\omega_1}$ becomes

$$\begin{aligned}\hat{\mathbf{x}} \cdot \mathbf{P}^{(2)}_{2\omega_1} &= \chi_{xxx}(E^2_{1x} - E^2_{1y}) \\ &\quad + E_{1x}[\chi_{xzx}E_{1z} + \chi_{xxy}E_{1y}]\end{aligned} \tag{8.35}$$

and in the second

$$\begin{aligned}\hat{\mathbf{x}} \cdot \mathbf{P}^{(2)}_{2\omega_1} &= \chi_{xxx}E^2_{1x} + \chi_{xyy}E^2_{1y} + \chi_{xzz}E^2_{1z} \\ &\quad + E_{1x}[\chi_{xzx}E_{1z} + \chi_{xxy}E_{1y}]\end{aligned} \tag{8.36}$$

which typically (in either case) will *not* remain unchanged when subject to the same modulation of driving-beam amplitudes. Clearly, the variation in this quantity arises purely from the symmetry broken by the chiral molecules, suggesting that some form of lock-in technique might fruitfully be employed to isolate it from the static background. Note, however, that so long as only linear polarisation is used in the driving beams, the magnitude of any such 'chiral' signal will indicate something of the degree to which the mirror symmetries of the clean surface have been broken but cannot distinguish between left- and right-handed enantiomers. For that, one must either look to circular dichroism or optical rotation in the SHG signal, as has been done in the two examples we shall now discuss.

8.4.3 Chiral SHG on Cu{111} and Au{110}

The earliest report of optical activity in SHG from a chiral overlayer on a metal surface was published by Kadodwala and co-workers [20] and subsequently expanded upon by the same authors in the following year [21]. In essence, their experiment made use of linearly polarised infrared light (1064 nm) incident upon a Cu{111} surface along one of its high-symmetry in-plane ⟨110⟩ directions, with (green) second harmonic outgoing light (532 nm) detected along the specular direction. In this geometry, no optical rotation due to chirality of the experimental geometry is to be expected, and intensity profiles recorded as a function of either (i) rotating the analyser applied to the second harmonic beam for fixed incident polarisation or (ii) rotating the incident polarisation for fixed analyser orientation matched well with expressions derived previously for these cases [22]. After adsorption of *S*- or *R*-2-butanol, no asymmetry was detected in either type of experiment, but when 1-(1-naphthyl)ethylamine (NEA) was used instead there was a marked difference between results obtained with the *R* and *S* enantiomers, albeit only in experiments of type (i). The insensitivity to variation in the incident polarisation was taken as evidence that the susceptibility of the system is dominated by the metal, rather than the overlayer, and the lack of any

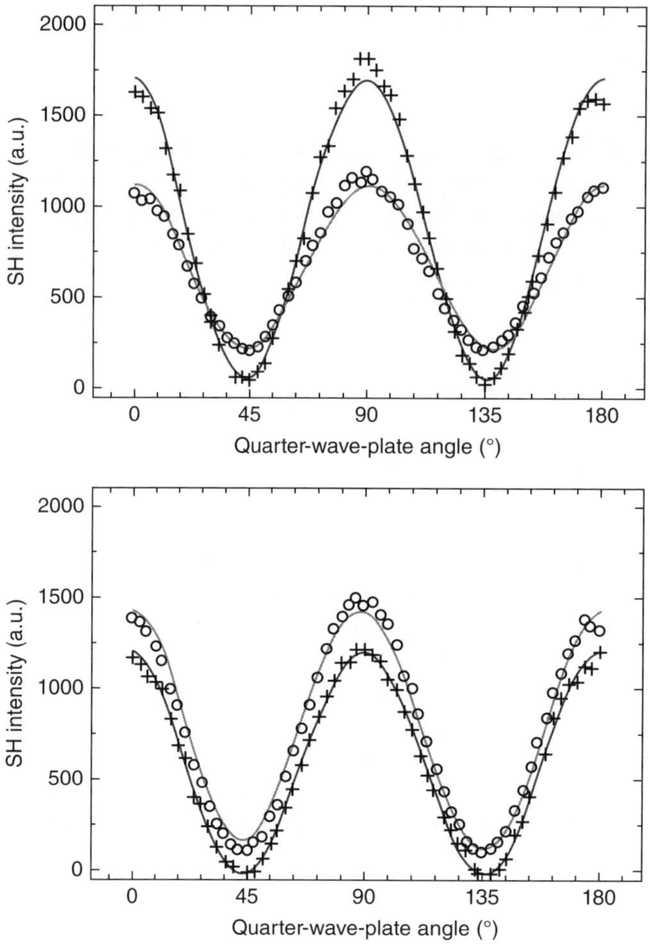

Figure 8.5 Asymmetry in second harmonic light generated from D- and L-cysteine ((a) and (b), respectively) adsorbed on Au{110}, as a function of incident polarisation. Circles denote data obtained in the presence of adsorbate, while crosses indicate data obtained from clean surfaces. Reproduced with permission from [23]. Copyright (2015) Wiley.

obvious surface reconstruction was interpreted as indicating a lowering of symmetry only from C_{3v} to C_3 upon adsorption; the authors therefore rationalised their observations in terms of a reversible perturbation of the surface electronic states that was induced by the large chiral footprint of NEA but not by the point-like contact of 2-butanol [21].

In a more recent work, reported by Persechini and McGilp [23], linearly polarised infrared light (800 nm) was passed through a quarter-wave plate[16] so that the polarisation of incident light could be continuously varied between linear and circular extremes, through a range of elliptical intermediate states.

16 That is, an optical element capable of retarding one polarisation component by a phase angle of $\pi/2$.

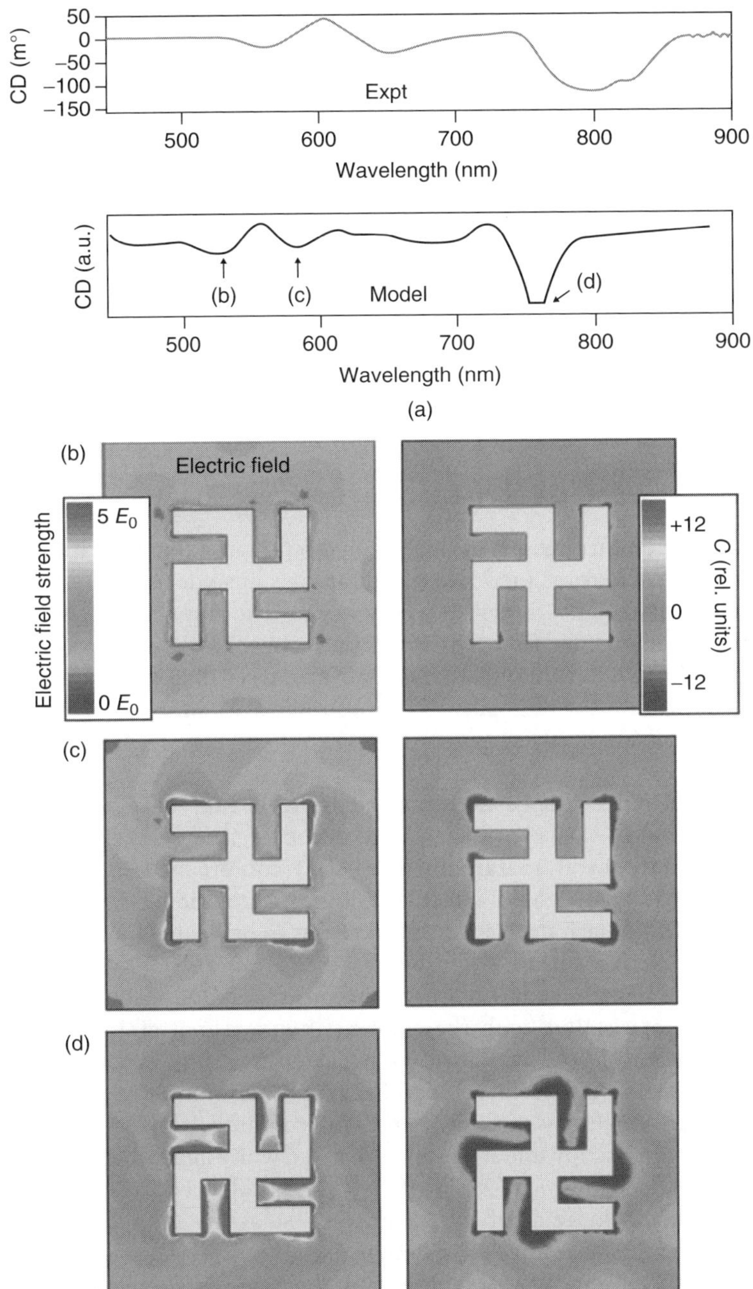

Figure 8.6 Simulated circular dichroism and electromagnetic chirality measure, *C*, for sub-micron gold gammadions. In panel (a) the calculated circular dichroism spectrum is compared with experiment. In panels (b)–(d) the left-hand images show (for three specific frequencies) calculated electric field strength, while the right-hand images show (for the same three frequencies) the calculated value of *C*. Reproduced with permission from [30]. Copyright (2010) Nature Publishing Group. (*See Plate IX for the colour representation of this figure.*)

Measuring the surface-parallel component of polarisation in the (violet) second harmonic beam (400 nm) from a Au{110} surface (angles of incidence and reflection both 45°) the authors noted a sinusoidal variation in the intensity as a function of incident polarisation; this changed upon adsorption of the amino acid cysteine, with the two enantiomers producing markedly different dichroic results (Fig. 8.5). Interpretation of the SHG data in this case is, however, complicated by evidence that the achiral gold substrate behaves anomalously with respect to the two enantiomers of cysteine, in the manner described by Kuhnle *et al.* [24] (and discussed in Section 4.3.4). Certainly, it is not clear that the response observed by Persechini and McGilp [23] for L- and D-cysteine shows the anticipated mirror-image behaviour. More work remains to be done in this area before a full understanding can be expected to emerge.

8.5 Near-Field Phenomena

The implicit assumption throughout the foregoing discussion is that light may be treated as a travelling electromagnetic wave, exhibiting sinusoidal oscillations in both time and space. In fact, however, this is only strictly true in spatially isotropic and homogeneous regions, far from any interfaces or surfaces – the so-called far-field optical regime. In the immediate vicinity of a material discontinuity – the near-field optical regime – the possibility of evanescent waves, sinusoidally oscillating in time but *decaying* in space, cannot be ruled out and may even dominate the phenomenology.

The first experimental evidence that near-field effects may have profound implications for chirality was provided by the group of Zheludev [25], who showed that a **planar chiral metamaterial** (PCM) comprising an array of micron-scale gammadions[17] could effect a significant rotation – up to as much as 30° – in the polarisation axis of a 632 nm light beam diffracted at angles such that no such rotation would be expected from an achiral grating; moreover, the initially linear polarisation of the beam was found to adopt a distinctly elliptical polarisation, with the ellipticity angle also attaining values up to around 30°, which again would not be expected from the experimental geometry in the absence of chirality. On subsequently examining a range of gammadion gratings, Zheludev's group found that coloured fringes could be observed in optical microscopy under linearly polarised white light when viewed through a cross-polarised analyser [26]. Strikingly, these coloured fringes break the fourfold rotational symmetry of the gammadion structure (i.e. bluish and yellowish fringes interchange upon a 90° rotation operation) and were said by the authors to imply broken time-reversal symmetry in the grating's interaction with polarised light, although this interpretation was later convincingly challenged

17 The term 'gammadion' is this field's euphemistic description of choice for a family of shapes that might otherwise be recognised as swastikas, presumably in the interests of good taste. In this instance, the gammadions in question were etched into a 100 nm thick gold layer sandwiched between 20 nm thick titanium layers, and ultimately supported on a polished silicon wafer.

on the basis of transmission experiments[18] [27]. The origin of both polarisation modulation and fringe colouration has been attributed to the coupling of incident light with surface plasmons, creating surface plasmon-polaritons that are sensitive to the chiral structure of individual gammadions [26], and this coupling is generally considered essential to near-field optical chirality.[19]

Common to the plasmon-mediated chirality generated by these PCM systems is an extremely high degree of optical activity, when compared to that of non-metamaterials, justifying the adoption of the term **superchiral** to describe the nature of the optical near-field behaviour. Notwithstanding our general admonition concerning the futility of quantifying chirality in an entirely absolute sense (see Section 1.6) Tang and Cohen [29] have shown that the quantity

$$C = \frac{\epsilon_0}{2}\mathbf{E} \cdot (\nabla \times \mathbf{E}) + \frac{1}{\mu_0}\mathbf{B} \cdot (\nabla \times \mathbf{B}) \tag{8.37}$$

is proportional to the enantiomeric excess in the rate constants of excitation for chiral molecules interacting with an electromagnetic field. Furthermore, they have shown (i) that for circularly polarised light, C takes the values $\pm 2U\omega/c$, where U is the time-averaged local electric energy density, and (ii) that for electromagnetic standing waves, C may exceed these values by multiple orders of magnitude. Making use of this figure of merit, Hendry *et al.* [31] have recently modelled the near-field properties of a simple PCM comprising asymmetrically offset pairs of nanoslits, not only finding values of C in excess of a hundred times the limit for circularly polarised light, but also mapping their distribution in the spatial and frequency domains. Similar calculations for a sub-micron gammadion-based PCM[20] reveal enhancement of C relative to circularly polarised light only up to around one order of magnitude (see Fig. 8.6) but such gratings are nevertheless capable of producing significant circular dichroism in reflection geometry, and this has been shown to respond in diasteroemeric manner to perturbation by adsorbed chiral molecules from amino acids to proteins [30]. The potential for PCM-induced optical superchirality as a tool for the investigation of surface chiral chemistry is evident, but as yet little explored.

References

1 J.D.F. Arago, *Mém. Inst. de France* **1**, 93 (1811).
2 J.-B. Biot, *Mém. Inst. de France* **13**, 1 (1812).
3 J.F.W. Herschel, *Trans. Cambridge Philos. Soc.* **1**, 43 (1819).
4 K.-H. Ernst, S. Baumann, C.P. Lutz, S. Seibel, L. Zoppi and A.J. Heinrich, *Nano Lett.* **15**, 5388 (2015).

18 The transmission experiments used sub-micron gold gammadions of 95 nm thickness supported on fused silica and coated with 23 nm of chromium.

19 We note in passing, however, that recent investigations, from the group of Kadodwala [28], have demonstrated significant optical activity in the SHG signal from an array of sub-micron gold gammadions at frequencies thought to be out of resonance with surface plasmons, demonstrating that plasmonic effects are not necessarily a prerequisite for chiral phenomena within the realm of non-linear optics.

20 In this instance, the gammadions are 100 nm in thickness, bonded to a glass substrate by a 5 nm layer of chromium.

5 L. Pasteur, *C. R. Seances Acad. Sci.* **26**, 535 (1848).
6 B. Kahr and O. Arteaga, *ChemPhysChem* **13**, 79 (2011).
7 R.C. Jones, *J. Opt. Soc. Am.* **31**, 488 (1941).
8 O. Plashkevych, V. Carravetta, O. Vahtras and H. Ågren, *Chem. Phys.* **232**, 49 (1998).
9 M. Polcik, F. Allegretti, D.I. Sayago, G. Nisbet, C.L.A. Lamont and D.P. Woodruff, *Phys. Rev. Lett.* **92**, 236103 (2004).
10 F. Allegretti, M. Polcik, D.I. Sayago, F. Demirors, S. O'Brien, G. Nisbet, C.L.A. Lamont and D.P. Woodruff, *New J. Phys.* **7**, 109 (2005).
11 N.A. Cherepkov, *Chem. Phys. Lett.* **87**, 344 (1982).
12 R.L. Dubs, S.N. Dixit and V. McKoy, *Phys. Rev. Lett.* **54**, 1249 (1985).
13 N. Chandra, *Phys. Rev. A* **39**, 2256 (1989).
14 J.W. Kim, M. Carbone, J.H. Dil, M. Tallarida, R. Flammini, M.P. Casaletto, K. Horn and M.N. Piancastelli, *Phys. Rev. Lett.* **95**, 107601 (2005).
15 G. Contini, S. Turchini, S. Sanna, D. Catone, J. Fujii, I. Vobornik, T. Prosperi and N. Zema, *Phys. Rev. B* **86**, 035426 (2012).
16 L. Ferrari, P. Moras, P. Gori, S. Turchini, N. Zema, A. Palma, J. Fujii, I. Vobornik, G. Alejandro, D. Catone, T. Prosperi and C. Carbone, *Phys. Rev. B* **91**, 085408 (2015).
17 M.O. Lorenzo, S. Haq, T. Bertrams, P. Murray, R. Raval and C.J. Baddeley, *J. Phys. Chem. B* **103**, 10661 (1999).
18 A.G. Lambert, P.B. Davies and D.J. Neivant, *Appl. Spectrosc. Rev.* **40**, 103 (2005).
19 R.W. Boyd, *Nonlinear Optics* 3rd Edition (Elsevier, 2008).
20 A. Mulligan, I. Lane, G.B.D. Rousseau, S.M. Johnston, D. Lennon and M. Kadodwala, *Angew. Chem. Int. Ed.* **44**, 1830 (2005).
21 A. Mulligan, I. Lane, G.B.D. Rousseau, S.M. Johnston, D. Lennon and M. Kadodwala, *J. Phys. Chem. B* **110**, 1083 (2006).
22 M.C. Schanne-Klein, F. Hache, A. Roy, C. Flytzanis and C. Payrastre, *J. Chem. Phys.* **108**, 9436 (1998).
23 L. Persechini and J.F. McGilp, *Phys. Status Solidi B* **252**, 95 (2015).
24 A. Kühnle, T.R. Linderoth and F. Besenbacher, *Top. Catal.* **54**, 1384 (2011).
25 A. Papakostas, A. Potts, D.M. Bagnall, S.L. Prosvirnin, H.J. Coles and N.I. Zheludev, *Phys. Rev. Lett.* **90**, 107404 (2003).
26 A.S. Schwanecke, A. Krasavin, D.M. Bagnall, A. Potts, A.V. Zayats and N.I. Zheludev, *Phys. Rev. Lett.* **91**, 247404 (2003).
27 M. Kuwata-Gonakami, N. Saito, Y. Ino, M. Kauranen, K. Jefimovs, T. Vallius, J. Turunen and Y. Svirko, *Phys. Rev. Lett.* **95**, 227401 (2005).
28 N. Abdulrahman, C.D. Syme, C. Jack, A. Karimullah, L.D. Barron, N. Gadegaard and M. Kadodwala, *Nanoscale* **5**, 12651 (2013).
29 Y. Tang and A.E. Cohen, *Phys. Rev. Lett.* **104**, 163901 (2010).
30 E. Hendry, T. Carpy, J. Johnston, M. Popland, R.V. Mikhaylovskiy, A.J. Lapthorn, S.M. Kelly, L.D. Barron, N. Gadagaard and M. Kadodwala, *Nat. Nanotechnol.* **5**, 783 (2010).
31 E. Hendry, R.V. Mikhaylovskiy, L.D. Barron, M. Kadodwala and T.J. Davis, *Nano Lett.* **12**, 3640 (2012).

9

Magnetic Consequences of Surface Chirality

And I'll begin to treat by what decree
Of Nature it came to pass that iron can be
By that stone drawn which Greeks the magnet call.
De Rerum Natura, Lucretius c.50 BC(tr. William E. Leonard, 1916)

It is an important but commonly overlooked fact that the phenomenon of magnetism is inherently chiral. That it *is* chiral ought, however, to be plainly evident when one considers that the mathematical description of magnetism at the microscopic scale follows closely that of angular momentum at the macroscopic. The very language of magnetism, employing words such as **spin** and **orbit**, connotes a sense of revolution that immediately implies the possibility of clockwise or anticlockwise rotation – mirror-image states of motion that relate to one another in enantiomeric fashion. Let us begin, therefore, by reviewing the origin of these two fundamental forms of magnetism from first principles.

9.1 Spin and Orbital Magnetism

The quantum mechanical description of a single particle begins and ends with its wavefunction. This mathematical ledger carries within itself the entirety of information that can, in principle, be discovered about the particle to which it belongs. Quantum mechanical operators, acting upon the wavefunction, are the lenses with which that information is brought into focus. Thus, the position operator, $\hat{\mathbf{r}}$, yields up the particle's position; the momentum operator, $\hat{\mathbf{p}}$, its momentum; the Hamiltonian operator, $\hat{H}$, its energy; and so on. The fundamental incompatibility of certain pairs of operators with one another is the root cause of quantum uncertainty – one cannot, for instance, measure both the position and momentum of a particle simultaneously with absolute accuracy – but the nature and evolution of the wavefunction is rigidly deterministic. The opaque quality of quantum mechanics derives, therefore, from the fact that the wavefunction itself is not an observable entity. Its square modulus, on the other hand, is eminently observable, corresponding to the probability distribution for finding the described particle at a given location at a specific time.

Chirality at Solid Surfaces, First Edition. Stephen J. Jenkins.

9.1.1 Fermions and the Dirac Equation

When dealing with multiple particles, the wavefunction necessarily becomes more complicated; in three-dimensional space, it will depend upon $3N$ coordinates, where N is the number of particles, and the many-body problem rapidly spirals beyond all hope of analytic solution for all but the smallest and/or most symmetric of systems. Where one *can* make progress, however, is in the case of two identical particles. Here, the two-body wavefunction will admittedly depend upon the coordinates of both particles, but the symmetry implied by their indistinguishability imposes strong constraints upon its possible form. Specifically, we know that simply swapping the particles ought not to change the square modulus of the wavefunction, since this would imply a measurable change in the probability distribution, contradicting the *ansatz* of identical particles. It follows, therefore, that swapping two identical particles cannot alter the magnitude of the wavefunction, only its phase. Taking the extremal cases, one predicts a class of particle whose wavefunctions remain entirely unchanged upon such an exchange, and another whose wavefunctions change sign upon exchange. The former category corresponds, of course, to the particles known as **bosons**, the latter to the **fermions**. One upshot of a wavefunction subject to sign-inversion upon exchange is that two identical fermions can never simultaneously occupy the same location in space (the Pauli exclusion principle) since this would imply a vanishing wavefunction. Electrons are fermions, and hence subject to this crucial constraint. Solution of the Schrödinger equation, however, yields electronic orbitals that must apparently be doubly occupied with electrons in order to achieve correspondence with experiment. The implication is that some property of the electrons, not explicitly included within the Schrödinger equation, must exist, so that each pair of electrons found within a single orbital may differ in this aspect, even if they are identical in all others; the exchange behaviour of fermions will then *not* cause the wavefunction of the orbital to vanish. This property is, of course, the electrons' spin.

The existence of spin may be anticipated, according to the argument presented above, on the basis of the Schrödinger equation, but the absence of any explicit reference to spin in the Schrödinger formalism means that the use of a word connoting rotational motion is not obviously justified – at this level, one might as well refer to the property of 'spin' by any name one wishes. The Stern–Gerlach experiment demonstrates, however, that free electrons possess an intrinsic magnetic moment approximately equal to $\pm e\hbar/2m$ (with e and m the electronic charge and mass, respectively) and it is natural both to attribute this to a similarly intrinsic spinning motion of the electron and to identify this with the property responsible for the Pauli exclusion principle. Indeed, it was Pauli himself who first modified the Schrödinger equation to include spin, based on this essentially phenomenological reasoning. Defining a vector, $\boldsymbol{\sigma}$, whose components were themselves the (2×2) matrices

$$\sigma_x = \begin{pmatrix} 0 & 1 \\ 1 & 0 \end{pmatrix} \tag{9.1}$$

$$\sigma_y = \begin{pmatrix} 0 & -i \\ i & 0 \end{pmatrix} \tag{9.2}$$

and

$$\sigma_z = \begin{pmatrix} 1 & 0 \\ 0 & -1 \end{pmatrix} \tag{9.3}$$

he showed that the Hamiltonian

$$\hat{H}_P = \frac{1}{2m}((\hat{\mathbf{p}} + e\mathbf{A})^2 + e\hbar\boldsymbol{\sigma}.\mathbf{B}) - e\phi \tag{9.4}$$

would correctly reproduce the observed behaviour of an electron with momentum $\mathbf{p}$ in an electromagnetic field described by scalar potential ϕ and vector potential $\mathbf{A}$ (the magnetic field $\mathbf{B}$ being simply equivalent to $\nabla \times \mathbf{A}$). The Hamiltonian is, of course, itself a (2×2) matrix, and thus must act upon a two-vector wavefunction (a so-called **spinor**) whose components may be interpreted as the spin-up and spin-down solutions of the Pauli equation (i.e. the Schrödinger equation with the Pauli Hamiltonian). That is, the Pauli equation takes the form

$$i\hbar\frac{\partial}{\partial t}\psi(\mathbf{r}, t) = \hat{H}_P\psi(\mathbf{r}, t) \tag{9.5}$$

with $\psi(\mathbf{r}, t)$ implicitly a two-component spinor.

Subsequently, Dirac demonstrated that one could derive a wave equation compatible with special relativity from first principles, obtaining (in the potential- and field-free case) the Hamiltonian

$$\hat{H}_D = \beta mc^2 + c\boldsymbol{\alpha} \cdot \hat{\mathbf{p}} \tag{9.6}$$

where $\boldsymbol{\alpha}$ is the matrix

$$\alpha = \begin{pmatrix} 0 & \boldsymbol{\sigma} \\ \boldsymbol{\sigma} & 0 \end{pmatrix} \tag{9.7}$$

and $\boldsymbol{\beta}$ the matrix

$$\beta = \begin{pmatrix} I & 0 \\ 0 & -I \end{pmatrix} \tag{9.8}$$

with I being the (2×2) identity matrix. Clearly, the Hamiltonian is now a (4×4) matrix, and must act upon a four-vector wavefunction (spinor). The corresponding Dirac equation (i.e. the Schrödinger equation with the Dirac Hamiltonian) is thus simply

$$i\hbar\frac{\partial}{\partial t}\psi(\mathbf{r}, t) = \hat{H}_D\psi(\mathbf{r}, t) \tag{9.9}$$

and since its free-space solutions divide naturally into two pairs with positive and negative energies, the former are interpreted as spin-up and spin-down electron solutions and the latter as spin-up and spin-down positron solutions. The spin of electrons (and positrons) is thus found to be a direct consequence of including special relativity in the formulation of quantum mechanics.[1]

The quantisation of electronic spin implies discrete values for the electron's intrinsic angular momentum and its associated magnetic moment. Specifically, the spin angular momentum operator, $\hat{\mathbf{S}}$, is related to the Pauli spin vector by

$$\hat{\mathbf{S}} = \hbar\boldsymbol{\sigma}/2 \tag{9.10}$$

1 From a historical perspective, the Dirac equation was also notable, of course, for prompting the search for antiparticles.

and the spin magnetic moment operator, $\hat{\boldsymbol{\mu}}_s$, is given by

$$\hat{\boldsymbol{\mu}}_s = -g_s \mu_B \hat{\mathbf{S}}/\hbar \tag{9.11}$$

where g_s is the **electron spin g-factor** (equal to 2.0023193043622) and where

$$\mu_B = \frac{e\hbar}{2m} \tag{9.12}$$

is the **Bohr magneton**.

The spin angular momentum is quantised such that the eigenvalues of $\hat{\mathbf{S}}^2$ are restricted to the single value

$$S^2 = \hbar^2 s(s+1) \tag{9.13}$$

with $s = 1/2$ for electrons. Meanwhile, the projection of the spin angular momentum vector onto an arbitrary direction (customarily chosen to be the z axis) results in eigenvalues for the z-component of spin angular momentum, $\hat{\mathbf{S}}_z$, given by

$$S_z = \hbar m_s \tag{9.14}$$

and for the z-component of spin magnetic moment correspondingly given by

$$(\mu_s)_z = -g_s m_s \mu_B \tag{9.15}$$

with m_s being the **spin quantum number** and taking the values $m_s = \pm 1/2$. The z-resolved spin magnetic moment of an electron is thus approximately $\pm\mu_B$.

In contrast to electronic spin, the concepts of *orbital* angular momentum and its associated magnetic moment *do* emerge naturally from the Schrödinger equation, without any necessity to invoke special relativity. The angular momentum operator for motion of an electron about the origin (as opposed to spinning about its own axis) is simply

$$\hat{\mathbf{L}} = \hat{\mathbf{r}} \times \hat{\mathbf{p}} \tag{9.16}$$

and it is quantised (in the case of motion within a spherically symmetric potential) such that $\hat{\mathbf{S}}^2$ has eigenvalues

$$L^2 = \hbar^2 l(l+1) \tag{9.17}$$

where l is the **orbital quantum number**. The corresponding orbital magnetic moment operator, $\hat{\boldsymbol{\mu}}_l$, is then given by

$$\hat{\boldsymbol{\mu}}_l = -g_l \mu_B \hat{\mathbf{L}}/\hbar \tag{9.18}$$

where g_l is the **electron orbital g-factor**. Since g_l is identically equal to unity, we may write

$$\hat{\boldsymbol{\mu}}_l = -\mu_B \hat{\mathbf{L}}/\hbar \tag{9.19}$$

without approximation. Projection of the orbital angular momentum vector onto the z axis then yields eigenvalues of $\hat{\mathbf{L}}_z$ conforming to

$$L_z = \hbar m_l \tag{9.20}$$

with corresponding z-components of orbital magnetic moment given by

$$(\mu_l)_z = -m_l \mu_B \tag{9.21}$$

where m_l is the **magnetic quantum number** and takes integer values in the range $\pm l$ inclusive. The z-resolved orbital magnetic moment of an electron thus takes values that are integer multiples of μ_B, within the range $\pm \mu_B l$.

Now, it is important to note that this discussion of orbital angular momentum invoked the notion of a spherically symmetric potential from the outset; only in such a situation will l and m_l be good quantum numbers for the system. Clearly, therefore, the concept of orbital angular momentum as outlined above is *not* easily transposed to the case of electrons in solids. Nevertheless, practical methods for the calculation of electronic structure in solids typically either ignore orbital angular momentum (and its associated magnetic moment) altogether, or include it only in the vicinity of the ion cores (where the potential may be approximated as spherical). That either approach turns out to be reasonable is testament to the fact that the overall contribution of orbital motion to the magnetism of solids is rather small, especially in the interstitial regions away from ion cores. With this observation in place, one can safely proceed on the basis of separable spin and orbital contributions, treating deviations from this within the perturbational framework of spin–orbit coupling.

9.1.2 Spin–Orbit Coupling

As described above, the spin and orbital components of magnetism in solids may be treated as separable, to zeroth order, and their coupling included via perturbation theory. Indeed, this coupling is crucial to understanding how the geometric chirality of a solid or its surface may influence magnetic properties to any measurable degree. The spin magnetic moment of a solid may be large (e.g. in a paramagnetic or a ferromagnetic material) but the spin of each electron is essentially a point property, insensitive to structure on the atomic scale. The orbital magnetic moment, on the other hand, derives from motion on the atomic scale, and is therefore potentially sensitive to the geometric chirality of the material, but the result of any such asymmetry will be to modulate a fundamentally weak phenomenon (i.e. diamagnetism). Only when coupling between spin and orbital components is important will it be possible for structurally sensitive orbital magnetism to be amplified into a measurable change in otherwise structurally insensitive spin magnetism.

At heart, the phenomenon of spin–orbit coupling derives from the self-interaction of an electron's spin magnetic moment with the magnetic field generated by that same electron's orbital angular motion. Working within the rest frame of this electron, one begins by using the Biot–Savart law to derive the magnetic field operator due to the relative motion of the nucleus

$$\hat{\mathbf{B}} = -\frac{Ze\mu_0}{4\pi r^3}(\hat{\mathbf{v}} \times \hat{\mathbf{r}}) = \frac{Ze\mu_0}{4\pi m r^3}(\hat{\mathbf{r}} \times \hat{\mathbf{p}}) = \frac{Ze\mu_0}{4\pi m r^3}\hat{\mathbf{L}} \tag{9.22}$$

where Z is the atomic number, $\hat{\mathbf{v}}$ the electron velocity operator in the rest frame of the nucleus (or minus the nuclear velocity in the electron rest frame) and r the expectation value of the orbital radius.

Naively, the energy of interaction between the electron spin and this magnetic field would be simply equal to the scalar product of $-\hat{\boldsymbol{\mu}}_s$ with $\hat{\mathbf{B}}$, but less naively an extra factor of 1/2 must be inserted to account for relativistic effects that cause the electron spin to precess as it circulates around the nucleus. Accordingly, the spin–orbit interaction contributes a first-order term to the Hamiltonian (the so-called **LS coupling** term) of the form

$$\Delta\hat{H}_{LS} = \beta\hat{\mathbf{L}}\cdot\hat{\mathbf{S}} \tag{9.23}$$

with

$$\beta = \frac{g_s Z e \mu_0 \mu_B}{8\pi m \hbar r^3} \tag{9.24}$$

where we have used Eqs. (9.10) and (9.11) to avoid explicit reference to the spin magnetic moment. In this form, the spin–orbit coupling is seen to depend on the scalar product of the spin and orbital angular momentum vector operators. Since g_s is positive by definition, it is clear that the spin–orbit coupling constant, β, is positive and hence that the energy will be minimised when the expectation values of $\hat{\mathbf{S}}$ and $\hat{\mathbf{L}}$ are antiparallel. Furthermore, for a hydrogenic orbital having principal quantum number n, the expectation value of the radius, r, is approximately proportional to Z/n, implying that β is proportional to Z^4/n^3; accordingly, spin–orbit coupling is maximised for heavy elements and core electrons, whilst it is minimised for light elements and valence electrons.

As we have already mentioned, one of the most significant consequences of spin–orbit coupling is that it renders the electronic spin sensitive to the structure of the solid. The LS term is sufficient to impose a distinct directionality upon the spin – a phenomenon known as **magnetocrystalline anisotropy (MCA)**, which we shall discuss at length in the next section. Before doing so, however, we should also note the existence of a second-order effect due to spin–orbit coupling, whereby the spin on one electron is coupled to the spin of another electron via their orbital motions (as distinct from the direct exchange interaction responsible for ferromagnetism). In this interaction (the so-called **Dzyaloshinskii–Moriya (DM) coupling** [1, 2]) the overall contribution to the Hamiltonian is given by

$$\Delta\hat{H}_{DM} = \sum_{ij} \mathbf{D}_{ij}\cdot(\hat{\mathbf{S}}_i \times \hat{\mathbf{S}}_j) \tag{9.25}$$

where $\hat{\mathbf{S}}_i$ and $\hat{\mathbf{S}}_j$ are the spin vector operators for a pair of electrons, and $\mathbf{D}_{ij}$ may, in principle, be obtained by solution of the Dirac equation. Note that a different $\mathbf{D}_{ij}$ is required for each pair of spins, but that its magnitude typically decays rapidly with distance so that relatively few need to be calculated in practice. Note also, more importantly, that this term implies the possibility that configurations involving **non-collinear spin textures** may become favoured over the collinear arrangements involved in ferromagnetism, ferrimagnetism and antiferromagnetism. When the symmetry of a system includes a centre of inversion, the time-averaged DM coupling term reduces strictly to zero, and non-collinear spin solutions can only take the form of travelling waves; for non-centrosymmetric solids, however, and crucially for *all* surfaces, inversion symmetry is absent and static non-collinear spin configurations may result. Compared with the LS coupling, this effect is usually rather small, but it *can* become important when thin ferromagnetic films are deposited upon otherwise non-ferromagnetic substrates. We shall discuss such cases at the end of this chapter.

9.2 Bulk Magnetocrystalline Anisotropy

Working towards a quantification of the MCA, let us first review a number of key terms relating to the magnetisation of solids, as summarised in Fig. 9.1. Firstly, we note that the application of a magnetising field, **H**, to an initially non-magnetised sample (obtained, e.g. by annealing in a field-free region) produces a magnetisation, **M**, that increases monotonically but not linearly. At sufficiently high field, the magnetisation saturates at a value M_s, known as the **saturation magnetisation**. If the field is gradually removed, the magnetisation decreases but does not ultimately vanish, reaching a value at zero field known as the **remanent magnetisation**, labelled M_r. Only upon applying a magnetising field opposed to the original does the magnetisation decrease further, reaching zero when the field reaches a value of H_c, known as the **coercive field**. Beyond this point, the sample eventually achieves saturation in the direction opposed to the original, and clearly the system can be driven around the resulting hysteresis loop *ad infinitum*. The width of the hysteresis loop is indicative of the coercive field necessary to demagnetise a previously magnetised sample, while its height indicates the degree of spin-polarisation that may be achieved at the highest applied fields. If spin and orbital contributions were entirely decoupled, the magnetic properties of bulk solids would be very nearly isotropic – the coercive field would be essentially independent of the axis along which it is measured, as would the saturation and remanent magnetisations.[2] As it happens, however, spin–orbit coupling ensures that there will exist one or more axes along which the coercive field is

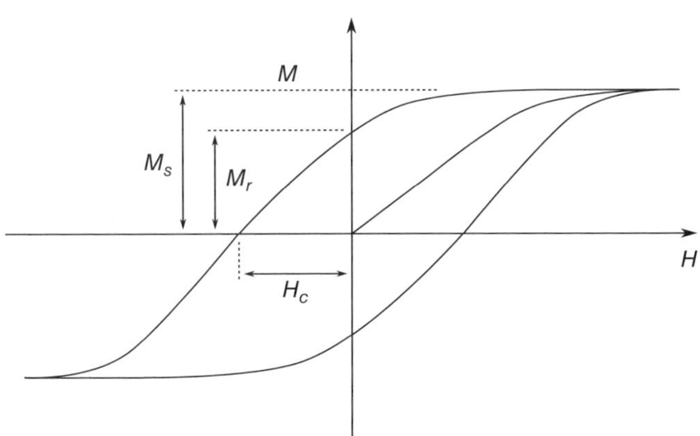

Figure 9.1 Schematic of magnetic hysteresis, showing saturation magnetisation (M_s), remanent magnetisation (M_r) and coercive field (H_c).

2 Long-range interactions between spin magnetic moments contribute a small degree of anisotropy dependent upon the macroscopic shape of the sample, but this will be neglected here.

relatively small (the **easy axes**) and one or more along which it is relatively large (the **hard axes**).

An alternative (but compatible) definition of the easy and hard axes references the energy per unit volume (E/V) required to magnetise a sample from a demagnetised state to saturation in a given direction – the **magnetic energy density**. In accordance with Neumann's principle, this property must display *at least* the symmetry of the point group (of the crystal's space group) so we might expect to be able to generate an expansion in powers of a small number of basis functions compatible with the same. In fact, macroscopic crystalline properties can *always* be expressed in the form

$$\sum_{i,j=0}^{\infty} K_{ij} s^i p^j + \sum_{k}^{3} \sum_{i,j=0}^{\infty} q_k K_{ijk} s^i p^j \tag{9.26}$$

where s, p and q_k are themselves functions of direction cosines (α_1, α_2 and α_3) taken relative to a set of Cartesian axes oriented to maximise their symmetry with respect to the lattice. Fortunately, Döring has tabulated just such a set of functions for each of the 32 crystallographic point groups [3].

Furthermore, when describing the MCA, we should restrict ourselves to only those terms of even overall order (e.g. α_1^2, $\alpha_1\alpha_2$, etc.) as a result of time-reversal symmetry.[3] As a result, the MCA of many crystals displays rather more symmetry than is dictated purely by their point group – specifically, the MCA will necessarily display inversion symmetry (and any other symmetries that this symmetry may imply) even if the point group does not. Of the 32 crystallographic point groups, only 11 contain inversion symmetry (the centrosymmetric point groups, mentioned in Section 8.4). Addition of inversion symmetry to any one of the other crystallographic point groups will result in its transformation into one of the centrosymmetric point groups. Thus, one may describe 11 so-called **Laue classes**, each consisting of one centrosymmetric point group (after which the class is named) together with the set of non-centrosymmetric point groups that would transform into it upon addition of inversion symmetry. The MCA for crystals conforming to *any* of the crystallographic point groups within a given Laue class will consequently itself display the full symmetry of the *one* centrosymmetric point group within that class. It suffices, therefore, to consider Eq. (9.26) explicitly only for the centrosymmetric point groups, confident that the same result would necessarily be obtained for all other point groups within the same Laue class.[4]

---

3 Reversal of the arrow of time results in a reversal of the spin direction, and hence the direction of magnetisation. Since the time-reversal operation must be a symmetry of the system, it follows that reversal of the sense of magnetisation must correspondingly be a symmetry of the system. Terms of odd overall order in the direction cosines (e.g. α_1, $\alpha_1^2\alpha_2$, etc.) would fail to satisfy this condition. It should, perhaps, be stressed that the DM interaction breaks time-reversal symmetry in systems without spatial inversion symmetry, but in this and the following section we consider only the effect of LS coupling.

4 Direct substitution of Döring's tabulated s, p and q_k functions into Eq. (9.26) for non-centrosymmetric point groups will sometimes lead to expressions that are not obviously identical with that obtained for the centrosymmetric member of the appropriate Laue class, but it will always be possible to demonstrate their equivalence by invoking the identity $\alpha_1^2 + \alpha_2^2 + \alpha_3^2 \equiv 1$.

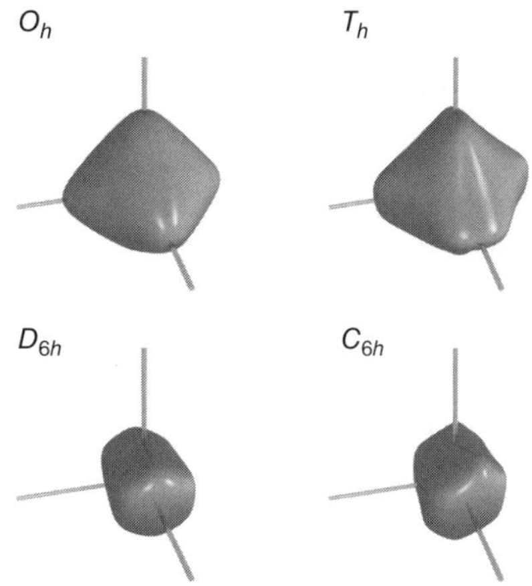

Figure 9.2 Magnetocrystalline anisotropy visualised in the four highest symmetry Laue classes. In each case, an isosurface of constant E/V is shown, together with a set of Cartesian axes; coefficients, K_n, defined in the text, have been chosen to emphasise the relevant symmetries. Magnetically hard and easy axes correspond respectively to directions in which the isosurface lies furthest or closest to the origin. Similar surfaces for the other seven Laue classes show progressively lower symmetry and are not visualised here.

9.2.1 Laue Class O_h (Cubic Crystal System: O_h, T_d and O)

In these cases, the direction cosines are measured relative to Cartesian axes aligned so that $\hat{\mathbf{x}} \equiv [100]$, $\hat{\mathbf{y}} \equiv [010]$ and $\hat{\mathbf{z}} \equiv [001]$. The functions appearing in Eq. (9.26) for the centrosymmetric O_h point group are $s = \alpha_1^2\alpha_2^2 + \alpha_2^2\alpha_3^2 + \alpha_3^2\alpha_1^2$ and $p = \alpha_1^2\alpha_2^2\alpha_3^2$ with $q_k = 0$ for all k (tabulated by Döring). Expanding to sixth order (and retaining only even-order terms) we have

$$E/V = K_0 + K_1(\alpha_1^2\alpha_2^2 + \alpha_2^2\alpha_3^2 + \alpha_3^2\alpha_1^2) + K_2\alpha_1^2\alpha_2^2\alpha_3^2 \tag{9.27}$$

which must hold for all three point groups within this Laue class. The actual direction of the easy and hard axes depends, however, upon the relative values of the K_1 and K_2 coefficients, and one representative example is depicted in Fig. 9.2.

9.2.2 Laue Class T_h (Cubic Crystal system: T_h and T)

Once again, the direction cosines for these crystallographic point groups are measured relative to Cartesian axes aligned so that $\hat{\mathbf{x}} \equiv [100]$, $\hat{\mathbf{y}} \equiv [010]$ and $\hat{\mathbf{z}} \equiv [001]$. The functions appearing in Eq. (9.26) for the centrosymmetric T_h point group are $s = \alpha_1^2\alpha_2^2 + \alpha_2^2\alpha_3^2 + \alpha_3^2\alpha_1^2$ and $p = \alpha_1^2\alpha_2^2\alpha_3^2$ with $q_1 = \alpha_1^4(\alpha_2^2 - \alpha_3^2) + \alpha_2^4(\alpha_3^2 - \alpha_1^2) + \alpha_3^4(\alpha_1^2 - \alpha_2^2)$ and $q_2 = q_3 = 0$. Expanding to sixth order we have

$$\begin{aligned} E/V = {} & K_0 + K_1(\alpha_1^2\alpha_2^2 + \alpha_2^2\alpha_3^2 + \alpha_3^2\alpha_1^2) + K_2\alpha_1^2\alpha_2^2\alpha_3^2 \\ & + K_3(\alpha_1^4(\alpha_2^2 - \alpha_3^2) + \alpha_2^4(\alpha_3^2 - \alpha_1^2) + \alpha_3^4(\alpha_1^2 - \alpha_2^2)) \end{aligned} \tag{9.28}$$

which must hold for both point groups within this Laue class. The final term lowers the symmetry relative to the other cubic Laue class (described above). One representative example is depicted in Fig. 9.2.

9.2.3 Laue Class D_{6h} (Hexagonal Crystal System: D_{6h}, D_{3h}, C_{6v} and D_6)

For these crystallographic point groups, the direction cosines must be measured relative to Cartesian axes aligned so that $\hat{\mathbf{x}} \equiv [2\bar{1}\bar{1}0]$, $\hat{\mathbf{y}} \equiv [01\bar{1}0]$ and $\hat{\mathbf{z}} \equiv [0001]$. The latter vector therefore lies along the sixfold rotational axis of the hexagonal lattice. Döring tabulates the functions appearing in Eq. (9.26) as $s = \alpha_3^2$ and

$p = \alpha_1^6 - 15\alpha_1^4\alpha_2^2 + 15\alpha_1^2\alpha_2^4 - \alpha_2^6$ for the D_{6h} point group (with $q_k = 0$ for all k) yielding

$$E/V = K_0 + K_1\alpha_3^2 + K_2\alpha_3^4 + K_3\alpha_3^6 + K_4(\alpha_1^6 - 15\alpha_1^4\alpha_2^2 + 15\alpha_1^2\alpha_2^4 - \alpha_2^6) \quad (9.29)$$

when expanded to sixth order. Again, this expression must hold for all four of the point groups within this Laue class.

For crystals in the hexagonal, trigonal, tetragonal or monoclinic systems, it is customary to recast the anisotropy in spherical polar coordinates. In the present case, one obtains

$$E/V = K_0 + K_1\cos^2\theta + K_2\cos^4\theta + K_3\cos^6\theta + K_4\sin^6\theta\cos 6\phi \quad (9.30)$$

by making the substitution $\alpha_1 \equiv \sin\theta\cos\phi$, $\alpha_2 \equiv \sin\theta\sin\phi$ and $\alpha_3 \equiv \cos\theta$, with θ and ϕ being, respectively, the polar angle relative to $\hat{\mathbf{z}}$ and the azimuthal angle relative to $\hat{\mathbf{x}}$. As above, one representative example of the anisotropies obtained with different relative values of these coefficients is depicted in Fig. 9.2.

9.2.4 Laue Class C_{6h} (Hexagonal Crystal System: C_{6h}, C_{3h} and C_6)

Once again, the direction cosines for these crystallographic point groups are to be measured relative to Cartesian axes aligned so that $\hat{\mathbf{x}} \equiv [2\bar{1}\bar{1}0]$, $\hat{\mathbf{y}} \equiv [01\bar{1}0]$ and $\hat{\mathbf{z}} \equiv [0001]$. The functions appearing in Eq. (9.26) for the D_{6h} point group are now $s = \alpha_3^2$ and $p = \alpha_1^6 - 15\alpha_1^4\alpha_2^2 + 15\alpha_1^2\alpha_2^4 - \alpha_2^6$ with $q_1 = \alpha_1\alpha_2(3\alpha_1^4 - 10\alpha_1^2\alpha_2^2 + 3\alpha_2^4)$ and $q_2 = q_3 = 0$, yielding

$$E/V = K_0 + K_1\alpha_3^2 + K_2\alpha_3^4 + K_3\alpha_3^6 + K_4(\alpha_1^6 - 15\alpha_1^4\alpha_2^2 + 15\alpha_1^2\alpha_2^4 - \alpha_2^6) + K_5\alpha_1\alpha_2(3\alpha_1^4 - 10\alpha_1^2\alpha_2^2 + 3\alpha_2^4) \quad (9.31)$$

when expanded to sixth order. As always, this expression must hold for all three of the point groups within this Laue class. In spherical polar coordinates, we then have

$$E/V = K_0 + K_1\cos^2\theta + K_2\cos^4\theta + K_3\cos^6\theta + K_4\sin^6\theta\cos 6\phi + K_5\sin 3\phi\cos 3\phi \quad (9.32)$$

in which the final term lowers the symmetry relative to the other hexagonal Laue class. One representative example of the anisotropies obtained with different relative values of these coefficients is depicted in Fig. 9.2.

9.2.5 Laue Class D_{3d} (Trigonal Crystal System: D_{3d}, C_{3v} and D_3)

These crystallographic point groups can be found in crystals having either hexagonal or rhombohedral Bravais lattices. In the latter case, however, it is common to reference directions with respect to the crystallographic axes of a non-primitive (i.e. conventional) hexagonal unit cell. We therefore define our Cartesian directions with respect to the hexagonal lattice in *both* cases. That is, we take $\hat{\mathbf{x}} \equiv [2\bar{1}\bar{1}0]$, $\hat{\mathbf{y}} \equiv [01\bar{1}0]$ and $\hat{\mathbf{z}} \equiv [0001]$, and define the direction cosines α_1, α_2 and

α_3 accordingly (as per the solely hexagonal cases above). Döring indicates $s = \alpha_3^2$ and $p = \alpha_1^6 - 15\alpha_1^4\alpha_2^2 + 15\alpha_1^2\alpha_2^4 - \alpha_2^6$ with $q_1 = \alpha_3(\alpha_1^3 - 3\alpha_1\alpha_2^2)$ and $q_2 = q_3 = 0$, so to sixth order we get

$$E/V = K_0 + K_1\alpha_3^2 + K_2\alpha_3^4 + K_3\alpha_3(\alpha_1^3 - 3\alpha_1\alpha_2^2) + K_4\alpha_3^6 + K_5(\alpha_1^6 - 15\alpha_1^4\alpha_2^2 + 15\alpha_1^2\alpha_2^4 - \alpha_2^6) + K_6\alpha_3^3(\alpha_1^3 - 3\alpha_1\alpha_2^2) \quad (9.33)$$

which must hold for all three point groups in this Laue class. Transforming to spherical polar coordinates, we have

$$E/V = K_0 + K_1\cos^2\theta + K_2\cos^4\theta + K_3\cos\theta\sin^3\theta\cos 3\phi + K_4\cos^6\theta + K_5\sin^6\theta\cos 6\phi + K_6\cos^3\theta\sin^3\theta\cos 3\phi \quad (9.34)$$

in which the K_3 and K_6 terms are evidently responsible for breaking the sixfold rotational symmetries found in the hexagonal Laue classes discussed above.

9.2.6 Laue Class S_6 (Trigonal Crystal System: S_6 and C_3)

Once again, we define our Cartesian directions with respect to the hexagonal unit cell (whether primitive or non-primitive) taking $\hat{\mathbf{x}} \equiv [2\bar{1}\bar{1}0]$, $\hat{\mathbf{y}} \equiv [01\bar{1}0]$ and $\hat{\mathbf{z}} \equiv [0001]$, and define the direction cosines α_1, α_2 and α_3 accordingly. The relevant functions for the S_6 group are $s = \alpha_3^2$ and $p = \alpha_1^6 - 15\alpha_1^4\alpha_2^2 + 15\alpha_1^2\alpha_2^4 - \alpha_2^6$ with $q_1 = \alpha_3(\alpha_1^3 - 3\alpha_1\alpha_2^2)$, $q_2 = \alpha_3(\alpha_1^3 - 3\alpha_1\alpha_2^2)$ and $q_3 = q_1q_2$. Expanding to sixth order, we obtain

$$E/V = K_0 + K_1\alpha_3^2 + K_2\alpha_3^4 + K_3\alpha_3(\alpha_1^3 - 3\alpha_1\alpha_2^2) + K_4\alpha_3(\alpha_2^3 - 3\alpha_1^2\alpha_2) + K_5\alpha_3^6 + K_6(\alpha_1^6 - 15\alpha_1^4\alpha_2^2 + 15\alpha_1^2\alpha_2^4 - \alpha_2^6) + K_7\alpha_3^3(\alpha_1^3 - 3\alpha_1\alpha_2^2) + K_8\alpha_3^3(\alpha_2^3 - 3\alpha_1^2\alpha_2) + K_9(\alpha_1^3 - 3\alpha_1\alpha_2^2)(\alpha_2^3 - 3\alpha_1^2\alpha_2) \quad (9.35)$$

which must hold for both point groups in this Laue class. Transforming to spherical polar coordinates, this becomes

$$E/V = K_0 + K_1\cos^2\theta + K_2\cos^4\theta + K_3\cos\theta\sin^3\theta\cos 3\phi + K_4\cos\theta\sin^3\theta\sin 3\phi + K_5\cos^6\theta + K_6\sin^6\theta\cos 6\phi + K_7\cos^3\theta\sin^3\theta\cos 3\phi + K_8\cos^3\theta\sin^3\theta\sin 3\phi + K_9\sin^6\theta\sin 3\phi\cos 3\phi \quad (9.36)$$

in which the K_4, K_8 and K_9 terms lower the symmetry relative to the other trigonal Laue class.

9.2.7 Laue Class D_{4h} (Tetragonal Crystal System: D_{4h}, D_{2d}, C_{4v} and D_4)

For these crystallographic point groups, we will define our Cartesian vectors parallel to the sides of the tetragonal unit cell, such that $\hat{\mathbf{x}} \equiv [100]$, $\hat{\mathbf{y}} \equiv [010]$ and $\hat{\mathbf{z}} \equiv [001]$ (with the latter corresponding to the direction of the fourfold rotational axis of the lattice). For the D_{4h} point group, the relevant functions of the corresponding direction cosines are $s = \alpha_3^2$ and $p = \alpha_1^2\alpha_2^2$ with $q_k = 0$ for all k. Expanding to fourth order[5] we obtain

$$E/V = K_0 + K_1\alpha_3^2 + K_2\alpha_3^4 + K_3\alpha_1^2\alpha_2^2 \tag{9.37}$$

which (as ever) must hold for all four point groups within this Laue class.

Transforming to spherical polar coordinates is not quite so straightforward in this case, but after some simplification we find

$$E/V = K_0' + K_1'\cos^2\theta + K_2'\cos^4\theta + K_3'\sin^4\theta\cos 4\phi \tag{9.38}$$

with

$$\begin{aligned}K_0' &= K_0 + K_3/8\\ K_1' &= K_1 - K_3/4\\ K_2' &= K_2 + K_3/8\\ K_3' &= -K_3/8\end{aligned} \tag{9.39}$$

in which fourfold rotational symmetry is evident in the K_3' term.

9.2.8 Laue Class C_{4h} (Tetragonal Crystal System: C_{4h}, S_4, & C_4)

Once again, we will define our Cartesian vectors parallel to the sides of the tetragonal unit cell, such that $\hat{\mathbf{x}} \equiv [100]$, $\hat{\mathbf{y}} \equiv [010]$ and $\hat{\mathbf{z}} \equiv [001]$. For the C_{4h} point group, the relevant functions of the corresponding direction cosines are $s = \alpha_3^2$ and $p = \alpha_1^2\alpha_2^2$ with $q_1 = \alpha_1\alpha_2(\alpha_1^2 - \alpha_2^2)$ and $q_2 = q_3 = 0$. Expanding to fourth order, we obtain

$$E/V = K_0 + K_1\alpha_3^2 + K_2\alpha_3^4 + K_3\alpha_1^2\alpha_2^2 + K_4\alpha_1\alpha_2(\alpha_1^2 - \alpha_2^2) \tag{9.40}$$

which holds true for all three point groups within this Laue class. Upon transforming to spherical polar coordinates, we find

$$E/V = K_0' + K_1'\cos^2\theta + K_2'\cos^4\theta + K_3'\sin^4\theta\cos 4\phi + K_4'\sin^4\theta\sin 2\phi\cos 2\phi \tag{9.41}$$

with

$$\begin{aligned}K_0' &= K_0 + K_3/8\\ K_1' &= K_1 - K_3/4\\ K_2' &= K_2 + K_3/8\\ K_3' &= -K_3/8\\ K_4' &= K_4/2\end{aligned} \tag{9.42}$$

in which the final term is responsible for lowering the symmetry relative to the other tetragonal Laue class.

9.2.9 Laue Class D_{2h} (Orthorhombic Crystal System: D_{2h}, C_{2v}, and D_2)

Here, the Cartesian axes must again be aligned with the sides of the primitive unit cell (orthorhombic in this case) so we have $\hat{\mathbf{x}} \equiv [100]$, $\hat{\mathbf{y}} \equiv [010]$ and

5 Sixth order becomes indulgent for lower symmetry systems.

$\hat{\mathbf{z}} \equiv [001]$ – all three vectors now lie along twofold rotational axes of the lattice. Using direction cosines defined with respect to these axes, we have, for the D_{2h} case, $s = \alpha_1^2$ and $p = \alpha_2^2$ with $q_k = 0$ for all k, and the fourth-order MCA hence takes the form

$$E/V = K_0 + K_1\alpha_1^2 + K_2\alpha_2^2 + K_3\alpha_1^4 + K_4\alpha_2^4 + K_5\alpha_1^2\alpha_2^2 \tag{9.43}$$

which at first glance *appears* to privilege the $\hat{\mathbf{x}}$ and $\hat{\mathbf{y}}$ directions over the $\hat{\mathbf{z}}$ direction (i.e. the equation features α_1 and α_2, but not α_3). This seems inappropriate, given that neither the D_{2h} point group nor the D_2 point group favours any one axis over any other. Using $\alpha_1^2 + \alpha_2^2 + \alpha_3^2 \equiv 1$, however, we may write expressions of equivalent form to Eq. (9.43) featuring either the pairing of α_2 with α_3 or of α_3 with α_1, simply by suitable redefinitions of the anisotropy parameters. Indeed, it is even possible to write

$$\begin{aligned} E/V = K_0' &+ K_1'\alpha_1^2 + K_2'\alpha_2^2 + K_3'\alpha_3^2 \\ &+ K_4'\alpha_1^4 + K_5'\alpha_2^4 + K_6'\alpha_3^2 \\ &+ K_6'\alpha_1^2\alpha_2^2 + K_8'\alpha_2^2\alpha_3^2 + K_9'\alpha_3^2\alpha_1^2 \end{aligned} \tag{9.44}$$

with

$$\begin{aligned} K_0' &= K_0 + (K_3 + K_4)/3 \\ K_1' &= (2K_1 - K_2 - 2K_4)/3 \\ K_2' &= (2K_2 - K_1 - 2K_3)/3 \\ K_3' &= -(K_1 + K_2 + 2(K_3 + K_4))/3 \\ K_4' &= (2K_3 + K_4)/3 \\ K_5' &= (2K_4 + K_3)/3 \\ K_6' &= (K_3 + K_4)/3 \\ K_7' &= K_5 \\ K_8' &= 2K_3/3 \\ K_9' &= 2K_4/3 \end{aligned} \tag{9.45}$$

which emphasises the full orthorhombic symmetry. Note, however, that the primed coefficients are necessarily subject to four linear constraints, which may be expressed as

$$\begin{aligned} K_1' + K_2' + K_3' + 2(K_8' + K_9') &= 0 \\ 3K_8' + 2K_9' - K_4' &= 0 \\ 3K_9' + 2K_8' - K_5' &= 0 \\ K_6' - (K_8' + K_9') &= 0 \end{aligned} \tag{9.46}$$

and together these ensure that there remain only six truly independent coefficients. Needless to say, the MCA must take the forms dictated by Eqs. (9.43) and (9.44) not only for the D_{2h} and D_2 point groups but also for the C_{2v} point group, since all belong to the same Laue class.

Transforming to spherical polar coordinates immediately prohibits a representation that explicitly reflects the full symmetry of the D_{2h} point group, so we must arbitrarily choose one axis from which to measure the polar angle and another from which to measure the azimuthal angle. Taking these to be the $\hat{\mathbf{z}}$ and $\hat{\mathbf{x}}$ axes respectively (as we have done for the various hexagonal, trigonal and tetragonal cases) we obtain

$$\begin{aligned} E/V = K_0'' + K_1''\cos^2\theta + K_2''\cos^4\theta \\ + K_3''\sin^2\theta\cos 2\phi + K_4''\sin^4\theta\cos 2\phi \\ + K_5''\sin^4\theta\cos 4\phi \end{aligned} \tag{9.47}$$

with

$$\begin{aligned} K_0'' &= (8K_0' + 4(K_1' + K_2') + 3(K_4' + K_5') + K_7')/8 \\ K_1'' &= (4K_3' - 2(K_1' + K_2') - 3(K_4' + K_5') - K_7' + 2(K_8' + K_9'))/4 \\ K_2'' &= (3(K_4' + K_5') + K_7' + 8K_6' - 4(K_8' + K_9'))/8 \\ K_3'' &= (K_1' - K_2' + K_9' - K_8')/2 \\ K_4'' &= (K_4' - K_5' + K_8' - K_9')/2 \\ K_5'' &= (K_4' + K_5' - K_9')/8 \end{aligned} \tag{9.48}$$

making the link to the symmetrised form expressed in Eq. (9.44). The presence of a twofold rotational axis oriented in the $\hat{\mathbf{z}}$ direction is immediately evident from the K_3'' and K_4'' terms – that of two further such axes parallel to $\hat{\mathbf{x}}$ and $\hat{\mathbf{y}}$ much less so, but they do exist.

9.2.10 Laue Class C_{2h} (Monoclinic Crystal System: C_{2h}, C_{1h}, and C_2)

In the monoclinic case, just one of the three lattice vectors spanning the highest symmetry primitive unit cell lies at right angles to the other two, and we shall choose to align the $\hat{\mathbf{z}}$ axis of our Cartesian coordinate system with this direction. The $\hat{\mathbf{x}}$ axis may then be chosen freely, within the plane perpendicular to $\hat{\mathbf{z}}$, but once this choice is made, the direction of $\hat{\mathbf{y}}$ is, of course, fully constrained. Measuring direction cosines relative to these axes, Döring provides for the C_{2h} point group the functions $s = \alpha_1$ and $p = \alpha_2$ with $q_k = 0$ for all k, so to fourth order one obtains the expression

$$\begin{aligned} E/V = K_0 + K_1\alpha_1^2 + K_2\alpha_2^2 + K_3\alpha_1\alpha_2 + K_4\alpha_1^4 + K_5\alpha_2^4 + K_6\alpha_1\alpha_2^3 \\ + K_7\alpha_1^2\alpha_2^2 + K_8\alpha_1^3\alpha_2 \end{aligned} \tag{9.49}$$

for the MCA, which must apply to all three point groups belonging to this Laue class. Transforming to spherical polar coordinates, we then have

$$\begin{aligned} E/V = K_0' + K_1'\cos^2\theta + K_2'\cos^4\theta \\ + K_3'\sin^2\theta\cos 2\phi + K_4'\sin^2\theta\sin 2\phi \\ + K_5'\sin^4\theta\cos 2\phi + K_6'\sin^4\theta\sin 2\phi \\ + K_7'\sin^4\theta\cos 4\phi + K_8'\sin^4\theta\sin 4\phi \end{aligned} \tag{9.50}$$

with

$$
\begin{aligned}
K_0' &= (8K_0 + 4(K_1 + K_2) + 3(K_4 + K_5) + K_7)/8 \\
K_1' &= -(2(K_1 + K_2) + 3(K_4 + K_5) + K_7)/4 \\
K_2' &= (3(K_4 + K_5) + K_7)/8 \\
K_3' &= (K_1 - K_2)/2 \\
K_4' &= K_3/2 \\
K_5' &= (K_4 - K_5)/2 \\
K_6' &= (K_6 + K_8)/4 \\
K_7' &= (K_4 + K_5 - K_7)/8 \\
K_8' &= (K_8 - K_6)/8
\end{aligned}
\tag{9.51}
$$

which displays the full symmetry of the C_{2h} point group, as expected.

9.2.11 Laue Class S_2 (Triclinic Crystal System: S_2 and C_1)

For the two crystallographic point groups of the triclinic crystal system, there is really no strong argument in favour of any particular alignment of the Cartesian axes needed to specify the MCA. With an eye towards the surface case, however, it will be useful to adopt the convention that $\hat{\mathbf{x}}$ and $\hat{\mathbf{y}}$ lie within the plane defined by two of the primitive triclinic lattice vectors (let us say $\mathbf{a}_1$ and $\mathbf{a}_2$) so that $\hat{\mathbf{z}} \cdot (\mathbf{a}_1 \times \mathbf{a}_2) \geq 0$. The necessary functions for expanding the MCA in the S_2 point group are then $s = \alpha_1^2$ and $p = \alpha_2^2$ with $q_1 = \alpha_1\alpha_2$, $q_2 = \alpha_2\alpha_3$ and $q_3 = \alpha_3\alpha_1$. To fourth order, we obtain

$$
\begin{aligned}
E/V = K_0 &+ K_1\alpha_1^2 + K_2\alpha_2^2 + K_3\alpha_1\alpha_2 + K_4\alpha_2\alpha_3 + K_5\alpha_3\alpha_1 \\
&+ K_6\alpha_1^4 + K_7\alpha_2^4 + K_8\alpha_1^2\alpha_2^2 + K_9\alpha_1^3\alpha_2 + K_{10}\alpha_1\alpha_2^3 \\
&+ K_{11}\alpha_1^2\alpha_2\alpha_3 + K_{12}\alpha_1\alpha_2^2\alpha_3 + K_{13}\alpha_1^3\alpha_3 + K_{14}\alpha_2^3\alpha_3
\end{aligned}
\tag{9.52}
$$

which must apply to both point groups within this Laue class. Converting to spherical polar coordinates, we find

$$
\begin{aligned}
E/V = K_0' &+ K_1'\cos^2\theta + K_2'\cos^4\theta + K_3'\sin\theta\cos\theta\sin\phi \\
&+ K_4'\sin\theta\cos\theta\cos\phi \\
&+ K_5'\sin^3\theta\cos\theta\sin\phi + K_6'\sin^3\theta\cos\theta\cos\phi \\
&+ K_7'\sin^2\theta\sin 2\phi + K_8'\sin^2\theta\cos 2\phi \\
&+ K_9'\sin^4\theta\sin 2\phi + K_{10}'\sin^4\theta\cos 2\phi \\
&+ K_{11}'\sin^3\theta\cos\theta\sin\phi\cos 2\phi + K_{12}'\sin^3\theta\cos\theta\cos\phi\cos 2\phi \\
&+ K_{13}'\sin^4\theta\sin 2\phi\cos 2\phi + K_{14}'\sin^4\theta\cos 4\phi
\end{aligned}
\tag{9.53}
$$

with

$$\begin{aligned}
K'_0 &= (8K_0 + 4(K_1 + K_2) + 3(K_6 + K_7) + K_8)/8 \\
K'_1 &= -(2(K_1 + K_2) + 3(K_6 + K_7) + K_8)/4 \\
K'_2 &= (3(K_6 + K_7) + K_8)/8 \\
K'_3 &= K_4 \\
K'_4 &= K_5 \\
K'_5 &= (K_{11} + K_{14})/2 \\
K'_6 &= (K_{12} + K_{13})/2 \\
K'_7 &= K_3/2 \\
K'_8 &= (K_1 - K_2)/2 \\
K'_9 &= (K_9 + K_{10})/4 \\
K'_{10} &= (K_6 - K_7)/2 \\
K'_{11} &= (K_{11} - K_{14})/2 \\
K'_{12} &= (K_{13} - K_{12})/2 \\
K'_{13} &= (K_9 - K_{10})/2 \\
K'_{14} &= K_6 + K_7 - K_8
\end{aligned} \tag{9.54}$$

which displays no symmetry other than inversion (present by default in the S_2 point group, but added to the MCA for the C_1 point group as a consequence of time-reversal symmetry).

9.2.12 General Comments

The expressions enumerated above provide an idealised description of the MCA in each of the 32 crystallographic point groups, but it is important to mention that the values of the various anisotropy parameters may often prove to be negligible (i.e. too small to measure). In such cases, the observed MCA might display even higher symmetry than that suggested by the relevant expansion, but it is never possible for it to show less.[6] Within the hexagonal, trigonal, tetragonal and monoclinic crystal systems, for example, the MCA is often found to display nearly cylindrical symmetry.

In addition, it ought also to be stressed that the anisotropy parameters are not, strictly speaking, constants of the material considered. They certainly vary as a function of temperature, and one would expect that they ought also to vary with strain and other perturbations, affecting the precise nature of the bulk MCA. It is quite possible, therefore, for the direction (and even the number) of bulk

6 It *is*, of course, possible for a crystal nominally conforming to one crystallographic point group to distort upon magnetisation to adopt a lower symmetry crystallographic point group. For instance, although the structure of iron is *bcc* (O_h) it distorts upon magnetisation into a tetragonal structure (D_{4h}). The distortion in this case is extremely slight, however, and customarily ignored.

easy/hard axes to change with context. It is equally likely that the same will be true of the surface.

9.3 Surface Magnetocrystalline Anisotropy

In the preceding section, we have seen how the symmetry of the bulk material, coupled with time-reversal symmetry, dictates the symmetry of the bulk MCA. Even for a very highly symmetric material, however, the alignment of the bulk MCA with respect to the surface plane may break some of that symmetry, for a sufficiently asymmetric choice of surface orientation. Furthermore, the MCA in the region close to the surface is actually dominated by purely surface effects, and in fact is often orders of magnitude more pronounced than in the bulk [4]. It makes sense, therefore, not to begin with the bulk MCA and consider the surface as a perturbation, but instead to directly determine the form of the surface MCA in its own right. To do this, we shall need first to identify the point group of the surface, and then to make use of the appropriate MCA expression for its Laue class. These expressions will take the same form as those presented above, but it will be necessary to pay careful attention to the orientation of Cartesian axes relative to which the direction cosines and/or spherical polar coordinates are defined.

As mentioned in Section 8.4 it turns out that only 10 of the crystallographic point groups are compatible with crystalline surfaces (C_6, C_4, C_3, C_2, C_1, C_{6v}, C_{4v}, C_{3v}, C_{2v} and C_{1h}) but these span 9 of the Laue classes (D_{6h}, C_{6h}, D_{3d}, S_6, D_{4h}, C_{4h}, D_{3h}, C_{2h} and S_2) so there is scope for considerable diversity in the MCA symmetry at surfaces in general. For the surfaces of any specific material, however, the range will be considerably more limited. Analysis of the manner in which the symmetry of a given point group is reduced upon the imposition of a surface allows the construction of a useful diagram showing which sub-groups may be created from the centrosymmetric point groups whose symmetry defines the Laue classes (Fig. 9.3). To make use of the diagram for a material having a particular bulk point group, one begins by identifying the Laue class to which that bulk point group belongs; the Laue classes reachable from that starting point comprise the full set of Laue classes to which the surface MCA could conceivably belong (depending upon orientation, reconstruction or adsorption). Thus, for example, we see that for *fcc* ferromagnets (point group and Laue class both O_h) the surface MCA may display D_{4h}, D_{3d}, D_{2h}, C_{4h}, C_{2h}, S_6 or S_2 symmetry. For *hcp* ferromagnets (point group and Laue class both D_{6h}) the permitted surface MCA symmetries are D_{6h}, D_{3d}, D_{2h}, C_{6h}, C_{2h}, S_6 and S_2. To take a more exotic example, the structure of MnSi[7] (point group T, Laue class T_h) may display surface MCA symmetries in the D_{2h}, C_{2h}, S_6 and S_2 classes. In all cases, however, the question of which MCA symmetry arises for which surface orientation requires one to systematically examine the various possible surface point groups individually. Let us consider the example of

7 The ferromagnetic state of MnSi is only metastable, relative to a more complex non-collinear magnetic order, but its properties are fascinating [5].

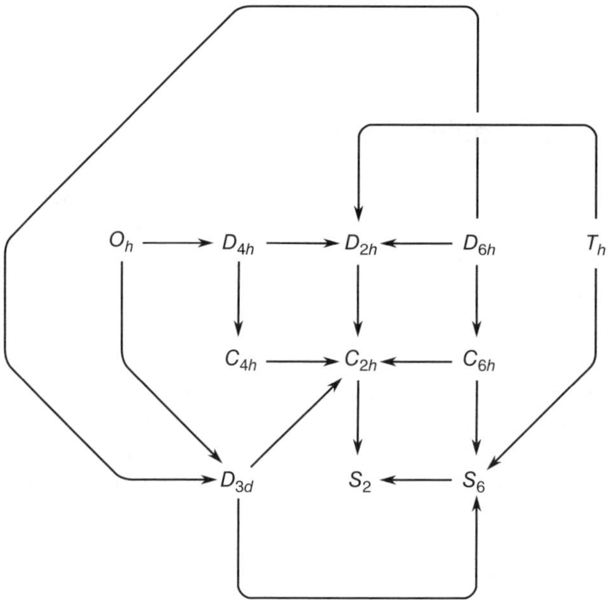

Figure 9.3 Relationships between bulk and surface Laue classes. Starting from the Laue class of the parent bulk material, that of the surface must either be identical (except in the O_h and T_h cases) or reachable by following the arrows (multiple steps are permitted). Precisely which Laue class a particular surface belongs to will depend both upon the point group of the bulk and the orientation of the surface plane.

fcc and *bcc* crystals in more detail, which should suffice to illustrate the necessary logic that must be employed.

9.3.1 Surface MCA of Face-Centred and Body-Centred Cubic Ferromagnets

For both the *fcc* and *bcc* crystal structure, the point group is O_h, and the analysis presented above dictates that the MCA at any one of their surfaces must conform to one of a limited set of possible point groups, namely: D_{4h}, D_{3d}, D_{2h}, C_{4h}, C_{2h}, S_6 or S_2. To determine for which surfaces these actually arise, we need first to ascertain the crystallographic point groups that occur for each type of surface.

The highest order crystallographic point group found amongst the surfaces of *fcc* and *bcc* crystals is C_{4v}, which occurs for surfaces of $\{100\}$ type (the 'quadruply reflexive surfaces' within the stereographic scheme discussed in Chapter 2). The Laue class of the C_{4v} point group is D_{4h}, so the surface MCA will display the tetragonal symmetry appropriate to this class. That is, we must have

$$(E/V)_{100} = K_0 + K_1\cos^2\theta + K_2\cos^4\theta + K_3\sin^4\theta\cos 4\phi \tag{9.55}$$

and furthermore the $\hat{\mathbf{z}}$ direction (from which the polar angle θ is measured) must be aligned with the surface normal, while the $\hat{\mathbf{x}}$ direction (from which the azimuthal angle ϕ is measured) must be aligned with one of the high-symmetry directions of the surface (either of $\langle 110\rangle$ or of $\langle 100\rangle$ type).

After the $\{100\}$ surfaces, the second most symmetric *fcc* and *bcc* surfaces are those of $\{111\}$ type (the 'triply reflexive surfaces') having point group C_{3v}.

This point group belongs to the D_{3d} Laue class, so the surface MCA will display trigonal symmetry of the form

$$(E/V)_{111} = K_0 + K_1\cos^2\theta + K_2\cos^4\theta + K_3\cos\theta\sin^3\theta\cos 3\phi + K_4\cos^6\theta + K_5\sin^6\theta\cos 6\phi + K_6\cos^3\theta\sin^3\theta\cos 3\phi \quad (9.56)$$

with $\hat{\mathbf{z}}$ defined to lie along the surface normal and $\hat{\mathbf{x}}$ directed along a high-symmetry surface axis (either of $\langle 110\rangle$ or of $\langle 211\rangle$ type).

The $\{110\}$ surfaces of *fcc* and *bcc* crystals (the 'doubly reflexive surfaces') conform to the point group C_{2v}, which belongs to the D_{2h} Laue class. Accordingly, the surface MCA will display orthorhombic symmetry, its magnetisation energy expressible as

$$(E/V)_{110} = K_0 + K_1\cos^2\theta + K_2\cos^4\theta + K_3\sin^2\theta\cos 2\phi + K_4\sin^4\theta\cos 2\phi + K_5\sin^4\theta\cos 4\phi \quad (9.57)$$

with $\hat{\mathbf{z}}$ lying normal to the surface and $\hat{\mathbf{x}}$ along a high-symmetry in-plane direction (either of $\langle 110\rangle$ or of $\langle 100\rangle$ type).

The three surfaces described above exhaust those possessing rotational symmetry, but there exists a whole family of surface types having a single mirror plane as their defining symmetry. These are the surfaces whose poles lie on a single mirror zone (the 'reflexive surfaces' of Chapter 2, having either two Miller indices of identical magnitude or one Miller index of zero magnitude) and all exhibit the C_{1h} point group, which belongs to the C_{2h} Laue class. The surface MCA will therefore display monoclinic symmetry, of the form

$$(E/V)_{hhk} \equiv (E/V)_{hk0} = K_0 + K_1\cos^2\theta + K_2\cos^4\theta + K_3\sin^2\theta\cos 2\phi + K_4\sin^2\theta\sin 2\phi + K_5\sin^4\theta\cos 2\phi + K_6\sin^4\theta\sin 2\phi + K_7\sin^4\theta\cos 4\phi + K_8\sin^4\theta\sin 4\phi \quad (9.58)$$

where it is *absolutely crucial* to note that $\hat{\mathbf{z}}$ now must lie within the surface plane, perpendicular to the sole mirror plane; the $\hat{\mathbf{x}}$ direction must then lie within the mirror plane, and one might suggest – but not insist – that parallel to the surface normal would be a sensible choice.

Finally, we turn to the chiral surfaces – those having three non-zero Miller indices of unequal magnitudes – for which the point group is always C_1. The appropriate Laue class is therefore S_2, and the surface MCA must be describable as

$$(E/V)_{hkl} = K_0 + K_1\cos^2\theta + K_2\cos^4\theta + K_3\sin\theta\cos\theta\sin\phi + K_4\sin\theta\cos\theta\cos\phi + K_5\sin^3\theta\cos\theta\sin\phi + K_6\sin^3\theta\cos\theta\cos\phi + K_7\sin^2\theta\sin 2\phi + K_8\sin^2\theta\cos 2\phi + K_9\sin^4\theta\sin 2\phi + K_{10}\sin^4\theta\cos 2\phi + K_{11}\sin^3\theta\cos\theta\sin\phi\cos 2\phi + K_{12}\sin^3\theta\cos\theta\cos\phi\cos 2\phi + K_{13}\sin^4\theta\sin 2\phi\cos 2\phi + K_{14}\sin^4\theta\cos 4\phi \quad (9.59)$$

where the $\hat{\mathbf{x}}$, $\hat{\mathbf{y}}$ and $\hat{\mathbf{z}}$ directions may be chosen without restriction.

Of the seven possible Laue classes that can arise at the surfaces of *fcc* and *bcc* crystals, therefore, we see that only five of these may be found when the surface is clean and unreconstructed; neither the C_{4h} nor the S_6 Laue class is represented.

9.3.2 Role of Adsorbates and Reconstruction

Before moving on from the surface MCA, we ought to consider briefly the role of adsorbates and/or reconstructions in modifying the surface symmetry and hence the symmetry of the anisotropy. Clearly, the adsorption of atoms or molecules cannot add to the symmetry of the surface, and neither can any conceivable reconstruction. It is quite possible, however, for adsorption or reconstruction to lower the symmetry, and this *may* result in a change in the Laue class of the surface. A neat example may be provided by the adsorption of benzene on Ni{111}, where a $(\sqrt{7} \times \sqrt{7})$R30° overlayer is formed at saturation coverage [6]. The rotation of this overlayer with respect to the substrate results in a lowering of the overall point group (i.e. that of the combined system of substrate plus overlayer) from C_{3v} to C_3. This changes the Laue class from D_{3d} to S_6, and one might therefore expect to find the surface MCA conforming to

$$\begin{aligned}(E/V)_{111'} = K_0 + K_1\cos^2\theta + K_2\cos^4\theta + K_3 \cos\theta\sin^3\theta \cos 3\phi \\ + K_4 \cos\theta\sin^3\theta \sin 3\phi + K_5\cos^6\theta + K_6\sin^6\theta \cos 6\phi \\ + K_7\cos^3\theta\sin^3\theta \cos 3\phi + K_8\cos^3\theta\sin^3\theta \sin 3\phi \\ + K_9\sin^6\theta \sin 3\phi \cos 3\phi \end{aligned} \tag{9.60}$$

rather than the higher symmetry form of Eq.(9.56). The situation is not quite so clear-cut, however, as one should expect the formation of domains corresponding both to clockwise *and* to anticlockwise rotation of the overlayer relative to the substrate. Whilst the form expressed in Eq. (9.60) ought indeed to hold true if measurements within a single domain were possible, we should expect measurements that average over multiple domains to include contributions from very nearly equal areas of both types of domain. Moreover, symmetry dictates that the anisotropy parameters K_4, K_8 and K_9 should each change sign between the clockwise and anticlockwise domains. The sum of clockwise and anticlockwise contributions would then yield an expression entirely equivalent to Eq. (9.56). The same ought to be true upon the adsorption of *any* achiral species onto an achiral surface, or upon the reconstruction of a surface in the absence of chiral adsorbates – individual domains may show lowered symmetry in the surface MCA, but averaging over multiple domains restores the apparent symmetry of the substrate.

In contrast, the adsorption of chiral adsorbates carries with it the possibility to lower the symmetry in a global fashion. Adsorption of the chiral molecule L-tartaric acid (as L-bitartrate) onto Ni{110}, for example, must break the mirror symmetry of the surface, changing its point group from C_{2v} to C_2 (if the twofold axis is preserved) or C_1 (if it is not). Although the adsorbate is *not* found to display any long-range order on this surface [7], which would strictly argue for the latter point group, it seems quite likely that the surface MCA will be dominated by sufficiently short-range effects that in this regard the surface should exhibit *approximate* C_2 symmetry. The Laue class accordingly changes from D_{2h} to C_{2h}, and the surface MCA should now adopt the form

$$\begin{aligned}(E/V)_{110'} = {} & K_0 + K_1\cos^2\theta + K_2\cos^4\theta \\ & + K_3\sin^2\theta\cos 2\phi + K_4\sin^2\theta\sin 2\phi \\ & + K_5\sin^4\theta\cos 2\phi + K_6\sin^4\theta\sin 2\phi \\ & + K_7\sin^4\theta\cos 4\phi + K_8\sin^4\theta\sin 4\phi \end{aligned} \tag{9.61}$$

in place of Eq. (9.57).

In summary, adsorbates and/or reconstructions can lower the symmetry sufficiently to change the Laue class, and hence materially alter the symmetry of the surface MCA, but in most cases this is limited to a single domain and the effect is cancelled when multiple domains are considered. The exception to this rule is when chiral molecules are involved, in which case it is possible for a change in symmetry of the surface MCA to take effect globally. It follows, therefore, that measurement of the surface MCA can, in principle, be considered as a probe for the presence of chiral molecules at otherwise achiral surfaces.[8] As ever, the absolute size of any effect, and hence its practical utility, will vary from system to system.

9.4 An Aside: Vectors and Pseudovectors

We have already remarked that the direction of magnetisation in a material changes sign upon reversing the arrow of time. Fundamentally, this is because the magnetic moment of an electron arises from its rotational motion, whether about its own axis in the case of spin, or about an external axis in the case of orbital magnetism. Either way, the reversal of time converts a clockwise motion to an anticlockwise one, and vice versa. A further important consequence of the rotational origin of magnetism, however, relates to the action of improper symmetry operations, such as reflection or inversion. Since these are key to the definition of chirality, it will be advisable to make a few observations at this point.

The first thing to note is the behaviour of what one might describe as 'ordinary' vectors under improper transformations. Unsurprisingly, such vectors reverse their direction upon inversion, and upon reflection across a mirror plane oriented perpendicular to their axis (Fig. 9.4a). They do *not* reverse direction upon reflection across a mirror plane oriented parallel to their axis (Fig. 9.4b). Vectors subject to these transformations under improper symmetry operations are known as **true vectors** or **polar vectors**.

In addition to these true vectors, however, we may define a second type of vector (which we shall describe as a **pseudovector** or **axial vector**) formed by the vector product of two true vectors. The direction of such a vector is determined by the right-hand rule, but this leads to strikingly unusual behaviour under

8 Since inherently chiral surfaces already display anisotropy conforming to the lowest symmetry Laue class, it is clear that the adsorption of chiral molecules cannot reduce the symmetry of the surface MCA any further.

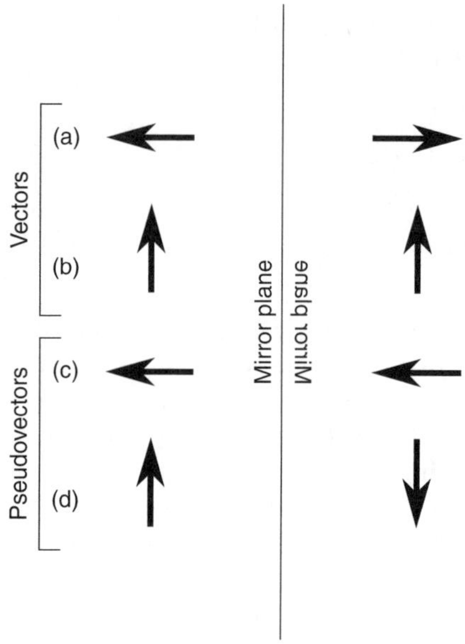

Figure 9.4 Behaviour of vectors and pseudovectors under the operation of mirror symmetry. Examples (a) and (b) show vectors, perpendicular and parallel to a mirror plane, while (c) and (d) show pseudovectors, similarly oriented.

improper symmetry operations. Consider first the result of inversion; since *both* the generating vectors change sign upon inversion, the pseudovector does not. Moreover, reflection in a plane perpendicular to the pseudovector leaves both generating vectors unchanged, so the pseudovector is also unchanged (Fig. 9.4c). Finally, reflection in a plane parallel to the pseudovector will reverse the relative orientation of the generating vectors and hence the direction of the pseudovector (Fig. 9.4d). In short, Cartesian components that would change sign under the action of a given improper symmetry operation upon a true vector will remain unaltered in the case of a pseudovector, while those that would remain unaltered for a true vector will change sign in the case of a pseudovector. An important consequence of this fact is that the action of any improper transformation upon a pseudovector may always be considered to be equivalent to a twofold rotation with a suitably chosen axis.

It will also be useful to note that, just as the vector product of two true vectors is a pseudovector, so too the vector product of two pseudovectors is also a pseudovector; conversely, the vector product of a true vector with a pseudovector is a true vector. On the other hand, the scalar product of two pseudovectors is a true scalar (unchanged under all transformations) just like the scalar product of two vectors; the scalar product of a true vector with a pseudovector, however, is a pseudoscalar (changing sign under improper transformations). It follows from this that any system that may be characterised in some sense by a single constant non-zero pseudoscalar (or by multiple constant non-zero pseudoscalars of the same sign) must necessarily be chiral, since all improper transformations will result in a change of pseudoscalar sign that cannot be reversed by any combination of proper transformations. A system that may only be characterised by multiple pseudoscalars of both signs, or by any number of pseudoscalars having time-varying sign, requires more detailed consideration, since certain proper transformations and/or temporal displacement may interchange the individual signs. Examples of both situations will arise in the following section.

The specific significance of the foregoing discussion to magnetism is twofold. Firstly, the spin and orbital magnetic moments of the electron may both be described as pseudovectors, deriving as they do from the angular momenta of the electron.[9] Secondly, any magnetic field, **B**, may be described as a pseudovector field, as it may be defined via the curl of the vector potential: $\mathbf{B} = \nabla \times \mathbf{A}$; both ∇ and **A** are true vectors, so their vector product will be a pseudovector. In the discussion of chirality that follows, it will be necessary to keep very much in mind the fact that magnetic moments and magnetic fields both behave as pseudovectors under improper transformations. Arrangements of spin that might *appear* to be chiral if one were to treat spin as a vector quantity may sometimes be revealed as achiral when spin is properly treated as a pseudovector, and the converse is, of course, equally true.

9.5 Spin Waves in Centrosymmetric Media

To this point, we have restricted our discussion to cases where all the electronic spin vectors of the system share a common axis. We earlier noted in passing, however, that the DM coupling term (Eq. (9.25)) can stabilise non-collinear spin configurations in non-centrosymmetric systems (e.g. surfaces and thin films). Even in centrosymmetric systems, non-collinear spin can exist as a perturbation, in the form of a travelling spin wave[10] (which may be quantised as a **magnon**). Since these may also exist at the surface, we shall begin by describing these travelling waves, considering the static solutions only later. In centrosymmetric systems, one can categorise travelling spin waves[11] into two types: those where the spin precesses around an axis perpendicular to the wavevector (the **spin spiral**) and those where it precesses around the wavevector itself (the **spin helix**). Here, we investigate the chirality, or otherwise, of these two types in turn.

9.5.1 Spin Helices

Let us begin with the spin helix shown schematically in Fig. 9.5a for an infinite and regularly spaced linear chain (oriented in the $\hat{\mathbf{x}}$ direction) of localised spins. The arrangement of spin sites clearly possesses inversion symmetry (centred on any one of the spin sites) and we assume that nothing in the surrounding medium breaks that symmetry. We can express the local spin orientation at the ith site as

$$\mathbf{S}_i = \hat{\mathbf{x}} \cos\theta + \hat{\mathbf{y}} \sin\theta \cos\phi_i + \hat{\mathbf{z}} \sin\theta \sin\phi_i \tag{9.62}$$

9 The orbital angular momentum is explicitly defined as the vector product of position and momentum operators; the spin angular momentum may be considered, at least conceptually, as the limiting case where the former tends to zero and the latter to infinity.

10 In effect, the wave's direction of travel defines a vector that breaks the inversion symmetry of the medium through which it travels. The DM coupling is subject only to the symmetry of the sub-group formed by removing inversion from the space group of the medium.

11 We exclude, here, the possibility of a spin *density* wave, where the spin is strictly collinear.

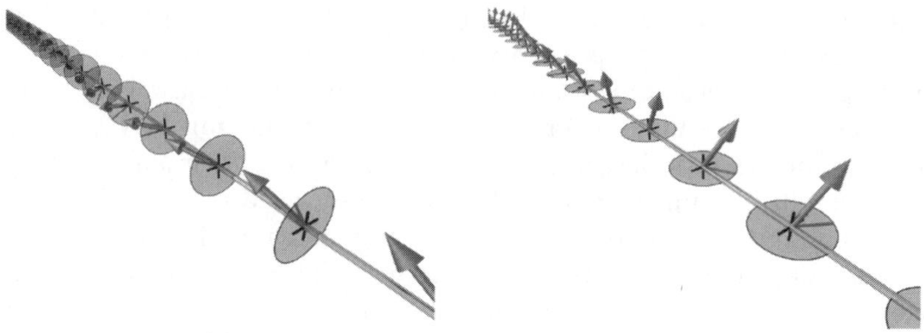

(a) (b)

Figure 9.5 Visualisation of spin waves, showing in (a) the spin helix, and in (b) the spin spiral cases.

where θ (in the interval $0 < \theta \leq \pi/2$) is the constant polar angle between all spin vectors and the $\hat{\mathbf{x}}$ axis (we must explicitly exclude the $\theta = 0$ limit, as this would imply straightforward ferromagnetism). The azimuthal angle ϕ_i varies from site to site and satisfies

$$\phi_i = kx_i - \omega t + \delta \tag{9.63}$$

with k the wavenumber, ω the frequency, δ a constant phase, and t the time. The wavevector, $\mathbf{k}$, is defined as $k\hat{\mathbf{x}}$, and we restrict its magnitude to the interval $-\pi/a \leq k \leq \pi/a$, with a being the one-dimensional lattice spacing between spins. The phase velocity is given by $\mathbf{v} = \mathbf{k}/\omega$ and the wavelength by $\lambda = 2\pi/k$. The special cases $k = 0$ and $\omega = 0$ must each be excluded from consideration at this stage, since these would not correspond to travelling-wave solutions; the special case with $k = \pi/a$ must similarly be excluded, as the direction of travel becomes ambiguous.

In order to determine whether such a wave is chiral, we must ask ourselves whether there exists an improper transformation (reflection or inversion) after which the transformed wave *cannot* be mapped back onto the original by some combination of proper transformations (rotation or translation). Only the failure to find *any* such improper symmetry operation will imply chirality. As per the preceding section, an efficient strategy to answer this question will be to attempt the construction of a pseudoscalar quantity, by the scalar multiplication of a true vector with a pseudovector.

The simplest pseudoscalar imaginable for this system may be constructed by taking the scalar product of the wavevector, $\mathbf{k}$, with the spin vector, $\mathbf{S}_i$. Denoting this $P_1(i, t)$, we find

$$P_1(i,t) \equiv \mathbf{k} \cdot \mathbf{S}_i = \begin{bmatrix} +k\cos\theta \\ -k\cos\theta \end{bmatrix} \tag{9.64}$$

for all i, where the upper solution corresponds to the wave described by Eq. (9.62), and the lower solution to a wave obtained by an improper transformation of that wave. We can, furthermore, define a second, slightly more complicated, pseudovector, which we shall denote $P_2(i, t)$, by taking the scalar product of $\mathbf{k}$ with the vector product $\mathbf{S}_{i+1} \times \mathbf{S}_i$. Evaluating this, we obtain

$$P_2(i,t) \equiv \mathbf{k} \cdot (\mathbf{S}_{i+1} \times \mathbf{S}_i) = \begin{cases} +k \sin^2\theta \sin ka \\ -k \sin^2\theta \sin ka \end{cases} \tag{9.65}$$

for all i, where once again the upper solution corresponds to the wave described by Eq. (9.62), and the lower solution to a wave obtained after an improper transformation. When $P_1(i,t)$ and $P_2(i,t)$ are independent of both i and t, as they are here, we shall refer to them simply as P_1 and P_2.

Now, so long as one or other of these pseudoscalars is non-zero, its sign will be sufficient to distinguish between a travelling spin helix and its mirror image. But P_1 and P_2 can only *both* vanish when we set $k = 0$ or $k = \pi/a$, and these cases we have already excluded as incompatible with a travelling-wave solution. It follows, therefore, that a travelling spin helix must *always* be chiral.

9.5.2 Spin Spirals

Where the spin precesses about a direction perpendicular to the wavevector (see Fig. 9.5b) the local spin orientation may be written in the form

$$\mathbf{S}_i = \hat{\mathbf{x}} \sin\theta \cos\phi_i + \hat{\mathbf{y}} \sin\theta \sin\phi_i + \hat{\mathbf{z}} \cos\theta \tag{9.66}$$

where $\hat{\mathbf{x}}$ is again the direction of travel, but where $\hat{\mathbf{z}}$ is the axis about which the spin precesses.[12] The azimuthal angle, ϕ_i, continues to satisfy Eq. (9.63), with the wavevector and related quantities defined as before. The quantities $P_1(i,t)$ and $P_2(i,t)$ are no longer, however, independent of i and t. In fact, we have

$$P_1(i,t) \equiv \mathbf{k} \cdot \mathbf{S}_i = \begin{cases} +k \sin\theta \cos\phi_i \\ -k \sin\theta \cos\phi_i \end{cases} \tag{9.67}$$

and

$$P_2(i,t) \equiv \mathbf{k} \cdot (\mathbf{S}_{i+1} \times \mathbf{S}_i) = \begin{cases} +k \sin\theta \cos\theta(\sin(\phi_i + ka) - \sin\phi_i) \\ -k \sin\theta \cos\theta(\sin(\phi_i + ka) - \sin\phi_i) \end{cases} \tag{9.68}$$

where, in each case, the upper solution corresponds to the wave described by Eq. (9.66) and the lower to a wave obtained from that via an improper transformation. It is *not* now possible to define a pseudoscalar whose value is constant in time and independent of the spin site at which it is calculated.

Now, it is clearly impossible for all instances of $P_1(i,t)$ and $P_2(i,t)$ to vanish simultaneously, so at first glance it would appear that they collectively provide a means by which the original travelling spin spiral and its mirror image may be distinguished. We must note, however, that the effect of an improper transformation is simply to reverse the sign of each and every instance of $P_1(i,t)$ and $P_2(i,t)$, but precisely the same thing would happen under the transformation $\phi_i \to \phi_i + \pi$, and this in turn merely corresponds to a change in the origin of t by π/ω. Since there is no absolute reference point for the time axis, we conclude that the spin spiral and its mirror image are, in fact, indistinguishable and hence achiral.

12 Clearly, we could interchange $\hat{\mathbf{y}}$ and $\hat{\mathbf{z}}$ without affecting our argument.

9.6 Spin Waves at a Featureless Surface

Having clarified the symmetry of travelling spin waves in centrosymmetric media, we turn now to the non-centrosymmetric case, where static solutions are also possible. It will be instructive to consider first the introduction of an entirely featureless surface to the scenario – we shall consider the effect of crystalline surfaces (of various symmetries) only after this minimal intervention has been fully discussed. For now, we shall represent the surface simply by means of its surface normal vector, $\hat{\mathbf{n}}$, which we shall take to lie in the $\hat{\mathbf{z}}$ direction; the presence of this vector precludes the existence of inversion symmetry, and will allow us to define some additional pseudoscalars in the analysis that follows, altering some (but by no means all) of our conclusions regarding the chirality or otherwise of spin waves.

9.6.1 Spin Helices

Returning to the case of a spin helix, described by Eq. (9.62), the first point to note is that the introduction of a featureless surface changes nothing in the analysis of travelling waves already provided. These will simply remain chiral, as a consequence of the pseudoscalars P_1 and P_2 (Eqs. (9.64) and (9.65)) which are not only independent of both i and t, but which cannot both vanish under the same set of circumstances. All we need do here, therefore, is consider the $k = 0$, $k = \pi/a$ and $\omega = 0$ cases, excluded from our previous discussion of centrosymmetric media but allowed here because the surface has broken the inversion symmetry.

Now, simply setting $\omega = 0$ on its own changes nothing in the previous analysis, since neither P_1 nor P_2 are functions of ω; the wave remains chiral, even when it is static, so long as k equals neither zero nor π/a. The other two special cases require more detailed consideration.

Setting $k = 0$, we find that the pseudoscalars

$$P_3(i,t)|_{k=0} \equiv \hat{\mathbf{n}} \cdot \mathbf{S}_i|_{k=0} = \begin{bmatrix} +\sin\theta\sin(\delta-\omega t) \\ -\sin\theta\sin(\delta-\omega t) \end{bmatrix} \tag{9.69}$$

and

$$P_4(i,t)|_{k=0} \equiv \hat{\mathbf{n}} \cdot (\mathbf{S}_i \times \partial\mathbf{S}_i/\partial\theta)|_{k=0} = \begin{bmatrix} +\cos(\delta-\omega t) \\ -\cos(\delta-\omega t) \end{bmatrix} \tag{9.70}$$

are independent of i (although evidently not of t). As usual, the upper solution in each case corresponds to the original wave defined in Eq. (9.62), and the lower to a wave obtained via an improper transformation of that wave. Given that we explicitly exclude $\theta = 0$ from consideration (as it is straightforwardly ferromagnetic) it is clear that $P_3(i,t)$ and $P_4(i,t)$ cannot be simultaneously zero. They do, however, both vary sinusoidally with t, meaning that the change of sign effected by an improper transformation is formally equivalent to a simple change in the origin of t by π/ω. This implies that the situation is achiral, so long as ω is non-zero. In the $\omega = 0$ limit, on the other hand, no amount of tinkering with the origin of t can mimic the effect of an improper transformation, so the situation becomes chiral once again.

Finally, setting $k = \pi/a$, we find that $P_3(i,t)$ and $P_4(i,t)$ become

$$P_3(i,t)|_{k=\pi/a} \equiv \hat{\mathbf{n}} \cdot \mathbf{S}_i|_{k=\pi/a} = \begin{bmatrix} +\sin\theta \sin(\pi x_i/a + \delta - \omega t) \\ -\sin\theta \sin(\pi x_i/a + \delta - \omega t) \end{bmatrix} \tag{9.71}$$

and

$$P_4(i,t)|_{k=\pi/a} \equiv \hat{\mathbf{n}} \cdot (\mathbf{S}_i \times \partial \mathbf{S}_i/\partial\theta)|_{k=\pi/a} = \begin{bmatrix} +\cos(\pi x_i/a + \delta - \omega t) \\ -\cos(\pi x_i/a + \delta - \omega t) \end{bmatrix} \tag{9.72}$$

which not only vary sinusoidally with t, but also alternate in sign from one spin to the next. That is, we have $P_3(i+1,t) = -P_3(i,t)$ and $P_4(i+1,t) = -P_4(i,t)$, meaning that the change of sign effected by an improper transformation can be replicated by a mere displacement of a along the $\hat{\mathbf{x}}$ axis. The case with $k = \pi/a$ is therefore achiral.

In summary, spin helices (whether static or dynamic) are generally chiral at featureless surfaces. The only exceptions are the cases of (i) a dynamic helix of infinite wavelength and (ii) a dynamic *or* static helix of wavelength equal to twice the lattice spacing between spins.

9.6.2 Spin Spirals

In the presence of a surface normal vector, the spin spiral class bifurcates into two possibilities: one where the axis around which the spin precesses lies along the surface normal and one where it lies perpendicular to the surface normal. We shall describe the former cases as a **vertical spin spiral** and the latter as a **lateral spin spiral**.

In the lateral case with non-zero frequency, the analysis remains unaltered from that presented above for centrosymmetric media, and the wave is always achiral. When we have $\omega = 0$, however, it is no longer possible to argue that the reversal of sign for $P_1(i,t)$ and $P_2(i,t)$ can be compensated by a simple change in the origin of t. Instead, we can point out that a displacement along the $\hat{\mathbf{x}}$ axis by some odd multiple of $\lambda/2$ will achieve exactly the same thing, so long as the distance involved is a multiple of the spin lattice spacing, a. That is, we must have $(2n+1)\lambda = 2ma$, or equivalently $(2n+1)\pi = mak$, with n and m integers, in order to claim the wave as achiral. Now, this condition will be met exactly for only certain wavevectors, k, but it will be met approximately *to arbitrary exactitude* for *all* wavevectors *not* satisfying $2n\pi = mak$; this latter expression therefore constitutes the condition under which a lateral spin spiral (in the static case only) becomes chiral.

In the vertical case, on the other hand, we continue to make use of Eq. (9.66) to define the wave, but in addition to $P_1(i,t)$ and $P_2(i,t)$, it will be convenient to consider

$$P_3(i,t) \equiv \hat{\mathbf{n}} \cdot \mathbf{S}_i = \begin{bmatrix} +\cos\theta \\ -\cos\theta \end{bmatrix} \tag{9.73}$$

and

$$P_5(i,t) \equiv \hat{\mathbf{n}} \cdot (\mathbf{S}_{i+1} \times \mathbf{S}_i) = \begin{bmatrix} +\sin^2\theta \sin ka \\ -\sin^2\theta \sin ka \end{bmatrix} \tag{9.74}$$

which are independent of both i and t. It is impossible for these to be simultaneously zero, except where $\theta = \pi/2$ and $k = 0$ or $k = \pi/a$, so the vertical spin spiral must generally be chiral. The exceptional situations correspond to spins revolving in the plane of the surface, either precisely in-phase or precisely out-of-phase from one site to the next. To deal with these, we consider another new pseudoscalar, defined as

$$P_6(i,t)|_{\theta=\pi/2} \equiv \hat{\mathbf{n}} \cdot (\mathbf{S}_i \times \partial\mathbf{S}_i/\partial\mathbf{t})|_{\theta=\pi/2} = \begin{bmatrix} +\omega \\ -\omega \end{bmatrix} \tag{9.75}$$

which can clearly only be zero for a static wave. The only potentially achiral form of a vertical spin spiral therefore corresponds to static ferromagnetic or antiferromagnetic ordering with in-plane spins, but only the cases where those spins line up parallel and/or antiparallel to the $\hat{\mathbf{x}}$ axis turn out actually to be achiral.

9.7 Spin Waves at Structured Surfaces

Thus far, we have talked about the wavevectors and frequencies of various spin waves as if they were entirely separable properties, assessing the chirality of different combinations without regard to which particular combinations may actually arise. In common with all excitations of a crystalline material, however, we should expect that the wavevector and frequency are in fact linked via a dispersion relation, suggesting the concept of a magnon bandstructure. Actually calculating or measuring the nature of this bandstructure lies beyond the purpose of this book, but we should nevertheless recognise that, for a given wavevector, only a discrete set of frequencies will be possible. Since these frequencies are linked to the energy of the spin wave ($\hbar\omega$) the symmetry of the bandstructure will determine which types of wave will be found at real surfaces, and in what orientations.

9.7.1 Spin Helices at Achiral Surfaces

It will be recalled from preceding sections that spin helices are *always* chiral as travelling waves. When found at an achiral surface, therefore, their behaviour may usefully be compared with that of chiral molecules at the same surface. One expects that both enantiomers ought to arise and that they ought to be degenerate. That is, for a given wavenumber, k, it should be possible to find both left- and right-handed spin helices at the same set of frequencies. Just as chiral molecules may segregate into different surface domains, however, it is entirely reasonable to anticipate that the frequencies of left- and right-handed spin helices may differ at any one specific wavevector, $\mathbf{k}$, so long as the anisotropy respects the point group of the surface. For wavevectors along high-symmetry directions, the degeneracy of frequencies must be strictly observed.

Turning away from travelling waves, static spin helices with any wavelength other than twice the spin lattice spacing are also chiral entities. The frequencies of such waves are, of course, zero, but instead we can speak of the energy associated with the static spin distortion involved. Once again, for any given wavenumber, k, degenerate distortions of both left- and right-handed type ought to be possible, but the degeneracy may be lifted for any specific low-symmetry wavevector, **k**. Overall, however, the anisotropy of spin helical energies must respect the point group of the surface. It follows, therefore, that chiral spin helices may be found in domains on an achiral surface, but that there must be equal areas of left- and right-handed helical domains to respect the global symmetry.

Achiral spin helices arise only for the special cases where the wavelength is infinite (non-zero frequency only) and where the wavelength is equal to twice the spin lattice spacing (any frequency). When such helices occur with the helical axis aligned along a high-symmetry direction in the surface, the global symmetry would be strictly preserved. It is not, however, impossible for an achiral spin helix to align its axis along a low-symmetry direction, thereby breaking the surface symmetry locally. In such circumstances, we ought again to expect that domains of locally broken symmetry are to be found in equal areas, respecting the global symmetry of the surface structure.

9.7.2 Spin Helices at Chiral Surfaces

When the point group of the surface is chiral, we no longer expect degeneracy in the bandstructures of chiral spin helices. For travelling waves, this means that it becomes possible for left-handed helices to have generally lower frequencies (and hence energies) than right-handed helices, or vice versa. In consequence, it would be possible to have a distinct imbalance in the number of left-handed magnons versus right-handed magnons. In the case of static waves, it would be possible for domains with left-handed helical spin distortion to differ in area from those exhibiting right-handed spin distortion. Achiral spin helices, on the other hand, would necessarily be insensitive to the chirality of the surface structure.

9.7.3 Spin Spirals at Achiral Surfaces

We have seen that vertical spin spirals are generally chiral, excepting those cases that actually reduce to ferromagnetic or antiferromagnetic ordering with spins aligned along a high-symmetry axis. As such, they behave at an achiral surface in much the same manner as the spin helices described above. As travelling waves, one expects that both left- and right-handed versions should be found at the same set of frequencies for a given wavenumber, k, but that asymmetry may arise at any specific low-symmetry wavevector, **k**. Overall, the anisotropy between left- and right-handed spin spirals must respect the point group of the surface. As static waves, on the other hand, domains of left- or right-handed spin distortion may be found, but these must have equal areas in order to respect the global symmetry. In contrast, lateral spin spirals are always achiral, so their behaviour matches that of the achiral spin helices described above.

9.7.4 Spin Spirals at Chiral Surfaces

As one might anticipate, when the point group of the surface is chiral, degeneracy between chiral spin spirals may be lifted. In the case of vertical spin spirals, therefore, we ought to expect left- and right-handed travelling magnons to occur in asymmetric proportion to one another, and for the corresponding static spin distortions to arise in larger domains of one handedness relative to smaller domains of the other. Such arguments do not apply, however, to lateral spin spirals, which remain stubbornly achiral and of limited interest in this context.

9.8 Surface Spin Spirals Observed

The foregoing discussion of spin waves at surfaces would clearly be of merely academic interest, were it not for the fact that such entities can be (and indeed have been) observed experimentally. Elegant spin-polarised scanning tunnelling microscopy (STM) studies from the group of Wiesendanger, supported by non-collinear density functional calculations by the group of Blügel, have demonstrated the existence of static spin spirals in monolayers of Mn deposited onto W{110} and W{100} substrates [8–10]. The method of detection involves use of a magnetic STM tip, whose polarity can be modulated by means of an externally applied field. The tunnel current passing between tip and sample is then found to vary depending upon whether the spin of the tip is aligned with that of the part of the sample above which it sits; this generates spin contrast, in addition to the usual topographic and electronic information supplied by the tunnel current. Although it is not clear whether it is the parallel or antiparallel spin configuration between tip and sample that maximises the current, the techniques is nevertheless capable of recording variations in the sample spin polarisation at atomic resolution.

In a Mn monolayer on the W{110} substrate, Bode *et al.* [8] report the existence of alternating stripes of opposed spin polarity, with wavelength 6 nm in the $[\bar{1}10]$ surface direction. Making use of symmetry rules provided by Moriya [2], the authors determine that the $\mathbf{D}_{ij}$ vector appearing in the DM coupling term (Eq. (9.25)) must lie always in the [001] direction, implying that the corresponding static spin wave must be a lateral spin spiral, with polar angle $\theta = \pi/2$. This situation, in which the spin tumbles end-over-end along the line of Mn atoms, may be described as **cycloidal** as the tips of the spin vector trace out a two-dimensional cycloid[13] (see Fig. 9.6). In fact, two possible cycloids may be imagined, according

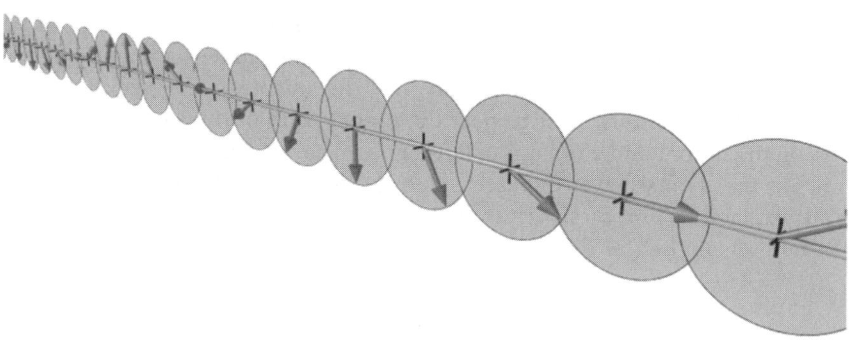

Figure 9.6 Visualisation of a cycloidal spin spiral.

13 We note, in passing, that non-cycloidal spin spirals have been observed by the same group in the case of a Mn bilayer on W{110}. Once again, these are lateral spin spirals, and hence achiral [11].

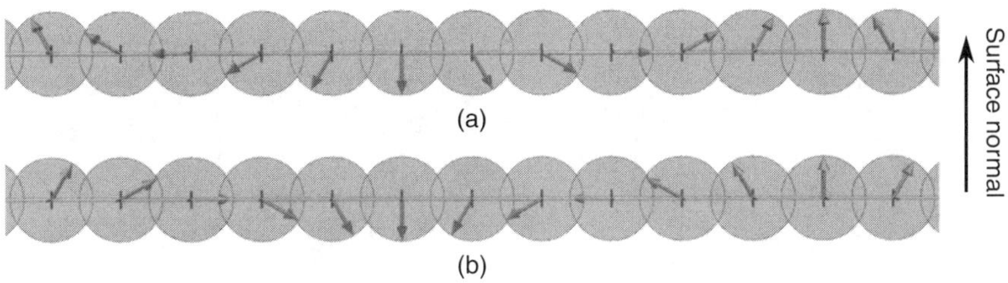

Figure 9.7 Non-degenerate cycloidal spin spirals, with equivalence between geometries (a) and (b) broken by the surface normal.

to the sense in which the spins tumble, and by carefully examining the change in the image when the tip polarisation direction is modified, the authors are able to distinguish that only one of these is actually found (Fig. 9.7a not Fig. 9.7b); spin–orbit coupled density functional calculations, reported in the same work, confirm that the two cycloids are non-degenerate. Unfortunately, the authors refer to this non-degeneracy as a consequence of chirality, when in fact (as we have demonstrated above) lateral spin spirals, including the cycloidal special case, are strictly achiral. Even if the spin spirals *were* chiral, it would be difficult to understand why the degeneracy between enantiomers should be lifted when they are found at a surface with an achiral point group. A very similar situation is found on the W{100} substrate, where Ferriani *et al.* [9, 10] report cycloidal spin spirals of wavelength 2.2 nm in the [110] surface directions, again with a preference for one type of cycloid over the other, attributed (incorrectly) to chirality.

What is going on here is simply that the two differing cycloidal spin spirals are, in fact, two non-degenerate achiral arrangements of spin, not enantiomers of one another. That is, each spiral is its own mirror image, and not that of the other, due to the reflection properties of pseudovectors (see Section 9.4). The DM vector (i.e. $\mathbf{D}_{ij}$) circulates in either a clockwise or an anticlockwise sense for one or other of the two spin spirals (as per the analysis of Hasselberg *et al.* [12]) but this is not in itself a physical field; indeed, as a pseudovector itself, the sense of its rotation is *not* reversed upon reflection. Granted that the two different cycloidal spin spirals are not enantiomerically related, the preference for one or other holds no mystery, and no global symmetry principles are violated.

9.9 Skyrmions, or How to Brush a Hedgehog

Having focused exclusively, in this chapter, on long-range manifestations of magnetism, we turn finally to a localised phenomenon – the quasiparticles known as **skyrmions**. First described by Skyrme in the context of nuclear physics [13] the concept has been generalised to include (dynamic, stable or metastable) excitations in the two-dimensional spin distribution of surfaces, interfaces or layered materials. In this case, a uniform spin field flips direction at a central point, while spins in the immediate vicinity must necessarily rotate to accommodate this flip, as per Fig. 9.8. Note that two types of skyrmion are

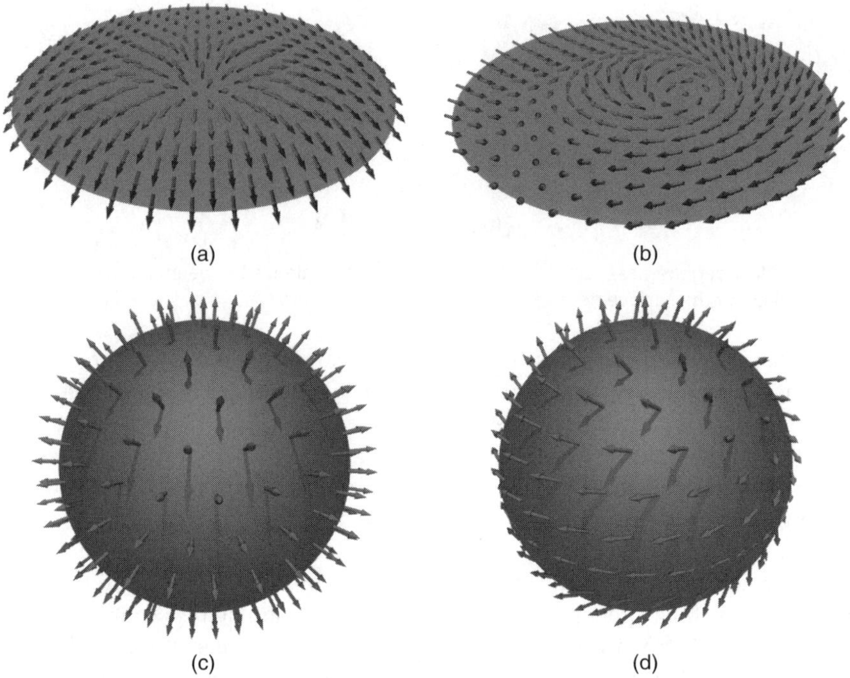

Figure 9.8 Relationship between hairy spheres and skyrmions. Panels (a) and (b) depict the spin distributions of a hedgehog and a vortex skyrmion, respectively. Panels (c) and (d) show the corresponding spherical spin distributions whose stereographic projections (not to scale) produce the two-dimensional spin distributions displayed above. Note that, despite appearances, the vortex skyrmion of panels (b) and (d) is its own mirror image, once the pseudovector nature of spin is taken into account, and hence achiral. The hedgehog skyrmion of panels (a) and (c) is, however, truly chiral.

broadly possible: the **hedgehog skyrmion** and the **vortex skyrmion**. Of these, the hedgehog variety is chiral[14] and in either case it is the familiar DM coupling that stabilises the non-collinear spin texture. Formally, skyrmions are usually considered to be stereographic projections of spins distributed on a sphere – the so-called 'hairy sphere' – pointing outwards to form a hedgehog, or brushed into a whorl to form the vortex [14]. Although skyrmions in bulk materials are possible in principle, broken symmetry in reduced-dimensionality environments is anticipated to promote their formation [15].

The presence of two-dimensional skyrmions in magnetic systems has recently been spectacularly confirmed in a number of studies based on bulk MnSi [16] and thin films of $Fe_{1-x}Co_xSi$ [17] or FeGe [18] (see Fig. 9.9). In each case, a triangular lattice of vortex skyrmions was inferred from neutron reflection measurements, or imaged by transmission electron microscopy (TEM). What all three systems have in common, however, is that the parent crystals take the so-called

14 The hedgehog skyrmion is often described as achiral, with chirality ascribed to the vortex variety. Whilst this would be accurate if the constituent spins were true vectors, the pseudovector nature of spin implies that the opposite is actually correct.

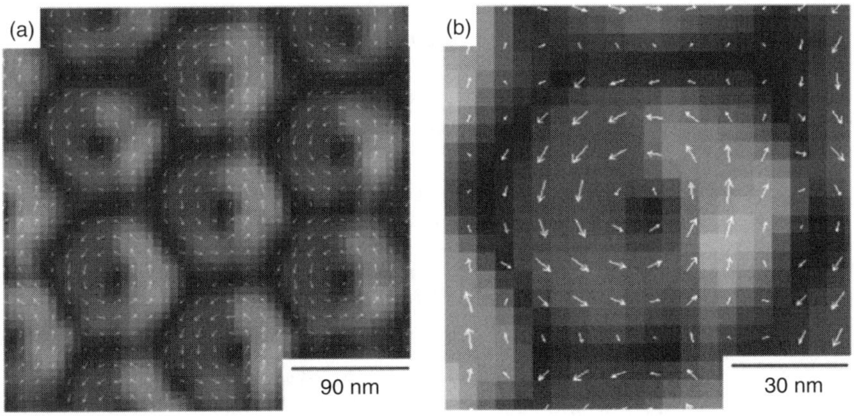

Figure 9.9 Skyrmions observed in a thin film of $Fe_{1-x}Co_xSi$ by means of Lorentz transmission electron microscopy, showing (a) the triangular lattice and (b) an individual skyrmion; arrows indicate in-plane spin components. Reproduced with permission from Ref. [17]. Copyright (2010) Nature Publishing Group. (*See Plate X for the colour representation of this figure.*)

B20 structure, exhibiting one of the Sohncke space groups and being, in consequence, chiral [19]. Indeed, in the TEM images obtained from $Fe_{1-x}Co_xSi$, only a single skyrmion 'enantiomer' was said to be observed [17], while in the FeGe case the skyrmion 'chirality' was said to switch across a grain boundary. In fact, however, as noted above, vortex skyrmions ought not to be thought of as truly chiral, despite the distribution of their spin vectors apparently making them so. What is going on in these systems, therefore, is simply a preference for one of two non-degenerate spin conformations that only appear to be mirror images of one another.

A further wrinkle to the skyrmion story has also been added by Heinze *et al.* [20] (following earlier speculation by the same group [21]) who demonstrated, via spin-polarised STM, the existence of what they term 'nanoskyrmions' in a monolayer of Fe on Ir{111} (see Fig. 9.10). Whereas the skyrmions reported in B20-structure materials were all of 20–90 nm diameter, the nanoskyrmions are less than 1 nm in diameter, packing into a square lattice that is incommensurate with the triangular parent overlayer. The skyrmion nature of the lattice is confirmed by evaluation of the **skyrmion number** integral

$$S = \frac{1}{4\pi} \int \hat{\mathbf{s}}. \left(\frac{\partial \hat{\mathbf{s}}}{\partial x} \times \frac{\partial \hat{\mathbf{s}}}{\partial y} \right) dxdy \tag{9.76}$$

where $\hat{\mathbf{s}}$ is a unit vector representing the local magnetisation; for non-skyrmion spin distributions, this integral is strictly zero when evaluated over a unit cell in the xy plane, while positive or negative values correspond to the existence of one or other skyrmion enantiomer. Moreover, the nanoskyrmions reported by Heinze *et al.* [20] are stable in the absence of an applied magnetic field (unlike those of the B20 materials which were not) and are of the hedgehog variety. Subsequent studies by Sonntag *et al.* [22] showed this texture to be stable up to 27.8 K, seemingly coincident with the Curie temperature of the overlayer.

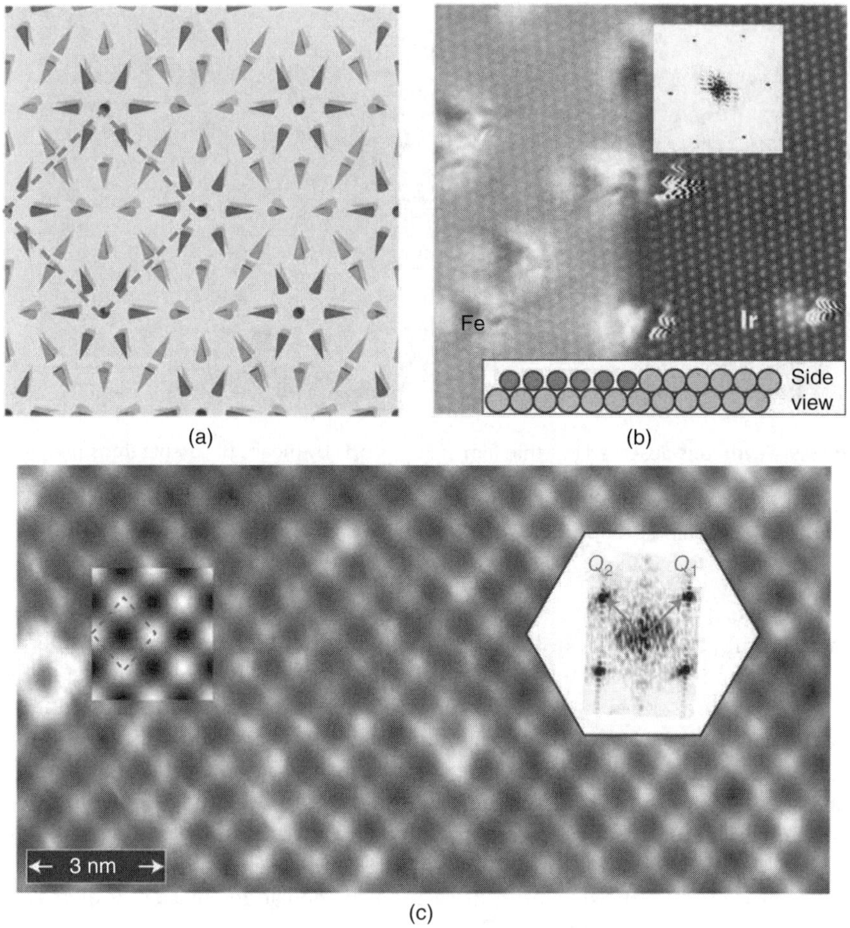

Figure 9.10 Skyrmions observed in a monolayer film of Fe on Ir{111} by means of spin-polarised scanning tunnelling microscopy. Panel (a) shows a schematic of hedgehog nanoskyrmions in a square lattice, while in panel (b) the hexagonal structure of the Fe monolayer (left-hand side) and underlying Ir{111} substrate (right-hand side) are imaged in spin-unpolarised mode. In panel (c) the square nanoskyrmion lattice is revealed in the spin-polarised image. Reproduced with permission from Ref. [20]. Copyright (2011) Nature Publishing Group.

Intriguingly, quite different behaviour was recently observed in ostensibly the same system by von Bergmann *et al.* [23] by focusing upon sub-monolayer Fe deposited at somewhat lower growth temperature. The result of these changes is that islands can be found in which the overlayer atoms sit in *hcp* rather than *fcc* hollow sites of the underlying Ir{111} substrate, subtly altering the coupling between neighbouring spins compared with the case described above. Under these circumstances, commensurate triangular skyrmion lattices are found – once again with sub-nanometre skyrmions (nanoskyrmions) of the hedgehog variety. The presence of both skyrmion enantiomers at zero applied field is, in effect, equivalent to the existence of counter-polarised magnetic domains, which are indeed observed.

Clearly, there remains much scope for investigating the properties of skyrmions at surfaces. Thus far, truly chiral hedgehog skyrmions have been observed on achiral surfaces, and essentially achiral vortex skyrmions on chiral surfaces; it would be interesting to see what effects may arise when chiral skyrmions exist on chiral surfaces. In the meantime, for an in-depth overview of this fascinating and important topic, the recent review by Wiesendanger [24] is highly recommended.

References

1 I.E. Dzyaloshinskii, *J. Phys. Chem. Solids* **4**, 241 (1958).
2 T. Moriya, *Phys. Rev.* **120**, 91 (1960).
3 W. Döring, *Ann. Phys.* **456**, 102 (1957).
4 P. Brune, Physical origins and physical models of magnetic anisotropy, in *Magnetismus von Festkörpern und Grenzflächen*, eds. P.H. Dederichs, P. Grünberg and W. Zinn (Furschungzentrum Jülich, 1993).
5 T. Jeong and W.E. Pickett, *Phys. Rev. B* **70**, 075114 (2004).
6 S. Yamagishi, S.J. Jenkins and D.A. King, *J. Chem. Phys.* **114**, 5765 (2001).
7 V. Humblot, S. Haq, C. Muryn, W.A. Hofer and R. Raval, *J. Am. Chem. Soc.* **124**, 503 (2002).
8 M. Bode, M. Heide, K. von Bergmann, P. Ferriani, S. Heinze, G. Bihlmayer, A. Kubetzka, O. Pietzsch, S. Blügel and R. Wiesendanger, *Nature* **447**, 190 (2007).
9 P. Ferriani, K. von Bergmann, E.Y. Vedmedenko, S. Heinze, M. Bode, M. Heide, G. Bihlmayer, S. Blügel and R. Wiesendanger, *Phys. Rev. Lett.* **101**, 027201 (2008).
10 P. Ferriani, K. von Bergmann, E.Y. Vedmedenko, S. Heinze, M. Bode, M. Heide, G. Bihlmayer, S. Blügel and R. Wiesendanger, *Phys. Rev. Lett.* **102**, 019901 (2009).
11 Y. Yoshida, S. Schröder, P. Ferriani, S. Serrate, A. Kubetzka, K. von Bergmann, S. Heinze and R. Wiesendanger, *Phys. Rev. Lett.* **108**, 087205 (2012).
12 G. Hasselberg, R. Yanes, D. Hinzke, P. Sessi, M. Bode, L. Szunyogh and U. Nowak, *Phys. Rev. B* **91**, 064402 (2015).
13 T.H. Skyrme, *Proc. R. Soc. London, Ser. A* **260**, 127 (1961).
14 C. Pfleiderer, *Nat. Phys.* **7**, 673 (2011).
15 A.N. Bogdanov and U.K. Rößler, *Phys. Rev. Lett.* **87**, 037203 (2001).
16 S. Mühlbauer, B. Binz, F. Jonietz, C. Pfliederer, A. Rosch, A. Neubauer, R. Georgii and P. Böni, *Science* **323**, 915 (2009).
17 X.Z. Yu, Y. Onose, N. Kanazawa, J.H. Park, J.H. Han, Y. Matsui, N. Nagaosa and Y. Tokura, *Nature* **465**, 901 (2010).
18 X.Z. Yu, N. Kanazawa, Y. Onose, K. Kimoto, W.Z. Zhang, S. Ishiwata, Y. Matsui and Y. Tokura, *Nat. Mater.* **10**, 106 (2010).
19 S.V. Grigoriev, D. Chernyshov, V.A. Dyadkin, V. Dmitriev, S.V. Malayev, E.V. Moskvin, D. Menzel, J. Schoenes and H. Eckerlebe, *Phys. Rev. Lett.* **102**, 037204 (2009).

20 S. Heinze, K. von Bergmann, M. Menzel, J. Brede, A. Kubetzka, R. Wiesendanger, G. Bihlmayer and S. Blügel, *Nat. Phys.* **7**, 713 (2011).
21 K. von Bergmann, S. Heinze, M. Bode, E.Y. Vedmedenko, G. Bihlmayer, S. Blügel and R. Wiesendanger, *Phys. Rev. Lett.* **96**, 167203 (2006).
22 A. Sonntag, J. Hermenau, S. Krause and R. Wiesendanger, *Phys. Rev. Lett.* **113**, 077202 (2014).
23 K. von Bergmann, M Menzel, A. Kubetzka and R. Wiesendanger, *Nano Lett.* **15**, 3280 (2015).
24 R. Wiesendanger, *Nat. Rev. Mater.* **1**, 16044 (2016).

10

Chiral Electronic States in Two Dimensions

The chemist has in general been content with what I will call the 'loafer' electron theory. He has imagined these electrons sitting around on dry goods boxes at every corner ready to shake hands with, or hold on to, similar electrons in other atoms. The physicist, on the other hand, has preferred to think of them as leading more active lives, playing ring-around-the-rosy, crack-the-whip and other interesting games.

Robert A. Millikan, Science 59, 473 (1924)

Perhaps the most fundamental message of this book is that chirality may be found in many phenomena related to surfaces – chemical, mechanical, optical and magnetic – but even so, it may at first seem odd to assert that a definition of chirality prevalent in the realm of particle physics is also highly relevant. Nevertheless, a variety of novel materials investigated over the last decade do indeed possess two-dimensional electronic structures that are characterised in terms of their chirality in a way that would be very familiar to a particle physicist. Let us begin with a few definitions.

The **helicity** of a fundamental particle (which is not, in general, the same thing as chirality) is defined in terms of its spin and its momentum, or more precisely their scalar product. That is, we must consider the term $\boldsymbol{\sigma} \cdot \mathbf{k}$, with $\boldsymbol{\sigma}$ the spin and $\mathbf{k}$ the wavevector of the particle. A particle for which this scalar product is positive is said to have right-handed helicity, while left-handed helicity implies that this scalar product is negative. Now, for a particle of finite mass, the momentum vector depends, of course, on the frame of reference of the observer, so the apparent helicity of a given particle is necessarily a frame-dependent property, not an intrinsic property of the particle. Particles of zero mass, however, move at the speed of light, so *their* momentum is *not* dependent on the observer's frame of reference. For such particles, the helicity is indeed an intrinsic property, and is then (and only then) referred to as their chirality.

It should be clear, therefore, that chirality in this sense is *not* to be found in the electronic structures of most materials, where the electron (and hole) effective masses may differ from their free-space values but are not, in general, zero. That is to say, the electrons (and holes) of most materials may be said to possess helicity, but not true frame-independent chirality. In a small number of materials, however, particular circumstances of symmetry and/or topology conspire to give

Chirality at Solid Surfaces, First Edition. Stephen J. Jenkins.

rise to massless excitations within the electronic structure, and for *these* it may be said both that true chirality exists and that it largely determines their behaviour at a fundamental level.

10.1 Dirac Cones in Graphene

The first case in which massless excitations were proposed is that of graphene – a material consisting of a single sheet of carbon atoms arranged in a two-dimensional honeycomb structure. The electronic structure of this material had been theoretically described as early as 1946 [1] but the results seem to have generally been viewed as of academic interest, or at best as an approximation to the electronic structure of graphite (comprising a stack of graphene sheets bound together by dispersion forces). Interest was only revived after Novoselov and Geim published, in 2004, their 'Scotch Tape' technique for the isolation of graphene[1] [2].

The honeycomb structure of graphene implies two carbon atoms per primitive unit cell in a trigonal planar bonding arrangement (Fig. 10.1). Each atom contributes three sp^2 hybrid orbitals to σ bonds/antibonds with its nearest neighbours, leaving a further p orbital to be incorporated into two π bands

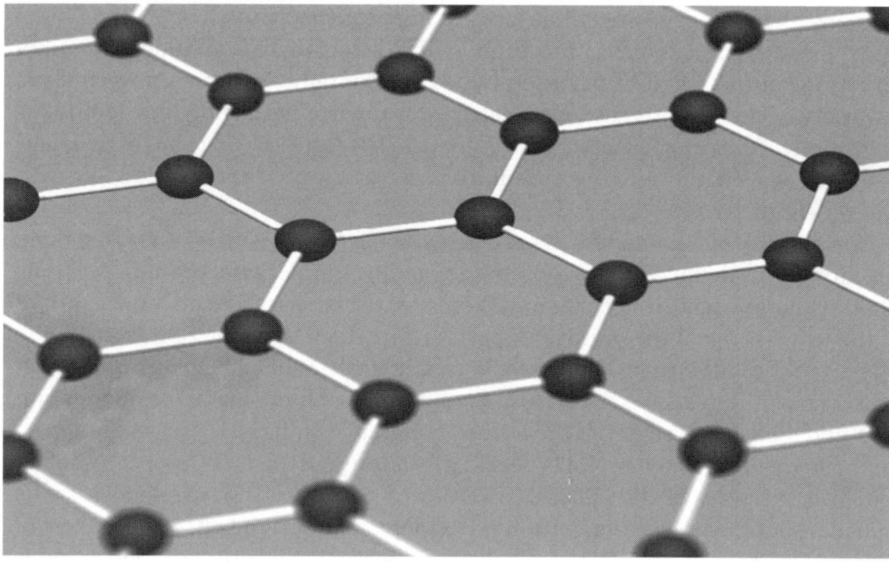

Figure 10.1 The honeycomb structure of graphene.

1 The isolation of graphene, and subsequent characterisation of its properties, ultimately prompted the award of a Nobel Prize to Novoselov and Geim in 2010. With hindsight, many instances of 'graphitic carbon' reported in the surface science literature throughout the 1970s, 1980s and 1990s probably amounted to adsorbed graphene, but were not generally discussed as anything other than a nuisance. As ever, recognising the significance of an observation is at least as important as making the observation itself.

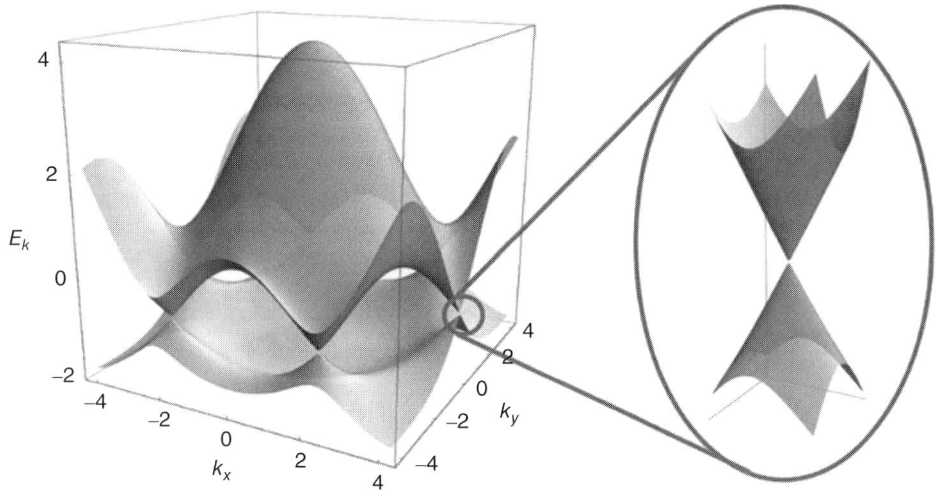

Figure 10.2 Schematic rendering of the π bands in graphene. Reproduced with permission from [3]. Copyright (2009) American Physical Society.

(bonding and antibonding) that span the entire two-dimensional crystal. It is the remarkable structure of these latter bands that gives graphene its unique electronic properties.

The first point to note about the π-derived bands is that they are degenerate at the corners of graphene's hexagonal Brillouin zone (see Fig. 10.2). It will turn out to be convenient to examine the dispersion of these bands in the vicinity of each corner, recognising from the outset that of the six corners only two need be considered to be independent. If we label one corner K and the opposite corner K' then all other corners may be seen to be identical to one or other of these to within a reciprocal lattice vector.

Let us begin in the vicinity of K, defining a relative wavevector $\mathbf{q}$ such that

$$\mathbf{k} = \mathbf{K} + \mathbf{q} \tag{10.1}$$

is the wavevector measured from the zone centre, with $\mathbf{K}$ being the wavevector of the K-point. In general, the eigenfunctions of the system, $\psi_{\mathbf{k}}(\mathbf{r})$, must satisfy the Bloch form

$$\psi_{\mathbf{k}}(\mathbf{r}) = u_{\mathbf{k}}(\mathbf{r})e^{i\mathbf{k}\cdot\mathbf{r}} \tag{10.2}$$

with $u_{\mathbf{k}}(\mathbf{r})$ being a Bloch function whose periodicity matches that of the real-space lattice.[2]

Now, at the K-point itself, a convenient representation of the degenerate eigenfunctions can be found in which the Bloch function of one, $u^A_{\mathbf{K}}(\mathbf{r})$, is localised entirely upon the sub-lattice labelled 'A' in Fig. 10.3, and the Bloch function of the other, $u^B_{\mathbf{K}}(\mathbf{r})$, entirely upon the sub-lattice labelled 'B'. In the vicinity of the K-point, we will assert that it ought to be possible to expand the Bloch functions

2 This requirement is, of course, a corollary of Bloch's theorem (or an alternate statement of that theorem, if one prefers).

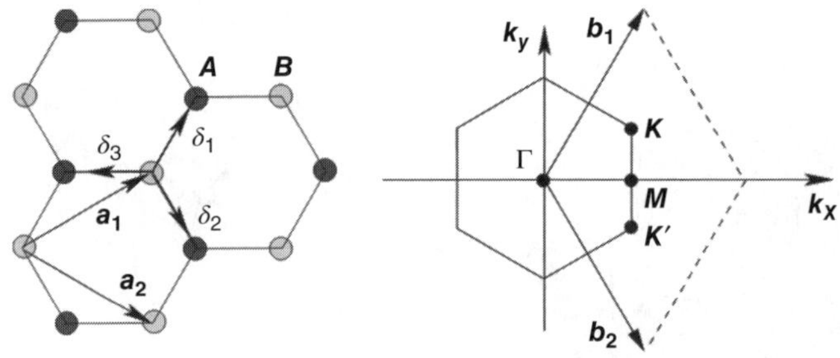

Figure 10.3 Labelling the sub-lattices and high-symmetry **k**-points in graphene. Reproduced with permission from [3]. Copyright (2009) American Physical Society.

of the eigenfunctions as linear combinations of similarly localised Bloch functions, $u_{\mathbf{k}}^A(\mathbf{r})$ and $u_{\mathbf{k}}^B(\mathbf{r})$. Tracking away from K along the $\hat{\mathbf{x}}$ axis, for instance, the Bloch functions take the form $(u_{\mathbf{k}}^A(\mathbf{r}) \pm u_{\mathbf{k}}^B(\mathbf{r}))/\sqrt{2}$, while tracking along the $\hat{\mathbf{y}}$ axis they take the form $(u_{\mathbf{k}}^A(\mathbf{r}) \pm iu_{\mathbf{k}}^B(\mathbf{r}))/\sqrt{2}$. More generally, we should expect

$$u_{\mathbf{k}}(\mathbf{r}) = \alpha u_{\mathbf{k}}^A(\mathbf{r}) + \beta u_{\mathbf{k}}^B(\mathbf{r}) \tag{10.3}$$

where α and β are complex coefficients satisfying

$$|\alpha|^2 + |\beta|^2 = 1 \tag{10.4}$$

whose particular values are to be determined.

If we now define a Hamiltonian matrix, $\hat{H}$, whose effect is simply to couple the Bloch functions localised on the A and B sub-lattices, we must have

$$\hat{H} = \epsilon \begin{pmatrix} 0 & e^{i\theta} \\ e^{-i\theta} & 0 \end{pmatrix} \tag{10.5}$$

where ϵ is a characteristic energy and θ is some phase angle. By inspection (i.e. by comparing the eigensolutions of $\hat{H}$ with those expected along the $\hat{\mathbf{x}}$ and $\hat{\mathbf{y}}$ axes) we conclude that θ must, in fact, simply be the anticlockwise angle between $\mathbf{q}$ and the $\hat{\mathbf{x}}$ axis. That is, we may rewrite the Hamiltonian matrix as

$$\hat{H} = \frac{\epsilon}{q} \begin{pmatrix} 0 & q_x + iq_y \\ q_x - iq_y & 0 \end{pmatrix} \tag{10.6}$$

which may further be simplified to the form

$$\hat{H} = \frac{\epsilon}{q} \boldsymbol{\sigma} \cdot \mathbf{q} \tag{10.7}$$

where $\boldsymbol{\sigma}$ is a vector whose components are the Pauli matrices

$$\sigma_x = \begin{pmatrix} 0 & 1 \\ 1 & 0 \end{pmatrix} \tag{10.8}$$

and

$$\sigma_y = \begin{pmatrix} 0 & -i \\ i & 0 \end{pmatrix} \tag{10.9}$$

familiar from the treatment of spin in three dimensions.[3] Here the spin is not the *true* spin of the electrons involved, but rather a **pseudospin** that transforms in the same way but describes an entirely different physical phenomenon (i.e. the coupling between states on the two sub-lattices of graphene).

Diagonalising the graphene Hamiltonian, one finds that the eigenvalues, ϵ, are simply proportional to q in the limit $q \to 0$, so that close to K we can set

$$\epsilon = \pm\hbar v q \tag{10.10}$$

and hence

$$\hat{H} = \hbar v \boldsymbol{\sigma} \cdot \mathbf{q} \tag{10.11}$$

with v being a constant speed. In the vicinity of the Brillouin zone corners, therefore, the bandstructure takes on a locally conical form, quite unfamiliar in the realm of condensed matter physics. Significantly, however, a Hamiltonian of precisely this form is readily found in the realm of fundamental particle physics, where the free-particle Hamiltonian of a massless Dirac fermion in two spatial dimensions[4] may be written as

$$\hat{H} = \hbar c \boldsymbol{\sigma} \cdot \mathbf{k} \tag{10.12}$$

with c the speed of light and **k** the wavevector. The corresponding eigenvalues are simply $\epsilon = \pm\hbar c k$, with c (of course) constant. Clearly, the constant v appearing in the graphene Hamiltonian plays the role of a renormalised speed of light, albeit one that turns out to be a factor of around 300 smaller than the actual speed of light. The existence of these **Dirac cones** in the electronic structure of graphene has been well established, through experimental techniques such as angle-resolved ultraviolet photoemission spectroscopy (ARUPS) (see, e.g. the work of Siegel *et al.* [4] in Fig. 10.4).

As mentioned in the introduction to this chapter, the adherence of electrons in the vicinity of the Dirac cones to a massless energy–wavevector dispersion curve implies that the product of the pseudospin vector, $\boldsymbol{\sigma}$, with the relative wavevector, $\mathbf{q}$, defines a true chirality, not merely a helicity. For the Dirac cones located at K, the eigenfunctions of the upper branch may be taken[5] to have chirality D, while those of the lower branch will have chirality L; these assignments are reversed for the Dirac cones located at K'. In a sense, therefore, the Dirac cones themselves may be thought of as chiral entities, with those located at K and K' representing the two enantiomers. Since the geometric structure of graphene displays no chirality, however, one should not be surprised to find that these

3 Since the pseudospin, $\boldsymbol{\sigma}$, is restricted to two dimensions, the third Pauli matrix is not required here.

4 Compared with the Dirac equation in three spatial dimensions, introduced in Chapter 9 as Eq. (9.6), the mass-dependent term is absent here, and the vector $\boldsymbol{\alpha}$ replaced by $\boldsymbol{\sigma}$; the first of these changes is because we wish to discuss massless particles, and the second is due to the reduced dimensionality. We have also replaced the momentum, $\mathbf{p}$, via the identity $\mathbf{p} \equiv \hbar\mathbf{k}$, in order to work directly with the wavevector.

5 The assignment of chirality labels here is, as ever, entirely arbitrary in the first instance; let us adopt, without loss of generality, the convention that the chirality label D corresponds to $\boldsymbol{\sigma} \cdot \mathbf{q} > 0$ and that chirality label L corresponds to $\boldsymbol{\sigma} \cdot \mathbf{q} < 0$.

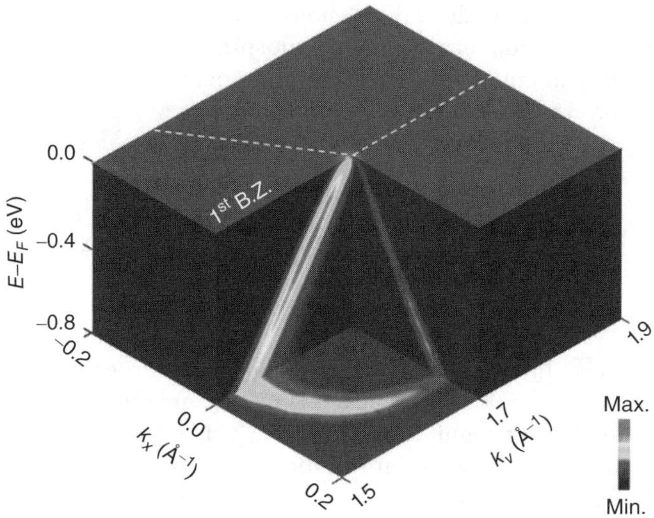

Figure 10.4 ARUPS data showing the lower half of a Dirac cone at the corner of the graphene Brillouin zone. Reproduced with permission from [4].

enantiomers arise in equal numbers (i.e. six partial Dirac cones encroach into the Brillouin zone – three that may be reconstructed by translation through reciprocal lattice vectors into a single cone at K, and three that may likewise be compiled into another cone at K'). The electronic structure of graphene is thus, in essence, racemic in nature.

One important consequence of the electronic chirality within each individual Dirac cone may be found in the phenomenon of **Klein tunnelling**, illustrated in Fig. 10.5 (and described fully by Katsnelson *et al.* [5]). The situation is essentially that of a lateral heterostructure, where one region of a graphene sheet is subject to an electrostatic potential that differs from that of the surrounding regions. This could be achieved, for example, by laying the graphene sheet across a

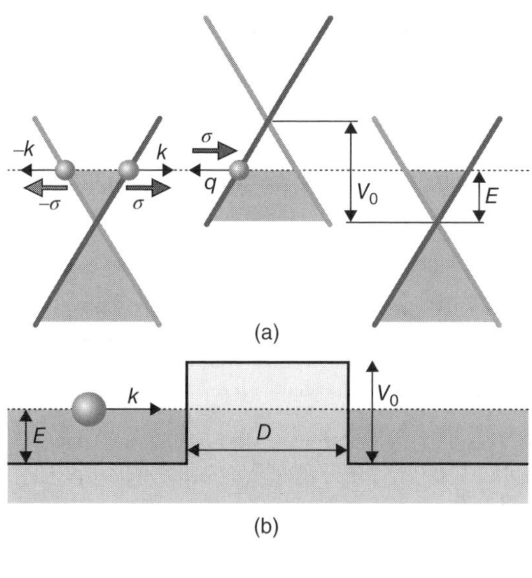

Figure 10.5 Schematic illustration of Klein tunnelling. Panel (a) shows Dirac cones on either side of, and within, the potential barrier shown in panel (b). Reproduced with permission from [5]. Copyright (2006) Nature Publishing Group.

patterned substrate whose surface domains have different work functions. The resulting offset of the Dirac cones on either side of each domain boundary implies that one ought to consider the transport of electrons on one side and holes on the other. The linear nature of the Dirac equation, however, ensures that for any electron-like wavepacket solution on the *outside* of the potential barrier in Fig. 10.5, there necessarily exists a time-reversed (i.e. hole-like) solution on the *inside*; these perfectly annihilate at the boundaries of the barrier region, implying a transmission coefficient of unity. Potential barriers hence become essentially transparent with respect to the transport of chiral particles.

10.2 Dirac Cones at Half-Metal Surfaces

The Dirac cones associated with graphene arise in a fundamentally two-dimensional material, but in principle the mathematical treatment that predicts their existence can potentially apply in any situation where electrons are confined within a two-dimensional sub-space. One example is to be found in the surface states of at least one half-metallic ferromagnet.

The intriguing phenomenon of half-metallicity was first proposed in theoretical work presented by De Groot *et al.* [6], who noted that the separate bandstructures occupied by spin-up and spin-down electrons within a ferromagnetic material might differ to such a degree that one could be metallic in nature while the other was semiconducting (or insulating). Several materials have subsequently been reported to show half-metallic behaviour, but the very first such material to be predicted in the original work was NiMnSb – a member of the class of inter-metallic compounds known as the semi-Heusler alloys.

The structure of a classic semi-Heusler alloy is based upon a face-centred cubic unit cell, in which the (0.00, 0.00, 0.00) position is occupied by a d-block element, the (0.25, 0.25, 0.25) position by Mn and the (0.75, 0.75, 0.75) position by a p-block element; a full-Heusler alloy would also find the (0.50, 0.50, 0.50) position occupied by the same d-block element found at the origin [7, 8]. Calculations for the spin-resolved bandstructure of bulk NiMnSb reveal a band gap of approximately 0.5 eV for the minority-spin channel, while the majority-spin channel is found to be metallic (i.e. zero band gap). As such, the material might be expected to function as an efficient spin valve, permitting the injection of majority-spin electrons while preventing the flow of minority-spin electrons, with promise for a variety of spintronic applications. A key question, however, is whether the half-metallic property is preserved right up to the surface of the material, or indeed its interface with any adjacent semiconductor. Even a thin layer of fully metallic material would likely lead to a significant degree of spin-flip scattering, compromising device performance.

Loss of half-metallicity at the surface or interface of a bulk half-metal can occur for one of two fundamental reasons. Firstly, and most obviously, the stoichiometry and/or structure of the material itself may differ at the surface relative to the bulk; clearly if the particular coincidence of stoichiometry, structure and symmetry that gives rise to half-metallicity in the bulk no longer pertains in the surface region, there is no reason to suppose that this desirable electronic property will itself survive. In the case of NiMnSb, however, experimental work from the group of Dowben indicates that it is possible to generate a non-reconstructed surface of {100} orientation, with only minimal segregation of Mn towards the surface [9]. A second reason for loss of half-metallicity at the surface may nonetheless be found in the existence of surface-localised electronic states; even when the stoichiometry and structure of the surface remain near-identical to that of the bulk, the three-dimensional (periodic) translational symmetry of the bulk crystal is necessarily broken, and this can give rise to electronic states that can only exist in the vicinity of the surface. Such states are familiar as Shockley states at the surfaces of nearly free-electron metals, for instance the {111} surfaces of coinage metals, and as Tamm states at the surfaces of covalent materials; the former retain essentially free-electron character in the two in-plane dimensions of the surface, whilst being strongly localised in the surface-normal dimension, and the latter are fully localised in the form of dangling bonds. In order to be strongly surface-localised, it is necessary for these bands to occupy regions of energy–wavevector space that are forbidden to bulk states – a situation that can occur across the entire Brillouin zone within the band gap of an insulator, but which may also arise in localised pockets within the Brillouin zone for both insulators *and* metals. Here, of course, the majority-spin channel is metallic, presenting only limited scope for surface-localised electronic states, but the minority-spin channel is insulating and a much more likely home for this kind of phenomenon.

Calculations of the stoichiometric NiMnSb{100} surface have confirmed the experimental observation that termination with a mixed MnSb layer is favoured over the alternative Ni termination, but also reveal the existence of two distinct surface-localised electronic states within the minority-spin channel [11]. From the perspective of spin-valve applications, these surface states present a real problem, since they imply that half-metallicity is indeed lost within the top-layer of the material. Once one looks beyond this disappointment, however, it becomes clear that the energy–wavevector dispersion of the surface states is potentially of great interest in its own right. Specifically, there exist two points within the surface Brillouin zone where the two surface states touch, and in the vicinity of these points the dispersion is found to take the form of a cone (see Fig. 10.6) [10]. Indeed, these features are, for practical purposes, analogous to the Dirac cones found in graphene. They occur at somewhat less symmetrical points within the surface Brillouin zone, and therefore display a notable tilt and a degree of elliptical deformation, but otherwise they satisfy the major criteria: the eigenfunctions comprising the upper and lower branches may be expressed as a linear combination of orthogonal components that transforms as one rotates around the central point in just the same way as for graphene. In graphene, the component eigenfunctions correspond to functions localised on two sub-lattices, whereas here they differ in their intrinsic symmetry rather than their physical location, but the result is essentially the same. Any chiral physics that can potentially be displayed by the Dirac electrons of graphene ought equally to be possible for the surface states of NiMnSb{100}.[6] As before, we can choose arbitrarily to identify

6 It seems likely that Dirac cones with similar properties ought to be present at the {100} surfaces of other half-metal semi-Heusler alloys too, although none have yet been reported.

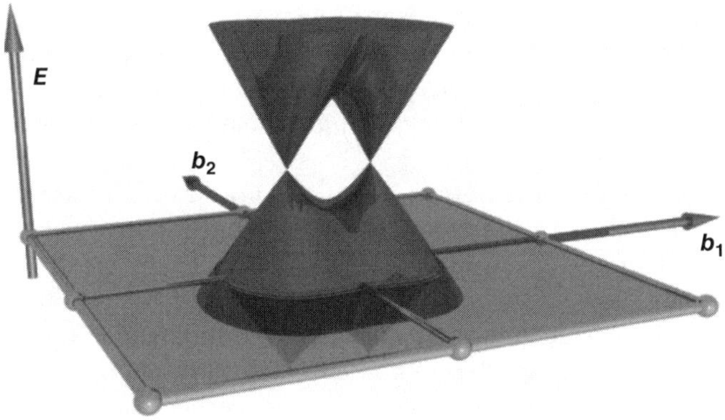

Figure 10.6 Minority-spin surface states of NiMnSb{100}. The horizontal square represents the surface Brillouin zone (spanned by reciprocal lattice vectors $\mathbf{b}_1$ and $\mathbf{b}_2$) positioned at the Fermi level on a vertical axis representing the band energy. Throughout the central region, there are two surface states at each wavevector, meeting at two distinct Dirac points. Reproduced with permission from [10]. Copyright (2010) American Physical Society.

the upper branch of one cone (let us choose the one centred at $\mathbf{k} = 0.087\mathbf{b}_1$) with the chirality label D and its lower branch with the label L. The other cone will, of course, have these labels reversed, ensuring that the surface is effectively racemic as far as its electronic chirality is concerned.

Interestingly, the Dirac cones found on the NiMnsb{100} surface turn out to be rather robust, so long as the symmetry of the surface is preserved. Calculations in which layers of Sb are progressively deposited upon the surface[7] show that although sub-monolayer coverages may disrupt the surface electronic structure significantly, integer multiples of a monolayer generally do not [12]. Although the position, tilt and ellipticity of the Dirac cones is altered by the presence of the overlayer, the fact of their existence is not compromised [10]. This suggests, at least in principle, the possibility of constructing vertical heterostructures in which two-dimensional interfaces featuring Dirac cones are separated by spacer layers of Sb; the interactions of Dirac cones across those layers would, of course, be subject to chiral constraints, in much the same way as described above in the case of Klein tunnelling through lateral heterostructures made from graphene.

10.3 Dirac Cones at the Surfaces of Topological Insulators

The examples discussed above share a common feature, in that the Dirac cones found in the electronic structures of graphene and the NiMnSb{100} surface exist because of very specific symmetries in the atomic geometry. Breaking

---

7 In growing stoichiometric surfaces of NiMnSb{100}, Dowben and co-workers found it beneficial to adopt a protocol where, in the final stage, deposition of Mn and Ni was halted while that of Sb continued, allowing a thick layer of Sb to build up on top of the surface [9]. This capping-layer could subsequently be removed by sputtering, revealing the stoichiometric surface beneath.

these symmetries, for instance, through strain or the adsorption of adspecies, lifts the degeneracy at the Dirac point and reduces the well-defined chirality of electrons in the ideal system to mere helicity; predictions based upon the concept of chirality hence become more and more approximate as the degree of symmetry-breaking perturbation increases.

In contrast, there exists a class of materials in which Dirac cones arise not through the symmetry properties of the atomic geometry, but rather through the topological properties of the electronic structure itself. These materials are known as **topological insulators**, and the Dirac cones are to be found in surface-localised electronic states at their surfaces. To understand these, however, we must first review the properties of bulk topological insulators.[8]

Imagine, then, an insulating material in which the valence band maximum and conduction band minimum approach one another very closely, and let us label (rather generically) the symmetries of these bands (at high-symmetry points within the Brillouin zone) either 'v' or 'c', as in Fig. 10.7a.[9] Such a material constitutes a 'trivial' insulator, albeit one with a rather narrow band gap (i.e. a semiconductor). Now, however, let us imagine that this insulator is subject to spin–orbit coupling, which typically induces a splitting of bands into spin-up and spin-down branches. If this splitting is small, the valence and conduction bands remain on their own side of the band gap (Fig. 10.7b) and nothing fundamentally changes, but if the splitting is sufficiently large the upshifted branch of the valence band is driven through the downshifted branch of the conduction band (Fig. 10.7c); interactions between the upshifted valence and downshifted

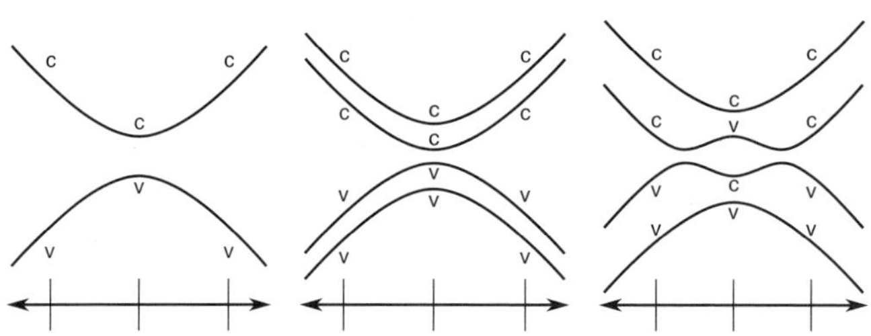

Figure 10.7 Schematic illustration of bandstructure topology. In panel (a) the valence and conduction bands of an insulator are separated by a band gap; symmetries characteristic of the valence and conduction band states are labelled generically 'v' or 'c' at high-symmetry **k**-points. In panel (b) the gap remains after spin–orbit splitting of each band; the symmetry character of bands remains unchanged. In panel (c) the spin–orbit splitting is sufficiently large to interchange states of valence and conduction character, in a manner that cannot be trivially undone without passing through an intermediate metallic situation.

8 A considerably more complete description, including bulk *and* surface phenomena, may be found in [13].

9 Electronic states must conform both to the symmetry of the material *and* that of the **k**-point at which they are found. The labels 'v' and 'c' may, therefore, indicate different symmetries at different points within the Brillouin zone, hence our description of them as 'generic'. All that matters for the discussion here is that the symmetries of conduction and valence bands be distinct from one another at the high-symmetry **k**-points.

conduction bands ensure that the states anticross rather than cross,[10] but the overall effect is that some states bearing the symmetry label 'v' end up *above* the band gap, and others labelled 'c' end up *below* it. Such a situation represents a distinctly different 'non-trivial' topology relative to the trivial case that pertained in the absence of spin–orbit splitting, and it is not possible to imagine reverting to that trivial case by any continuous process without closing and reopening the band gap.

The topological argument presented qualitatively above may be placed on a firm quantitative footing by introducing the concept of a **topological invariant** that flips between one value and another according to whether the topology is trivial or non-trivial. Fu and Kane [14] achieved this with reference to the parity of occupied eigenfunctions at those special **k**-points within the Brillouin zone where time-reversal symmetry is synonymous with translational symmetry (i.e. $\mathbf{k}_i \equiv -\mathbf{k}_i \pm \mathbf{G}$, with $\mathbf{G}$ a reciprocal lattice vector). At such points it is useful to define an overall parity for occupied states as

$$\delta_i = \prod_j \xi_{2j}(\mathbf{k}_i) \tag{10.13}$$

where $\xi_{2j}(\mathbf{k}_i)$ is the parity of the $(2j)$th occupied eigenfunction at wavevector $\mathbf{k}_i$; the product includes only every other eigenfunction, due to the degeneracy implicit at these special **k**-points.

The topological invariant itself, ν, is then defined as

$$(-1)^{\nu} = \prod_i \delta_i \tag{10.14}$$

and takes values of either zero (for trivial topology) or unity (for non-trivial topology).[11] It is not possible for a system to be transformed smoothly from $\nu = 0$ to $\nu = 1$ (or vice versa) without closing and reopening the band gap. A number of bulk materials displaying non-trivial electronic topology have been identified in recent years, generally involving relatively heavy elements conducive to strong spin–orbit coupling, including examples such as Bi_2Se_3, Bi_2Te_3, Sb_2Te_3 and possibly SmB_6.

The significance of topological insulators for electronic chirality only becomes apparent when considering the properties of states existing at their surfaces and/or interfaces. Here, the topological invariant evaluated *inside* the topological insulator clearly differs from the invariant evaluated *outside*, so something distinctly unusual *must* happen at the location where the topology changes. Reverting to our qualitative picture, it is necessary for the system to abruptly

---

10 When two bands of well-defined but differing symmetry approach one another in energy–wavevector space, they may literally cross through one another, sharing precisely the same energy at the crossing point and carrying their distinctive characters through from one side of this point to the other. Conversely, if the symmetries of two such bands are compatible, they must necessarily interact with one another, leading to an avoidance of crossing. In such a scenario, the characters of the two bands interchange in passing through the 'anticrossing' point, just as if they *had* in fact crossed, but there is no actual wavevector at which the two energies coincide; the bands possess mixed character in the immediate vicinity of the anticrossing point.

11 In view of its being capable of only two possible numerical values, akin to an integer expressed modulo-two, this topological invariant is often referred to as a Z_2 invariant.

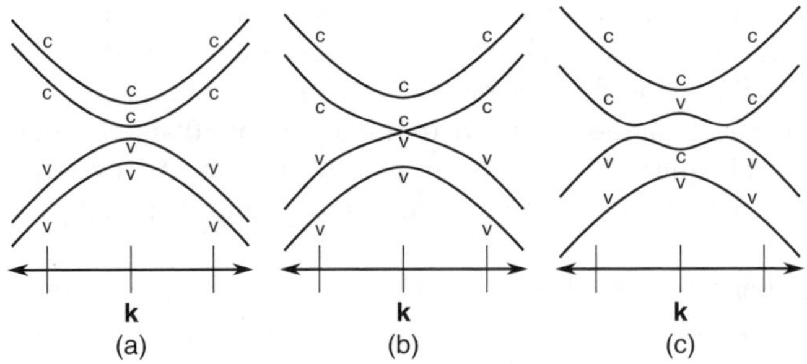

Figure 10.8 Schematic illustration of an interfacial Dirac cone resulting from bandstructure topology. In panel (a) the bandstructure on one side of the interface is that of a normal insulator, while in panel (c) it is that of a topological insulator; in panel (b) one observes that there must necessarily exist an intermediate situation in which the valence and conduction band labels ('v' and 'c') swap at a point.

convert from a bandstructure of the form shown in Fig. 10.7c to one of the form shown in Fig. 10.7b. This can only be achieved by passing through a transitional bandstructure in which the 'v' states above the band gap and the 'c' states below it cross back to their more usual places, as shown in Fig. 10.8. The resulting bandstructure, which exists only at the surface/interface itself, necessarily features surface/interface-localised electronic states that cross the Fermi level. Moreover, these metallic states evidently form a Dirac cone (see Fig. 10.9) albeit of a type that turns out to be subtly different from those found in the graphene and half-metal surface cases discussed above. The difference turns out to be largely connected with chirality.

In the non-topological cases, it will be recalled that chirality was defined with respect to the relative orientation of the electronic wavevector and a pseudospin vector – the latter transforming in a manner akin to the true spin of the electron, but in actual fact representing either the localisation (for graphene) or the symmetry (for the half-metal) of the electronic wavefunction. Here, in contrast, the non-trivial topology of the bulk insulator is a direct consequence of the *actual* electronic spin, and it is with respect to this true spin that any chiral properties ought to be defined. It turns out that whereas the pseudospin *radiates* towards and away from the Dirac point in the non-topological case (Fig. 10.10a) the true spin *circulates* around the Dirac point of a topologically mandated Dirac cone (Fig. 10.10b). The Hamiltonian for this latter case is, correspondingly, defined by the vector product of the relative wavevector and the spin

$$\hat{H} = \hbar v(\sigma \times \mathbf{q}) \tag{10.15}$$

as opposed to the scalar product involved in the non-topological cases. This distinction immediately suggests that topological Dirac cones may not be chiral, since the scalar product of spin vector with relative wavevector must always be zero. Indeed, correctly treating the true spin as a pseudovector (see Section 9.4) one can demonstrate that each such cone is its own mirror image

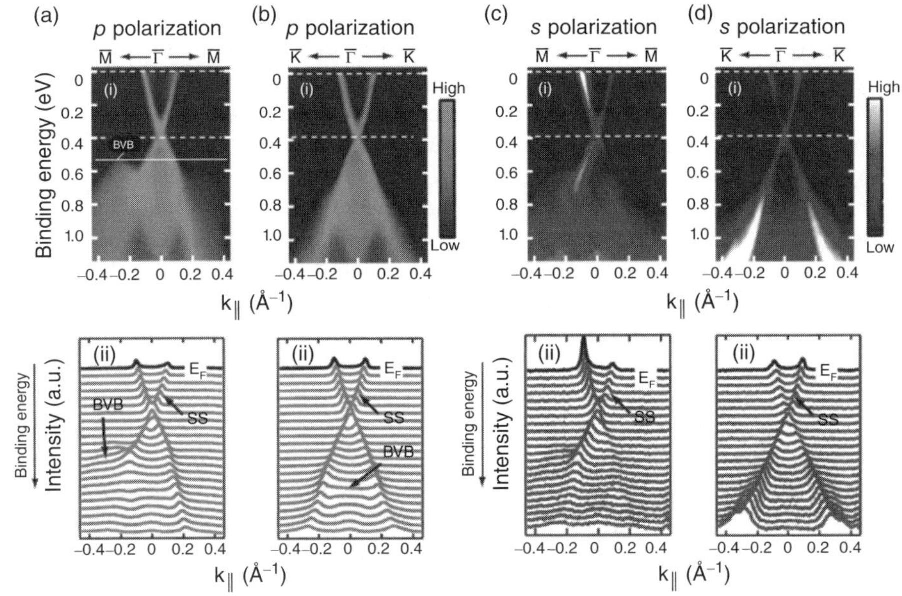

Figure 10.9 A Dirac cone revealed by ARUPS on the {111} surface of the topological insulator $TlBiSe_2$. Panels (a)–(d) show different cuts through the Brillouin zone and different polarisation of the ultraviolet light used in the experiments. Reproduced with permission from [15]. Copyright (2010) American Physical Society.

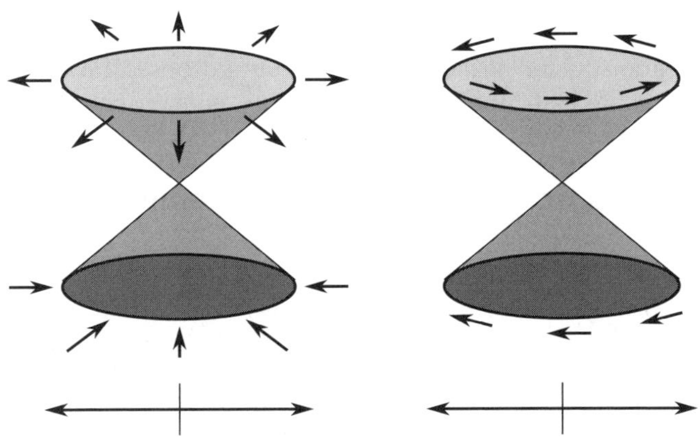

Figure 10.10 Schematic illustration of (a) pseudopin radiating towards and away from a non-topological Dirac cone and (b) true spin circulating clockwise and anticlockwise around a topological Dirac cone.

and hence overall achiral.[12] A cone with spins reversed relative to the example in Fig. 10.10b constitutes a non-degenerate alternative spin configuration that is not enantiomerically related to the original.

In fact, it ought to come as something of a relief that topological Dirac cones are achiral, in light of a particularly important distinction between the topological and non-topological cases: the surface states of a topological insulator necessarily give rise to an odd number of Dirac cones, whereas for the non-topological cases the number of cones must always be even. Accordingly, if the topologically mandated cones were chiral, we would be faced with an unexplained (and potentially inexplicable) symmetry-breaking in the electronic structure. As things stand, the preference of a particular surface for one sense of circulation over the other implies no such mysterious behaviour, so long as spin is treated correctly as a pseudovector.

It is worth emphasising, once again, that the Dirac cones found at the surfaces of topological insulators are mandated by the nature of the bulk electronic structure. They are thus robust against perturbations of the surface atomic geometry, either through clean-surface reconstructions or interactions with adsorbed species. This contrasts with the non-topological cases of graphene and half-metals, where persistence of the Dirac cones is contingent upon the system maintaining sub-lattice localisation or surface state symmetry, respectively. In this respect, one could argue that topological insulators offer a superior environment in which to study the behaviour of Dirac electrons, relatively unhampered by the effect of these perturbations. On the other hand, materials thus far predicted to display non-trivial topology are also, typically, non-trivial to handle experimentally, and calculations involving spin–orbit coupling are also rather challenging. It seems likely, therefore, that both types of Dirac cone will continue to be of interest for the foreseeable future. For our purposes, however, it bears repeating that only the non-topological instances are actually chiral.

10.4 Prospects for Electronic Chirality in the Chemical Context

In this chapter, we have briefly reviewed the state of knowledge relating to chirality within the electronic structure of otherwise non-chiral two-dimensional systems. To date, the author is unaware of any studies aimed specifically at understanding how electronic chirality may interact with the other forms of chirality described elsewhere in this book. How, for example, might the adsorption of chiral molecules on graphene modify its electronic structure, and could this be exploited in some form of sensor technology? Might the same thing be possible at a half-metal surface? And how might any intrinsic chirality associated with

12 It is worthwhile comparing this situation with the skyrmions discussed in Section 9.9. There, the vortex skyrmions appear to be chiral at first glance, and the hedgehog skyrmions achiral, based on the visual representation of spin as a vector. Correct treatment of spin as a pseudovector reversed this assessment, however, and a similar reversal occurs here.

the structure of the substrate alter things? Could chirality somehow be bestowed, either by adsorbate or substrate, upon the robust but achiral electronic structure of a topological insulator?

As a first step towards addressing this last question, some indications may be gleaned from recent ARUPS studies on the electronic structure of a stepped Bi_2Se_3 surface [16]. By growing a thin film of the topological insulator on a slightly miscut InP{100} substrate, the authors achieved a surface orientation identified as {221}. Whilst not chiral, such a surface orientation displays only a single mirror symmetry, meaning that its properties should, in principle, be anisotropic. The ARUPS results clearly show the existence of a Dirac cone, consistent with the prediction that one ought to be present at *all* topological insulator surfaces, not only those with the highest symmetry. Unlike previous observation of Dirac cones on the {111} surfaces of topological insulators, the cone reported in this case showed a distinct ellipticity, which may imply a chiral contribution to its symmetry; the close resemblance of these results to theoretical predictions by Moon *et al.* [17] is quite remarkable, and unequivocally demonstrates how sensitive this aspect of electronic structure is to the substrate symmetry. There is evidently no fundamental reason why adsorbates might not induce similar effects, albeit the magnitudes of any such phenomena remain to be determined. In any case, the utility of electronic structure asymmetry for sensor applications certainly rests upon designing a set of circumstances in which transport measurements, rather than spectroscopic measurements, are capable of detecting the effect in question; as yet, the potential of such transport experiments remains unexplored.

References

1 P.R. Wallace, *Phys. Rev.* **71**, 622 (1946).

2 K.S. Novoselov, A.K. Geim, S.V. Morozov, D. Jiang, Y. Zhang, S.V. Dubonos, I.V. Grigorieva and A.A. Firsov, *Science* **306**, 666 (2004).

3 A.H. Castro Neto, F. Guinea, N.M.R. Peres, K.S. Novoselov and A.K. Geim, *Rev. Mod. Phys.* **81**, 109 (2009).

4 D.A. Siegel, C.-H. Park, C. Hwang, J. Deslippe, A.V. Fedorov, S.G. Louie and A. Lanzara *Proc. Natl. Acad. Sci. U.S.A.* **108**, 11365 (2011).

5 M.I. Katsnelson, K.S. Novoselov and A.K. Geim, *Nat. Phys.* **2**, 620 (2006).

6 R.A. de Groot, F.M. Mueller, P.G. van Engen and K.H.J. Buschow, *Phys. Rev. Lett.* **50**, 2024 (1983).

7 F. Heusler, *Verhandl. Deuts. Ges.* **5**, 219 (1903).

8 F. Heusler, W. Stark and E. Haupt, *Verhandl. Deuts. Ges.* **5**, 220 (1903).

9 D. Ristoiu, J.P. Nozieres, C.N. Borca, T. Komesu, H.-K. Jeong and P.A. Dowben, *Europhys. Lett.* **49**, 624 (2000).

10 S.J. Jenkins, *Phys. Rev. B* **82**, 020403R (2010).

11 S.J. Jenkins and D.A. King, *Surf. Sci.* **494**, L793 (2001).

12 S.J. Jenkins and D.A. King, *Surf. Sci.* **501**, L185 (2002).

13 F. Ortmann, S. Roche and S.O. Valenzuela, eds. *Topological Insulators* (Wiley-VCH Verlag GmbH, 2015).
14 L. Fu and C.L. Kane, *Phys. Rev. B* **76**, 045302 (2007).
15 K. Kuroda, M. Ye, A. Kimura, S.V. Eremeev, E.E. Krasovskii, E.V. Chulkov, Y. Ueda, K. Miyamoto, T. Okuda, K. Shimada, H. Namatame and M. Taniguchi, *Phys. Rev. Lett.* **105**, 146801 (2010).
16 Z. Xu, X. Guo, M. Yao, H. He, L. Miao, L. Jiao, H. Liu, J. Wang, D. Qian, J. Jia, W. Ho and M. Xie, *Adv. Mater.* **25**, 1557 (2013).
17 C.-Y. Moon, J. Han, H. Lee and H.J. Choi, *Phys. Rev. B* **84**, 195425 (2011).

11

Postscript

Then felt I like some watcher of the skies
When a new planet swims into his ken;
Or like stout Cortez when with eagle eyes
He star'd at the Pacific – and with all his men
Look'd at each other with a wild surmise –
Silent, upon a peak in Darien.

On First Looking into Chapman's Homer, John Keats, 1816

I ended the preface to this book with a gratuitous metaphor likening efforts in the study of surface chirality to the exploration of new and fertile lands holding great promise for the future. It seems, therefore, only appropriate to return to that metaphor and take stock of where this journey of exploration has collectively taken the surface chirality community. I believe that our position is not entirely dissimilar to that which Keats imagined for stout Cortez.[1] We have jointly hacked our way through a jungle of complicated phenomena to a point where, abruptly, things seem to make some kind of sense. What lies ahead, however, has the potential to be even more exciting and vastly challenging. Here, for what it is worth, I offer my own wild surmise.

First, some trail notes from the jungle, by which I mean, of course, that tangled morass of interwoven phenomenology, related to the chiral chemistry of surfaces, through which we have already passed. When serious efforts to understand the origins of asymmetry in these systems began, around two decades ago, it is probable that even those who saw some potential in the field will not have fully appreciated the range of complex behaviours they, and those who followed, were going to uncover. Now, with the benefit of hindsight, the possibilities of kinetic and thermodynamic chiral amplification are increasingly evident, the mechanisms of modifier action in some classic cases of asymmetric heterogeneous catalysis have largely been revealed, and the catalogue of basic data pertaining to hydrogen bonding and surface-mediated intermolecular interactions continues to grow apace.

It would be wrong to suggest, however, that little of note remains to be done here. It is one thing to beat a path through a jungle in a race to the other side, and

1 Apparently, twas not Cortez who first discovered the Pacific (or discovered it for Spain, at any rate) but Vasco Núñez de Balboa.

Chirality at Solid Surfaces, First Edition. Stephen J. Jenkins.

quite another to tame that wilderness. We have, at the present time, a good idea of the territory and its riches; all the better to plot how to exploit those riches in the pursuit of academic achievement and public prosperity. There are, meanwhile, still plenty of gaps in our knowledge, and not all of these are small. It will not have escaped even the most casual of readers, for example, that the set of intrinsically chiral surfaces deemed worthy of study ranges from the predictable to the wildly eclectic. In some cases, the choice has been motivated by sound reasoning – surface orientations known to be stabilised by certain chiral molecules upon facetting are an obvious example – but in others the reasoning is less clear and one might be forgiven for imagining that Miller indices have been chosen more or less at random.[2] Furthermore, when one is dealing with chirality, it seems odd that a rigorous distinction between those aspects of the system that are symmetrically necessary and those that are structurally contingent has often been lacking;[3] the distinction between matters of symmetry (which are absolute) and matters of structure (which are not) deserves more attention in general than it usually receives. Surely, the more systematic an approach we take in this regard, the more rewards will accrue.

Now, turning once again to the waters ahead of us, charting the hidden currents that will steer research in surface chirality over the coming decades remains decidedly difficult. We are, however, most certainly ready for the next big leap into the unknown, which in my view will involve the coupling of chiral surface chemistry with chiral surface physics. I believe that, somewhere across a metaphorical ocean of opportunity, the great undiscovered continent awaiting us is a land in which asymmetric physical stimuli can be used to effect control over asymmetric surface chemistry (and *vice versa*) – a concept that we might wish to call 'surface asymmetric transduction'.[4] I imagine the scientists that live on this happy continent enjoy the benefits of sensor technology capable of readily translating enantiomeric excess in chemical systems into a digital readout, not to mention heterogeneous catalysts whose ability to synthesise a desirable chiral product can be switched from one enantiomer to its mirror image at the flick of a switch.

Undoubtedly, the physical aspect of this endeavour is at a much more primitive stage than the chemical, although the use of optical techniques based on

2 In my own group, work over the last several years has focused upon the Cu{311} surface precisely because, although it seemed clear to us that its symmetry and structure were of similar importance to that of Cu{110}, it had been almost entirely neglected as a subject of study. Similarly, the literature is currently replete with studies of the Cu{531} surface, but the Cu{421} surface is essentially absent; the former has the distinction of having the highest density of chiral kink sites amongst all surfaces of face-centred cubic materials, but the latter is barely less distinguished on that score; importantly, however, whilst the oft-studied orientation is the simplest chiral surface to possess a rhombic unit cell, the less popular choice is the simplest with a rectangular unit cell.
3 Indeed, in writing this book, it has been striking to note the number of occasions on which systems based on rhombic or rectangular two-dimensional lattices – whether intrinsically chiral substrates or adsorbed overlayers – have been labelled in matrix notation in such a manner as to obscure their lattice symmetry. It seems to me that it is important to highlight whether a system is chiral (i) because of the inherently oblique nature of its lattice, (ii) because its otherwise achiral lattice does not align with another lattice in the system or (iii) because the contents of the lattice break its symmetry.
4 Admittedly, this is less catchy than, say, 'Australia', but it captures the essence of an interplay between different forms of chirality quite neatly.

circular polarisation, near-field and non-linear phenomena have shown real promise. The recent observation of chiral skyrmions at magnetic surfaces also underlines the possibilities of spin as a probe of two-dimensional asymmetry. As yet, however, the study of two-dimensional systems exhibiting chirality in their electronic structure, such as graphene and the surfaces of half-metals, appears disconnected from the possibility that those systems might one day interact with others showing chirality in their geometric structure. Similarly, on the mechanical front, whilst anisotropy in surface stress has commonly been invoked to explain surface reconstructions and the adoption of particular overlayer geometries in achiral systems, the consequences of stress at chiral surfaces have barely been acknowledged, let alone addressed.

It is to be hoped that future efforts will see not only the development of new physical probes of surface chirality, but also a concerted effort to apply these to systems of chemical interest. Only then will our metaphorical continent hove into view. Between ourselves and that distant goal lies an archipelago of scattered islands – disconnected outcrops of phenomenology relating to aspects of mechanical, optical, magnetic and electronic manifestations of chirality at surfaces – across which we must hop, in an unpredictable and (dare one say) asymmetric path. Doubtless we shall spend much of our time paddling in the shallows, and much of it out of our depth, but with perseverance we shall reach our destination in the end.

A

List of Abbreviations

2-MBA	2-Methylbutanoic acid
3-MCHO	3-Methylcyclohexanone
ARUPS	Angle-resolved UPS
ATR	Attenuated total internal reflection
CD	Cinchonidine or circular dichroism
CDAD	Circular dichroism in the angular distribution
CV	Cyclic voltammetry
DBA-OC*n*	Dehydrobenzo[12]annulene
DCSB	4,4′-Dicarboxystilbene
DFG	Difference-frequency generation
DFT	Density functional theory
DM	Dzyaloshinskii–Moriya
DNA	Deoxyribosenucleic acid
EP	Ethyl pyruvate
FWHM	Full-width at half maximum
HCD	10,11-Dihydrocinchonidine
HOPG	Highly Ordered pyrolytic graphite
IET	Inelastic electron tunnelling
LEED	Low-energy electron diffraction
M3HB	Methyl-3-hydroxybutyrate
MAA	Methyl acetoacetate
MCA	Magnetocrystalline anisotropy
MEIS	Medium-energy ion scattering
ML	Monolayers
MM	Molecular mechanics
MP	Methyl pyruvate
NEA	1-(1-Naphthyl)ethylamine
NEXAFS	Near-edge X-ray absorption fine structure
NMR	Nuclear magnetic resonance
OPV*n*T	Oligo (*p*-phenylenevinylene) diaminotriazine
PCM	Planar chiral metamaterial
PhD	Photoelectron diffraction
PM-RAIRS	Polarisation modulation RAIRS
PO	Propylene oxide

Chirality at Solid Surfaces, First Edition. Stephen J. Jenkins.

PVBA	2-(4-Pyridyl)-4-vinylbenzoic acid
RAIRS	Reflection absorption infrared spectroscopy
SFG	Sum-frequency generation
SHG	Second-harmonic generation
STM	Scanning tunnelling microscopy
TPD	Temperature programmed desorption
UPS	Ultraviolet photoemission spectroscopy
XPS	X-ray photoemission spectroscopy

B

Rules for Overlayer Periodicity Assignment

The periodicity of an adsorbed overlayer may readily be described by means either of Wood's notation or the more general Matrix notation. In each case, primitive lattice vectors of the overlayer are effectively defined in terms of primitive lattice vectors pertaining to the ideal substrate. Problems arise, however, because there are always infinitely many ways to choose each set of primitive lattice vectors; the same overlayer may, therefore, be reasonably referred to in different terms by different individuals. Whilst certain 'unwritten rules' exist within the surface science community, such that commonly occurring cases are typically treated consistently, the more esoteric overlayers and substrates found amongst chiral systems cry out for some 'written rules' to be agreed upon. A few attempts have been made along these lines, most recently by Merz and Ernst [1]. Although unambiguous, however, their approach retains a fundamentally 'right-handed' philosophy, so that the notation for 'left-handed' surfaces is simply that of their mirror image, augmented by a symbol to indicate that reflection should be applied. Here, in contrast, we shall make use of an alternative scheme, outlined below, in which right- and left-handed lattices (whether overlayer or substrate) imply the use of right- and left-handed coordinate systems, respectively. At the same time, the method ensures that both the substrate and overlayer lattices are always described by primitive lattice vectors spanning cells whose shapes display the maximum possible symmetry for the relevant lattice type. Once the primitive lattice vectors of the substrate and overlayer are firmly defined in this manner, Wood's and/or Matrix notation may be employed as usual, but freed from any element of user choice.

B.1 Substrate Lattice

S1. In the case of a square, triangular, rectangular or rhombic substrate lattice, identify the primitive unit cell of highest symmetry. For the triangular lattice, there will be three alternative ways to draw this, whereas for the square, rectangular and rhombic lattice types there will be only one. In the case of an oblique lattice, select the primitive unit cell for which the ratio of smallest-to-largest internal angle is closest to unity; again, there should only be one.

Chirality at Solid Surfaces, First Edition. Stephen J. Jenkins.

S2. Identify all possible combinations of lattice vectors $\mathbf{a}_1$ and $\mathbf{a}_2$ that span the specific primitive cell (or cells, in the triangular case) identified in the previous step. For the triangular case, there should be 24 such combinations in total and for all other lattice types, just 8.

S3. In the triangular and rhombic cases, eliminate those combinations of lattice vectors for which the angle between $\mathbf{a}_1$ and $\mathbf{a}_2$ is acute. Likewise, in the rectangular and oblique cases, eliminate those combinations for which $\mathbf{a}_1$ exceeds $\mathbf{a}_2$. This should leave 12 remaining combinations for the triangular case, 8 for the square case and 4 in any of the other cases.

S4. In the case of a substrate bearing the chiral label D, eliminate those combinations of lattice vectors which form a left-handed set of axes with the surface normal. In the case of a substrate labelled L, eliminate those combinations which form a right-handed set of axes with the surface normal. Treat achiral substrates as for the D case in this respect. This should leave six remaining combinations for the triangular case, four in the square case and two in any of the other cases.

S5. Owing to the twofold rotational symmetry possessed by *all* two-dimensional lattices, one may now select *either* of the two remaining combinations of $\mathbf{a}_1$ and $\mathbf{a}_2$ when dealing with rectangular, rhombic or oblique lattices, without loss of generality. In the case of the triangular lattice, one may select one combination from each of three pairs related by a threefold rotation, but the three chosen combinations must all be carried forward for consideration. Similarly, in the case of the square lattice, one may select one combination from each of two pairs related by a fourfold rotation, but the two chosen combinations must both be carried forward.

S6. Define reference vectors for each combination of $\mathbf{a}_1$ and $\mathbf{a}_2$ as follows. In the square, rectangular and oblique cases, set $\mathbf{v}_1 = \mathbf{a}_1$ and $\mathbf{v}_2 = \mathbf{a}_2$. In the triangular and rhombic cases, set $\mathbf{v}_1 = \mathbf{a}_1 + \mathbf{a}_2$ and $\mathbf{v}_2 = \mathbf{a}_1 - \mathbf{a}_2$. In the triangular case, there will be three possible combinations of $\mathbf{v}_1$ and $\mathbf{v}_2$; in the square case there will be two possible combinations; in all other cases there will be just one.

B.2 Overlayer Lattice

O1–O3. As for S1–S3 above, but applied to the overlayer lattice and denoting the overlayer lattice vectors $\mathbf{a}'_1$ and $\mathbf{a}'_2$.

O4. Define reference vectors for each combination of $\mathbf{a}'_1$ and $\mathbf{a}'_2$ as follows. In the square, rectangular and oblique cases, set $\mathbf{v}'_1 = \mathbf{a}'_1$ and $\mathbf{v}'_2 = \mathbf{a}'_2$. In the triangular and rhombic cases, set $\mathbf{v}'_1 = \mathbf{a}'_1 + \mathbf{a}'_2$ and $\mathbf{v}'_2 = \mathbf{a}'_1 - \mathbf{a}'_2$. In the case of a triangular overlayer lattice, there will be 12 possible combinations of $\mathbf{v}'_1$ and $\mathbf{v}'_2$, while for a square overlayer lattice there will be 8 and in all other cases just 4.

O5. Of the various substrate and overlayer combination still in play, make a selection that maximises the scalar product $\mathbf{v}'_1 \cdot \mathbf{v}_1$. If more than one combination satisfies this criterion, make the selection from amongst these that maximises $\mathbf{v}'_1 \cdot \mathbf{v}_2$.

O6. If $(\mathbf{a}'_1 \times \mathbf{v}_1) \cdot \hat{\mathbf{n}} \geq 0$ insist that $\mathbf{a}'_1$, $\mathbf{a}'_2$ and $\hat{\mathbf{n}}$ must form a right-handed set. Otherwise, let $\mathbf{a}'_1$, $\mathbf{a}'_2$ and $\hat{\mathbf{n}}$ form a left-handed set.

O7. At this point, the choices of $\mathbf{a}_1$, $\mathbf{a}_2$, $\mathbf{a}'_1$ and $\mathbf{a}'_2$ should be fully constrained (to within a trivial twofold rotation) and the Wood's and/or Matrix notation for the overlayer may be unambiguously assigned.

B.3 Illustrative Examples

In order better to motivate and explain the convention outlined above, let us consider a few concrete examples. In the first instance, we turn to the overlayer (described in Section 4.3.3) formed upon prolonged exposure of Cu{110} to L-lysine. In the original report [2] the overlayer was given the matrix notation ($\bar{2}$, $\bar{2}$; 5, $\bar{1}$), implying the primitive unit cell shown in the lower part of Fig. B.1a, which is clearly oblique in shape. Application of the rules proposed by Merz and Ernst would lead unambiguously to the same primitive cell, but demands a different choice for the primitive lattice vectors (as marked) resulting in the notation (2, 2; $\bar{5}$, 1). Considering the mirror-image overlayer[1] shown in Fig. B.1b, however, the Merz–Ernst rules dictate a primitive cell that is *not* merely the reflection of the original across either of the two substrate mirror planes, and which consequently gains a notation that does not simply permute and/or negate the indices of the original, to whit (2, $\bar{2}$; 3, 3). Application of an additional set of rules defined within the Merz–Ernst convention identifies this latter notation as the 'master matrix' for the system, so that the overlayer of Fig. B.1b would indeed be denoted as (2, $\bar{2}$; 3, 3), while that of Fig. B.1a may, if desired, be denoted as (2, $\bar{2}$; 3, 3)^M rather than (2, 2; $\bar{5}$, 1), the superscript indicating a mirror operation. The advantage of the master matrix approach is that mirror-image overlayers gain notations that emphasise their symmetrical relationship, but at the expense that the matrix elements in the superscripted case do not straightforwardly yield the primitive lattice vectors of the overlayer itself, only of its mirror image. One major drawback to the Merz–Ernst convention, however, is that it does not always recognise the symmetry of the overlayer, which in the example chosen here is actually rhombic, not oblique.

Following the rules proposed here, in contrast, the notation ($\bar{3}$, 3; 5, $\bar{1}$) is dictated for the situation shown in Fig. B.1a, reflecting the rhombic symmetry of the overlayer lattice through the shape of the primitive cell shown in the upper part of the panel. For the mirror-image overlayer shown in Fig. B.1b, the same rules naturally dictate a notation of (3, 3; $\bar{5}$, $\bar{1}$) and a corresponding primitive cell that is automatically simply the mirror image of the original; the symmetrical affinity between the two overlayers is, moreover, reflected in the fact that their notations differ only in the signs of some matrix elements. These attractive features of the scheme are achievable only because we have relaxed the usual insistence upon an exclusively right-handed system of axes; however, the matrix corresponding to Fig. B.1a has a negative determinant, signalling that the overlayer vectors, $\mathbf{a}'_1$ and $\mathbf{a}'_2$, form a left-handed set of axes with the surface normal vector, $\hat{\mathbf{n}}$, whilst

1 One might, of course, expect the mirror-image overlayer to form if D-lysine were used in place of L-lysine, but this experiment seems not yet to have been performed.

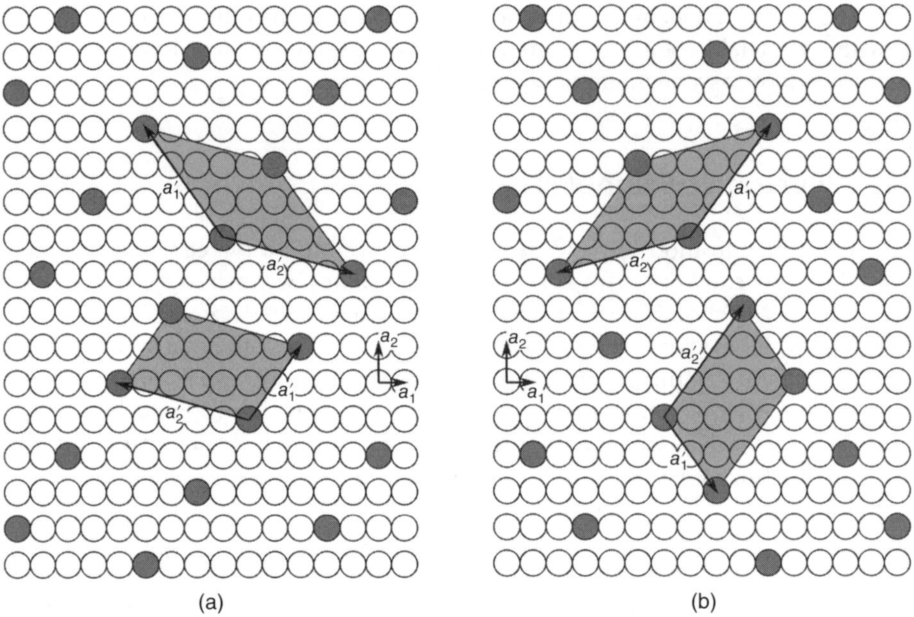

Figure B.1 Schematic illustration of the overlayer formed at high exposure by lysine on Cu{110}, showing only the top-layer substrate atoms, and marking a subset of these at the appropriate periodicity: (a) corresponding to the use of L-lysine, and (b) corresponding to the use of D-lysine. The primitive unit cell shaded in the lower part of (a) corresponds to the (2, 2; $\bar{5}$, 1) matrix notation dictated by the Merz–Ernst convention, while that in the upper part corresponds to the ($\bar{3}$, 3; 5, $\bar{1}$) matrix notation obtained within the convention adopted here. In the lower part of (b) the primitive cell corresponds to the (2, $\bar{2}$; 3, 3) matrix notation dictated by the Merz–Ernst convention, while once again the upper part shows the primitive cell obtained within the convention adopted in this book, denoted (3, 3; $\bar{5}$, $\bar{1}$).

the opposite is true for the matrix corresponding to Fig. B.1b. In this way, the notational scheme provides a natural and unambiguous way to assign left- or right-handed status to overlayers displaying either intrinsic or (as in this case) extrinsic lattice chirality.

For our second example, we take the overlayers formed upon adsorption of racemic heptahelicene on Cu{111} (described in Section 6.2.2). Here, mirror-image domains are observed, and the original report [3] assigns these the matrices (4, 5; $\bar{1}$, 8) and (8, $\bar{1}$; 5, 4). The Merz–Ernst convention, on the other hand, assigns a total of six different matrices, owing to the existence of a threefold rotational symmetry in addition to reflection of the overlayer; not all of the primitive unit cells implied by these matrices are of the same shape, however, with three being very nearly rectangular, two being distinctly oblique and one being very nearly rhombic. As before, one of these may be singled out (by means of the Merz–Ernst rules) as the master matrix and in this case the appropriate notation for this is (5, 4; $\bar{3}$, 5); the other five overlayers can then be indicated with reference to this matrix by means of a superscript to indicate reflection, and a subscript to indicate rotation. For example, one may speak of the $(5, 4; \bar{3}, 5)^{M}_{120}$ overlayer, meaning that overlayer obtained from that described by the master matrix via reflection followed by anticlockwise rotation through 120°. Once again, the advantage of a single matrix to define all the overlayers is counterbalanced by the fact that the master matrix does not always directly indicate either the primitive unit cell of the overlayer or the primitive lattice vectors that span it.

Application of the rules adopted in this book to this case illustrates an important feature, whereby the choice of substrate lattice vectors is, to a degree, dependent upon the overlayer when rotational degeneracy is present. Specifically, we find that the overlayer originally defined as $(4, 5; \overline{1}, 8)$ is now represented as $(4, 5; 5, \overline{3})$ in all three of its rotational variants (see Fig. B.2) while the mirror-image overlayer, originally defined as $(8, \overline{1}; 5, 4)$ is now indicated by $(5, 4; \overline{3}, 5)$ in all of its rotational variants (not shown). The matrix notation derived in this way is rather

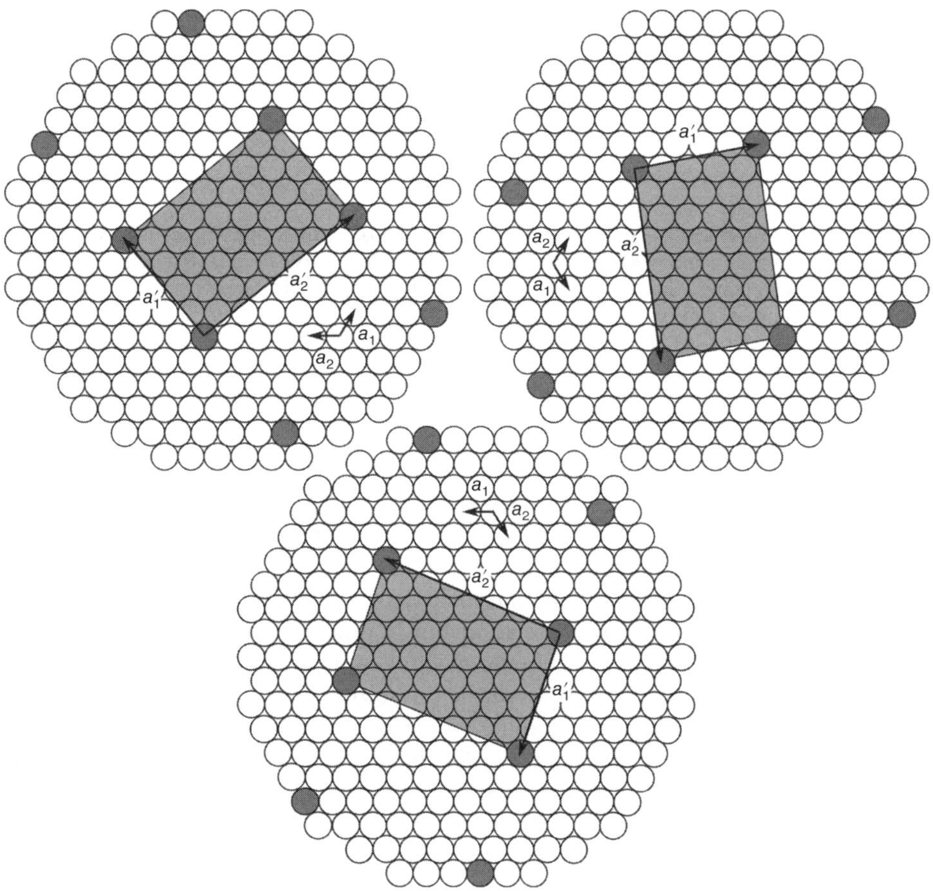

Figure B.2 Schematic illustration of the overlayer formed upon adsorption of racemic heptahelicene on Cu{111}, showing only the top-layer substrate atoms, and marking a subset of these at the appropriate periodicity. The three images show that threefold rotation results in degenerate overlayers, and the convention adopted in this book dictates that the substrate lattice vectors should be considered to rotate along with the overlayer, at least for the purposes of assigning a canonical matrix notation; accordingly, all three overlayers are denoted $(4, 5; 5, \overline{3})$.

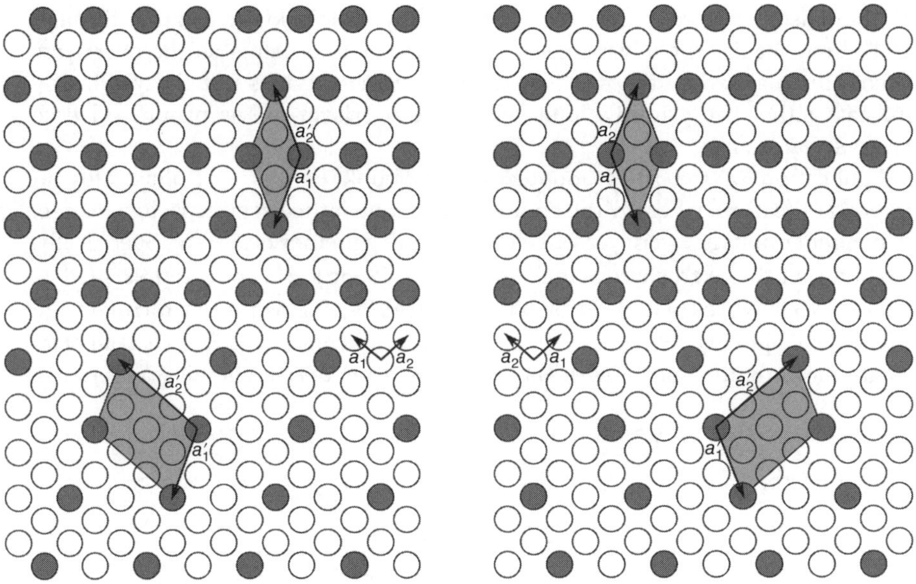

(a) (b)

Figure B.3 Schematic illustration of the overlayer formed upon low-coverage adsorption of serine on Cu{531}, showing only the top-layer substrate atoms, and marking a subset of these at the appropriate periodicity. Panel (a) shows the Cu{531}-L surface, with D-serine adsorption at the top and L-serine adsorption at the bottom; panel (b) shows the Cu{531}-D surface, with L-serine adsorption at the top and D-serine adsorption below. In both cases where the chiral label of molecules and substrate match (top), the rules adopted in this book produce the matrix notation $(\bar{1}, \bar{2}; 3, 0)$, while in both cases where the chiral labels do not match (bottom), the same rules produce the matrix notation $(\bar{1}, \bar{2}; 2, 1)$.

more closely linked to the actual overlayer lattice than was the case in the master matrix approach, albeit the rotational degeneracy is subsumed into the multiplicity of substrate lattice vector pairs. Note, however, that the distinction between mirror images is automatically maintained, as in the previous example on the {110} surface, and again the relationship between matrix elements is simple (here, their order is permuted upon reflection, but the signs remain the same).

Finally, we turn to an example where the substrate itself is chiral, namely the low-coverage adsorption of serine on Cu{531} (described in Section 5.3.2). In this case, we shall focus only upon the rules presented here, but we will consider adsorption of each molecular enantiomer on both the Cu{531}-L *and* the Cu{531}-D surface. Starting with Cu{531}-L, depicted in Fig. B.3a, the adsorption of D-serine leads to an overlayer described in the original report [4] with the matrix notation $(\bar{1}, 1; \bar{1}, \bar{2})$ relative to a set of substrate axes that span an oblique primitive cell. In fact, the substrate lattice of Cu{531} displays rhombic symmetry, and there is no very good reason not to reflect this in our notation. Defining the substrate primitive lattice vectors accordingly, the overlayer lattice now takes on the matrix notation $(\bar{1}, \bar{2}; 2, 1)$. When using L-serine, on the other hand, the original notation of $(\bar{1}, 1; \bar{2}, \bar{4})$ becomes $(\bar{1}, \bar{2}; 3, 0)$ relative to our symmetrically defined substrate lattice vectors. Note that our rules require that $\mathbf{a}_1$, $\mathbf{a}_2$ and $\hat{\mathbf{n}}$ form a left-handed set of axes for this left-handed substrate. Conversely, for the Cu{531}-D surface, shown in Fig. B.3b, we must adopt a right-handed set. Now, the adsorption of L-serine gives rise to an overlayer denoted $(\bar{1}, \bar{2}; 2, 1)$, while the use of D-serine produces an overlayer denoted $(\bar{1}, \bar{2}; 3, 0)$. That is, the matrix notation remains identical when both the molecular and substrate chirality are swapped simultaneously, changing only when one or other is changed alone.

References

1 L. Merz and K.-H. Ernst, *Surf. Sci.* **604**, 1049 (2010).
2 V. Humblot *et al.*, *Surf. Sci.* **601**, 4189 (2007).
3 R. Fasel, M. Parschau and K.-H. Ernst, *Nature* **439**, 449 (2006).
4 T. Eralp *et al.*, *J. Am. Chem. Soc.* **134**, 9615 (2012).

C

Further Reading

The following selection of review articles is by no means exhaustive, but nevertheless provides a flavour of the field as it stands today.

Intrinsically Chiral Surfaces

'Naturally Chiral Surfaces'
J.D. Horvath and A.J. Gellman, *Top. Catal.* **25**, 9 (2003).

'Beyond the Surface Atlas: A Roadmap and Gazetteer for Surface Symmetry and Structure'
S.J. Jenkins and S.J. Pratt, *Surf. Sci. Rep.* **62**, 373 (2007).

Surface Stress

The Role of Surface Stress in Reconstruction, Epitaxial Growth and Stabilization of Mesoscopic Structures
H. Ibach, *Surf. Sci. Rep.* **29**, 193 (1997).

Chiral Adsorption and Catalysis

'Complex Organic Molecules at Metal Surfaces: Bonding, Organisation and Chirality'
S.M. Barlow and R. Raval, *Surf. Sci. Rep.* **50**, 201 (2003).

'Chiral Modification of Solid Surfaces: A Molecular Level View'
F. Zaera, *J. Phys. Chem. C* **112**, 16196 (2008).

'Chiral Surfaces: Accomplishments and Challenges'
A.J. Gellman, *ACS Nano* **4**, 5 (2010).

'Fundamental Investigations of Enantioselective Catalysis'
C.J. Baddeley, T.E. Jones, A.G. Trant and K.E. Wilson, *Top. Catal.* **54**, 1348 (2011).

'Recognition and Ordering at Surfaces: The Importance of Handedness and Footedness'
A.G. Mark, M. Forster and R. Raval, *ChemPhysChem* **12**, 1474 (2011).

'Molecular Chirality in Surface Science'
K.-H. Ernst, *Surf. Sci.* **613**, 1 (2013).

Chirality at Solid Surfaces, First Edition. Stephen J. Jenkins.

'Crucial Aspects in the Design of Chirally Modified Noble Metal Catalysts for Asymmetric Hydrogenation of Activated Ketones'
A. Baiker, *Chem. Soc. Rev.* **44**, 7449 (2015).

'Surface Chemistry for Enantioselective Catalysis'
A.J. Gellman, W.T. Tysoe and F. Zaera, *Catal. Lett.* **145**, 220 (2015).

'Stereochemical Recognition of Helicenes on Metal Surfaces'
K.-H. Ernst, *Acc. Chem. Res.* **49**, 1182 (2016).

'Simple Rules and the Emergence of Complexity in Surface Chirality'
M. Forster and R. Raval, *Chem. Commun.* **52**, 14075 (2016).

Optical, Magnetic and Electronic Properties

'Implementing the Theory of Sum Frequency Generation Vibrational Spectroscopy: A Tutorial Review'
A.G. Lambert, P.B. Davies and D.J. Neivandt, *Appl. Spectrosc. Rev.* **40**, 102 (2005).

'The Electronic Properties of Graphene'
A.H. Castro Neto, F. Guinea, N.M.R. Peres, K.S. Novoselov and A.K. Geim, *Rev. Mod. Phys.* **81**, 109 (2009).

'Nanoscale Magnetic Skyrmions in Metallic Films and Multilayers: A New Twist for Spintronics'
R. Wiesendanger, *Nat. Rev. Mater.* **1**, 16044 (2016).

Index

1,2,4-Trichlorobenzene, *see* Solvents
1,3-Butanediol, *see* Alcohols
1-(1-Naphthyl)ethylamine, *see* Alkaloids
1-Phenyl-1-octanol, *see* Solvents
1-Phenyloctane, *see* Solvents
10,11-Dihydrocinchonidine, *see* Alkaloids
2,3-Butanediol, *see* Alcohols
2-(4-Pyridyl)-4-vinylbenzoic acid, *see* Stilbenes
2-Butanol, *see* Alcohols
2-Butanone, *see* Ketones
2-Chloropropanol, *see* Alcohols
2-Methylbutanoic acid, *see* Alkyl acids
2-Octanol, *see* Solvents
3-Methylcyclohexanone, *see* Ketones
4,4′-Dicarboxystilbene, *see* Stilbenes

a

Acetic acid, *see* Solvents
Alanine, *see* Amino acids
Alaninol, *see* Alcohols
Alcohols
  1,3-butanediol 229
  2,3-butanediol 229, 264–266
  2-butanol 145, 185–188, 191, 229, 273–274
  2-chloropropanol 229
  alaninol 191, 266–267
  glycidol 190–191
Aldehydes
  glyceraldehyde 7
Alkaloids
  1-(1-naphthyl)ethylamine 189, 244, 246–247, 273–274
  10,11-dihydrocinchonidine 240
  cinchonidine 236–237, 240–247
  cinchonine 236
  lepidine 243
  quinidine 236
  quinine 236
  quinoline 243–244
Alkyl acids
  2-methylbutanoic acid 188–189
Amino acids xiv–xv, 277
  alanine 80–81, 112–120, 147–149, 151, 157–158, 171–174, 184, 189–192, 206, 262–265
  aspartic acid 183–184, 192, 205, 215–217, 247, 250–253
  cysteine 128–135, 158, 174–176, 262, 274–276
  glutamic acid 140–144, 152–153, 247, 249–250, 252
  glycine 102–107, 112–113, 150–151, 157–159, 169–172, 174, 191, 206
  histidine 120
  homoalanine 189–190
  isoleucine 189–190
  leucine xv, 189–190
  lysine 125–128, 151, 153–154, 159, 174, 182–184, 191–192
  norvaline 189–190

Amino acids (*contd.*)
phenylalanine 184
proline 120–125, 154–156, 158, 226
serine 125–126, 128, 158, 159, 173–174, 184, 191–192, 206, 262
tryptophan 120
valine 189–190, 262
Antiferromagnetism 284, 306
Antimony telluride 325
Antonius, Marcus xiii
Arabinose, *see* Sugars
Aspartic acid, *see* Amino acids
Asymmetric chemistry
homogeneous xv
heterogeneous 19, 197, 225–226
Attard convention 10, 31–32
Axial vector, *see* Pseudovector

b

Benzene 95–96, 99, 157
Biot–Savart law 283
Bismuth nitrate 229
Bismuth selenide 325
Bismuth telluride 325
Bloch's theorem 317
Bosons 280
Brillouin zone
of graphene 317–320
of NiMnSb{100} 322–323
of $TlBiSe_2$ {111} 327

c

Caesar, Gaius Julius xiii
Cahn-Ingold-Prelog convention 6–7
Carbon (graphitic) 200, 202, 316
Carbon dioxide 200, 202
Carbon monoxide 94–96, 99, 200, 202
Carbon tetrachloride, *see* Solvents
Castles xiv
Centrosymmetry 270, 286, 301–305
Chemical potential 65
Chiral amplification 197–221, 236
Chiral centre (definition) 6
Chiral coercivity 217
Chiral conflict 209
Chiral conventions
D/L convention for surfaces 23–24, 32
R/S convention for surfaces 31–32
Chiral imprinting/facetting 138–140, 149–156, 159
Chiral quantification
futility of 15–17
Chiral recognition 171
induced-fit process 190, 225, 246, 249
lock-and-key process 190, 225, 250
one-to-one model 190, 225, 237, 243, 253
template model 225, 248–250, 253
Chirality 4
in bulk crystals 8–9
in heraldry xv–xvii
in particle physics 315–316
in pharmaceuticals xv
in the animal kingdom xiv–xv
in the human realm xiii–xiv
lattice 12–13
molecular 5–8, 11–12
substrate 12–14
Chirality conventions
D/L convention for molecules 7
D/L convention for surfaces 10–11
R/S convention for molecules 6–7
R/S convention for surfaces 10
Chloroauric acid 229
Chromium 277
Cinchonidine, *see* Alkaloids
Cinchonine, *see* Alkaloids
Circular dichroism 261–267, 273–276
in the angular distribution 263–267
Clock reconstruction 94
Collagen xiv
Compound surfaces
Bi_2Se_3 {221} 329
InP{100} 329
NiMnSb{100} 321–323
$TlBiSe_2$ {111} 326
Coordination numbers
body-centred cubic 37–39
diamond structure 59–62
face-centred cubic 25–27, 30–31

hexagonal close-packed 51–52, 55–56
Copper surfaces
Cu{100} 101, 150–151, 159, 182–183, 191–192, 212, 266–267
Cu{110} 80–81, 101–105, 107–115, 119–122, 125–132, 135–140, 157, 158, 190–191, 200–202, 206–209, 262–264, 267
Cu{111} 168, 177, 179–180, 191, 210–217, 236, 257, 273–274
Cu{11 7 5} 167
Cu{13 9 1} 181–182
Cu{17 5 1} 203–205
Cu{211} 98
Cu{221} 180
Cu{311} 105–107, 115–121, 123–125, 157–158, 168, 206
Cu{3 1 17} 150–151, 182–184, 192
Cu{410} 165–166
Cu{531} 165–174, 176, 180–181, 191–192, 203–205
Cu{532} 167–168
Cu{553} 180
Cu{643} 167, 176–180, 203–204, 235
Cu{651} 181–182, 201–203, 205
Cu{653} 180
Cu{762} 172
Cu{831} 172
Cricket xiii
Cyclohexanone, *see* Ketones
Cysteine, *see* Amino acids

d

DBA-OC*n* 219–221
Desorption rate
explosive 199
non-explosive 198–199
Diamagnetism 283
Diastereoisomerism (definition) 14–15
Dicarboxylic acids
fumaric acid 109–111
malic acid 138, 202, 209
succinic acid 107–111, 158, 183, 200, 202, 206–209
tartaric acid 7–9, 111, 135–140, 146–147, 158, 176, 183, 190–191, 200–209, 247–250, 252, 257, 267, 298
Dichloromethane, *see* Solvents
Dirac cones 319, 322–324, 326–328
Dirac equation 281
Dirac Hamiltonian 281
DNA xiv, 98
Domains 12
heterochiral 100–101, 104–105
homochiral 99–101, 105, 125–126, 133, 135, 138, 157
Dzyaloshinskii-Moriya (DM) coupling 284–286, 301, 308–309

e

Eccentricity 74, 77, 79
Egyptology xiii
Electro-oxidation 227–229
Electromagnetic wave equations 258
Electron orbital g-factor 282
Electron spin g-factor 282
Ellipticity angle 261, 276
Enantiomeric excess
in the rate constants 18–19, 277
in the rates 17
of chiral entities 17
Enantiomeric ratio 17
Enantiomers (definition) 6
Enantioselectivity
chiral modification 226, 236–253
intrinsic chirality 226–236
Entropy 26, 65–66, 100
Epoxides
glycidol 190–191
propylene oxide 176–179, 186–191, 226
Equilibium crystallite shape 89
Ethanol, *see* Solvents
Ethyl lactate see Hydroxyesters
Ethyl pyruvate, *see* Ketoesters
Ethylene 185

f

Fermions 280
Ferrimagnetism 284

Ferromagnetic analogy 216–217
Ferromagnetism 283–284, 306
Ferromagnets 295–298, 321–323
Flat surfaces
Ag{100} 143–144
Ag{111} 213–214
Au{111} 80, 95–96, 98–101, 145, 152–156, 158, 213–214, 226
Cu{100} 101, 150–151, 159, 182–183, 191–192, 212, 266–267
Cu{110} 262–264
Cu{111} 168, 177, 179–180, 191, 210–217, 236, 257, 273–274
Fe{110} 77–78
Ir{111} 311–313
Mo{110} 77–78
Ni{100} 93–94
Ni{111} 95, 99, 146–147, 190, 248–250
Pd{111} 147–149, 158, 185–191, 226
Pt{111} 187–189, 191, 227, 237–241, 243–244, 246
W{110} 77–78, 308–309
Four-index notation 42–43
Full-Heusler alloys 321
Fumaric acid, *see* Dicarboxylic acids

g
Gammadions 275–277
Gastropods xiv
Germanium surfaces
Ge{001} 81–82
Gibbs free energy 67–68
Glass 277
Glucose, *see* Sugars
Glutamic acid, *see* Amino acids
Glyceraldehyde, *see* Aldehydes
Glyceraldehyde convention 7
Glycidol, *see* Alcohols, Epoxides
Glycine, *see* Amino acids
Gold (polycrystalline) 276–277
Gold surfaces
Au{110} 132–135, 158, 274–276
Au{111} 80, 95–96, 98–101, 145, 152–156, 158, 213–214, 226
Au{17 11 9} 174–176
Graphene
electronic structure 317–323, 326, 328
structure and bonding 316–317
Graphite 316
Greek xiii
Guanine, *see* Ketones

h
Half-metals 321–323, 326, 328
Helicity 315, 319, 324
Helmholtz free energy 67–68, 86
Heptahelicene 8, 210–214, 257
Heptanol, *see* Solvents
Herringbone reconstruction 80, 98, 145, 152–153
Highly ordered pyrolytic graphite (HOPG) 217–221
Histidine, *see* Amino acids
Homoalanine, *see* Amino acids
Hydrogen
adsorbed 235, 237–238
evolved 200, 202
Hydroxyacids
lactic acid 5–6
Hydroxyesters
ethyl lactate 236
methyl lactate 235–236, 245, 247
methyl-3-hydroxybutyrate 247, 250

i
Inelastic electron tunnelling 98
Internal energy 65–66
Iridium surfaces
Ir{111} 311–313
Iron
bulk 294
monolayer 311–313
Iron germanide 310–311
Iron surfaces
Fe{110} 77–79
Fe{211} 77–79
Fe{321} 77–79
Fe{431} 77–79
Fe{532} 77–79
Iron-cobalt silicide 310–311

Isoleucine, *see* Amino acids
Izumi reaction 247–253

j
Jones vectors (definition) 259

k
Ketoacids
  pyruvic acid 5–6
Ketoesters
  ethyl pyruvate 236–237, 239–240, 242, 244
  methyl pyruvate 235–240, 244–247
  methylacetoacetate 247–253
Ketones
  2-butanone 185–187
  3-methylcyclohexanone 177, 179–181, 191
  cyclohexanone 167
  guanine 98–101, 158
Kinked surfaces
  Au{17 11 9} 174–176
  Cu{11 7 5} 167
  Cu{13 9 1} 181–182
  Cu{17 5 1} 203–205
  Cu{3 1 17} 150–151, 182–184, 192
  Cu{410} 165–166
  Cu{531} 165–174, 176, 180–181, 191–192, 203–205
  Cu{532} 167–168
  Cu{643} 167, 176–180, 182, 203–204, 235
  Cu{651} 181–182, 202–203, 205
  Cu{653} 180
  Cu{762} 172
  Cu{831} 172
  Ge{001} 81–82
  Pt{11 7 1} 229
  Pt{321} 229
  Pt{431} 229
  Pt{531} 166, 229
  Pt{643} 227–229
  Pt{976} 229
  Re{$11\bar{2}1$} 79–80
  Si{001} 264–265
  W{100} 308–309
Kipling, Rudyard 225
Klein tunnelling 320, 323

l
Lactic acid, *see* Hydroxyacids
Latin xiii, 7
Laue classes 286–298
Lepidine, *see* Alkaloids
Leucine, *see* Amino acids
Light
  angular momentum 261–262
  circularly polarised 259–261, 268, 277
  elliptically polarised 260–261
  linearly polarised 7, 15, 259–261, 268
  planar 258–262
  twisted 259, 261–262
LS coupling 284, 286
Lysine, *see* Amino acids

m
Machiavelli, Niccolò 221
Magnetic hysteresis 285–286
Magnetically easy axes 286, 294
Magnetically hard axes 286, 294
Magnetisation
  coercive field 285
  remanent 285
  saturation 285
Magnetocrystalline Anisotropy 284
  bulk 285–295
  surface 295–299
Magnons 301, 306
Majority rule 206, 209
Malic acid, *see* Dicarboxylic acids
Manganese
  bilayer 308
  monolayer 308–309
Manganese silicide 310–311
Mannose, *see* Sugars
Massless Dirac fermions 319
Mercator projection 27
Meso isomer (definition) 8
Metallorganic frameworks 151–156, 159

Methyl lactate, *see* Hydroxyesters
Methyl pyruvate, *see* Ketoesters
Methyl-3-hydroxybutyrate, *see* Hydroxyesters
Methylacetoacaetate, *see* Ketoesters
Mirror domains 95
Missing row reconstruction 94
Moiré pattern 214
Molybdenum surfaces
  Mo{110} 77–79
  Mo{211} 77–79
  Mo{321} 77–79
  Mo{431} 77–79
  Mo{532} 77–79
Motifs
  butterfly 101
  cloverleaf 211, 214
  comb-like 143–144
  flower-like 143–144
  herringbone 101
  parquet 101
  pinwheel 96, 211–212
  rosettes 217–218
  row-like 143–144
  square-like 143–144
  zig-zag 211, 213–214
Mystery asymmetry 133

n

Nanoskyrmions 311–312
Nanoslits 277
Neumann's principle 76, 286
Nickel catalyst (Raney) 247
Nickel lysinate 154, 159
Nickel prolinate 156, 159
Nickel pyroglutamate 152–153, 159
Nickel surfaces
  Ni{100} 93–94
  Ni{110} 298
  Ni{111} 95, 99, 146–147, 190, 248–250
Nickel-Manganese Antimonide 321–323
Nitrogen (adsorbed) 93–94
Nitrogen dioxide 80, 95–96, 99
Non-collinear spin 284, 295, 301
Non-linear optics 267–276
  difference-frequency generation 270, 272
  second harmonic generation 270, 272–277
  sum-frequency generation 270, 272
Norvaline, *see* Amino acids

o

Optical activity, *see* Optical rotation
Optical rotation 7, 15, 257–258, 260–261, 273–274
Optics
  far-field 276
  near-field 276–277
OPV*n*T 217–219
Orbital angular momentum 282–284, 301
Orbital magnetic moment 282–283, 299, 301
Orito reaction 236–247
Oxygen (adsorbed) 165–167

p

Palladium surfaces
  Pd{111} 147–149, 158, 185–191, 226
Paramagnetism 283
Parity 325
Pauli equation 281
Pauli exclusion principle 280
Pauli Hamiltonian 281
Pauli matrices 280–281, 318–319
Pauli spin vector 280–281
Phase angle 260, 261, 274
Phenylalanine, *see* Amino acids
Planar chiral metamaterial 276–277
Platinum catalyst (supported) 241
Platinum foil (polycrystalline) 241
Platinum surfaces
  Pt{110} 94–95
  Pt{111} 187–189, 191, 227, 237–241, 243–244, 246
  Pt{11 7 1} 229
  Pt{211} 227
  Pt{321} 229
  Pt{332} 227
  Pt{431} 229
  Pt{531} 166, 229

Pt{643} 227–229
Pt{976} 229
Platypus (duck-billed) xiv
Point group 3–4
bulk 270–272, 286–295
surface 270–272, 295–297
Poles (definition) 22
Porphyrin derivatives 221
Positrons 281
Pratt-Jenkins convention 10–11, 23–24, 32
Principal stress axes 71–79, 84
Principal stress components 71–77, 81–82, 84
Prochiral (definition) 6
Projection antipole 27
Projection pole 27–28
Proline, *see* Amino acids
Propene 98
Propylene (propene) 177
Propylene oxide
see Epoxides
Proteins xiv–xv, 277
Pseudoscalar (vs. scalar) 300–301
Pseudospin 319, 326
Pseudovector (vs. vector) 299–301
Pyramid-and-pit 88–89, 165–166, 168
Pyrrole-2-carboxylate 124
Pyruvic acid, *see* Ketoacids

q

Quartz 9, 257
Quinidine, *see* Alkaloids
Quinine, *see* Alkaloids

r

Racemisation 115, 157
Remanent chirality 217
Rhenium surfaces
Re{$11\bar{2}1$} 79–80
Ridge-and-furrow 86–87, 94–95, 140, 165–167
Rule of the road xiv

s

Samarium boride 325
Scalar (vs. pseudoscalar) 300–301
Schrödinger equation 280–282
Science-Fiction replicator 1
Semi-Heusler alloys 321
Sergeants and corporals 221
Sergeants and soldiers 206–209, 218–221
Serine, *see* Amino acids
Shakespeare, William xiii
Shockley states 322
Shuttleworth equation 68–69
Silica (fused) 277
Silicon surfaces
Si{001} 264–265
Silicon wafer 276
Silver perchlorate 229
Silver surfaces
Ag{100} 143–144
Ag{110} 140–143
Ag{111} 213–214
Skyrmions
hedgehog 309–313
vortex 309–311
Sodium chloride 168
Solvents
1,2,4-trichlorobenzene 219
1-phenyl-1-octanol 218
1-phenyloctane 217–219
2-octanol 219, 221
acetic acid 242
carbon tetrachloride 241–242, 244
dichloromethane 241
ethanol 240
heptanol 221
Space group 3
chiral 9
Sohncke 9, 311
Special relativity 281–282
Specific surface energy 66, 68, 86, 89
Spherical representation 21–27
Spin 280–282, 315
Spin angular momentum 281–282, 284, 301
Spin helices 301–307
Spin magnetic moment 280, 282–284, 299, 301
Spin precession 284

Spin spirals 303, 305–309
Spin valve (definition) 321
Spin waves 301–309
Spin–orbit coupling 324–325
Spin-flip scattering 321
Spin-orbit coupling 283–285
Spintronics 321
Stepped surfaces
 Ag{110} 140–143
 Au{110} 132–135, 140–143, 158, 274–276
 Cu{110} 80–81, 101–105, 107–115, 119–122, 125–132, 135–140, 157–158, 190–191, 200–202, 206–209, 267
 Cu{211} 98
 Cu{221} 180
 Cu{311} 105–107, 115–121, 123–125, 157–158, 168, 206
 Cu{553} 180
 Cu{651} 201–202
 Fe{211} 77–78
 Fe{321} 77–78
 Fe{431} 77–78
 Fe{532} 77–78
 Mo{211} 77–78
 Mo{321} 77–78
 Mo{431} 77–78
 Mo{532} 77–78
 Ni{110} 298
 Pt{110} 94–95
 Pt{211} 227
 Pt{332} 227
 W{211} 77–78
 W{321} 77–78
 W{431} 77–78
 W{532} 77–78
Stereographic projection 27–28
Stern–Gerlach experiment 280
Stilbenes
 2-(4-pyridyl)-4-vinylbenzoic acid 101
 4,4′-dicarboxystilbene 101
Stress components
 normal 70
 shear 70
Stress visualisations
 normal stress conic 71–73
 shear stress quartic 72–73
 stress ellipse 72, 74–75
Succinic acid, *see* Dicarboxylic acids
Succinimide 252
Sugars
 arabinose 229
 glucose 227–229
 mannose 229
 xylose 229
Superchiral light 277
Surface classification
 body-centred cubic crystals 36–41
 diamond structure crystals 56–62
 face-centred cubic crystals 29–36
 hexagonal close-packed crystals 42–56
Surface explosions 197–205
Surface plasmon-polaritons 277
Surface plasmons 277
Surface strain 68–69, 79
Surface stress 68–86
Surface stress measurement
 cantilever deformation 85–86
 contact angle 82–85
 island shape 81–82
Surface types (structure)
 flat 25–27, 29–30, 33, 37–38, 49, 52
 geminal 49, 52, 55, 59–61
 interrupted flat 50–53
 kinked 25–27, 29, 31–34, 37–40, 49, 52, 54–55, 59–62
 meandering row 49–50, 52, 54, 59–60
 stepped 25–27, 29, 31, 33–34, 37–41, 49, 52–54
Surface types (symmetry)
 bayonet 46–48, 52, 59
 chiral 31–32, 38–41, 44
 chiral, doubly 46–48, 52, 55, 58, 62
 chiral, singly 46–48, 52, 55, 58, 61–62
 glissadic 44, 46–48, 52, 54, 58, 61–62
 racemic 44–48, 52, 54–55, 58, 61
 reflexive 29, 31, 39, 44, 47, 54, 61

reflexive, doubly 29, 31, 38, 44, 47, 53
reflexive, quadruply 29–30, 38
reflexive, triply 29–30, 38, 60
reflexive-glissadic 44, 47, 52, 59–60, 62
uniterminated-reflexive 46–48, 53–54, 58, 60–62
Surface-localised electronic states 321–324, 328
Susceptibility tensors 267–268
Symmetry 2
improper 4
proper 4
Symmetry operations
glide 43
inversion 2
reflection 2, 23, 46
reflexotruncation 46
rotation 2, 46, 55
rototruncation 46
translation 2
truncation 46

t

Tamm states 322
Tartaric acid, *see* Dicarboxylic acids
Termination degeneracy 45–46
Thermal roughening 25–27, 37–38
Time-reversal symmetry 286, 295
Titanium 276
Tools xiii
Topological insulators 324–328
Topological invariant 325
Tryptophan, *see* Amino acids
Tungsten surfaces
W{100} 308–309
W{110} 77–79, 308–309
W{211} 77–79
W{321} 77–79
W{431} 77–79
W{532} 77–79
Two-dimensional lattice classification 34–36

u

Unitermination 46

v

Valine, *see* Amino acids
Vector (vs. pseudovector) 299–301
Volume 65, 66

w

Water (evolved) 200, 202
Work
done in cleavage 66
done in stretching 68
Writing systems xiv
Wulff construction 89–91

x

Xylose, *see* Sugars

y

Young equation 83

z

Zisman plot 83
Zones (structural)
interrupted primary 49, 58–59
primary 24–25, 37, 49
secondary 32–33, 39–40
tertiary 33–34, 40
Zones (symmetry)
glide 43–44, 57
mirror 22–23, 37, 43, 56–57